BIOCHEMICAL CALCULATIONS

How to Solve Mathematical Problems in General Biochemistry

BIOCHEMICAL CALCULATIONS

How to Solve Mathematical Problems in General Biochemistry

Second Edition

IRWIN H. SEGEL

Department of Biochemistry and Biophysics
University of California
Davis, California

JOHN WILEY & SONS

New York • Chichester • Brisbane • Toronto • Singapore

Library of Congress Cataloging in Publication Data

Segel, Irwin H   1935–
    Biochemical calculations.

    Includes index.
    1.   Biological chemistry--Problems, exercises,
etc.   I.   Title.
QD415.3.S4   1975   574.1'92'0151   75-23140
ISBN   0-471-77421-9

Printed in the United States of America
30 29 28 27 26

This book is dedicated with much love to my sons Jonathan and Daniel

PREFACE

In the six years since the publication of the first edition, *Biochemical Calculations* has been used as a supplementary text at many colleges and universities in the United States, and has been translated into Japanese and Spanish. This new edition is the result of suggestions that I received from instructors and students as well as my own desire to give a broader and more balanced picture of quantitative biochemistry. My objectives remain the same: to introduce students to the mathematical basis of biochemical discoveries and to show that modern biochemistry is more than the memorization of structures and pathways.

The present edition differs from the first in several ways. First, the descriptive material (formerly in the appendixes) has been incorporated into the body of the text so that the problems immediately follow the corresponding theory. The main emphasis of the book is still on numerical problem solving, and there is very little duplication of descriptive material found in standard biochemistry textbooks. Second, a greater variety of problems has been included. For example, the amount of material on aqueous solutions and acid-base chemistry has been markedly reduced, but the coverage has been expanded to include new subjects such as blood buffers. A new chapter, "Chemistry of Biological Molecules," has been added. The material on biochemical energetics now includes entropy and enthalpy changes, activation energy, and the energetics of membrane transport and photosynthesis. These problems do not require previous exposure to physical chemistry. Chapter 4, "Enzymes," includes new material on equilibrium binding studies, inhibitors, enzyme units and assays, the effects of pH and temperature, and the kinetics of allosteric enzymes. Kinetic mechanisms of bisubstrate enzymes are introduced without complicated mathematics. Chapter 5, "Spectrophotometry and Other Optical Methods," includes new problems on protein determination, fluorometry, and optical rotation. New problems on biological half-life, precursor-product relationships, dual-label scintillation counting, and counting errors have been added to Chapter 6, "Isotopes in Biochemistry." Although the scope of this edition has been broadened, the subject matter and problems remain at a suitable level for a modern, introductory, general biochemistry course.

I thank Prof. Francis W. Sayre of University of the Pacific, Prof. Kermit L. Carraway of Oklahoma State University, and Prof. Roland R. Rueckert of University of Wisconsin for their critical and extensive reviews of the manuscript.   I also thank my colleague Prof. Jerry L. Hedrick for his advice on Chapter 2, my graduate student, Mr. John Cuppoletti, for checking the calculations, and my wife, Dr. Leigh D. Segel, for her tireless editorial assitance, proofreading, and artwork.   A special thanks goes to the people at John Wiley & Sons who put it all together: Mr. Gary R. Carlson, my editor; Ms. Maria Colligan and Ms. Regina Malone, who supervised the production; and Ms. Eileen Thaxton, the designer and cover artist.

Irwin H. Segel

Davis, California
September 1975

PREFACE TO THE FIRST EDITION

Biochemistry is a quantitative science. Yet too often the subject is taught in a purely descriptive manner. This is understandable because many students in biological sciences have had only one course in general chemistry and one in introductory organic chemistry as prerequisites to general biochemistry. *Biochemical Calculations* was written to introduce students to some of the mathematical aspects of biochemistry that should be discussed in any introductory course. The problems discussed need no knowledge of mathematics beyond that required for general chemistry. In the very few problems where elementary calculus is employed, an alternate algebraic approximation is also given. In order for this book to be as useful as possible to the student, almost every problem is solved completely. There are no "rearranging terms and solving for X" statements. This book is meant to be used in conjunction with a standard textbook of general biochemistry. Nevertheless, a substantial amount of descriptive background (including mathematical derivations) is given in the appendixes.

A mimeographed edition of *Biochemical Calculations* was used by more than 800 students on the Davis campus during 1966–1967. Most of the students were enrolled in our general biochemistry course. Many others were graduate students studying for their M.S. and Ph.D. examinations. It was the response of these students that prompted me to submit the book for publication and general distribution.

I wish to express my appreciation to Dr. Wayne W. Luchsinger and Dr. Ronald S. Watanabe for their advice and critical review of the original manuscript.

I thank Dr. Eric E. Conn and Dr. Paul K. Stumpf for their suggestions and encouragement during the writing of this book. I am especially grateful to Miss Leigh Denise Albizati for all her help.

Irwin H. Segel

October 1968
Davis, California

CONTENTS

BIOCHEMICAL CALCULATIONS

How to Solve Mathematical Problems in General Biochemistry

1

AQUEOUS SOLUTIONS AND ACID-BASE CHEMISTRY

A. AQUEOUS SOLUTIONS

The majority of reactions studied by biochemists occur in solution. Consequently, it is appropriate to begin our mathematical survey by reviewing the various ways of expressing and interconverting concentrations of solutions.

CONCENTRATIONS BASED ON VOLUME

Concentrations based on the amount of dissolved solute per unit volume are the most widely used in biochemistry laboratories. The most common conventions are defined below.

$$\text{\textit{Molarity (M)}} = \text{the number of moles of solute per liter of solution} \tag{1}$$

Molar concentrations are usually given in square brackets, for example, $[H^+]$ = molarity of H^+ ion. To calculate M, we need to know the weight of dissolved solute and its molecular weight, MW.

$$\frac{\text{wt}_g}{\text{MW}} = \textbf{moles} \tag{2}$$

Dilute solutions are often expressed in terms of millimolarity, micromolarity, and so on, where:

$$1 \text{ mmole} = 10^{-3} \text{ moles}$$
$$1 \text{ } \mu\text{mole} = 10^{-6} \text{ moles}$$
$$1 \text{ nmole} = 1 \text{ m}\mu \text{ mole} = 10^{-9} \text{ moles}$$
$$1 \text{ pmole} = 1 \text{ } \mu\mu\text{mole} = 10^{-12} \text{ moles}$$

Therefore:

$$1 \text{ m}M = 10^{-3} \text{ } M = 1 \text{ mmole/liter} = 1 \text{ } \mu\text{mole/ml}$$
$$1 \text{ } \mu M = 10^{-6} \text{ } M = 1 \text{ } \mu\text{mole/liter} = 1 \text{ nmole/ml}$$
$$1 \text{ n}M = 10^{-9} \text{ } M = 1 \text{ nmole/liter} = 1 \text{ pmole/ml}$$

1

A 1 M solution contains one Avogadro's number of molecules per liter.

$$
\begin{aligned}
\text{Avogadro's number} &= \text{number of molecules per g-mole} \\
&= \text{number of atoms per g-atom} \\
&= \text{number of ions per g-ion} \\
&= 6.023 \times 10^{23}
\end{aligned}
$$

In general practice, one Avogadro's number of particles (i.e., 1 g-mole or 1 g-atom or 1 g-ion) is frequently called a "mole" regardless of whether the substance is ionic, monoatomic, or molecular in nature. For example, 35.5 g of Cl^- ions may be called a "mole" instead of a "gram-ion."

$$
\boxed{\textit{Activity (a)} = \text{the effective or apparent molarity of a solute} } \tag{3}
$$

Activity and actual molarity are related by:

$$
\boxed{a = \gamma[M]} \tag{4}
$$

where γ = activity coefficient (i.e., the fraction of the actual concentration that is active). Because of interactions between solute molecules that prevent their full expression, γ is usually less than unity. For example, HCl in a 0.1 M solution is fully ionized, yet the solution behaves as if it contains only 0.086 M H^+. Thus, $\gamma = 0.86$.

$$
\boxed{\textit{Normality (N)} = \text{the number of equivalents of solute per liter of solution} } \tag{5}
$$

To calculate N, we need to know the weight of dissolved solute and its equivalent weight, EW.

$$
\boxed{\frac{\text{wt}_\text{g}}{\text{EW}} = \text{equivalents}} \tag{6}
$$

One equivalent (i.e., the EW) of an acid or base is the weight that contains 1 g-atom (1 mole) of replaceable hydrogen, or 1 g-ion (1 mole) of replaceable hydroxyl. The EW of a compound involved in an oxidation-reduction reaction is the weight that provides or accepts 1 faraday (1 mole) of electrons. In general:

$$
\boxed{\text{EW} = \frac{\text{MW}}{n}} \tag{7}
$$

where n = the number of replaceable H^+ or OH^- per molecule (for acids and bases)

or n = the number of electrons lost or gained per molecule (for oxidizing and reducing agents).

The molarity and normality are related by:

$$N = nM \qquad (8)$$

For example, a 0.01 M solution of H_2SO_4 is 0.02 N.

> *Weight/Volume Percent (% w/v)* = the weight in g of a solute per 100 ml of solution   (9)

Weight/volume percent is often used for routine laboratory solutions where exact concentrations are not too important.

> *Milligram Percent (mg %)* = the weight in mg of a solute per 100 ml of solution   (10)

Milligram percent is often used in clinical laboratories.   For example, a clinical blood sugar value of 225 means 225 mg of glucose per 100 ml of blood serum.

> *Osmolarity* = the molarity of particles in a solution   (11)

A 1 M solution of a nondissociable solute is also 1 Osmolar.   (The solution contains 6.023×10^{23} particles per liter.)   A 1 M solution of a dissociable salt is n Osmolar, where n is the number of ions produced per molecule.   Thus, a 0.03 M solution of KCl is 0.06 Osmolar.   Osmolarity is often considered in physiological studies where tissues or cells must be bathed in a solution of the same osmolarity as the cytoplasm in order to prevent the uptake or release of water.   Blood plasma is 0.308 Osmolar.   Thus, red blood cells suspended in a 0.308 Osmolar NaCl solution (0.154 M) would neither shrink nor swell.   The 0.154 M NaCl solution is said to be *isotonic* with respect to the red blood cells.

· **Problem 1-1**

(a) How many grams of solid NaOH are required to prepare 500 ml of a 0.04 M solution?   (b) Express the concentration of this solution in terms of N, g/liter, % w/v, mg %, and osmolarity.

Solution

(a)
$$\text{liters} \times M = \text{number of moles NaOH required}$$
$$0.5 \times 0.04 = 0.02 \text{ mole NaOH required}$$

$$\text{number of moles} = \frac{\text{wt}_g}{\text{MW}} \qquad 0.02 = \frac{\text{wt}_g}{40}$$

$$\boxed{\textbf{wt = 0.8 g}}$$

∴ Weigh out 0.8 g, dissolve in water, and dilute to 500 ml.

(b) NaOH contains one OH per molecule.

∴ $M = N$ and the solution is $\boxed{\textbf{0.04 } N}$

The solution contains 0.8 g/500 ml, or $\boxed{\textbf{1.6 g/liter}}$

% (w/v) = g per 100 ml   1.6 g/liter = 0.16 g/100 ml = $\boxed{\textbf{0.16\%}}$

mg % = mg per 100 ml   0.16 g/100 = 160 mg/100 ml = $\boxed{\textbf{160 mg \%}}$

NaOH yields two particles (Na^+ and OH^-)

∴ osmolarity $= 2 \times M = \boxed{\textbf{0.08 Osmolar}}$

· Problem 1-2

How many milliliters of 5 M H_2SO_4 are required to make 1500 ml of a 0.002 M H_2SO_4 solution?

Solution

The number of moles of H_2SO_4 in the dilute solution equals the number of moles of H_2SO_4 taken from the concentrated solution.

$$\text{liters} \times M \text{ (dilute solution)} = \text{liters} \times M \text{ (concentrated solution)}$$
$$1.5 \times 0.002 = \text{liters} \times 5$$
$$\frac{1.5 \times 0.002}{5} = \text{liters concentrated solution required}$$
$$\frac{3 \times 10^{-3}}{5} = 0.6 \times 10^{-3} \text{ liters} = \boxed{\textbf{0.6 ml}}$$

∴ Take 0.6 ml of the concentrated solution and dilute to 1.5 liters.

$$\text{Ionic Strength } \left(\frac{\Gamma}{2}\right) = 1/2 \, \Sigma \, M_i Z_i^2$$

where M_i = the molarity of the ion

Z_i = the net charge of the ion (regardless of sign) (12)

Σ = a symbol meaning "the sum of"

Ionic strength measures the concentration of charges in solution. As the ionic strength of a solution increases, the activity coefficient of an ion decreases. The relationship between the ionic strength and the molarity of a solution of ionizable salt depends on the number of ions produced and their net charge, as summarized below.

Salt		Ionic Strength
Type	Example	
1:1	KCl, NaBr	M
2:1	$CaCl_2$, Na_2HPO_4	$3 \times M$
2:2	$MgSO_4$	$4 \times M$
3:1	$FeCl_3$, Na_3PO_4	$6 \times M$
2:3	$Fe_2(SO_4)_3$	$15 \times M$

"Type" refers to the net charge on the ions. Thus $MgSO_4$, which yields Mg^{2+} and SO_4^{2-}, is called a 2:2 salt. Na_2HPO_4, which yields HPO_4^{2-} and Na^+ ions, is called a 2:1 salt.

Only the *net* charge on an ion is used in calculating ionic strength. Thus, un-ionized compounds (e.g., un-ionized acetic acid) or species carrying an equal number of positive and negative charges (e.g., a neutral amino acid) do not contribute toward the ionic strength of a solution.

· **Problem 1-3**

Calculate the ionic strength of a 0.02 M solution of $Fe_2(SO_4)_3$.

Solution

$$\frac{\Gamma}{2} = \tfrac{1}{2} \sum M_i Z_i^2 = \tfrac{1}{2}[M_{Fe^{3+}} Z_{Fe^{3+}}^2 + M_{SO_4^{2-}} Z_{SO_4^{2-}}^2]$$

The 0.02 M $Fe_2(SO_4)_3$ yields 0.04 M Fe^{3+} and 0.06 M SO_4^{2-}.

$$\frac{\Gamma}{2} = \frac{(0.04)(3)^2 + (0.06)(-2)^2}{2} = \frac{(0.04)(9) + (0.06)(4)}{2}$$

$$= \frac{(0.36) + (0.24)}{2} = \frac{0.60}{2}$$

$$\boxed{\frac{\Gamma}{2} = 0.30}$$

Or, from the precalculated relationship for $2:3$ salts:

$$\frac{\Gamma}{2} = 15 \times M = (15)(0.02) = \boxed{\textbf{0.30}}$$

CONCENTRATIONS BASED ON WEIGHT

> *Weight / Weight Percent (% w/w)* = the weight in g of a solute per 100 g of solution   (13)

The concentrations of many commercial acids are given in terms of % w/w. In order to calculate the volume of the stock solution required for a given preparation, we must know its *density* or *specific gravity* where:

ρ = density = weight per unit volume

SG = specific gravity = density relative to water. Since the density of water is 1 g/ml, specific gravity is numerically equal to density.

· **Problem 1-4**

Describe the preparation of 2 liters of 0.4 M HCl starting with a concentrated HCl solution (28% w/w HCl, SG = 1.15).

Solution

liters $\times M$ = number of moles $\boxed{\textbf{2} \times \textbf{0.4} = \textbf{0.80 mole HCl needed}}$

wt_g = number of moles $\times$ MW $wt_g = 0.80 \times 36.5$

$$\boxed{\textbf{wt}_g = \textbf{29.2 g pure HCl needed}}$$

The stock solution is not pure HCl but only 28% HCl by weight.

$$\therefore \ \frac{29.2}{0.28} = \boxed{\textbf{104.3 g stock solution needed}}$$

Instead of weighing out 104.3 g of stock solution, we can calculate the *volume* required.

$$vol_{ml} = \frac{wt_g}{\rho_{g/ml}} = \frac{104.3}{1.15} = \boxed{\textbf{90.7 ml stock solution needed}}$$

$\therefore$ Measure out 90.7 ml of stock solution and dilute to 2 liters with water.

All of the above relationships (between weight, density, and percent w/w) can be combined into a single expression.

$$\text{wt}_g = \text{vol}_{ml} \times \rho_{g/ml} \times \% \text{ (as decimal)}$$

where wt_g = weight of *pure* substance required in g

vol_{ml} = volume of stock solution needed in ml

$\%$ = fraction of total weight that is pure substance

$$\therefore \quad \text{vol} = \frac{\text{wt}}{\rho \times \%} = \frac{29.2}{1.15 \times 0.28} = 90.7 \text{ ml}$$

As an alternate method, we can calculate the molarity of the stock solution.   First calculate the weight of pure HCl in 1 liter of stock solution.

$$\text{wt} = \text{vol} \times \rho \times \%$$
$$\text{wt}_g = 1000 \text{ ml} \times 1.15 \text{ g/ml} \times 0.28$$
$$\text{wt} = 322 \text{ g}$$

In other words, 1000 ml (1 liter) of stock solution contains 322 g of pure HCl.

$$\text{number of moles} = \frac{\text{wt}_g}{\text{MW}} = \frac{322}{36.5} = 8.82$$

$\therefore$ The concentrated stock solution is 8.82 M.
We need 0.80 mole.

$$\text{liters} = \frac{\text{number of moles}}{M} = \frac{0.80}{8.82} = 0.0907 \text{ liter}$$

$\therefore$ Take 0.0907 liter (90.7 ml) of stock and dilute to 2 liters.

> *Molality (m)* = the number of moles of solute per 1000 g of solvent   (14)

Molality is used in certain physical chemical calculations (e.g., calculations of boiling-point elevation and freezing-point depression).   For dilute aqueous solutions, m and M will be quite close.   In order to interconvert m and M, we need to know % w/w.

> *Mole Fraction* = the fraction of the total number of moles present represented by the compound in question   (15)

For example, in a solution containing n_1 moles of compound 1, n_2 moles of compound 2, and n_3 moles of compound 3, the mole fraction of compound 2,

MF_2, is given by:

$$MF_2 = \frac{n_2}{n_1 + n_2 + n_3}$$

The mole fraction of a compound is important in certain physical chemical calculations, but is not often used in biochemistry.

· Problem 1-5

Calculate   (a) the molality of the concentrated stock HCl solution described in Problem 1-4.   (b) Calculate the mole fraction of HCl in the solution.

Solution

(a)   The solution contains 28% w/w HCl, or 28 g HCl per 100 g total, or 28 g HCl per $(100 - 28) = 72$ g water.

$$\frac{28 \text{ g HCl}}{72 \text{ g H}_2\text{O}} \times 1000 = 388.9 \text{ g HCl}/1000 \text{ g H}_2\text{O}$$

$$\frac{\text{wt}_g}{\text{MW}} = \text{moles} \qquad \frac{388.9}{36.5} = 10.65 \text{ moles HCl}/1000 \text{ g H}_2\text{O}$$

$\therefore$ | **the solution is 10.65 m** |

(b)   In 100 g of solution, for example, we have:

$$\frac{28 \text{ g HCl}}{36.5 \text{ g/mole}} = 0.767 \text{ moles of HCl}$$

and

$$\frac{72 \text{ g H}_2\text{O}}{18 \text{ g/mole}} = 4.0 \text{ moles of H}_2\text{O}$$

$$MF_{HCl} = \frac{n_{HCl}}{n_{HCl} + n_{H_2O}} = \frac{0.767}{4.767}$$

| **$MF_{HCl} = 0.161$** |

CONCENTRATION BASED ON DEGREE OF SATURATION

Proteins are often purified by differential precipitation with neutral salts.   Ammonium sulfate is the most common salt used for this purpose, although occasionally NaCl is used.   The concentration of ammonium sulfate used to "salt out" proteins is almost always expressed in terms of "percent saturation."

Percent Saturation = the concentration of salt in solution as a percent of the maximum concentration possible at the given temperature

(16)

In order to take into account volume changes that occur when a large amount of salt is added to an aqueous solution we need to know the *specific volume* of the salt, where:

$$\bar{v} = \text{specific volume} = \text{volume occupied by 1 g of salt (ml/g)}$$
$$= \text{the reciprocal of the density}$$

· Problem 1-6

The specific volume of solid ammonium sulfate is 0.565 ml/g. The solubility of ammonium sulfate at 0°C is 706 g/1000 g water*. Calculate (a) the concentration of ammonium sulfate in a saturated solution at 0°C and (b) the amount of solid ammonium sulfate that must be added at 0°C to 500 ml of a "40% saturated" solution to bring it to "60% saturation."

Solution

(a) A saturated solution at 0°C prepared by adding 706 g of ammonium sulfate (AS) to 1000 g of H_2O occupies:

$$1000 \text{ ml} + (706)(0.565) \text{ ml} = 1399 \text{ ml}$$

The concentration of AS in the solution is:

$$\frac{706}{1399} = \boxed{0.505 \text{ g/ml}} = \boxed{505 \text{ g/liter}}$$

The MW of AS is 132.14. Therefore the molarity of the solution is:

$$\frac{505}{132.14} = \boxed{3.82 \; M}$$

(b) We can derive a simple equation giving the amount of solid AS that must be added to 1000 ml of solution at an initial degree of saturation, S_1, in order to bring the solution to a different degree of saturation, S_2. Knowing that 0.505 g/ml = 100% saturation, or 1.00 saturation:

$$S_2 = \frac{(\text{initial wt}_\text{g} \text{ of AS present}) + (\text{wt}_\text{g} \text{ AS added})}{(\text{final vol. of solution in ml})(0.505 \text{ g/ml})}$$

$$= \frac{(1000 \text{ ml})(0.505 \text{ g/ml})(S_1) + (\text{wt}_\text{g})}{[1000 \text{ ml} + 0.565 (\text{wt}_\text{g})]0.505 \text{ g/ml}}$$

$$\boxed{\text{wt}_\text{g} = \frac{505(S_2 - S_1)}{1 - 0.285 S_2}} \tag{17}$$

---

* Different sources give slightly different values. For example, Appendix II gives 697 g/liter at 0°C; Appendix III gives 706.8 g/liter.

where wt_g = wt of solid AS to be added to 1000 milliliters

S_2 = the final saturation (as a decimal)

S_1 = the initial saturation (as a decimal)

505 = g AS per 1000 ml for 100% saturation

$$\therefore \quad wt_g = \frac{505(0.60 - 0.40)}{1 - 0.285(0.60)} = \frac{101}{0.829}$$

$$= 121.8 \, g/liter \quad \text{or} \quad \boxed{\textbf{60.9 g/500 ml}}$$

· Problem 1-7

How many milliliters of a saturated ammonium sulfate solution must be added to 40 ml of a 20% saturated solution to make the final solution 70% saturated?   Assume that the volumes are additive.

Solution

$$(40 \text{ ml})(0.20) + (X \text{ ml})(1.00) = (40 + X \text{ ml})(0.70)$$

$$8 + X = 28 + 0.70X$$

$$0.3X = 20$$

$$\boxed{\textbf{X = 66.7 ml}}$$

In general, the volume of saturated ammonium sulfate solution to be added to 100 ml of solution at saturation S_1 (as a decimal) to produce a final saturation, S_2, is given by:

$$\boxed{vol_{ml} = \frac{100(S_2 - S_1)}{(1 - S_2)}} \tag{18}$$

(Usually, the tables shown in Appendices II and III are used for enzyme purification by fractional precipitation with ammonium sulfate.)

B. EQUILIBRIUM CONSTANTS

A great many reactions that occur in nature are *reversible* and do not proceed to completion.   Instead, they come to an apparent halt or *equilibrium* at some point between 0 and 100% completion.   At equilibrium, the net velocity is zero because the absolute velocity in the forward direction exactly equals the absolute velocity in the reverse direction.   The position of equilibrium is conveniently described by an *equilibrium constant*, K_{eq}.   For example, consider the dissociation of a weak acid (which we will examine in more detail in the following pages).

$$HA \underset{k_{-1}}{\overset{k_1}{\rightleftharpoons}} H^+ + A^-$$

The forward velocity, v_f, is proportional to the concentration of HA:

$$v_f \propto [HA] \qquad \text{or} \qquad v_f = k_1[HA]$$

where k_1 is a proportionality constant, known as a *rate constant* (specifically, a *first-order rate constant* because v_f is proportional to the concentration of a single substance raised to power one). The reverse velocity, v_r, is proportional to the concentration of H^+ and A^-, and, therefore, to the products of the concentration of A^- and H^+:

$$v_r \propto [H^+] \qquad \text{and} \qquad v_r \propto [A^-]$$
$$\therefore \quad v_r \propto [H^+][A^-] \qquad \text{or} \qquad v_r = k_{-1}[H^+][A^-]$$

where k_{-1} is a *second-order rate constant*. Thus, doubling $[H^+]$ doubles v_r. Doubling $[A^-]$ doubles v_r. Doubling both $[H^+]$ and $[A^-]$ increases v_r fourfold. At equilibrium:

$$v_f = v_r$$

or

$$k_1[HA] = k_{-1}[H^+][A^-] \qquad \text{or} \qquad \frac{k_1}{k_{-1}} = \frac{[H^+][A^-]}{[HA]}$$

The ratio of the two constants k_1/k_{-1} is itself a constant and is defined as K_{eq}:

$$\boxed{K_{eq} = \frac{[H^+][A^-]}{[HA]}} \qquad (19)$$

In this particular case, K_{eq} is an acid dissociation constant and would be indicated as K_a.

If the reaction in question is $A + B \underset{k_{-1}}{\overset{k_1}{\rightleftharpoons}} 2C$, then:

$$v_f = k_1[A][B] \qquad \text{and} \qquad v_r = k_{-1}[C][C] = k_{-1}[C]^2$$

$$\therefore \quad \boxed{K_{eq} = \frac{[C]^2}{[A][B]}}$$

The dimensions of K_{eq} depend on the number of components in the system.

Strictly speaking, it is not the concentrations of the reaction components that are considered, but instead, their *activities* or *effective* or *apparent* concentrations. For most of the calculations in the following chapters, we will assume that $\gamma = 1$, that is, activity is equivalent to molar concentration. This assumption is reasonably valid for the dilute aqueous solutions of monovalent and divalent ions employed in biochemical studies. Appendix V lists some activity coefficients.

· **Problem 1-8**

Consider the reaction $E + S \xrightleftharpoons[k_{-1}]{k_1} ES$.   What are the units of (a) k_1, (b) k_{-1}, and (c) K_{eq}?

Solution

(a) $v_f = k_1[E][S]$
Let v_f = moles of ES formed per liter per minute
$= \text{moles} \times \text{liter}^{-1} \times \text{min}^{-1} = M \times \text{min}^{-1}$
and [E] and [S] = $\text{moles} \times \text{liter}^{-1} = M$

$$\therefore \quad k_1 = \frac{v_f}{[E][S]} = \frac{M \times \text{min}^{-1}}{(M)^2} = \frac{\text{min}^{-1}}{M}$$

or
$$\boxed{k_1 = \text{min}^{-1} \times M^{-1}}$$

(b) $v_r = k_{-1}[ES]$

$$k_{-1} = \frac{v_r}{[ES]} = \frac{M \times \text{min}^{-1}}{M} \quad \therefore \quad \boxed{k_{-1} = \text{min}^{-1}}$$

(c) $K_{eq} = \dfrac{k_1}{k_{-1}} = \dfrac{\text{min}^{-1} \times M^{-1}}{\text{min}^{-1}} \quad \therefore \quad \boxed{K_{eq} = M^{-1}}$

or

$$K_{eq} = \frac{[ES]}{[E][S]} = \frac{M}{(M)^2} = M^{-1}$$

Additional problems on chemical equilibria are found in Chapter 3.

C. ACIDS AND BASES

An understanding of acid-base chemistry is essential if we are to appreciate the properties of biological molecules.   A great many of the low-molecular-weight metabolites and macromolecular components of living cells are acids and bases, and thus, have the potential to ionize.   The electrical charges on these molecules are important factors in the rate of enzyme-catalyzed reactions, the stability and conformation of proteins, the interactions of macromolecules with each other and with small ions, and the analytical and purification techniques used in the laboratory.

BRONSTED CONCEPT OF CONJUGATE ACID-CONJUGATE BASE PAIRS

The most useful way of discussing acids and bases in general biochemistry is to define an "acid" as a substance that donates protons (hydrogen ions) and a "base" as a substance that accepts protons. This concept is generally referred to as the Bronsted concept of acids and bases. When a Bronsted acid loses a proton, a Bronsted base is produced. The original acid and resulting base are referred to as a conjugate acid-conjugate base pair. The substance that accepts the proton is a different Bronsted base; by accepting the proton, another Bronsted acid is produced. Thus, in every ionization of an acid or base, two conjugate acid-conjugate base pairs are involved.

$$HA \quad + \quad B^- \quad \rightleftharpoons \quad A^- \quad + \quad HB$$

[conjugate acid]$_1$ [conjugate base]$_2$ [conjugate base]$_1$ [conjugate acid]$_2$

IONIZATION OF STRONG ACIDS AND BASES

A "strong" acid is a substance that ionizes almost 100% in aqueous solution. For example, HCl in solution is essentially 100% ionized to H_3O^+ and Cl^-:

$$HCl + H_2O \longrightarrow H_3O^+ + Cl^-$$

H_3O^+ (the hydronium ion, or conjugate acid of water) is the actual form of the hydrogen ion (proton) in solution. The ionization of HCl could just as easily be represented as a simple dissociation:

$$HCl \longrightarrow H^+ + Cl^-$$

Thus, for all practical purposes, H_3O^+ and H^+ mean the same thing. We will use the two conventions for "hydrogen ion" interchangeably, depending on which is the more convenient.

A "strong" base is a substance that ionizes extensively in solution to yield OH^- ions. Sodium and potassium hydroxides are examples of strong inorganic bases.

$$KOH \longrightarrow K^+ + OH^-$$

IONIZATION OF WATER

The ionization of water itself can be considered in two ways: (1) as a simple dissociation to yield H^+ and OH^- ions and (2) in terms of Bronsted conjugate acid-conjugate base pairs. In either case, it is obvious that water is amphoteric—it yields both H^+ and OH^- ions; it can both donate and accept protons.

The ionization of water can be described by a "dissociation constant," K_d, an "ionization constant," K_i, and a specific constant for water, K_w, as shown below.

Simple Dissociation	Conjugate Acid-Conjugate Base
$HOH \rightleftharpoons H^+ + OH^-$	$HOH + HOH \rightleftharpoons H_3O^+ + OH^-$
$K_d = \dfrac{[H^+][OH^-]}{[HOH]}$	$K_i = \dfrac{[H_3O^+][OH^-]}{[HOH]^2}$

Note that water produces two conjugate acid-conjugate base pairs: HOH/OH^- and H_3O^+/HOH.

For every mole of H^+ (or H_3O^+), 1 mole of OH^- is produced. In pure water $[H^+] = 10^{-7} M$. $\therefore$ $[OH^-] = 10^{-7} M$. The molarity of HOH can be calculated as follows:

$$M = \frac{\text{number of moles } H_2O}{\text{liter}} \qquad \text{number of moles} = \frac{wt_g}{MW}$$

A liter of water weighs 1000 g.   The MW of H_2O is 18.

$$\therefore \quad M = \frac{1000/18\,g}{1\ \text{liter}} = 55.6$$

The M of H_2O is actually 55.6 M (original concentration), minus $10^{-7}\,M$ (the amount that ionized).   However, this amount is so close to 55.6 that we may neglect the $10^{-7}\,M$.   We can now substitute the above values into the K_d and K_i expressions.

$$K_d = \frac{(10^{-7})(10^{-7})}{55.6} = \frac{10^{-14}}{55.6}$$

$$\boxed{K_d = 1.8 \times 10^{-16}}$$

$$K_i = \frac{(10^{-7})(10^{-7})}{(55.6)^2} = \frac{10^{-14}}{3.09 \times 10^3}$$

$$\boxed{K_i = 3.24 \times 10^{-18}}$$

The molarity of H_2O is essentially constant in the dilute solutions considered in most biochemical problems.   Consequently, we can define a new constant for the dissociation or ionization of water, K_w, which combines the two constants (K_d and $[H_2O]$ or K_i and $[H_2O]^2$).

$$K_w = K_d \times [H_2O]$$
$$K_w = (1.8 \times 10^{-16})(55.6)$$

$$\boxed{K_w = 1 \times 10^{-14} = [H^+][OH^-]}$$

$$K_w = K_i \times [H_2O]^2$$
$$K_w = (3.24 \times 10^{-18})(55.6)^2$$
$$= (3.24 \times 10^{-18})(3.09 \times 10^3)$$

$$\boxed{K_w = 1 \times 10^{-14} = [H_3O^+][OH^-]}$$

$$(20)$$

pH *AND* pOH

pH is a shorthand way of designating the hydrogen ion activity of a solution.   By definition pH is the negative logarithm of the hydrogen ion activity.   Similarly, pOH is the negative logarithm of the hydroxyl ion activity.

$$\text{pH} = -\log a_{H^+} = \log \frac{1}{a_{H^+}}$$
$$= -\log \gamma_{H^+}[H^+]$$
$$= \log \frac{1}{\gamma_{H^+}[H^+]}$$

$$\text{pOH} = -\log a_{OH^-} = \log \frac{1}{a_{OH^-}}$$
$$= -\log \gamma_{OH^-}[OH^-]$$
$$= \log \frac{1}{\gamma_{OH^-}[OH^-]}$$

In dilute solutions of acids and bases and in pure water, the activities of H^+ and OH^- may be considered to be the same as their concentrations.

$$\boxed{\text{pH} = -\log[H^+] = \log \frac{1}{[H^+]}}$$

$$\boxed{\text{pOH} = -\log[OH^-] = \log \frac{1}{[OH^-]}}$$

$$(21)$$

In all aqueous solutions the equilibrium for the ionization of water must be satisfied, that is, $[H^+][OH^-] = K_w = 10^{-14}$. Thus, if $[H^+]$ is known, we can easily calculate $[OH^-]$. Furthermore, we can derive the following relationship between pH and pOH:

$$[H^+][OH^-] = K_w$$

Taking logarithms:

$$\log [H^+] + \log [OH^-] = \log K_w$$
$$-\log [H^+] - \log [OH^-] = -\log K_w$$
$$-\log [H^+] = pH \qquad -\log [OH^-] = pOH \qquad -\log K_w = pK_w$$
$$\therefore \quad pH + pOH = pK_w$$
$$K_w = 10^{-14} \qquad pK_w = -\log 10^{-14} = +14$$

$$\therefore \quad \boxed{\mathbf{pH + pOH = 14}} \tag{22}$$

Thus, if any one of the values $[H^+]$, $[OH^-]$, pH, or pOH is known, the other three can be calculated easily. At concentrations of H^+ and OH^- greater than 0.1 M, the activity coefficients must be taken into account.

· **Problem 1-9**

What are the (a) H^+ ion concentration, (b) pH, (c) OH^- ion concentration, and (d) pOH of a 0.001 M solution of HCl?

Solution

(a)   HCl is a "strong" inorganic acid; that is, it is essentially 100% ionized in dilute solution. Consequently, when 0.001 mole of HCl is introduced into 1 liter of H_2O, it immediately dissociates into 0.001 M H^+ and 0.001 M Cl^-.

Note that when we are dealing with strong acids, the H^+ contribution from the ionization of water is neglected.

(b) $\quad pH = -\log [H^+] \qquad$ or $\qquad pH = \log \dfrac{1}{[H^+]}$

$\qquad\qquad = -\log 10^{-3} \qquad\qquad\qquad\qquad = \log \dfrac{1}{10^{-3}}$

$\qquad\qquad = -(-3) = +3 \qquad\qquad\qquad\qquad = \log 10^3$

$\qquad\qquad \boxed{\mathbf{pH = 3}} \qquad\qquad\qquad\qquad\qquad \boxed{\mathbf{pH = 3}}$

(c) $\qquad\qquad [H^+][OH^-] = K_w \qquad [OH^-] = \dfrac{K_w}{[H^+]}$

$$[OH] = \frac{1 \times 10^{-14}}{1 \times 10^{-3}} \qquad \boxed{\mathbf{[OH^-] = 1 \times 10^{-11}}}$$

(d) $\qquad$ $pOH = -\log[OH^-]$ $\qquad$ or $\qquad$ $pOH = \log\dfrac{1}{[OH^-]}$

$$= -\log(10^{-11})$$

$$= \log\dfrac{1}{10^{-11}}$$

$$= -(-11)$$

$$= \log 10^{11}$$

$$\boxed{pOH = 11}$$ $\qquad$ $$\boxed{pOH = 11}$$

or:

$$pH + pOH = 14, \qquad pOH = 14 - pH$$

$$pOH = 14 - 3 \qquad \boxed{pOH = 11}$$

· **Problem 1-10**

What are the (a) $[H^+]$, (b) $[OH^-]$, (c) pH, and (d) pOH of a $0.002\ M$ solution of HNO_3?

Solution

(a) HNO_3 is a strong inorganic acid.

$$\boxed{[H^+] = 0.002\ M = 2 \times 10^{-3}\ M}$$

(b) $\qquad$ $[H^+][OH^-] = 1 \times 10^{-14}$

$$[OH^-] = \dfrac{1 \times 10^{-14}}{2 \times 10^{-3}} = 0.5 \times 10^{-11}$$

$$\boxed{[OH^-] = 5 \times 10^{-12}\ M}$$

(c) $\qquad$ $pH = \log\dfrac{1}{[H^+]}$

$$= \log\dfrac{1}{2 \times 10^{-3}}$$

$$= \log 0.5 \times 10^3 \qquad\qquad\quad \text{or} \qquad\qquad pH = \log 500$$

$$= \log 5 \times 10^2 \qquad\qquad\qquad\qquad\qquad\qquad \log 500 = 2.699$$

$$= \log 5 + \log 10^2$$

$$= 0.699 + 2 \qquad\qquad\qquad\qquad\qquad \boxed{pH = 2.699}$$

$$\boxed{pH = 2.699}$$

where $2 =$ the number of places between the first significant figure and
the decimal point and $0.699 = \log$ of
"5."

Check:

$$10^{-2} \, M \, [H^+] = pH \, 2$$
$$10^{-3} \, M \, [H^+] = pH \, 3$$
$$\therefore \quad 2 \times 10^{-3} \, M \, [H^+] = pH \text{ between 2 and 3}$$

(d) $pH + pOH = 14$ or $pOH = \log \dfrac{1}{[OH^-]}$

$$pOH = 14.000 - 2.699$$

$$\boxed{pOH = 11.301}$$

$$= \log \frac{1}{5 \times 10^{-12}}$$
$$= \log 0.2 \times 10^{12}$$
$$= \log 2 \times 10^{11}$$
$$= \log 2 + \log 10^{11}$$
$$= 0.301 + 11$$

$$\boxed{pOH = 11.301}$$

· **Problem 1-11**

What is the concentration of HNO_3 in a solution that has a pH of 3.4?

Solution

$$pH = \log \frac{1}{[H^+]} = 3.4 \qquad \text{or} \qquad [H^+] = 10^{-pH}$$
$$= 10^{-3.4}$$
$$= 10^{-4} \times 10^{+0.6}$$

where 3 = number of places between first significant figure and the decimal point.   Look up antilog of "4."

Look up antilog of 0.6

    antilog of 0.6 = "398"
           = 3.98

    antilog of 4 = "2512"

$$\therefore \quad [H^+] = \frac{1}{2512}$$

$$\boxed{[H^+] = 3.98 \times 10^{-4} \, M}$$

$$= \frac{1}{2.512 \times 10^3}$$
$$= 0.398 \times 10^{-3}$$

$$\boxed{[H^+] = 3.98 \times 10^{-4} \, M}$$

$$\therefore \quad HNO_3 = 3.98 \times 10^{-4} \, M \text{ assuming 100\% ionization}$$

Check:

$$pH \, 3 = 10^{-3} \, M \, [HNO_3]$$
$$pH \, 4 = 10^{-4} \, M \, [HNO_3]$$
$$\therefore \quad pH \, 3.4 = [HNO_3] \text{ between } 10^{-4} \text{ and } 10^{-3} \, M$$

· **Problem 1-12**

How many (a) H^+ ions and (b) OH^- ions are present in 250 ml of a solution of pH 3?

Solution

(a)
$$pH = 3$$
$$\therefore \quad [H^+] = 10^{-3} \, M \, (10^{-3} \, \text{g-ions/liter})$$
$$1 \, \text{g-ion/liter} = 6.023 \times 10^{23} \, \text{ions/liter}$$
$$\therefore \quad 10^{-3} \, \text{g-ions/liter} = 6.023 \times 10^{20} \, \text{ions/liter}$$

$$\therefore \quad \frac{6.023 \times 10^{20}}{4} = \boxed{\textbf{1.506} \times \textbf{10}^{\textbf{20}} \textbf{ ions/250 ml}}$$

(b)
$$pH + pOH = 14$$
$$pOH = 14 - 3 = 11$$
$$[OH^-] = 10^{-11} \, M \text{ or } 10^{-11} \, \text{g-ions/liter}$$
$$10^{-11} \, \text{g-ions/liter} \times 6.023 \times 10^{23} \, \text{ions/g-ion} = 6.023 \times 10^{12} \, \text{ions/liter}$$

$$\frac{6.023 \times 10^{12}}{4} = \boxed{\textbf{1.506} \times \textbf{10}^{\textbf{12}} \textbf{ ions/250 ml}}$$

· **Problem 1-13**

What is the pH of a $10^{-8} \, M$ solution of HCl?

Solution

The first tendency of many students is to say "pH = 8." This is obviously incorrect. No matter how much one dilutes a strong acid, the solution will never become alkaline. In this dilute solution, the contribution of H^+ ions from H_2O is actually greater than the amount contributed by HCl. As a *first approximation*, therefore, the H^+ ions from the HCl may be neglected. The pH then is around 7.

As a *second approximation*, we can solve for pH while taking into account the H^+ ions from both sources.

$$pH = -\log [H^+]$$
$$[H^+] = 10^{-7} \text{ (from } H_2O) + 10^{-8} \text{ (from HCl)}$$
$$pH = -\log (1 \times 10^{-7} + 0.1 \times 10^{-7})$$
$$= \log \frac{1}{1.1 \times 10^{-7}} = \log 0.909 \times 10^7$$
$$= \log 9.09 \times 10^6 = \log 9.09 + \log 10^6$$

$$= 0.959 + 6 \qquad \boxed{\textbf{pH} = \textbf{6.959}}$$

The above solution is still not completely correct. It assumes that the contribution of H^+ ions from water is still $10^{-7} \, M$ in the presence of $10^{-8} \, M$ HCl. Actually, the slight increase in H^+ ions from HCl tends to depress the ionization of H_2O, that is, shift the equilibrium of the $HOH \leftrightharpoons H^+ + OH^-$

reaction back to the left.   An exact solution to the problem can be obtained in the following manner: Both HOH and HCl ionize to form H^+ ions.

$$HOH \leftrightharpoons H^+ + OH^-$$
$$HCl \rightarrow H^+ + Cl^-$$

Let $X = [H^+]$ from H_2O

$$\therefore \quad [OH^-] = X$$

$[H^+]$ from $HCl = 10^{-8} M$

$$\therefore \quad [H^+] = X + 10^{-8}, [OH^-] = X$$
$$[H^+][OH^-] = 10^{-14}$$
$$(X + 10^{-8})(X) = 10^{-14}$$
$$X^2 + 10^{-8} X = 10^{-14}$$
$$X^2 + 10^{-8} X - 10^{-14} = 0$$

The above equation can be solved by substituting into the general solution for a quadratic equation:

$$X = \frac{-b \pm \sqrt{b^2 - 4ac}}{2a}$$

where $a = 1$, $b = 10^{-8}$ and $c = -10^{-14}$.

$$X = \frac{-10^{-8} \pm \sqrt{(10^{-8})^2 - 4(-10^{-14})}}{2} = \frac{-10^{-8} \pm \sqrt{10^{-16} + 4 \times 10^{-14}}}{2}$$

$$= \frac{-10^{-8} \pm \sqrt{4.01 \times 10^{-14}}}{2} = \frac{-10^{-8} \pm 2.0025 \times 10^{-7}}{2}$$

$$= \frac{-10^{-8} \pm 20.025 \times 10^{-8}}{2}$$

$$= \frac{19.025 \times 10^{-8}}{2} \quad \text{and} \quad \frac{-21.025 \times 10^{-8}}{2}$$

$X = 9.5125 \times 10^{-8}$ (neglecting the negative value)

$[H^+] = X + 10^{-8}$

$\quad = 9.5125 \times 10^{-8} + 10^{-8} = 10.5125 \times 10^{-8}$

$$pH = \log \frac{1}{10.5125 \times 10^{-8}}$$

$\quad = \log 0.09512 \times 10^8 = \log 9.512 \times 10^6$

$\quad = \log 9.512 + \log 10^6 = 0.978 + 6$

$$\boxed{pH = 6.978}$$

· **Problem 1-14**

What are the (a) a_{H^+} and (b) γ_{H^+} in a 0.010 M solution of HNO_3 if the pH is 2.08?

Solution

In this problem we can no longer assume that $a_{H^+} = [H^+]$.   It is obvious that if $a_{H^+} = [H^+]$ and $pH = -\log [H^+]$, the pH would be 2.0 and not 2.08. $\therefore \gamma \neq 1$.

(a)

$$pH = \log \frac{1}{a_{H^+}} \qquad \text{or} \qquad a_{H^+} = 10^{-pH}$$

$$2.08 = \log \frac{1}{a_{H^+}} \qquad\qquad a_{H^+} = 10^{-2.08}$$

$$\text{antilog of } 0.08 = \text{``120''} \qquad\qquad = 10^{-3} \times 10^{+0.92}$$

$$\frac{1}{a_{H^+}} = 120 \qquad\qquad = 10^{-3} \times 8.3$$

$$a_{H^+} = \frac{1}{120}$$

$$\boxed{a_{H^+} = 8.3 \times 10^{-3}}$$

$$a_{H^+} = 0.0083$$

$$\boxed{a_{H^+} = 8.3 \times 10^{-3}}$$

(b)

$$a_{H^+} = \gamma_{H^+}[H^+] \qquad \text{where } \gamma = \text{the activity coefficient}$$
$$0.0083 = \gamma_{H^+}(0.010)$$

$$\boxed{\gamma_{H^+} = 0.83}$$

Although the *actual* concentration of HNO_3 is 0.01 M, the solution behaves as if only 83% of the HNO_3 molecules are dissociated; the *effective* or *apparent* concentration (a_{H^+}) is 0.0083 M. The HNO_3 is actually 100% ionized. Interactions of ion clouds, however, prevent full expression of the H^+ ions; that is, the shielding effects of NO_3^- ions surrounding the H^+ ions make it seem as if some of the H^+ ions are not there.

NEUTRALIZATION AND TITRATION OF STRONG ACIDS AND BASES

· Problem 1-15

(a) How many milliliters of 0.025 M H_2SO_4 are required to neutralize exactly 525 ml of 0.06 M KOH?   (b) What is the pH of the "neutralized" solution?

Solution

(a)   number of moles (equivalents) of H^+ required =
number of moles (equivalents) of OH^- present

$$\text{liters} \times N = \text{number of equivalents}$$

$$\text{liters}_{acid} \times N_{acid} = \text{liters}_{base} \times N_{base}$$

$$H_2SO_4 = 0.025 \ M = 0.05 \ N$$

$$\text{liters}_{acid} \times 0.05 = 0.525 \times 0.06$$

$$\text{liters}_{acid} = \frac{0.525 \times 0.06}{0.05} = 0.63$$

$$\boxed{\textbf{acid required = 630 ml}}$$

(b) The neutralized solution contains only K_2SO_4 that, being a salt of a strong acid and strong base, has no effect on pH.

$$\therefore \quad \boxed{\text{pH} = 7}$$

· Problem 1-16

How many milliliters of 0.05 N HCl are required to neutralize exactly 8.0 g of NaOH?

Solution

At the equivalence point, the number of moles H^+ added equals the number of moles OH^- present.

$$\text{liters}_{\text{acid}} \times N_{\text{acid}} = \text{number of moles (equivalents) of } H^+ \text{ added}$$

$$\frac{\text{wt}_{\text{gNaOH}}}{\text{MW}_{\text{NaOH}}} = \text{number of moles of NaOH (and } OH^-\text{) present}$$

$$\text{liters} \times N = \frac{\text{wt}_g}{\text{MW}} \qquad \text{liters} \times 0.05 = \frac{8.0}{40}$$

$$\text{liters} = \frac{8.0}{40 \times 0.05} = \frac{8.0}{2}$$

$$= 4.0 \text{ liters} = \boxed{\textbf{4000 ml}}$$

· Problem 1-17

Calculate the appropriate values and draw the curve for the titration of 500 ml of 0.01 N HCl with 0.01 N KOH.

Solution

A titration curve is a plot of pH versus milliliters (or equivalents or moles) of standard titrant added.   For the titration of a given amount of acid, the curve is a plot of pH versus milliliters (or equivalents) of base added.   The pH at any position up to the equivalence point is calculated from the concentration of excess (untitrated) H^+ remaining (taking the increased volume into account).   At the equivalence point, the solution contains only KCl, a salt of a strong acid and strong base that has no effect on the pH.   Therefore, pH = 7.0.   The pH at positions beyond the equivalence point is calculated from the concentration of excess OH^-.   The titration curve is shown in Figure 1-1.

IONIZATION OF WEAK ACIDS

In an aqueous solution a weak acid ionizes to a limited extent as follows:

$$\text{HA} \quad + \quad H_2O \quad \rightleftharpoons \quad H_3O^+ \quad + \quad A^-$$

[conjugate acid]₁   [conjugate base]₂   [conjugate acid]₂   [conjugate base]₁

The proton released from HA is accepted by water to form the hydronium ion H_3O^+.   The reversible ionization reaction can be described by an equilib-

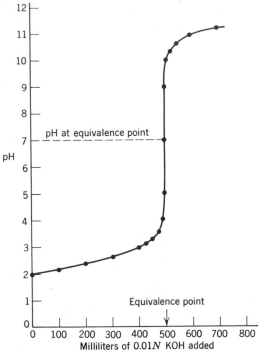

Figure 1-1 Titration of a strong acid (e.g., HCl) with a strong base (e.g., KOH).

rium constant, K_i :

$$K_i = \frac{[H_3O^+][A^-]}{[HA][H_2O]}$$

Because $[H_2O]$ is itself a constant, we can define a new constant, K_a, that combines K_i and $[H_2O]$.

$$K_i[H_2O] = K_a = \frac{[H_3O^+][A^-]}{[HA]}$$

Because $[H_3O^+]$ is the same as the "hydrogen ion concentration," $[H^+]$, the K_a expression is usually written as shown below.

$$K_a = \frac{[H^+][A^-]}{[HA]}$$

It is not surprising that the above K_a expression is identical to the expression that we would obtain if we assumed that the weak acid dissociates directly to yield H^+ and A^-:

$$HA \rightleftharpoons H^+ + A^-$$

IONIZATION OF WEAK BASES

In an aqueous solution, inorganic bases yield OH^- ions directly by dissociation.

$$KOH \rightarrow K^+ + OH^-$$

Organic bases such as amines $R\text{-}NH_2$, contain no OH to dissociate. However, if we assume that the $R\text{-}NH_2$ reacts with H_2O to form "$R\text{-}NH_3OH$," then we can consider the "dissociation" of organic bases to yield OH^- ions directly just as we do for inorganic bases. In fact, this is frequently done when we consider aqueous ammonia, NH_3; we assume that the dissociable substance present is "NH_4OH."

$$\text{"}R\text{-}NH_3OH\text{"} \rightleftharpoons R\text{-}NH_3^+ + OH^-$$
$$\text{"}NH_4OH\text{"} \rightleftharpoons NH_4^+ + OH^-$$

We should bear in mind that "$R\text{-}NH_3OH$" refers to the sum of $R\text{-}NH_2$ plus any small amount of $R\text{-}NH_3OH$ that might exist.

It usually makes little difference whether we consider the ionization of an acid as a simple dissociation or as the true ionization involving water as a conjugate base. However, in dealing with organic bases, it is usually more fruitful to consider the ionization as it actually occurs.

$$R\text{-}NH_2 \quad + \quad HOH \quad \rightleftharpoons \quad R\text{-}NH_3^+ \quad + \quad OH^-$$

[conjugate base]$_1$ [conjugate acid]$_2$ [conjugate acid]$_1$ [conjugate base]$_2$

The two conjugate acid-conjugate base pairs involved are $R\text{-}NH_3^+/R\text{-}NH_2$ and HOH/OH^-. The ionization can be described by an ionization constant, K_i.

$$K_i = \frac{[R\text{-}NH_3^+][OH^-]}{[R\text{-}NH_2][H_2O]}$$

Again, we can define a new constant, K_b, that combines the two constants, K_i and $[H_2O]$.

$$K_i[H_2O] = K_b = \frac{[R\text{-}NH_3^+][OH^-]}{[R\text{-}NH_2]}$$

The K_b expression is exactly the same as that which we obtain if we assume that $R\text{-}NH_2$ is actually $R\text{-}NH_3OH$, which dissociates directly to yield $R\text{-}NH_3^+$ and OH^-.

RELATIONSHIP BETWEEN K_a AND K_b FOR WEAK ACIDS AND BASES

When a weak acid, HA, is dissolved in water, it ionizes as shown previously to form H_3O^+ and the corresponding conjugate base, A^-. A K_a expression can be written for the ionization.

$$HA + H_2O \rightleftharpoons H_3O^+ + A^- \qquad K_a = \frac{[H_3O^+][A^-]}{[HA]}$$

If we *start* with the conjugate base, A^-, and dissolve it in water, it ionizes as a typical base; it accepts a proton from H_2O to form OH^- and the corresponding conjugate acid, HA. A K_b expression can be written for this ionization.

$$A^- + HOH \rightleftharpoons HA + OH^- \qquad K_b = \frac{[HA][OH^-]}{[A^-]}$$

Solving the K_a and K_b expressions for $[H_3O^+]$ and $[OH^-]$:

$$[H_3O^+] = \frac{[HA]K_a}{[A^-]} \qquad [OH^-] = \frac{[A^-]K_b}{[HA]}$$

Substituting into $[H_3O^+][OH^-] = K_w$:

$$\frac{[HA]K_a}{[A^-]} \times \frac{[A^-]K_b}{[HA]} = K_w$$

or
$$\boxed{K_a \times K_b = K_w} \qquad (23)$$

Taking logarithms:

$$\log K_a + \log K_b = \log K_w$$
$$-\log K_a - \log K_b = -\log K_w$$

Just as $-\log[H^+]$ has been defined as pH, we can define $-\log K_a$ as pK_a, $-\log K_b$ as pK_b, and $-\log K_w$ as pK_w (which equals 14).

$$\therefore \quad \boxed{pK_a + pK_b = 14} \qquad (23a)$$

For polyprotic acids, the K_a values are numbered in order of decreasing acid strength (K_{a_1}, K_{a_2}, etc.). The K_b values are numbered in order of decreasing base strength. However, remember that the conjugate base of the strongest acid group is the weakest basic group, and vice versa. The K_a and K_b values are numbered accordingly as shown below.

$$H_2A \underset{K_{b_2}}{\overset{K_{a_1}}{\rightleftharpoons}} H^+ + HA^-$$
$$K_{a_2} \big\Updownarrow K_{b_1}$$
$$H^+ + A^{2-}$$

We must be sure to use the K_a and K_b (or pK_a and pK_b) of *the same ionization.* For the diprotic acid illustrated above, the correct expressions are as follows:

$$K_{a_1} \times K_{b_2} = K_w \qquad pK_{a_1} + pK_{b_2} = pK_w$$
$$K_{a_2} \times K_{b_1} = K_w \qquad pK_{a_2} + pK_{b_1} = pK_w$$

pH OF SOLUTIONS OF WEAK ACIDS

The dissociation of a weak monoprotic acid, HA, yields H^+ and A^- in equal concentrations. If K_a and the initial concentration of HA are known, H^+ can be calculated easily:

$$K_a = \frac{[H^+][A^-]}{[HA]} = \frac{[H^+]^2}{[HA]}$$

$$[H^+] = \sqrt{K_a[HA]} \qquad (24)$$

Equation 24 assumes that the degree of ionization is small so that [HA] remains essentially unchanged. To obtain an expression giving pH, we put the above equation in logarithmic form:

$$\log[H^+] = \tfrac{1}{2}\log K_a[HA] = \tfrac{1}{2}(\log K_a + \log[HA])$$

$$-\log[H^+] = \frac{-\log K_a - \log[HA]}{2}$$

or

$$pH = \frac{pK_a + p[HA]}{2} \qquad (25)$$

where p[HA] is the negative logarithm of the HA concentration. Similar relationships can be derived for weak bases, A^- or RNH_2.

$$[OH^-] = \sqrt{K_b[A^-]} \qquad \text{and} \qquad pOH = \frac{pK_b + p[A^-]}{2} \qquad (26)$$

· Problem 1-18

The weak acid, HA, is 0.1% ionized (dissociated) in a 0.2 M solution. (a) What is the equilibrium constant for the dissociation of the acid (K_a)? (b) What is the pH of the solution? (c) How much "weaker" is the active acidity of the HA solution compared to a 0.2 M solution of HCl? (d) How many milliliters of 0.1 N KOH would be required to neutralize completely 500 ml of the 0.2 M HA solution?

Solution

(a)	HA	$\rightleftharpoons$	H^+	+	A^-
Start:	0.2 M		0		0
Change:	$-(0.1\%$ of 0.2 $M) =$				
	$-2 \times 10^{-4}\ M$		$+2 \times 10^{-4}\ M$		$+2 \times 10^{-4}\ M$
Equilibrium:	$0.2 - 2 \times 10^{-4}\ M$		$2 \times 10^{-4}\ M$		$2 \times 10^{-4}\ M$

$$K_a = \frac{[H^+][A^-]}{[HA]} = \frac{(2 \times 10^{-4})(2 \times 10^{-4})}{0.2 - 2 \times 10^{-4}}$$

When the amount of HA that has dissociated is small compared to the original concentration of HA, (e.g., 10% or less) the K_a expression may be simplified by ignoring the subtraction in the denominator.

$$K_a = \frac{(2 \times 10^{-4})(2 \times 10^{-4})}{0.2} = \frac{4 \times 10^{-8}}{2 \times 10^{-1}}$$

$$K_a = 2 \times 10^{-7}$$

(b)
$$\text{pH} = \log \frac{1}{[H^+]} = \log \frac{1}{2 \times 10^{-4}}$$
$$= \log 0.5 \times 10^4 = \log 5 \times 10^3$$
$$= \log 5 + \log 10^3 = 0.7 + 3$$

$$\boxed{\textbf{pH} = \textbf{3.7}}$$

(c)   A 0.2 M solution of HCl would be 100% ionized and yield 0.2 M H^+.

$$\text{pH} = \log \frac{1}{[H^+]} = \log \frac{1}{2 \times 10^{-1}} = \log 0.5 \times 10^1 = \log 5$$

$$\boxed{\textbf{pH} = \textbf{0.7}}$$ (assuming $\gamma = 1$)

$\therefore$ The weak acid is 3 pH units less acid than a comparable HCl solution.   Remember the pH scale is a *logarithmic* scale, not a linear scale.   The HA then is 10^3 or 1000 times less acid than the HCl (*not* 3 times).

(d)   Although the *active* acidity $[H^+]$ of the weak acid is 1000 times less than that of the HCl solution, the *total* acidity (free H^+ plus the undissociated hydrogen in HA) is the same.   When OH^- is added, it reacts with the free H^+ to form H_2O.   Some HA then immediately dissociates to H^+ and A^- to reestablish the equilibrium.   This H^+ is also neutralized by further additions of OH^- and so on until all the HA is neutralized.   Neutralization calculations for weak acids then may be conducted in the same manner as for strong acids.

number of moles of OH^- required = total number of moles H^+ available

$$\text{liters}_{\text{base}} \times N_{\text{base}} = \text{liters}_{\text{acid}} \times N_{\text{acid}}$$

HA is monoprotic.

$$\therefore \quad N = M$$
$$\text{liters}_{\text{base}} \times 0.1 = 0.5 \times 0.2$$
$$\text{liters}_{\text{base}} = \frac{0.5 \times 0.2}{0.1} = \frac{0.1}{0.1} = \boxed{\textbf{1 liter of base required}}$$

· Problem 1-19

The K_a for a weak acid, HA, is 1.6×10^{-6}.   What are the (a) pH and (b) degree of ionization of the acid in a 10^{-3} M solution?   (c) Calculate pK_a and pK_b.

Solution

(a)   Let $x = M$ of HA that dissociates. $\therefore$ $x = M$ of H^+ and also M of A^- produced.

	HA	$\rightleftharpoons$	H^+	$+$	A^-
Start:	10^{-3} M		0		0
Change:	$-x$ M		$+x$ M		$+x$ M
Equilibrium:	$10^{-3} - x$ M		x M		x M

$$K_a = \frac{[H^+][A^-]}{[HA]} = \frac{(x)(x)}{10^{-3} - x} = 1.6 \times 10^{-6}$$

First calculate x assuming that x is very much smaller than the concentration of un-ionized acid, that is, assuming that the acid is less than 10% ionized. The denominator of the K_a expression may then be simplified.

$$1.6 \times 10^{-6} = \frac{x^2}{10^{-3}}$$

$$x^2 = 1.6 \times 10^{-9} = 16 \times 10^{-10}$$

$$x = \sqrt{16 \times 10^{-10}} = \sqrt{16} \times \sqrt{10^{-10}} = 4.0 \times 10^{-5}$$

$$\boxed{[H^+] = 4 \times 10^{-5}\ M}$$

or using equation 24:

$$[H^+] = \sqrt{K_a[HA]} = \sqrt{(1.6 \times 10^{-6})(10^{-3})}$$

$$= \sqrt{16 \times 10^{-10}} = 4 \times 10^{-5}\ M$$

$$pH = \log\frac{1}{[H^+]}$$

$$= \log\frac{1}{4 \times 10^{-5}} = \log 0.25 \times 10^5 = \log 2.5 \times 10^4$$

$$= \log 2.5 + \log 10^4 = 0.398 + 4$$

$$\boxed{pH = 4.398}$$

(b) $$\text{degree of ionization} = \frac{[H^+]}{[HA]_{orig}} \times 100$$

$$= \frac{4 \times 10^{-5}}{10^{-3}} \times 100 = \frac{4 \times 10^{-3}}{10^{-3}} = \boxed{4\%}$$

The acid is indeed less than 10% ionized. Therefore, the simplification of the denominator term in the expression for K_a is reasonably valid.

(c) The pK_a is the negative logarithm of K_a.

$$pK_a = -\log K_a = \log\frac{1}{K_a}$$

$$pK_a = \log\frac{1}{1.6 \times 10^{-6}} = \log 6.25 \times 10^5$$

$$pK_a = \log 6.25 + \log 10^5 = 0.796 + 5$$

$$\boxed{pK_a = 5.796}$$

$$pK_a + pK_b = 14 \qquad pK_b = 14 - pK_a = 14 - 5.796$$

$$\boxed{pK_b = 8.204}$$

· Problem 1-20

Calculate (a) the H^+ ion concentration in a 0.02 M solution of a moderately strong acid, HA, where $K_a = 3 \times 10^{-2}$ M, and (b) the degree of dissociation of the acid.

Solution

(a)   Let: $x = M$ of HA dissociated
               $= M$ of H^+ produced

	HA	$\rightleftharpoons$	H^+	+	A^-
Start:	0.02 M		0		0
Change:	$-x\,M$		$+x\,M$		$+x\,M$
Equilibrium:	$0.02 - x\,M$		$x\,M$		$x\,M$

$$K_a = \frac{[H^+][A^-]}{[HA]} \qquad 3 \times 10^{-2} = \frac{(x)(x)}{0.02 - x}$$

The relatively large $K_a (>10^{-3})$ suggests that the acid is more than 10% dissociated.   Therefore, the denominator term in the K_a expression should not be simplified.

$$(3 \times 10^{-2})(0.02 - x) = x^2$$
$$6 \times 10^{-4} - 3 \times 10^{-2}x = x^2$$
$$x^2 + 3 \times 10^{-2}x - 6 \times 10^{-4} = 0$$

Solve for x using the general solution for quadratic equations.

$$x = \frac{-b \pm \sqrt{b^2 - 4ac}}{2a}$$

where $a = 1$ $b = 3 \times 10^{-2}$ $c = -6 \times 10^{-4}$.

$$x = \frac{-3 \times 10^{-2} \pm \sqrt{(3 \times 10^{-2})^2 - 4(-6 \times 10^{-4})}}{2}$$

$$= \frac{-3 \times 10^{-2} \pm \sqrt{(9 \times 10^{-4}) + 24 \times 10^{-4}}}{2}$$

$$= \frac{-3 \times 10^{-2} \pm \sqrt{33 \times 10^{-4}}}{2} = \frac{-3 \times 10^{-2} \pm 5.74 \times 10^{-2}}{2}$$

$$= \frac{+2.74 \times 10^{-2}}{2} \qquad \text{(neglecting the negative answer)}$$

$$\boxed{[H^+] = 1.37 \times 10^{-2} \ M}$$

(b) $$\text{degree of dissociation} = \frac{[H^+]}{[HA]_{orig}} \times 100$$

$$= \frac{0.0137}{0.0200} \times 100 \qquad \boxed{= 68.5\%}$$

· Problem 1-21

What is the pH of a 3.5×10^{-2} M solution of an amine with a pK_a of 9.6?

Solution

$$pK_b = pK_w - pK_a$$
$$= 14 - 9.6$$
$$pK_b = 4.4$$

$$p[RNH_2] = -\log [RNH_2]$$
$$= \log \frac{1}{3.5 \times 10^{-2}}$$
$$= \log 28.57$$
$$p[RNH_2] = 1.456$$

$$pOH = \frac{pK_b + p[RNH_2]}{2} = \frac{4.4 + 1.456}{2}$$
$$= \frac{5.86}{2} = 2.93$$

$$pH = 14 - pOH = 14 - 2.93$$

$$\boxed{\mathbf{pH = 11.07}}$$

· Problem 1-22

Calculate the ionic strength of a $0.1\,M$ solution of butyric acid. $K_a = 1.5 \times 10^{-5}$.

Solution

Butyric acid is only partially ionized. The undissociated molecules have no effect on the ionic strength of the solution. First calculate M_{H^+} and $M_{butyrate^-}$.

$$K_a = \frac{[H^+][butyrate^-]}{[butyric\ acid]} = \frac{(X)(X)}{0.1 - X} = \frac{X^2}{0.1} = 1.5 \times 10^{-5}$$

$$X^2 = 1.5 \times 10^{-6}$$

$$X = \sqrt{150 \times 10^{-8}} = 12.25 \times 10^{-4}$$

$$[H^+] = 1.225 \times 10^{-3}\,M, \quad [butyrate^-] = 1.225 \times 10^{-3}\,M$$

$$\frac{\Gamma}{2} = \frac{1}{2} \sum M_i Z_i^2 = \frac{1}{2}[M_{H^+} Z_{H^+}^2 + M_{butyrate^-} Z_{butyrate^-}^2]$$

$$= \frac{(1.225 \times 10^{-3})(1)^2 + (1.225 \times 10^{-3})(-1)^2}{2}$$

$$\boxed{\frac{\Gamma}{2} = 1.225 \times 10^{-3}}$$

EFFECT OF CONCENTRATION ON DEGREE OF DISSOCIATION

As a consequence of the law of mass action, the degree of dissociation of a weak acid varies with concentration. A dissociating system of the type $HA \rightleftharpoons H^+ + A^-$ has an unequal number of particles on the two sides of the equilibrium. As the total concentration of all species increases, the equilibrium shifts to the left (i.e., to the side with the fewer particles). As the total concentration of all species decreases, the equilibrium shifts to the right (i.e., to the side with the greater number of particles). The equilibrium constant remains constant. Only the relative proportions of the various species change.

A simple example will illustrate why the relative proportions must change upon dilution if K_{eq} is to remain constant. Consider the reaction $A \rightleftharpoons B + C$. At equilibrium, the concentrations of A, B, and C are related by:

$$K_{eq} = \frac{[B][C]}{[A]} = 1.0$$

Let the letters A, B, and C stand for some equilibrium concentrations, for example, 1 M. Now dilute the solution tenfold. If the relative proportions of A, B, and C remain constant:

New concentration of A = 0.1 [A] = 0.1 M

New concentration of B = 0.1 [B] = 0.1 M

New concentration of C = 0.1 [C] = 0.1 M

But $\dfrac{(0.1)(0.1)}{0.1}$ does *not* equal 1.0

Therefore, the relative concentrations must change to reestablish equilibrium. B and C increase while A decreases.

A mathematical relationship between concentration, K_a, and the degree of dissociation can be derived easily.

	HA	$\rightleftharpoons$	H^+	+	A^-
Start:	C		0		0
Change:	$-nC$		$+nC$		$+nC$
Equilibrium:	$C - nC$		nC		nC

where C = the original total concentration of HA
          n = the fraction dissociated (as a decimal)

$$K_a = \frac{(nC)(nC)}{(C - nC)} \qquad\qquad K_aC - K_anC = n^2C^2$$

$$K_a - K_an = n^2C \qquad\qquad\qquad K_a(1 - n) = n^2C$$

$$\boxed{C = \frac{1 - n}{n^2} K_a} \qquad\qquad (27)$$

If nC is very small compared to C, the expression for K_a simplifies to:

$$K_a = \frac{(nC)(nC)}{C} = \frac{n^2C^2}{C} = n^2C$$

Solving for C or n:

$$\boxed{C = \frac{K_a}{n^2}} \qquad \text{and} \qquad \boxed{n = \sqrt{\frac{K_a}{C}}} \qquad (28)$$

· **Problem 1-23**

At what concentration (in terms of K_a) of a weak acid, HA, will the acid be (a) 10% dissociated, (b) 50% dissociated, and (c) 90% dissociated?

Solution

(a)
$$C = \frac{1-n}{n^2} K_a = \frac{1-0.1}{(0.1)^2} K_a = \frac{0.9}{0.01} K_a \qquad \boxed{C = 90K_a}$$

(b)
$$C = \frac{1-0.5}{(0.5)^2} K_a = \frac{0.5}{0.25} K_a \qquad \boxed{C = 2K_a}$$

(c)
$$C = \frac{1-0.9}{(0.9)^2} K_a = \frac{0.1}{0.81} K_a \qquad \boxed{C = 0.123K_a}$$

Thus, we see that the degree of dissociation increases as the initial concentration of HA decreases.

"HYDROLYSIS" OF SALTS OF WEAK ACIDS AND BASES

Salts of weak acids (the conjugate base anion of weak acids) react with water to produce the weak parent acid (conjugate acid) and OH^- ions.

$$A^- + HOH \rightleftharpoons HA + OH^-$$

We can see that the "hydrolysis" is nothing more than the ionization of the conjugate base as described earlier. The "hydrolysis constant," K_h, is identical to K_b.

$$K_{eq} = \frac{[HA][OH^-]}{[A^-][HOH]} \qquad K_{eq}[HOH] = K_h = \frac{[HA][OH^-]}{[A^-]} = K_b$$

Similarly, salts of weak bases (the conjugate acid of weak bases) react with water to produce the weak parent base and H^+ ions.

$$R\text{-}NH_3^+ + HOH \rightleftharpoons R\text{-}NH_2 + H_3O^+$$
$$NH_4^+ + HOH \rightleftharpoons NH_3 + H_3O^+$$

or

$$NH_4^+ + HOH \rightleftharpoons \text{"}NH_4OH\text{"} + H^+$$

Again, we see that the "hydrolysis" is nothing more than the usual ionization of the conjugate acid. In this instance, the K_h that is defined is identical to K_a for the conjugate acid.

$$K_{eq} = \frac{[R\text{-}NH_2][H_3O^+]}{[R\text{-}NH_3^+][HOH]} \qquad K_{eq}[HOH] = K_h = \frac{[R\text{-}NH_2][H_3O^+]}{[R\text{-}NH_3^+]} = K_a$$

· **Problem 1-24**

(a) Calculate the pH of a 0.1 M solution of NH_4Cl. The K_b for NH_4OH is 1.8×10^{-5}. (b) What is the degree of hydrolysis of the salt?

Solution

(a)   NH₄Cl is a salt of a weak base and a strong acid.   Therefore a solution of NH₄Cl will be acidic because of "hydrolysis" of the NH_4^+ ion.

$$NH_4^+ + HOH \rightleftharpoons NH_4OH + H^+$$

$$K_h = K_a = \frac{K_w}{K_b} = \frac{1 \times 10^{-14}}{1.8 \times 10^{-5}} = 5.56 \times 10^{-10}$$

$$K_h = \frac{[NH_4OH][H^+]}{[NH_4^+]} = 5.56 \times 10^{-10}$$

Let

$$y = M \text{ of } NH_4OH \text{ produced upon hydrolysis}$$

$$\therefore \quad y = M \text{ of } H^+ \text{ produced upon hydrolysis}$$

$$\frac{(y)(y)}{(0.1 - y)} = 5.56 \times 10^{-10}$$

Simplifying:

$$y^2 = 5.56 \times 10^{-11} \qquad y = \sqrt{55.6} \times \sqrt{10^{-12}}$$

$$[H^+] = 7.46 \times 10^{-6} \qquad pH = \log \frac{1}{7.46 \times 10^{-6}}$$

$$\boxed{pH = 5.13}$$

(b) $$\text{degree of hydrolysis} = \frac{[H^+]}{[NH_4^+]} \times 100\% = \frac{7.46 \times 10^{-6}}{10^{-1}} \times 100$$

$$= \frac{7.46 \times 10^{-4}}{10^{-1}} = \boxed{7.46 \times 10^{-3}\%}$$

Because the NH_4^+ is less than 10% hydrolyzed (ionized), our substitution of 0.1 M for $0.1 - y\ M$ is reasonably valid.

HENDERSON-HASSELBALCH EQUATION

A useful expression relating the (a) K_a of a weak acid, HA, and the pH of a solution of the weak acid or (b) K_b of a weak base and the pOH of a solution of the weak base can be derived.

(a) $$K_a = \frac{[H^+][A^-]}{[HA]}$$

Rearranging terms:

$$[H^+] = K_a \frac{[HA]}{[A^-]}$$

Taking logarithms of both sides:

$$\log [H^+] = \log K_a + \log \frac{[HA]}{[A^-]}$$

Multiplying both sides by -1:

$$-\log [H^+] = -\log K_a - \log \frac{[HA]}{[A^-]}$$

$$pH = pK_a - \log \frac{[HA]}{[A^-]}$$

$$\boxed{pH = pK_a + \log \frac{[A^-]}{[HA]}} \tag{29}$$

(b)

$$K_b = \frac{[M^+][OH^-]}{[MOH]}$$

$$[OH^-] = K_b \frac{[MOH]}{[M^+]}$$

$$\log [OH^-] = \log K_b$$
$$+ \log \frac{[MOH]}{[M^+]}$$

$$-\log [OH^-] = -\log K_b$$
$$- \log \frac{[MOH]}{[M^+]}$$

$$pOH = pK_b - \log \frac{[MOH]}{[M^+]}$$

$$\boxed{pOH = pK_b + \log \frac{[M^+]}{[MOH]}}$$

$$K_b = \frac{[R\text{-}NH_3^+][OH^-]}{[R\text{-}NH_2]}$$

$$[OH^-] = K_b \frac{[R\text{-}NH_2]}{[R\text{-}NH_3^+]}$$

$$\log [OH^-] = \log K_b$$
$$+ \log \frac{[R\text{-}NH_2]}{[R\text{-}NH_3^+]}$$

$$-\log [OH^-] = -\log K_b$$
$$- \log \frac{[R\text{-}NH_2]}{[R\text{-}NH_3^+]}$$

$$pOH = pK_b - \log \frac{[R\text{-}NH_2]}{[R\text{-}NH_3^+]}$$

$$\boxed{pOH = pK_b + \log \frac{[R\text{-}NH_3^+]}{[R\text{-}NH_2]}}$$

$$\tag{30}$$

Note that when the concentrations of conjugate acid and conjugate base are equal, $pH = pK_a$ and $pOH = pK_b$. This same relationship can be seen from the original K_a or K_b expressions; when $[A^-] = [HA]$, $[H^+] = K_a$ and when $[R\text{-}NH_2] = [R\text{-}NH_3^+]$, $[OH^-] = K_b$.

According to the above equations, the pH of a solution containing HA and A^- is independent of concentration; the pH is established solely by the *ratio* of conjugate base to conjugate acid. This is not quite true, as we will see. For the moment, we will assume that the conjugate base/conjugate acid ratio is the determining factor. The assumption is valid as long as $[A^-]$ and $[HA]$ are high compared to K_a, but not so high as to warrant corrections for activity coefficients. Under the usual laboratory conditions, the concentrations might be 0.1 M or less with K_a values of 10^{-3} or less, so this condition is met.

D. LABORATORY BUFFERS

TITRATION OF A WEAK ACID

When a strong acid is titrated with a strong base the pH at any point is determined solely by the concentration of untitrated acid or excess base (Problem 1-17). The conjugate base that is formed (e.g., Cl⁻) has no effect on pH.   The situation is quite different when a weak acid is titrated with a strong base.   A weak acid dissociates in an aqueous solution to yield a small amount of H^+ ions.

1. $$HA \rightleftharpoons H^+ + A^-$$

When OH^- ions are added, they are neutralized by the H^+ ions to form H_2O.

2. $$OH^- + H^+ \rightarrow H_2O$$

The removal of H^+ ions disturbs the equilibrium between the weak acid and its ions.   Consequently,   more   HA   ionizes   to   reestablish   the   equilibrium.   The newly produced H^+ ions can then be neutralized by more OH^- and so on until all of the hydrogen originally present is neutralized.   The overall result, the sum of reactions 1 and 2, is the titration of HA with OH^-.

3. $$HA + OH^- \rightleftharpoons H_2O + A^-$$

The number of equivalents of OH^- required equals the total number of equivalents of hydrogen present (as H^+ plus HA).

The pH at the exact end (equivalence) point of the titration is not 7 but higher because of the hydrolysis of the A^- ion; that is, because reaction 3 itself is an equilibrium reaction.   In the absence of any remaining HA, the A^- ion reacts with H_2O to produce OH^- ions and the undissociated weak acid, HA.   Because equilibrium conditions must always be satisfied in solutions of weak acids and bases, the H^+ ion concentration and pH during the titration can be calculated from the K_a expression or from the Henderson-Hasselbalch equation, provided the concentration of conjugate acid and conjugate base (or the ratio of their concentrations) is known.   When calculating the values for [HA] and $[A^-]$ during a titration, it is safe to assume that moles HA_{remain} = moles HA_{orig} − moles $HA_{titrated}$, and moles A^- = moles $HA_{titrated}$ throughout *most* of the titration curve.   Significant errors (resulting from hydrolysis of the salt) arise only when an equivalence point is approached.   The weaker the acid (in terms of K_a as well as original concentration), the sooner (in terms of percent of the original acid titrated) anomalous answers result from ignoring hydrolysis.

In other words, if A^- is 10% ionized (to $HA + OH^-$), then the Henderson-Hasselbalch equation cannot be used to estimate the pH of a solution of HA that has been 99% titrated.

· Problem 1-25

Calculate the appropriate values and draw the curve for the titration of 500 ml of 0.1 M weak acid, HA, with 0.1 M KOH. $K_a = 10^{-5}$ ($pK_a = 5.0$).

Solution

The titration curve is shown in Figure 1-2.   The values were calculated as shown below.

(a)   At the start, the pH depends solely on the concentration of HA and the value of K_a.

$$pH = \frac{pK_a + p[HA]}{2} = \frac{5+1}{2} = 3.0$$

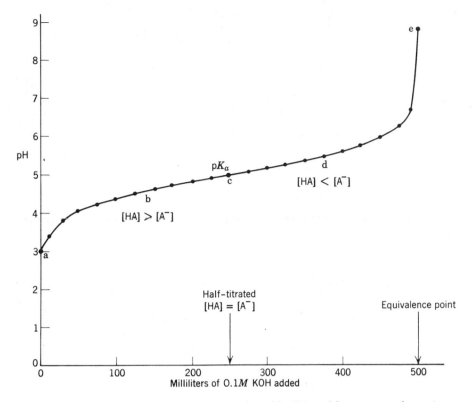

Figure 1-2 Titration of a weak monoprotic acid, HA, with a strong base (e.g., KOH). $pK_a = 5.0$; [HA] at the start $= 0.1\ M$.

(b) At any point during the titration the pH can be calculated from:

$$pH = pK_a + \log \frac{[A^-]}{[HA]}$$

For example, after adding 100 ml of 0.1 M KOH:

$\qquad$ 0.1 liter $\times$ 0.1 $M = 0.01$ moles of OH^- have been added

$\therefore\quad$ 0.01 moles of HA have been converted to 0.01 moles of A^-

$\qquad$ moles HA remaining = moles HA originally present
$\qquad\qquad\qquad\qquad\qquad$ $-$ moles of HA titrated to A^-
$\qquad\qquad\qquad\qquad$ $= (0.5\ \text{liter} \times 0.1\ M) - 0.01$
$\qquad\qquad\qquad\qquad$ $= 0.05 - 0.01 = 0.04$ moles

The volume has changed, but the ratio of moles A^-/moles HA is the same as the ratio of $[A^-]/[HA]$.

$$pH = 5.0 + \log \frac{0.01}{0.04} = 5.0 + \log 0.25$$

To avoid dealing with the log of a number smaller than unity, the Henderson-Hasselbalch equation can be written as

$$pH = pK_a - \log\frac{[HA]}{[A]} = 5.0 - \log\frac{0.04}{0.01}$$
$$= 5.0 - \log 4 = 5.0 - 0.602$$

$$\boxed{pH = 4.40}$$

(c)   When 250 ml of 0.1 M KOH has been added, the original HA is half titrated.

$$pH = pK_a + \log\frac{[A^-]}{[HA]}$$

$$[A^-] = [HA] \qquad \frac{[A^-]}{[HA]} = 1 \qquad \log 1 = 0$$

$$\therefore \quad pH = pK_a \qquad or \qquad \boxed{pH = 5.0}$$

(d)   Beyond the halfway point, the pH is still given by:

$$pH = pK_a + \log\frac{[A^-]}{[HA]}$$

For example, after adding 375 ml of 0.1 M KOH:

$$0.375 \text{ liter} \times 0.1 \ M = 0.0375 \text{ moles OH}^- \text{ added}$$
$$\therefore \quad 0.0375 \text{ moles of HA have been converted to } 0.0375 \text{ moles of A}^-$$
$$\text{HA remaining} = 0.0500 - 0.0375 = 0.0125$$
$$pH = 5.0 + \log\frac{0.0375}{0.0125} = 5.0 + \log 3$$

$$= 5.0 + 0.477 \qquad \boxed{pH = 5.48}$$

Note that when the acid is less than half titrated, the pH is less than pK_a.   When the acid is exactly half titrated, the pH equals pK_a.   When the acid is more than half titrated, the pH is greater than pK_a.

(e)   When exactly 500 ml of 0.1 M KOH have been added, we have theoretically titrated all the HA to A$^-$.   However, the pH at the endpoint is not 7 because A$^-$, the salt or conjugate base of HA, ionizes:

$$A^- + HOH \rightleftharpoons HA + OH^-$$

Note that the ionization equation read backwards is the equation for the titration of HA with OH$^-$.   Thus, the titration is an equilibrium reaction that does not go to completion.   The addition of 1 mole of OH$^-$ to 1 mole of HA does not produce exactly 1 mole of A$^-$.   The pH at the endpoint can be calculated from K_b.   First note that the concentration of A$^-$ is $5 \times 10^{-2} \ M$ at the endpoint (500 ml of 0.1 M KOH have been added to 500 ml of 0.1 M HA yielding 1 liter of solution containing 0.05 moles of A$^-$).

$$K_b = \frac{[HA][OH^-]}{[A^-]} = \frac{K_w}{K_a} = \frac{10^{-14}}{10^{-5}} = 10^{-9}$$

$$pOH = \frac{pK_b + p[A^-]}{2} = \frac{9 + \log\frac{1}{0.05}}{2}$$

$$= \frac{9 + 1.30}{2} = \frac{10.30}{2}$$

$$\boxed{pOH = 5.15}$$

$$pH = pK_w - pOH \qquad pH = 14 - 5.15$$

$$\boxed{pH = 8.85}$$

If we started with A^- and titrated with HCl, the curve would be essentially identical to that shown in Figure 1-2 with the horizontal axis reading (from right to left) "ml of 0.1 M HCl added."

WHAT IS A BUFFER? HOW DOES A BUFFER WORK?

A "buffer" is something that resists change. In common chemical usage, a pH buffer is a substance, or mixture of substances, that permits solutions to resist *large changes* in pH upon the addition of small amounts of H^+ or OH^- ions. To put it another way, a buffer helps maintain a *near constant* pH upon the addition of small amounts of H^+ or OH^- ions to a solution.

Common buffer mixtures contain two substances, a conjugate acid and a conjugate base. An "acidic" buffer contains a weak acid and a salt of the weak acid (conjugate base). A "basic" buffer contains a weak base and a salt of the weak base (conjugate acid). Together the two species (conjugate acid plus conjugate base) resist large changes in pH by *partially* absorbing additions of H^+ or OH^- ions to the system. If H^+ ions are added to the buffered solution, they react partially with the conjugate base present to form the conjugate acid. Thus, some H^+ ions are taken out of circulation. If OH^- ions are added to the buffered solution, they react partially with the conjugate acid present to form water and the conjugate base. Thus, some OH^- ions are taken out of circulation. Buffered solutions *do* change in pH upon the addition of H^+ or OH^- ions. However, the change is much less than that which would occur if no buffer were present. The amount of change depends on the strength of the buffer and the $[A^-]/[HA]$ ratio (see Buffer Capacity). The solution obtained by titrating HA with KOH (Problem 1-25) is a buffer. If we examine the titration curve, we see that in the region of pK_a the pH changes only slightly as OH^- is added. Thus $HA + A^-$ provides good buffering action in the neighborhood of pH 5. However, this particular weak acid plus its salt is a poor buffer at pH 7. If we wished to prepare a solution that would buffer at pH 7, we would use a weak acid whose pK_a is around 7.

· **Problem 1-26**

(a) Describe the components of an "acetate" buffer.   (b) Show the reactions by which an acetate buffer resists changes in pH upon the addition of OH^- and H^+ ions.

Solution

(a)   An "acetate" buffer contains un-ionized acetic acid (HOAc) as the conjugate acid and acetate ions (OAc^-) as the conjugate base.   The OAc^- may be provided directly by NaOAc, KOAc, and the like, or by neutralizing a portion of the HOAc with KOH or NaOH.

(b)   In a solution containing a weak acid such as HOAc, a certain condition must be met—namely, the product of $[H^+][OAc^-]$ divided by [HOAc] must be constant:

$$K_a = \frac{[H^+][OAc^-]}{[HOAc]}$$

A change in the concentration of any one of the three components of the K_a expression causes the concentrations of the other two to alter appropriately so that $[H^+][OAc^-]$ divided by [HOAc] is still the same constant value (K_a).

For example, if OH^- ions are added to the system, they react with the H^+ ions present to form H_2O.

$$\boxed{OH^- + H^+ \rightarrow H_2O}$$

The reduction in $[H^+]$ disturbs the equilibrium momentarily.   Consequently, more HOAc dissociates to reestablish the equilibrium condition.

$$\boxed{HOAc \rightleftharpoons H^+ + OAc^-}$$

The net result (as well as the sum of the above two reactions) is as if the OH^- ions react directly with the conjugate acid of the acetate buffer to yield H_2O plus more conjugate base $[OAc^-]$.

$$\boxed{OH^- + HOAc \rightleftharpoons H_2O + OAc^-}$$

All of this, of course, happens almost instantaneously.

Similarly, if H^+ ions are added to the system, the equilibrium again shifts.   This time the conjugate base $[OAc^-]$ reacts with some of the excess H^+ ions to form un-ionized HOAc.

$$\boxed{H^+ + OAc^- \rightleftharpoons HOAc}$$

It should be emphasized that the excess H^+ or OH^- ions are not *completely* neutralized by the buffer; that is, the pH does not remain absolutely constant upon addition of H^+ or OH^- ions to a buffer. The reactions by which H^+ and OH^- ions are absorbed are themselves equilibrium reactions and do not go to completion.

PREPARATION OF BUFFERS

· Problem 1-27

What are the concentrations of HOAc and OAc⁻ in a 0.2 M "acetate" buffer, pH 5.00? The K_a for acetic acid is 1.70×10^{-5} ($pK_a = 4.77$).

Solution

A "0.2 M acetate" buffer contains a *total* of 0.2 mole of "acetate" per liter. Some of the total acetate is in the conjugate acid form, HOAc, and some is in the conjugate base form, OAc⁻. The proportions (hence, the concentrations) of each form may be solved by using either the K_a expression or the Henderson-Hasselbalch equation.

$$K_a = \frac{[H^+][OAc^-]}{[HOAc]}$$

Let
$$y = M \text{ of } OAc^-$$
$$\therefore \quad 0.2 - y = M \text{ of } HOAc$$

$$pH = 5$$

$$\therefore \quad [H^+] = 10^{-5}$$

$$1.7 \times 10^{-5} = \frac{(10^{-5})(y)}{(0.2 - y)}$$

$$3.4 \times 10^{-6} - 1.7 \times 10^{-5}y$$

$$= 1 \times 10^{-5}y$$

$$3.4 \times 10^{-6} = 2.7 \times 10^{-5}y$$

$$y = \frac{3.4 \times 10^{-6}}{27 \times 10^{-6}}$$

$$y = 0.126$$

$$\boxed{[OAc^-] = 0.126 \ M}$$

$$[HOAc] = 0.200 - 0.126$$

$$\boxed{[HOAc] = 0.074 \ M}$$

$$pH = pK_a + \log \frac{[OAc^-]}{[HOAc]}$$

Let
$$y = M \text{ of } OAc^-$$
$$\therefore \quad 0.2 - y = M \text{ of } HOAc$$

$$5.00 = 4.77 + \log \frac{y}{0.2 - y}$$

$$0.23 = \log \frac{y}{0.2 - y}$$

$$\frac{y}{0.2 - y} = \text{antilog of } 0.23$$

$$\frac{y}{0.2 - y} = 1.70$$

$$0.34 - 1.70y = y$$

$$0.34 = 2.7y$$

$$y = 0.126$$

$$\boxed{[OAc^-] = 0.126 \ M}$$

$$[HOAc] = 0.2 - y \ M$$

$$[HOAc] = 0.2 - 0.126$$

$$\boxed{[HOAc] = 0.074 \ M}$$

or $\qquad \mathrm{pH} = \mathrm{p}K_a + \log \dfrac{[\mathrm{OAc^-}]}{[\mathrm{HOAc}]} \qquad 5 = 4.77 + \log \dfrac{[\mathrm{OAc^-}]}{[\mathrm{HOAc}]}$

$$0.23 = \log \frac{[\mathrm{OAc^-}]}{[\mathrm{HOAc}]} \qquad \frac{[\mathrm{OAc^-}]}{[\mathrm{HOAc}]} = \text{antilog of } 0.23$$

$$\frac{[\mathrm{OAc^-}]}{[\mathrm{HOAc}]} = 1.70 = \frac{1.70}{1}$$

$\therefore \quad \dfrac{1.70}{2.70}$ of total $= \mathrm{OAc^-} \qquad$ and $\qquad \dfrac{1.00}{2.70}$ of total $= \mathrm{HOAc}$

$\dfrac{1.70}{2.7} \times 0.2\ M = 0.126 \qquad$ $\boxed{[\mathrm{OAc^-}] = 0.126\ M}$

$\dfrac{1.00}{2.70} \times 0.2\ M = 0.074 \qquad$ $\boxed{[\mathrm{HOAc}] = 0.074\ M}$

Check: The pH is higher than the $\mathrm{p}K_a$. $\therefore$ The solution should contain more conjugate base than conjugate acid. Conjugate base $= 0.126\ M$. Conjugate acid $= 0.074\ M$.

· Problem 1-28

Describe the preparation of 3 liters of a 0.2 M acetate buffer, pH 5.00, starting from solid sodium acetate trihydrate (MW 136) and a 1 M solution of acetic acid.

Solution

First calculate the molarities of $\mathrm{OAc^-}$ and HOAc present. Any of the three methods shown in Problem 1-27 may be used to obtain $[\mathrm{OAc^-}] = 0.126\ M$ and $[\mathrm{HOAc}] = 0.074\ M$. We need 3 liters of the 0.2 M buffer.

$$3 \text{ liters} \times 0.2\ M = 0.6 \text{ mole } total \text{ (HOAc plus OAc}^-\text{)}$$

The total of 0.6 mole is obtained from two sources:

$$3 \text{ liters} \times 0.126\ M = 0.378 \text{ mole } \mathrm{OAc^-}$$
$$3 \text{ liters} \times 0.074\ M = 0.222 \text{ mole HOAc}$$

The 0.378 mole of $\mathrm{OAc^-}$ comes from solid NaOAc.

$$\text{number of moles} = \frac{\mathrm{wt_g}}{\mathrm{MW}} \qquad 0.378 = \frac{\mathrm{wt_g}}{136}$$

$$\boxed{\mathrm{wt_g} = 51.4\ g}$$

The 0.222 mole of HOAc comes from a 1 M stock solution.

$$\text{number of moles} = \text{liters} \times M \qquad 0.222 = \text{liters} \times 1$$

$$\text{liters} = 0.222 \qquad \text{or} \qquad \boxed{222\ \mathrm{ml}}$$

Therefore, to prepare 3 liters of the buffer, dissolve 51.4 g of the sodium acetate in some water, add 222 ml of the 1 M acetic acid, and then dilute to a total final volume of 3.0 liters.

· **Problem 1-29**

Describe the preparation of 5 liters of a 0.3 M acetate buffer, pH 4.47, starting from a 2 M solution of acetic acid and a 2.5 M solution of KOH.

Solution

As in the previous problem, first calculate the proportions of the two acetate species present.

$$pH = pK_a + \log \frac{[OAc^-]}{[HOAc]} \qquad 4.47 = 4.77 + \log \frac{[OAc^-]}{[HOAc]}$$

$$-0.30 = \log \frac{[OAc^-]}{[HOAc]} \quad \text{or} \quad +0.30 = \log \frac{[HOAc]}{[OAc^-]}$$

$$\frac{[HOAc]}{[OAc^-]} = \text{antilog of } 0.3 = 2 = \frac{2}{1} \text{ ratio}$$

∴ $\frac{2}{3}$ of the total acetate is present as HOAc and $\frac{1}{3}$ of the total acetate is present as OAc⁻. The final solution contains:

$$\frac{2}{3} \times 0.3 \ M = 0.2 \ M \ HOAc \ (1 \text{ mole in 5 liters})$$

$$\frac{1}{3} \times 0.3 \ M = 0.1 \ M \ OAc^- \ (0.5 \text{ mole in 5 liters})$$

In this buffer, *all* of the acetate must be provided by the HOAc. The buffer is prepared by converting the *proper proportion* of the HOAc to OAc⁻ by adding KOH. We need 5 liters × 0.3 M = 1.5 moles *total* acetate. Calculate how much stock 2 M HOAc is needed to obtain 1.5 moles.

$$\text{liters} \times M = \text{number of moles}$$

$$\text{liters} \times 2 = 1.5 \qquad \text{liters} = \frac{1.5}{2} = 0.75$$

∴ 750 ml of the 2 M HOAc is required.

Next, convert $\frac{1}{3}$ of the 1.5 moles to OAc⁻ by adding the proper amount of 2.5 M KOH.

$$\frac{1}{3} \times 1.5 \text{ moles} = 0.5 \text{ mole KOH needed}$$

$$\text{liters} \times M = \text{number of moles}$$

$$\text{liters} \times 2.5 = 0.5 \text{ mole}$$

$$\text{liters} = \frac{0.5}{2.5} = 0.2 \text{ liter}$$

∴ Add 200 ml of 2.5 M KOH.

The solution now contains 1 mole of HOAc and 0.5 mole of OAc⁻. Finally, add sufficient water to bring the volume up to 5 liters. The final solution contains 0.2 M HOAc and 0.1 M OAc⁻.

pH *CHANGES IN BUFFERS*

In general, a buffer is used to maintain the pH relatively constant during the course of a reaction that produces or utilizes H⁺ ions. As we shall see in a following section, the ability of a buffer to maintain a near-constant pH increases as the concentration of the buffer increases. However, it is not always possible to use a relatively concentrated buffer. The enzyme, tissue, or cells under investigation may be sensitive to high ionic strength, or the assay may require that the pH be adjusted easily to some higher or lower value at the end of the reaction. Thus, a compromise is necessary. The concentration and pH of the buffer are chosen so that the pH will remain as near constant as possible without introducing complications due to high ionic strength. There are some circumstances where we want the pH to change significantly (e.g., when the extent of a reaction is measured by the pH change). In this case, we would use the lowest concentration of buffer possible without allowing the pH to move out of a range optional for the reaction under investigation.

· **Problem 1-30**

Show mathematically why an acetate buffer cannot maintain an absolutely constant pH upon the addition of H^+.

Solution

Suppose we have a buffer containing $0.01 \ M$ HA and $0.01 \ M$ A^-. Assume that the K_a of the weak acid is also 10^{-5}. Consequently, the H^+ ion concentration must also be $10^{-5} \ M$.

$$\frac{[H^+][A^-]}{[HA]} = K_a \qquad \frac{(10^{-5})(10^{-2})}{(10^{-2})} = 10^{-5}$$

Now suppose $10^{-3} \ M$ H^+ is added to the buffer. If *all* of the H^+ reacts with A^- to yield HA (thus maintaining $[H^+]$ at $10^{-5} \ M$), the new concentration of $[HA]$ would be $1.1 \times 10^{-2} \ M$ and the new concentration of $[A^-]$ would be $0.9 \times 10^{-2} \ M$. Substituting these values into the K_a expression, we can see that the $[H^+][A^-]$ divided by $[HA]$ is *not* constant and equal to 10^{-5}.

$$\frac{(10^{-5})(0.9 \times 10^{-2})}{(1.1 \times 10^{-2})} \neq 10^{-5}$$

· **Problem 1-31**

Consider a $0.002 \ M$ acidic buffer containing $10^{-3} \ M$ HA and $10^{-3} \ M$ A^-. The $pH = pK_a = 5$ ($K_a = 10^{-5}$). Suppose that 5×10^{-4} moles of H^+ are added to 1 liter of the buffer (assume that the volume remains at 1 liter). (a) Calculate the *exact* concentrations of A^- and HA and the pH of the solution after addition of the HCl. (b) Calculate the concentrations of A^- and HA and the pH of the solution assuming that the increase in the amount of HA (and the decrease in the amount of A^-) is equal to the amount of H^+ added.

Solution

The added H^+ is partially utilized by reacting with A^- to form un-ionized HA.

$$H^+ + A^- \rightleftharpoons HA$$

(a)   Let

$$y = M \text{ of } H^+ \text{ utilized by the buffer}$$

$$[A^-] = 10^{-3} - y \, M \qquad [HA] = 10^{-3} + y \, M$$

$$[H^+] = 10^{-5} + 5 \times 10^{-4} - y = 51 \times 10^{-5} - y$$

$$K_a = \frac{[H^+][A^-]}{[HA]} = \frac{(51 \times 10^{-5} - y)(10^{-3} - y)}{(10^{-3} + y)} = 10^{-5}$$

Cross multiplying:

$$10^{-8} + 10^{-5}y = 51 \times 10^{-8} - 51 \times 10^{-5}y - 10^{-3}y + y^2$$

Rearranging and collecting terms:

$$y^2 - 10^{-3}y - 51 \times 10^{-5}y - 10^{-5}y + 51 \times 10^{-8} - 10^{-8} = 0$$

$$y^2 - 100 \times 10^{-5}y - 51 \times 10^{-5}y - 1 \times 10^{-5}y + 50 \times 10^{-8} = 0$$

$$y^2 - 152 \times 10^{-5}y + 50 \times 10^{-8} = 0$$

$$y = \frac{-b \pm \sqrt{b^2 - 4ac}}{2a}$$

where $a = 1 \qquad b = -152 \times 10^{-5} \qquad c = 50 \times 10^{-8}$.

$$y = \frac{+152 \times 10^{-5} \pm \sqrt{(-15.2 \times 10^{-4})^2 - 4(50 \times 10^{-8})}}{2}$$

$$y = \frac{15.2 \times 10^{-4} \pm \sqrt{231 \times 10^{-8} - 200 \times 10^{-8}}}{2}$$

$$y = \frac{15.2 \times 10^{-4} \pm \sqrt{31 \times 10^{-8}}}{2} = \frac{15.20 \times 10^{-4} \pm 5.57 \times 10^{-4}}{2}$$

$$y = \frac{20.77 \times 10^{-4}}{2} \quad \text{and} \quad \frac{9.63 \times 10^{-4}}{2}$$

$$y = 10.39 \times 10^{-4} \quad \text{and} \quad 4.815 \times 10^{-4} \, M$$

The higher value is obviously incorrect because only $5 \times 10^{-4} \, M \, H^+$ was added.

$$\therefore \quad y = 4.815 \times 10^{-4} \, M$$

Thus, of the $5 \times 10^{-4} \, M \, H^+$ originally added, $4.815 \times 10^{-4} \, M$ was utilized by the buffer.   The final H^+ ion concentration was increased by $0.185 \times 10^{-4} \, M$.

$$[H^+]_{\text{final}} = (1 \times 10^{-5}) + (1.85 \times 10^{-5}) \qquad \boxed{[H^+]_{\text{final}} = 2.85 \times 10^{-5} \, M}$$

$$[A^-]_{\text{final}} = (10 \times 10^{-4}) - (4.82 \times 10^{-4}) \qquad \boxed{[A^-]_{\text{final}} = 5.18 \times 10^{-4} \, M}$$

$$[HA]_{\text{final}} = (10 \times 10^{-4}) + (4.82 \times 10^{-4}) \qquad \boxed{[HA]_{\text{final}} = 14.82 \times 10^{-4} \, M}$$

$$pH_{final} = \log \frac{1}{[H^+]_{final}} = \log \frac{1}{2.85 \times 10^{-5}}$$
$$= \log 0.351 \times 10^5 = \log 3.51 \times 10^4$$

$$\boxed{pH = 4.545}$$

In other words, the pH decreased by 0.455 unit.

(b)   If we assume that virtually *all* the H^+ reacts with A^- to form HA, in order to simplify the calculations:

$$[A^-]_{final} = (10 \times 10^{-4}) - (5 \times 10^{-4}) \qquad \boxed{\mathbf{[A^-]_{final} = 5.0 \times 10^{-4}\ M}}$$

$$[HA]_{final} = (10 \times 10^{-4}) + (5 \times 10^{-4}) \qquad \boxed{\mathbf{[HA]_{final} = 15 \times 10^{-4}\ M}}$$

The estimated new $[H^+]$ is that which is in equilibrium with $15 \times 10^{-4}\ M$ HA and $5 \times 10^{-4}\ M$ A^-.   The estimated value will be slightly high because, as shown in part a, the $[HA]/[A^-]$ ratio is actually a little less than $3:1$.

$$K_a = \frac{[H^+][A^-]}{[HA]} \qquad\qquad pH = pK_a + \log \frac{[A^-]}{[HA]}$$

Let $y = M$ of H^+ present $\qquad\qquad pH = 5.00 + \log \dfrac{(5 \times 10^{-4})}{(15 \times 10^{-4})}$

$$10^{-5} = \frac{(y)(5 \times 10^{-4})}{(15 \times 10^{-4})} \qquad\qquad pH = 5.00 - \log \frac{(15 \times 10^{-4})}{(5 \times 10^{-4})}$$

$$15 \times 10^{-9} = (5 \times 10^{-4})y \qquad\qquad pH = 5.00 - \log 3$$

$$y = \frac{15 \times 10^{-9}}{5 \times 10^{-4}} \qquad\qquad pH = 5.00 - 0.477$$

$$y = 3 \times 10^{-5}$$

$$\boxed{\mathbf{[H^+] = 3 \times 10^{-5}\ M}} \qquad\qquad \boxed{\mathbf{pH = 4.523}}$$

The calculated H^+ ion concentration increase is $2 \times 10^{-5}\ M$ (compared to the true value of $1.85 \times 10^{-5}\ M$). | The calculated pH decrease is 0.477 unit (compared to the true value of 0.455).

We can see that the error introduced by assuming that the buffer reacts *completely* with the added H^+ is small.   In the above problem, the buffer is relatively weak and the amount of H^+ ion added is of the same order of magnitude as the original A^- concentration.   In practice, the concentration of the buffer employed would be high compared to the expected change in H^+ (or OH^-) ion concentration.   Consequently, buffer calculations may be simplified greatly, as shown in part b above, without undue error.

· Problem 1-32

An enzyme-catalyzed reaction was carried out in a 0.2 M "Tris" buffer, pH 7.8. As a result of the reaction, 0.03 mole/liter of H^+ was produced. (a) What was the ratio of $Tris^+$ (conjugate acid)/$Tris^0$ (conjugate base) at the start of the reaction? (b) What are the concentrations of $Tris^+$ and $Tris^0$ at the start of the reaction? (c) Show the reaction by which the buffer maintained a near constant pH. (d) What were the concentrations of $Tris^0$ and $Tris^+$ at the end of the reaction? (e) What was the pH at the end of the reaction? The pK_a of Tris is 8.1. (f) What would the final pH be if no buffer were present?

Solution

(a)
$$pH = pK_a + \log \frac{Tris^0}{Tris^+} \qquad 7.8 = 8.1 + \log \frac{Tris^0}{Tris^+}$$

$$-0.3 = \log \frac{Tris^0}{Tris^+} \qquad \text{or} \qquad +0.3 = \log \frac{Tris^+}{Tris^0}$$

$$\frac{Tris^+}{Tris^0} = \text{antilog of } 0.3 = 2 \qquad \boxed{\frac{\mathbf{Tris^+}}{\mathbf{Tris^0}} = \frac{\mathbf{2}}{\mathbf{1}}}$$

(b)

$$\tfrac{2}{3} \times 0.2\,M = \boxed{\textbf{0.133 } \textbf{\textit{M}} \textbf{ Tris}^+} \qquad \text{and} \qquad \tfrac{1}{3} \times 0.2\,M = \boxed{\textbf{0.067 } \textbf{\textit{M}} \textbf{ Tris}^0}$$

   Check: The pH is less than the pK_a; $\therefore$ [conjugate acid] > [conjugate base]; 0.133 M > 0.067 M.

(c) The conjugate base reacts with the excess H^+.

$$\boxed{\textbf{Tris}^0 + \textbf{H}^+ \rightarrow \textbf{Tris}^+}$$

(d) As a result of the reaction, the amounts of $Tris^+$ and $Tris^0$ change as shown below.

$$[Tris^+] = 0.133 + 0.030 = \boxed{\textbf{0.163 } \textbf{\textit{M}}}$$

$$[Tris^0] = 0.067 - 0.030 = \boxed{\textbf{0.037 } \textbf{\textit{M}}}$$

(e)
$$pH = pK_a + \log \frac{Tris^0}{Tris^+} = 8.1 + \log \frac{0.037}{0.163}$$

$$= 8.1 - \log \frac{0.163}{0.037} = 8.1 - \log 4.4$$

$$pH = 8.1 - 0.644 \qquad \boxed{\textbf{pH} = \textbf{7.456}}$$

(f)   If no buffer were present, the production of 0.03 M H^+ would bring the pH to:

$$pH = \log \frac{1}{[H^+]} = \log \frac{1}{0.03} = \log 33.33$$

$$\boxed{pH = 1.52}$$

(The enzyme would very likely be denatured before the pH decreased to 1.52.)

BUFFER CAPACITY—THEORETICAL AND PRACTICAL

The ability of a buffer to resist changes in pH is referred to as the "buffer capacity." "Buffer capacity" can be defined in two ways: (1) the number of moles per liter of H^+ or OH^- required to cause a given change in pH (e.g., 1 unit), or (2) the pH change that occurs upon addition of a given amount of H^+ or OH^- (e.g., 1 mole/liter).   The first definition is better because it can be applied to buffers of any concentration.

   An expression for instantaneous buffer capacity, β, can be derived using calculus.   Essentially, β is the reciprocal of the slope of the titration curve at any point.   Starting with the Henderson-Hasselbalch equation:

$$pH = pK_a + \log \frac{[A^-]}{[HA]} = pK_a + \log [A^-] - \log [HA]$$
$$= pK_a + \log [A^-] - \log ([C] - [A^-])$$
$$= pK_a + \frac{\ln [A^-]}{2.3} - \frac{\ln ([C] - [A^-])}{2.3}$$

where C = the total concentration of buffer components
$$= [A^-] + [HA]$$

Differentiating with respect to $[A^-]$:

$$\frac{d\,pH}{d[A^-]} = \frac{1}{2.3[A^-]} + \frac{1}{2.3([C] - [A^-])} = \frac{[C]}{2.3[A^-]([C] - [A^-])}$$

$d[A^-]$ is the same as $d[H^+]$ or $d[OH^-]$ because for every mole of H^+ added a mole of A^- is utilized; for every mole of OH^- added a mole of A^- is produced.   Substituting and inverting:

$$\frac{d[H^+]}{d\,pH} = \frac{d[OH^-]}{d\,pH} = \frac{2.3[A^-]([C] - [A^-])}{[C]} = \beta$$

or
$$\boxed{\beta = \frac{2.3[A^-][HA]}{[A^-] + [HA]}} \qquad (31)$$

Further substitution from the expression for K_a yields:

$$\beta = \frac{2.3\,K_a[H^+][C]}{(K_a + [H^+])^2} \quad \text{and} \quad \beta_{max} = 0.575\,[C] \qquad (32)$$

where $[H^+]$ = the hydrogen ion concentration of the buffer.

We see that β increases as the concentration of the buffer increases. We might have arrived at this conclusion intuitively. It seems logical that a $0.25\,M$ buffer should resist a pH change better than a $0.01\,M$ buffer. It can also be shown (by calculus or by trial and error) that β will be maximum when $[A^-] = [HA]$ or $[H^+] = K_a$, that is, the slope of the titration curve is minimal at $pH = pK_a$. Also, when $[H^+] = K_a$, $\beta = 2.3[H^+]^2[C]/(2[H^+])^2 = 2.3[H^+]^2 [C]/4[H^+]^2$ or $\beta = 0.575[C]$.

Since β is related to the slope of the titration curve at one point, its value is the same whether H^+ or OH^- is added to the buffer. A more practical definition of buffer capacity is:

$$\begin{aligned}
\text{Buffer capacity}_a &= \text{the number of moles of } H^+ \text{ that must be} \\
&\quad \text{added to one liter of the buffer in order to} \\
&\quad \text{decrease the pH by 1 unit.} \\
&= \text{the buffer capacity in the acid direction.}
\end{aligned}$$

$$\begin{aligned}
\text{and} \quad \text{Buffer capacity}_b &= \text{the number of moles of } OH^- \text{ that must be} \\
&\quad \text{added to one liter of the buffer in order to} \\
&\quad \text{increase the pH by 1 unit.} \\
&= \text{the buffer capacity in the alkaline direction.}
\end{aligned} \qquad (33)$$

Biochemical reactions seldom produce OH^- ions. However, many reactions consume H^+ ions. The utilization of n moles/liter of H^+ ions during a reaction has the same effect on a buffer as the addition of n moles/liter of OH^- ions.

• Problem 1-33

Calculate (a) the instantaneous and (b) the practical buffer capacity in both directions of a $0.05\,M$ Tricine buffer, pH 7.5. Tricine is N-tris-(hydroxymethyl)-methylglycine. $pK_a = 8.15$. ($K_a = 7.08 \times 10^{-9}$.)

Solution

The titration curve for Tricine (base) showing the position of the buffer is sketched in Figure 1-3. The curve is shown for the titration of Tricine conjugate base with H^+. The curve for the titration of Tricine conjugate acid with OH^- would be the mirror image of that shown. The pH of the buffer is less than pK_a. Therefore, $[\text{Tricine}^+] > [\text{Tricine}^0]$, as shown below.

$$pH = pK_a + \log\frac{[\text{Tricine}^0]}{[\text{Tricine}^+]} \qquad\qquad 7.50 = 8.15 + \log\frac{[\text{Tricine}^0]}{[\text{Tricine}^+]}$$

$$7.50 = 8.15 - \log\frac{[\text{Tricine}^+]}{[\text{Tricine}^0]} \qquad\qquad 0.65 = \log\frac{[\text{Tricine}^+]}{[\text{Tricine}^0]}$$

$$\frac{[\text{Tricine}^+]}{[\text{Tricine}^0]} = \frac{4.47}{1}$$

$$[\text{Tricine}^+] = \frac{4.47}{5.47} \times 0.05 = 0.041 \ M$$

$$[\text{Tricine}^0] = \frac{1}{5.47} \times 0.05 = 0.009 \ M$$

At pH 7.5, $[H^+] = 3.16 \times 10^{-8} \ M$

(a)   or

$$\beta = \frac{2.3 \ K_a [H^+][C]}{(K_a + [H^+])^2}$$

$$= \frac{(2.3)(7.08 \times 10^{-9})(3.16 \times 10^{-8})(0.05)}{(7.08 \times 10^{-9} + 3.16 \times 10^{-8})^2}$$

$$= \frac{2.57 \times 10^{-17}}{(3.87 \times 10^{-8})^2}$$

$$= \frac{2.57 \times 10^{-17}}{1.50 \times 10^{-15}}$$

$$\boxed{\beta = 0.017 \ M}$$

$$\beta = \frac{2.3[\text{Tricine}^0][\text{Tricine}^+]}{[\text{Tricine}^0] + [\text{Tricine}^+]}$$

$$= \frac{2.3[\text{Tricine}^0][\text{Tricine}^+]}{C}$$

$$= \frac{(2.3)(0.041)(0.009)}{0.05}$$

$$= \frac{8.49 \times 10^{-4}}{5 \times 10^{-2}}$$

$$\boxed{\beta = 0.017 \ M}$$

(b)   We can see from the titration curve that the practical buffer capacity in the alkaline direction is greater than the practical buffer capacity in the acid direction.   To calculate BC_a, we start by calculating the concentrations of Tricine^+ and Tricine^0 at pH 6.5 (one pH unit less than the pH of the buffer).

$$pH = pK_a + \log \frac{[\text{Tricine}^0]}{[\text{Tricine}^+]} \qquad 6.50 = 8.15 + \log \frac{[\text{Tricine}^0]}{[\text{Tricine}^+]}$$

$$1.65 = \log \frac{[\text{Tricine}^+]}{[\text{Tricine}^0]} \qquad \frac{[\text{Tricine}^+]}{[\text{Tricine}^0]} = \frac{44.7}{1}$$

$$\therefore \quad \text{at pH 6.5:} \quad [\text{Tricine}^+] = \frac{44.7}{45.7} \times 0.05 = 0.049 \ M$$

$$[\text{Tricine}^0] = \frac{1}{45.7} \times 0.05 = 0.001 \ M$$

Next, calculate the $[H^+]$ required to change the original concentrations of conjugate acid and conjugate base to the final concentrations.

$$[\text{Tricine}^+]_{\text{final}} = [\text{Tricine}^+]_{\text{orig}} + [H^+]$$
$$0.049 = 0.041 + [H^+]$$

$$[H^+] = 0.008 \ M \qquad \therefore \quad \boxed{BC_a = 0.008 \ M}$$

or
$$[\text{Tricine}^0]_{\text{final}} = [\text{Tricine}^0]_{\text{orig}} - [H^+]$$
$$0.001 = 0.009 - [H^+]$$

$$[H^+] = 0.008 \ M \qquad \therefore \quad \boxed{BC_a = 0.008 \ M}$$

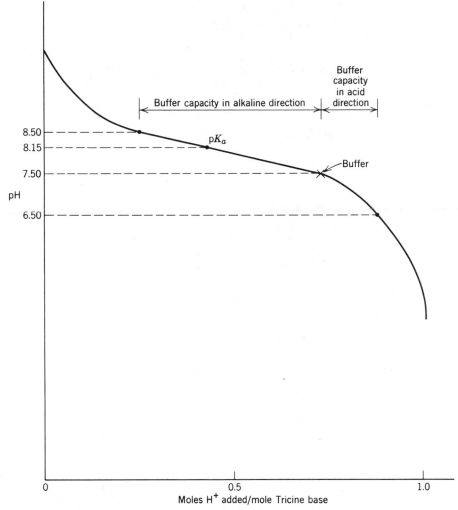

Figure 1-3   Titration of a weak monoprotic base (e.g., Tricine) with a strong acid (e.g., HCl). pK_a of Tricine = 8.15.

To calculate BC_b, we proceed in the same manner to obtain the concentrations of Tricine^0 and Tricine$^+$ at pH 8.5.

$$8.5 = 8.15 + \log \frac{[\text{Tricine}^0]}{[\text{Tricine}^+]}$$

$$0.35 = \log \frac{[\text{Tricine}^0]}{[\text{Tricine}^+]} \qquad \frac{[\text{Tricine}^0]}{[\text{Tricine}^+]} = \frac{2.24}{1}$$

$$[\text{Tricine}^0] = \frac{2.24}{3.24} \times 0.05 \; M = 0.035 \; M$$

$$[\text{Tricine}^+] = \frac{1}{3.24} \times 0.05 \; M = 0.015 \; M$$

Now, calculate the amount of OH$^-$ that must be added (or H$^+$ that must be removed) to change the original concentrations to the final concentrations.

$$[\text{Tricine}^0]_{\text{final}} = [\text{Tricine}^0]_{\text{orig}} + [\text{OH}^-]$$
$$0.035 = 0.009 + [\text{OH}^-]$$

$$[\text{OH}^-] = 0.026 \ M \qquad \therefore \quad \boxed{\textbf{BC}_b = \textbf{0.026} \ \textbf{\textit{M}}}$$

or
$$[\text{Tricine}^+]_{\text{final}} = [\text{Tricine}^+]_{\text{orig}} - [\text{OH}^-]$$
$$0.015 = 0.041 - [\text{OH}^-]$$

$$[\text{OH}^-] = 0.026 \ M \qquad \therefore \quad \boxed{\textbf{BC}_b = \textbf{0.026} \ \textbf{\textit{M}}}$$

Thus, the practical buffer capacity in the alkaline direction is more than three times the practical buffer capacity in the acid direction. At pH 7.5, $BC_b > \beta$, but $BC_a < \beta$. If the pH of the buffer were 8.15 (i.e., $\text{pH} = pK_a$):

$$BC_a = BC_b \simeq 0.020 \ M$$
and
$$\beta \simeq 0.029 \ M$$

That is, the slope of the line between pK_a and $pK_a + 1$ (or pK_a and $pK_a - 1$) is greater than the slope of the line drawn tangent to the titration curve at $\text{pH} = pK_a$.

Instead of proceeding stepwise as outlined above, we can derive a general equation for practical buffer capacity. For example, for BC_a where pH_1 is the initial pH and pH_2 is the final (lower) pH:

$$\text{pH}_1 = pK_a + \log\frac{[A^-]_1}{[HA]_1} \qquad \text{and} \qquad \text{pH}_2 = pK_a + \log\frac{[A^-]_2}{[HA]_2}$$

$$\therefore \qquad \text{pH}_1 - \text{pH}_2 = 1 = \left(pK_a + \log\frac{[A^-]_1}{[HA]_1}\right) - \left(pK_a + \log\frac{[A^-]_2}{[HA]_2}\right)$$

or
$$1 = \log\frac{[A^-]_1}{[HA]_1} - \log\frac{[A^-]_2}{[HA]_2}$$

or
$$1 = \log\frac{[A^-]_1[HA]_2}{[HA]_1[A^-]_2} \qquad \therefore \quad \frac{[A^-]_1[HA]_2}{[HA]_1[A^-]_2} = 10$$

Substituting $[HA]_2 = [HA]_1 + [H^+]$ and $[A^-]_2 = [A^-]_1 - [H^+]$, multiplying out, separating terms, and solving for $[H^+]$:

$$\boxed{[\textbf{H}^+] = \textbf{BC}_a = \frac{9[\textbf{HA}]_1[\textbf{A}^-]_1}{10[\textbf{HA}]_1 + [\textbf{A}^-]_1}} \qquad (34)$$

Similarly, we can derive the equation for BC_b:

$$\boxed{[\textbf{OH}^-] = \textbf{BC}_b = \frac{9[\textbf{HA}]_1[\textbf{A}^-]_1}{10[\textbf{A}^-]_1 + [\textbf{HA}]_1}} \qquad (35)$$

For Tricine, $HA = \text{Tricine}^+$, $A^- = \text{Tricine}^0$.

· Problem 1-34

In the preceding problem, we saw that it takes 8×10^{-3} moles H^+/liter to decrease the pH of a 0.05 M Tricine buffer, pH 7.5, by one unit. (a) Will the addition of 4×10^{-3} M H^+ decrease the pH by 0.5 unit? (b) What is the ΔpH caused by adding 4×10^{-3} M H^+?

Solution

(a) A titration curve is not a straight line and the pH scale is logarithmic, not linear. Hence, there is no direct proportionality between H^+ added and ΔpH. Thus, it does *not* take half as much H^+ to decrease the pH by 0.5 unit as it took to decrease the pH by 1.0 unit. The exact amount of H^+ required can easily be calculated as shown in the previous problem but substituting pH 7.0 for the final pH. We will find that at pH 7.0:

$$[\text{Tricine}^+] = 0.047 \ M \qquad \therefore \quad [H^+]_{added} = 0.006 \ M$$

Thus, more than half as much H^+ is required to reduce the pH by 0.5 unit as it took to reduce the pH by 1.0 unit.

(b)

$$pH = 8.15 + \log \frac{0.009 - 0.004}{0.041 + 0.004}$$

$$= 8.15 + \log \frac{0.005}{0.045} = 8.15 - \log 9 = 8.15 - 0.95$$

$$pH = 7.2 \qquad \boxed{\Delta \textbf{pH} = \textbf{0.3}}$$

Thus, half as much H^+ decreased the pH by less than 0.5 unit.

· Problem 1-35

Lactic dehydrogenase catalyzes the reversible reaction shown below.

$$\underset{\text{pyruvate}}{CH_3 - \underset{\underset{O}{\|}}{C} - COO^-} + NADH + H^+ \rightleftharpoons \underset{\text{lactate}}{CH_3 - \underset{\underset{OH}{|}}{\overset{\overset{H}{|}}{C}} - COO^-} + NAD^+$$

Suppose the enzyme that you are interested in has a relatively flat pH plateau (optimum) between pH 7.9 and pH 8.3. Beyond these limits, the reaction rate decreases markedly. The enzyme is also rapidly inactivated at high ionic strength. You wish to assay the enzyme in the direction of lactate production. The reaction will be allowed to proceed until 0.05 M lactate is produced. In order to minimize inactivation of the enzyme you must use the lowest concentration of buffer possible that will still maintain the pH within the limits of 7.9 to 8.4. You decide to use Tris as a buffer.

Describe in detail the characteristics of your buffer; that is, indicate the starting pH of the assay mixture, the final pH, the concentrations of conjugate acid and conjugate base at the beginning and end of the reaction, and the total buffer concentration (conjugate acid plus conjugate base). The pK_a value of Tris is 8.1.

Solution

The reaction utilizes H^+ ions.   Therefore, we would start at the lowest pH permissible (7.9).   The pH will rise as the reaction proceeds, but the final pH cannot be greater than 8.4.   The problem is to calculate the concentration of Tris buffer, C, that will maintain the pH within these limits.   Thus, C is unknown.   At the start:

$$pH = pK_a + \log \frac{[Tris^0]_0}{[Tris^+]_0} \qquad 7.9 = 8.1 + \log \frac{[Tris^0]_0}{[Tris^+]_0}$$

$$\frac{[Tris^+]_0}{[Tris^0]_0} = \text{antilog } 0.2$$

$$\frac{[Tris^+]_0}{[Tris^0]_0} = \frac{1.585}{1} \qquad \text{or} \qquad [Tris^+]_0 = \frac{1.585}{2.585}[C] = 0.613\,[C]$$

and

$$[Tris^0]_0 = \frac{1}{2.585}[C] = 0.387\,[C]$$

At the end:

$$pH = pK_a + \log \frac{[Tris^0]_f}{[Tris^+]_f} \qquad 8.4 = 8.1 + \log \frac{[Tris^0]_f}{[Tris^+]_f}$$

$$\frac{[Tris^0]_f}{[Tris^+]_f} = \text{antilog } 0.3$$

$$\frac{[Tris^0]_f}{[Tris^+]_f} = \frac{2}{1} \qquad \text{or} \qquad [Tris^0]_f = \frac{2}{3}[C] = 0.667\,[C]$$

$$\text{and} \qquad [Tris^+]_f = \frac{1}{3}[C] = 0.333\,[C]$$

$$[Tris^+]_f = [Tris^+]_0 - 0.05$$

$$\therefore \quad 0.333[C] = 0.613[C] - 0.05$$

$$0.05 = 0.28\,[C] \qquad [C] = \frac{0.05}{0.28}$$

$$\boxed{[C] = 0.1786\ M}$$

Thus, at the start, the buffer contains:

$$[Tris^+] = (0.613)(0.1786) = \boxed{0.1095\ M}$$

$$[Tris^0] = (0.387)(0.1786) = \boxed{0.0691\ M}$$

At the end of the reaction, the buffer contains:

$$[\text{Tris}^+] = (0.333)(0.1786) = \boxed{\textbf{0.0595} \textit{M}}$$

$$[\text{Tris}^0] = (0.667)(0.1786) = \boxed{\textbf{0.1191} \textit{M}}$$

POLYPROTIC ACIDS

A polyprotic acid ionizes in successive steps:

$$H_2A \underset{K_{a_1}}{\rightleftharpoons} H^+ + HA^-$$

$$\Big\Updownarrow K_{a_2}$$

$$H^+$$
$$+$$
$$A^{2-}$$

$$K_{a_1} = \frac{[H^+][HA^-]}{[H_2A]}$$

$$K_{a_2} = \frac{[H^+][A^{2-}]}{[HA^-]}$$

Note that HA^- is the conjugate base of H_2A, but the conjugate acid of A^{2-}. For most common weak diprotic acids, K_{a_1} is at least one order of magnitude greater than K_{a_2}. Consequently, the pH of a solution of H_2A is established almost exclusively by the first ionization.

Titration and buffer calculations for weak diprotic and triprotic acids are done exactly as shown earlier for weak monoprotic acids. The only new consideration is which K_a or pK_a value to use. Very simply, we use the appropriate constant that describes the equilibrium between the species we are dealing with. For example, Figure 1-4 shows the titration of a weak diprotic acid with OH^- (p$K_{a_1} = 4$, p$K_{a_2} = 7$). The pH at any point along the titration curve is given by:

$$pH = pK_{a_1} + \log \frac{[HA^-]}{[H_2A]} \quad \text{and} \quad pH = pK_{a_2} + \log \frac{[A^{2-}]}{[HA^-]}$$

Thus, if we know the exact ratio of either conjugate base-conjugate acid pair, we can calculate the pH. Alternately, if we know the pH, we can calculate the ratios of $[HA^-]/[H_2A]$ and $[A^{2-}]/[HA^-]$, hence, the ratio of all three species.

AMPHOTERIC SALTS—INTERMEDIATE IONS OF POLYPROTIC ACIDS

An intermediate ion of a polyprotic acid when dissolved in water undergoes both ionization as an acid (reaction 1) and ionization as a base or hydrolysis (reaction 2).

1. $$HA^- + H_2O \rightleftharpoons A^{2-} + H_3O^+ \qquad K_{eq_1} = K_{a_2}$$

2. $$HA^- + H_2O \rightleftharpoons H_2A + OH^- \qquad K_{eq_2} = K_{h_1} = K_{b_2} = \frac{K_w}{K_{a_1}}$$

If reaction 1 proceeds further to the right (that is, has a greater K_{eq}) than reaction 2, the solution is acidic. If reaction 2 proceeds further to the right

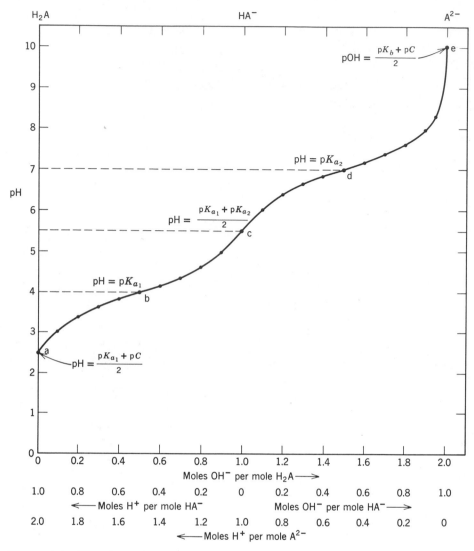

Figure 1-4 Titration of a weak diprotic acid, H_2A, with a strong base (e.g., KOH). $pK_{a_1} = 4.0$; $pK_{a_2} = 7.0$; $[H_2A]$ at the start $= 0.1\ M$.

than reaction 1, the solution is basic. However, the acidity or alkalinity is not as great as we might expect judging from the relative values of K_{eq_1} and K_{eq_2} because of a further compensating reaction that takes place. If reaction 1 goes further to the right than reaction 2, then some of the excess H_3O^+ produced reacts with unreacted HA^- according to reaction 3.

3. $$HA^- + H_3O^+ \rightleftharpoons H_2A + H_2O \qquad K_{eq_3} = \frac{1}{K_{a_1}}$$

If, however, reaction 2 proceeds further than reaction 1, some of the excess OH^- produced reacts with HA^- according to reaction 4.

4. $$HA^- + OH^- \rightleftharpoons A^{2-} + H_2O \qquad K_{eq_4} = \frac{1}{K_{h_2}} = \frac{1}{K_{b_1}} = \frac{K_{a_2}}{K_w}$$

Thus, the two major reactions taking place in the solution are either 1 plus 3 or 2 plus 4.   The sum of either pair of reactions is identical and is called a "disproportionation" reaction 5.   The equilibrium constant for a disproportionation reaction always is the ratio of the K_a for the next acid dissociation to the K_a for the preceding acid dissociation (K_{a_2}/K_{a_1}), as shown below.

1. $$HA^- + H_2O \rightleftharpoons A^{2-} + H_3O^+ \qquad K_{eq_1} = K_{a_2}$$

3. $$HA^- + H_3O^+ \rightleftharpoons H_2A + H_2O \qquad K_{eq_3} = \frac{1}{K_{a_1}}$$

Sum: 5. $$2HA^- \rightleftharpoons A^{2-} + H_2A \qquad K_{eq_5} = K_{eq_1} \times K_{eq_3} = K_{dis}$$

$$K_{dis} = \frac{K_{a_2}}{K_{a_1}}$$

or:

2. $$HA^- + H_2O \rightleftharpoons H_2A + OH^- \qquad K_{eq_2} = \frac{K_w}{K_{a_1}}$$

4. $$HA^- + OH^- \rightleftharpoons A^{2-} + H_2O \qquad K_{eq_4} = \frac{K_{a_2}}{K_w}$$

Sum: 5. $$2HA^- \rightleftharpoons A^{2-} + H_2A \qquad K_{eq_5} = K_{eq_2} \times K_{eq_4} = K_{dis}$$

$$K_{dis} = \frac{K_w}{K_{a_1}} \times \frac{K_{a_2}}{K_w}$$

$$K_{dis} = \frac{K_{a_2}}{K_{a_1}}$$

We can arrive at the same disproportionation reaction by assuming that 1 molecule of HA^- releases a proton and another molecule accepts the proton.

$$HA^- \rightleftharpoons H^+ + A^{2-} \qquad K_{eq} = K_{a_2}$$

$$HA^- + H^+ \rightleftharpoons H_2A \qquad K_{eq} = \frac{1}{K_{a_1}}$$

Sum: 5. $$2HA^- \rightleftharpoons H_2A + A^{2-} \qquad K_{eq} = \frac{K_{a_2}}{K_{a_1}} = K_{dis}$$

The disproportionation reaction tends to equalize the concentrations of the two ionic forms on either side of the intermediate ion.   When writing reaction 5, we assume that the component reactions proceed to the same extent (we cancel H_2O and OH^- or H_2O and H_3O^+).   Actually, reaction 1 or reaction 2 proceeds slightly further than reaction 3 or 4; that is why the solution is acidic or basic.   However, the actual *amounts* of OH^- or H_3O^+ produced (by reaction 1 or 2) and utilized (by reaction 3 or 4) are much greater than the *difference between* the amounts produced and utilized.   Consequently, we can safely use reaction 5 as a basis for calculating the concentrations of all ionic forms (H_2A, HA^-, and A^{2-}) in the solution.

The H^+ ion concentration and subsequently the pH of the solution may then be calculated from any K_{eq} expression containing the above components:

$$K_{a_1} = \frac{[\text{H}^+][\text{HA}^-]}{[\text{H}_2\text{A}]} \qquad\qquad K_{a_2} = \frac{[\text{H}^+][\text{A}^{2-}]}{[\text{HA}^-]}$$

$$K_{b_2} = \frac{[\text{H}_2\text{A}][\text{OH}^-]}{[\text{HA}^-]} = \frac{K_w}{K_{a_1}} \qquad K_{b_1} = \frac{[\text{HA}^-][\text{OH}^-]}{[\text{A}^{2-}]} = \frac{K_w}{K_{a_2}}$$

We can also derive equations from which the H^+ ion concentration and the pH may be determined directly from the K_{a_2} and K_{a_1} values.

$$K_{\text{dis}} = \frac{[\text{H}_2\text{A}][\text{A}^{2-}]}{[\text{HA}^-]^2} = \frac{K_{a_2}}{K_{a_1}}$$

$$[\text{H}_2\text{A}] = [\text{A}^{2-}]$$

$$\therefore \quad \frac{[\text{H}_2\text{A}]^2}{[\text{HA}^-]^2} = \frac{K_{a_2}}{K_{a_1}} \quad \text{or} \quad \frac{[\text{H}_2\text{A}]}{[\text{HA}^-]} = \frac{\sqrt{K_{a_2}}}{\sqrt{K_{a_1}}}$$

Substituting the $[\text{H}_2\text{A}]/[\text{HA}^-]$ ratio into the K_{a_1} expression:

$$K_{a_1} = \frac{[\text{H}^+][\text{HA}^-]}{[\text{H}_2\text{A}]} \qquad\qquad [\text{H}^+] = \frac{K_{a_1}[\text{H}_2\text{A}]}{[\text{HA}^-]} = \frac{K_{a_1}\sqrt{K_{a_2}}}{\sqrt{K_{a_1}}}$$

$$\frac{K_{a_1}}{\sqrt{K_{a_1}}} = \sqrt{K_{a_1}} \qquad\qquad \therefore \quad [\text{H}^+] = \sqrt{K_{a_1}}\sqrt{K_{a_2}}$$

$$\boxed{[\text{H}^+] = \sqrt{K_{a_1}K_{a_2}}} \qquad\qquad (36)$$

$$\log[\text{H}^+] = \tfrac{1}{2}(\log K_{a_1} + \log K_{a_2})$$

$$-\log[\text{H}^+] = -\tfrac{1}{2}(\log K_{a_1} + \log K_{a_2})$$

$$-\log[\text{H}^+] = \frac{-\log K_{a_1} - \log K_{a_2}}{2}$$

$$\boxed{\text{pH} = \frac{\text{p}K_{a_1} + \text{p}K_{a_2}}{2}} \qquad\qquad (37)$$

Intermediate ions are the predominant species at intermediate equivalence points. Therefore, we can predict that the pH at an intermediate equivalence point during the titration of a polyprotic acid (or amino acid) is the average of the $\text{p}K_a$ values on either side of the equivalence point. For amino acids the pH is designated $\text{p}I$ (isoelectric point) if the intermediate ion in question is the one that carries no *net* charge. At the $\text{p}I$ the major ionic species present is AA^0. However, because of the disproportionation reaction, small (and essentially equal) amounts of AA^{-1} and AA^{+1} are also present. The pH is designated pH_m if the major ionic species present carries the maximum number of charges, regardless of sign. By the same reasoning we can show that the pH of a solution of a salt of a weak acid and a weak base is the average of the two $\text{p}K_a$ values. For example, the pH of a solution of NH_4OAc is given by:

$$\text{pH} = \frac{\text{p}K_{a_{\text{NH}_4^+}} + \text{p}K_{a_{\text{HOAc}}}}{2}$$

· Problem 1-36

Describe the preparation of 10 liters of 0.045 M potassium phosphate buffer, pH 7.5.

Solution

The pH of this buffer is a little above the pK_{a_2} of H_3PO_4 as shown in the titration curve in Figure 1-5. Consequently, the two major ionic species present are $H_2PO_4^-$ (conjugate acid) and HPO_4^{2-} (conjugate base) with the HPO_4^{2-} predominating.

The buffer can be prepared in any one of several ways: (1) by mixing KH_2PO_4 and K_2HPO_4 in the proper proportions, (2) by starting with H_3PO_4 and converting it to KH_2PO_4 plus K_2HPO_4 by adding the proper amount of KOH, (3) by starting with KH_2PO_4 and converting a portion of it to K_2HPO_4 by adding KOH, (4) by starting with K_2HPO_4 and converting a portion of it to KH_2PO_4 by adding a strong acid such as HCl, (5) by starting with K_3PO_4 and converting it to KH_2PO_4 plus K_2HPO_4 by adding HCl, and (6) by mixing K_3PO_4 and KH_2PO_4 in the proper proportions. Regardless of which method is used, the first step involves calculating the proportion and amounts of the two ionic species in the buffer.

The buffer contains a total of 10 liters $\times$ 0.045 M = 0.45 mole of phosphate.

$$pH = pK_{a_2} + \log \frac{[HPO_4^{2-}]}{[H_2PO_4^-]}$$

$$7.5 = 7.2 + \log \frac{[HPO_4^{2-}]}{[H_2PO_4^-]} \qquad 0.3 = \log \frac{[HPO_4^{2-}]}{[H_2PO_4^-]}$$

$$\frac{[HPO_4^{2-}]}{[H_2PO_4^-]} = \text{antilog of } 0.3 = 2 = \frac{2}{1} \text{ ratio}$$

$\therefore$ $\frac{2}{3} \times 0.45$ mole = 0.30 mole of HPO_4^{2-} is needed and $\frac{1}{3} \times 0.45$ mole = 0.15 mole of $H_2PO_4^-$ is needed.

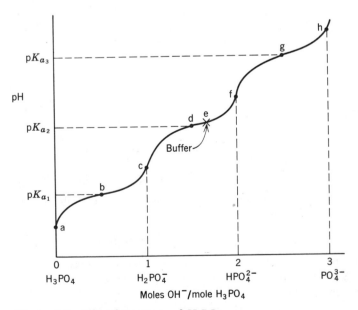

Figure 1-5 Titration curve of H_3PO_4.

1. *From KH_2PO_4 and K_2HPO_4*—Weigh out 0.30 mole of K_2HPO_4 (52.8 g) and 0.15 mole of KH_2PO_4 (20.4 g) and dissolve in sufficient water to make 10 liters final volume. Or, if stock solutions of the two phosphates are available, measure out the appropriate volumes of each and dilute to 10 liters. In practice, we might prepare 0.045 M K_2HPO_4 and 0.045 M KH_2PO_4 and simply mix the two until the pH (as measured with a pH meter) is 7.5. Since both stock solutions are 0.045 M, the total phosphate concentration will remain 0.045 M regardless of what volumes of each are added.

2. *From H_3PO_4 and KOH*—Start with 0.45 mole of H_3PO_4 (position a in Figure 1-5) and add sufficient KOH to titrate *completely* 1 hydrogen (to position c) and $\frac{2}{3}$ of the second hydrogen (to position e).

$$H_3PO_4 \xrightarrow{\text{OH}^-} H_2PO_4^- \xrightarrow{\text{OH}^-} HPO_4^{2-}$$

For example, suppose we have available only stock concentrated (15 M) H_3PO_4 and a standard solution of 1.5 M KOH. We need 0.45 mole of H_3PO_4.

$$\text{liters} \times M = \text{number of moles}$$

$$\text{liters} \times 15 = 0.45 \text{ mole}$$

$$\text{liters} = \frac{0.45}{15} = 0.03 \text{ liter}$$

$\therefore$ Take 30 ml of H_3PO_4. Add 0.45 mole of KOH to convert *all* the H_3PO_4 to $H_2PO_4^-$; then add another $\frac{2}{3} \times 0.45 = 0.30$ mole of KOH to convert 0.3 mole of $H_2PO_4^-$ to HPO_4^{2-}. In other words, a total of 0.75 mole of KOH is required.

Since we have available 1.5 M KOH, we can calculate how much of this solution to add.

$$\text{liters} \times M = \text{number of moles}$$

$$\text{liters} = \frac{\text{number of moles}}{M} = \frac{0.75}{1.5} = 0.500$$

$\therefore$ Add 500 ml of KOH to the 30 ml of concentrated H_3PO_4; then add sufficient water to bring the final volume to 10 liters.

3. *From KH_2PO_4 and KOH*—We can start with KH_2PO_4 (position c) and add sufficient KOH to convert $\frac{2}{3}$ of the $H_2PO_4^-$ to HPO_4^{2-} (position e).

$$H_2PO_4^- \xrightarrow{\text{OH}^-} HPO_4^{2-}$$

For example, suppose we have available only solid KH_2PO_4 and KOH. We need 0.45 mole of KH_2PO_4.

$$\frac{\text{wt}_\text{g}}{\text{MW}} = \text{number of moles}$$

$$\text{wt}_\text{g} = (0.45)(136) = 61.2 \text{ g}$$

Dissolve the KH_2PO_4 in some water; then add 0.30 mole of KOH (solid or dissolved in some water).

$$\text{wt}_\text{g} = (0.30)(56) = 16.8 \text{ g of KOH}$$

Next, add sufficient water to bring the volume to 10 liters.

4. *From K_2HPO_4 and HCl*—The HPO_4^{2-} (position f) may be converted to $H_2PO_4^-$ by adding HCl.

$$HPO_4^{2-} \xrightarrow{\text{H}^+} H_2PO_4^-$$

Because we want to end up with an $HPO_4^{2-}/H_2PO_4^-$ of $\frac{2}{1}$, we want to convert only $\frac{1}{3}$ of the HPO_4^{2-} to $H_2PO_4^-$ (to reach position e). Suppose we have available solid K_2HPO_4 and a 2 M solution of HCl. Weigh out 0.45 moles of K_2HPO_4.

$$wt_g = \text{number of moles} \times MW$$
$$wt_g = (0.45)(174) = 78.3 \text{ g}$$

Dissolve the KH_2PO_4 in some water; then add $\frac{1}{3} \times 0.45 = 0.15$ mole of HCl.

$$\text{liters} \times M = \text{number of moles}$$
$$\text{liters} = \frac{\text{number of moles}}{M} = \frac{0.15}{2} = 0.075 \text{ liter}$$

$\therefore$ Add 75 ml of 2 M HCl. Then add sufficient water to bring the volume to 10 liters.

5. *From K_3PO_4 and HCl*—Start with 0.45 mole of K_3PO_4 (position h) and add sufficient HCl to convert all the PO_4^{3-} to HPO_4^{2-} (position f). Then add additional HCl to convert $\frac{1}{3}$ of the HPO_4^{2-} to $H_2PO_4^-$ (position e).

$$PO_4^{3-} \xrightarrow{\;H^+\;} HPO_4^{2-} \xrightarrow{\;H^+\;} H_2PO_4^-$$

We need 0.45 mole of K_3PO_4.

$$wt_g = \text{number of moles} \times MW$$
$$wt_g = (0.45)(212) = 95.4 \text{ g}$$

Dissolve the K_3PO_4 in water. Add 0.45 mole of HCl to convert all the PO_4^{3-} to HPO_4^{2-}. Then add another $\frac{1}{3} \times 0.45 = 0.15$ mole of HCl to convert 0.15 mole of HPO_4^{2-} to $H_2PO_4^-$. The final solution then contains 0.15 mole of $H_2PO_4^-$ and 0.30 mole of HPO_4^{2-}. Now add sufficient water to make 10 liters.

6. *From KH_2PO_4 and K_3PO_4*—The KH_2PO_4 and K_3PO_4 react to form K_2HPO_4. The $H_2PO_4^-$ acts as an acid and the PO_4^{3-} acts as a base.

$$H_2PO_4^- + PO_4^{3-} \rightleftharpoons 2HPO_4^{2-}$$

The reaction is the reverse of the disproportionation reaction. Note that each mole of $H_2PO_4^-$ and PO_4^{3-} yields 2 moles of HPO_4^{2-}. Thus, to produce 0.30 mole of HPO_4^{2-}, 0.15 mole of $H_2PO_4^-$ and 0.15 mole of PO_4^{3-} are required. But, in addition to the 0.30 mole of HPO_4^{2-}, the final solution contains 0.15 mole of $H_2PO_4^-$. Therefore, dissolve 0.30 mole of KH_2PO_4 and 0.15 mole of K_3PO_4 in water. Of the original 0.30 mole of KH_2PO_4, 0.15 mole reacts with the PO_4^{3-} to produce 0.30 mole of HPO_4^{2-}, leaving 0.15 mole as $H_2PO_4^-$. Then add sufficient water to make 10 liters.

· **Problem 1-37**

Calculate the concentration of all ionic species of succinate present· in a solution (buffer) of 0.1 M succinate, pH 4.59.

Solution

We can see from the titration curve sketched in Figure 1-6 that the major ionic species present at pH 4.59 are H_2succinate and Hsuccinate$^-$ with the latter predominating. However, because pK_{a_2} is close to pK_{a_1}, an appreciable amount of succinate^{2-} is also present.

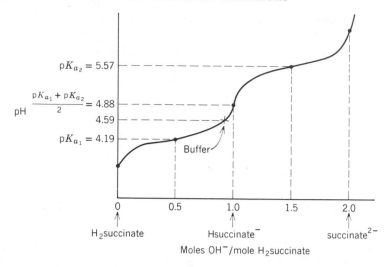

Figure 1-6   Titration curve of succinic acid.

First calculate the ratio of Hsuccinate$^-$/H$_2$succinate using the Henderson-Hasselbalch equation.

$$pH = pK_{a_1} + \log \frac{[\text{Hsuccinate}^-]}{[\text{H}_2\text{succinate}]} \qquad 4.59 = 4.19 + \log \frac{[\text{Hsuccinate}^-]}{[\text{H}_2\text{succinate}]}$$

$$0.40 = \log \frac{[\text{Hsuccinate}^-]}{[\text{H}_2\text{succinate}]} \qquad \frac{[\text{Hsuccinate}^-]}{[\text{H}_2\text{succinate}]} = \text{antilog of } 0.40 = 2.51$$

∴ The Hsuccinate$^-$/H$_2$succinate ratio is 2.51.

Next calculate the Hsuccinate$^-$/succinate^{2-} ratio at pH 4.59.

$$pH = pK_{a_2} + \log \frac{[\text{succinate}^{2-}]}{[\text{Hsuccinate}^-]} \qquad 4.59 = 5.57 + \log \frac{[\text{succinate}^{2-}]}{[\text{Hsuccinate}^-]}$$

$$-0.98 = \log \frac{[\text{succinate}^{2-}]}{[\text{Hsuccinate}^-]} \quad \text{or} \quad +0.98 = \log \frac{[\text{Hsuccinate}^-]}{[\text{succinate}^{2-}]}$$

$$\frac{[\text{Hsuccinate}^-]}{[\text{succinate}^{2-}]} = \text{antilog of } 0.98 = 9.55$$

∴ The Hsuccinate$^-$/succinate^{2-} ratio is 9.55 : 1.
The three ionic species are in the following proportions:

H$_2$succinate		Hsuccinate$^-$	succinate^{2-}
1	:	2.51	
		9.55	:	1

The two ratios must be expressed relative to a common component such as Hsuccinate$^-$. So, if 1 part succinate^{2-} is present for every 9.55 parts of Hsuccinate$^-$, calculate how much succinate^{2-} is present for every 2.51 parts of Hsuccinate$^-$.

$$\frac{1}{9.55} = \frac{y}{2.51} \qquad 9.55y = 2.51$$

$$y = \frac{2.51}{9.55} = 0.263$$

That is, the ratio of the three ionic species is:

H₂succinate		Hsuccinate⁻		succinate²⁻	total
1	:	2.51	:	0.263	3.773 parts

The *total* succinate concentration is 0.1 M.

$$\frac{1}{3.773} \times 0.1\ M = \boxed{\textbf{0.0265 } M \textbf{ H}_2\textbf{succinate}}$$

$$\frac{2.51}{3.773} \times 0.1\ M = \boxed{\textbf{0.0665 } M \textbf{ Hsuccinate}^-}$$

$$\frac{0.263}{3.773} \times 0.1\ M = \boxed{\textbf{0.00697 } M \textbf{ succinate}^{2-}}$$

· **Problem 1-38**

(a) What is the pH of a 0.1 M solution of monosodium succinate? (b) What are the concentrations of un-ionized succinic acid, Hsuccinate⁻ and succinate²⁻ in the solution ($pK_{a_1} = 4.19$ and $pK_{a_2} = 5.57$)?

Solution

Monosodium succinate (Hsuccinate⁻) is an intermediate ion of a polyprotic acid. The pH of a 0.1 M solution and the concentrations of all three ionic forms of succinic acid may be calculated as shown.

(a)
$$pH = \frac{pK_{a_1} + pK_{a_2}}{2} = \frac{4.19 + 5.57}{2} = \frac{9.76}{2}$$

$$\boxed{\textbf{pH} = \textbf{4.88}}$$

(b) The concentrations of all species present can be calculated in two ways. First, considering the disproportionation reaction:

$$2\text{Hsuccinate}^- \rightarrow \text{H}_2\text{succinate} + \text{succinate}^{2-}$$

$$K_{dis} = \frac{[\text{H}_2\text{succinate}][\text{succinate}^{2-}]}{[\text{Hsuccinate}^-]^2} = \frac{K_{a_2}}{K_{a_1}} = \frac{2.69 \times 10^{-6}}{6.46 \times 10^{-5}} = 4.16 \times 10^{-2}$$

Let

$$y = M \text{ of Hsuccinate}^- \text{ that disappears}$$

$$\therefore \frac{y}{2} = M \text{ of H}_2\text{succinate produced}$$

and

$$\frac{y}{2} = M \text{ of succinate}^{2-} \text{ produced}$$

$$\frac{(y/2)(y/2)}{(0.1-y)^2} = 4.16 \times 10^{-2} \qquad \sqrt{\frac{y^2}{4(0.1-y)^2}} = \sqrt{4.16 \times 10^{-2}}$$

$$\frac{y}{2(0.1-y)} = 2.04 \times 10^{-1} \qquad \frac{y}{(0.2-2y)} = 0.204$$

$$y = 0.0408 - 0.408y \qquad 1.408y = 0.0408$$

$$y = \frac{0.0408}{1.408} = 0.0290 \qquad \frac{y}{2} = 0.0145$$

> **H$_2$succinate = 0.0145 M**
>
> **succinate^{2-} = 0.0145 M**
>
> **Hsuccinate$^-$ = 0.100 − 0.029 = 0.071 M**

Alternatively, the concentrations can be calculated using the Henderson-Hasselbalch equation:

$$pH = pK_{a_1} + \log \frac{[\text{Hsuccinate}^-]}{[\text{H}_2\text{succinate}]} \qquad 4.88 = 4.19 + \log \frac{[\text{Hsuccinate}^-]}{[\text{H}_2\text{succinate}]}$$

$$0.69 = \log \frac{[\text{Hsuccinate}^-]}{[\text{H}_2\text{succinate}]} \qquad \frac{[\text{Hsuccinate}^-]}{[\text{H}_2\text{succinate}]} = \text{antilog of } 0.69 = 4.9$$

∴ The ratio of Hsuccinate$^-$/H$_2$succinate is 4.9:1.
Next calculate the ratio of Hsuccinate$^-$/succinate^{2-}.

$$pH = pK_{a_2} + \log \frac{[\text{succinate}^{2-}]}{[\text{Hsuccinate}^-]} \qquad 4.88 = 5.57 + \log \frac{[\text{succinate}^{2-}]}{[\text{Hsuccinate}^-]}$$

$$-0.69 = \log \frac{[\text{succinate}^{2-}]}{[\text{Hsuccinate}^-]} \quad \text{or} \quad +0.69 = \log \frac{[\text{Hsuccinate}^-]}{[\text{succinate}^{2-}]}$$

$$\frac{[\text{Hsuccinate}^-]}{[\text{succinate}^{2-}]} = \text{antilog of } 0.69 = 4.9$$

∴ The ratio of Hsuccinate$^-$/succinate^{2-} is also 4.9:1. The concentrations of all three ionic species are in the ratio of 1:4.9:1.

$$\text{H}_2\text{succinate} \rightleftharpoons \text{Hsuccinate}^- \rightleftharpoons \text{succinate}^{2-}$$
$$1 \qquad\qquad 4.9 \qquad\qquad 1$$

The *total* concentration of succinate is 0.1 M.

$$\frac{1}{6.9} \times 0.1 \ M = \text{H}_2\text{succinate concentration}$$

$$\frac{4.9}{6.9} \times 0.1 \ M = \text{Hsuccinate}^- \text{ concentration}$$

$$\frac{1}{6.9} \times 0.1 \ M = \text{succinate}^{2-} \text{ concentration}$$

> ∴ **[H$_2$succinate] = 0.0145 M**
>
> **[Hsuccinate$^-$] = 0.0710 M**
>
> **[succinate^{2-}] = 0.0145 M**

DILUTIONS OF BUFFERS

· Problem 1-39

According to the Henderson-Hasselbalch equation, the pH of a buffer depends only on the *ratio* of conjugate base activity to conjugate acid activity.   Explain then why the pH of a buffer changes when it is diluted.

Solution

The pH of a buffer changes with dilution for several reasons:

1.   *Changes in Activity Coefficients*—The activity coefficients of different ions are not the same at any given concentration and do not change in an identical manner with a given change in concentration.   For example, we can see from Appendix V that $\gamma_{HPO_4^{2-}}$ is 0.445 in a 0.1 M solution and 0.903 in a 0.001 M solution.   The activity coefficient of its conjugate acid ($\gamma_{H_2PO_4^-}$) is 0.744 in a 0.1 M solution and 0.928 in a 0.001 M solution.   In general, dilution results in an increase in γ; γ approaches unity at infinite dilution.   The greater the charge on the ion, the greater is the change in its activity coefficient for a given change in concentration.   Consider a "0.2 M phosphate buffer" containing equal molar amounts of HPO_4^{2-} and $H_2PO_4^-$ (0.1 M of each ionic species).   The exact pH of the solution can be calculated taking into account the activity coefficients of the two ions.   We can also calculate the pH of the solution after it is diluted 100-fold, taking into account the change in activity coefficients.

$$pH = pK_{a_2} + \log\frac{a_{HPO_4^{2-}}}{a_{H_2PO_4^-}} = pK_{a_2} + \log\frac{\gamma_{HPO_4^{2-}}[HPO_4^{2-}]}{\gamma_{H_2PO_4^-}[H_2PO_4^-]}$$

0.2 M Buffer	0.002 M Buffer
$pH = 7.2 + \log\dfrac{(0.445)(0.1)}{(0.744)(0.1)}$	$pH = 7.2 + \log\dfrac{(0.903)(0.001)}{(0.928)(0.001)}$
$= 7.2 + \log 0.598$	$= 7.2 + \log 0.973$
$= 7.2 - 0.22$	$= 7.2 - 0.01$
pH = 6.98	**pH = 7.19**

In general, the log a_{A^-}/a_{HA} term of "acidic" buffers increases upon dilution, resulting in an increase in pH.   In "basic" buffers, the log $a_{R-NH_2}/a_{R-NH_3^+}$ term decreases upon dilution, resulting in a decrease in pH.

2.   *Changes in the Degree of Dissociation of HA*—As shown earlier, the degree of dissociation of a weak acid increases as the solution is diluted.   In a solution of a weak acid alone (no added conjugate base), the acid is 10% dissociated when $[HA]_{orig} = 100\,K_a$ and 50% dissociated when $[HA]_{orig} = 2\,K_a$.   Thus, the log A^-/HA term increases as the solution is diluted.   In a buffer solution (weak acid plus added conjugate base), the A^- tends to suppress the dissociation of HA.   Consequently, in a buffer solution containing $[HA] = 2\,K_a$, the HA is somewhat less than 50% dissociated.

For example, consider a 0.02 M succinate buffer, prepared by dissolving 0.01 mole of succinic acid and 0.01 mole of monosodium succinate in sufficient water to make 1 liter final volume. The monosodium succinate ionizes completely and the succinic acid ionizes partially. Let

$$y = M \text{ of succinic acid that dissociates}$$

$$\therefore \quad y = M \text{ of } H^+ \text{ produced upon dissociation}$$

and

$$y = M \text{ of } HA^- \text{ produced upon dissociation of the acid}$$

$$\therefore \quad [HA^-] = 0.01 + y \qquad [H_2A] = 0.01 - y \qquad \text{and} \qquad [H^+] = y$$

$$K_a = \frac{[H^+][HA^-]}{[H_2A]} = \frac{(y)(0.01 + y)}{(0.01 - y)} = 6.46 \times 10^{-5}$$

Because the concentrations of H_2A and HA^- are more than 100 times the K_{a_1} value, y is small compared to 0.01 and may be neglected.

$$\therefore \quad [H^+] = K_{a_1} \qquad pH = pK_{a_1} = 4.19$$

Now let us dilute the above buffer 100 times.

$$[HA^-] = 10^{-4} + y \qquad [H_2A] = 10^{-4} - y \qquad \text{and} \qquad [H^+] = y$$

Now the concentration of buffer components is of the same order of magnitude as the K_a value. Thus, H_2A is more than 10% dissociated and y is not small compared to 10^{-4}. Consequently, we must solve for y exactly.

$$K_a = \frac{(y)(10^{-4} + y)}{(10^{-4} - y)} = 6.46 \times 10^{-5}$$

$$6.46 \times 10^{-9} - 6.46 \times 10^{-5}y = 10^{-4}y + y^2$$

$$y^2 + 10 \times 10^{-5}y + 6.46 \times 10^{-5}y - 6.46 \times 10^{-9} = 0$$

$$y^2 + 16.46 \times 10^{-5}y - 6.46 \times 10^{-9} = 0$$

$$y = \frac{-b \pm \sqrt{b^2 - 4ac}}{2a}$$

where $a = 1$ $b = 16.46 \times 10^{-5}$ $c = 6.46 \times 10^{-9}$.

$$y = \frac{-16.46 \times 10^{-5} \pm \sqrt{(16.46 \times 10^{-5})^2 - 4(-6.46 \times 10^{-9})}}{2}$$

$$y = \frac{-16.46 \times 10^{-5} \pm \sqrt{271 \times 10^{-10} + 258 \times 10^{-10}}}{2}$$

$$y = \frac{-16.46 \times 10^{-5} \pm \sqrt{529 \times 10^{-10}}}{2}$$

$$y = \frac{-16.46 \times 10^{-5} \pm 23.0 \times 10^{-5}}{2} = \frac{6.54 \times 10^{-5}}{2}$$

$$y = 3.27 \times 10^{-5}$$

$$[H^+] = 3.27 \times 10^{-5} M \qquad \therefore \quad pH = 4.49$$

$$[H_2A] = (10 \times 10^{-5}) - (3.27 \times 10^{-5}) = 6.73 \times 10^{-5} M$$

$$[HA^-] = (10 \times 10^{-5}) + (3.27 \times 10^{-5}) = 13.27 \times 10^{-5} M$$

Note that the $[HA^-]/[H_2A]$ ratio that is essentially 1 in the $10^{-2} M$ buffer changes to about 2 when the buffer is diluted 100-fold.

3. Finally, as the buffer is diluted extensively, its contribution toward the H^+ or OH^- ion concentration of the solution approaches that of water and the pH approaches 7.

· **Problem 1-40**

Suppose that you prepare a buffer by dissolving 0.10 mole per liter of a weak acid, HA, and 0.10 mole per liter of its sodium salt, A^-. Assume that $pK_a = 3$. (a) What is the pH of the buffer? (b) How much does the buffer have to be diluted for the pH to increase by 1 unit? Neglect changes in activity coefficients.

Solution

(a) Let $y = M$ of HA that dissociates

$$\therefore \quad [HA] = 0.10 - y \qquad [A^-] = 0.10 + y \qquad \text{and} \qquad [H^+] = y$$

$$K_a = \frac{[H^+][A^-]}{[HA]} = \frac{(y)(0.10 + y)}{(0.10 - y)} = 10^{-3}$$

Because the concentrations of HA and A^- are much larger than K_a, y is small and may be neglected in the HA and A^- terms.

$$K_a = \frac{(y)(0.10)}{(0.10)} = 10^{-3}$$

$$y = 10^{-3} = [H^+]$$

$$\boxed{\textbf{pH = 3}}$$

(b) If the pH increases by 1 unit upon dilution (to pH 4.0), then $[H^+] = 10^{-4} M = y$. Calculate the new "original" concentrations of HA and A^- that we must start with (or dilute the original buffer to) so that when the addition and subtraction of y are made (when the HA dissociates) the ratio of A^-/HA is 10:1.

Let $C =$ new "original" M of HA and A^-.

$$K_a = \frac{[H^+][A^-]}{[HA]} \qquad 10^{-3} = \frac{(10^{-4})(C + 10^{-4})}{(C - 10^{-4})}$$

$$10^{-3} C - 10^{-7} = 10^{-4} C + 10^{-8}$$

$$0.9 \times 10^{-3} C = 1.1 \times 10^{-7}$$

$$C = \frac{1.1 \times 10^{-7}}{0.9 \times 10^{-3}} = 1.22 \times 10^{-4}$$

In other words, if we start by dissolving 1.22×10^{-4} moles per liter each of HA and A^-, $1 \times 10^{-4} M$ HA dissociates producing $2.22 \times 10^{-4} M$ A^- and leaving $0.22 \times 10^{-4} M$ HA.

Check:

$$10^{-3} = \frac{(10^{-4})(2.22 \times 10^{-4})}{(0.22 \times 10^{-4})} = (10^{-4})(10)$$

We will obtain exactly the same result by diluting the 0.2 M buffer to the above new concentrations.

$$\left.\begin{array}{c}0.10 \ M \ \text{HA} \\ + \\ 0.10 \ M \ \text{A}^-\end{array}\right\} \xrightarrow{\text{dilution...}} \begin{array}{c} 1.22 \times 10^{-4} \ M \ \text{HA} \\ + \\ 1.22 \times 10^{-4} \ M \ \text{A}^- \end{array}$$

$$\downarrow \begin{array}{l} \text{...results in} \\ \text{further dissociation} \\ \text{of HA} \end{array}$$

$$\begin{array}{c} 0.22 \times 10^{-4} \ M \ \text{HA} \\ + \\ 2.22 \times 10^{-4} \ M \ \text{A}^- \end{array}$$

> $\therefore$ **The original buffer must be diluted 820-fold.**
> $\frac{1}{820} \times 0.10 \ M = 1.22 \times 10^{-4} \ M$

GENERAL RULE

As a general rule for solving buffer problems, we can assume that the concentrations of conjugate acid and conjugate base present in solution are the same as the concentrations of each originally added to the solution (or produced by partial titration of one or the other).   This general rule does not hold when the buffer is extremely dilute (when the concentrations of buffer components are in the region of the K_a value).   In such dilute buffers, the changes (y) are large compared to the original concentrations.

CORRECTIONS FOR TOTAL IONIC STRENGTH

Ions (in addition to those of the buffer components) influence the ionic strength and affect the activity coefficients of the buffer components.   Thus, even in a very dilute solution containing equimolar concentrations of, for example, HPO_4^{2-} and $H_2PO_4^-$, the pH will not equal pK_a if a relatively high concentration of NaCl is present.   Instead of correcting the activity coefficients of the buffer components for the total ionic strength, it is simpler to define a new, *apparent* or *effective* pK_a that relates the pH (i.e., $-\log a_{H^+}$) to the actual concentrations of buffer components at a given total ionic strength.   The apparent pK_a is designated pK_a'.

$$K_a' = \frac{a_{H^+}[A^-]}{[HA]} \qquad \text{and} \qquad pH = pK_a' + \log \frac{[A^-]}{[HA]} \qquad (38)$$

The effect of total ionic strength on K_a or pK_a can be quantitatively predicted from the Debye-Hückel equation.   Thus, at any total ionic strength, the effective pK_a is given by:

$$pK_a' = pK_a + \Delta pK_a \qquad (39)$$

The values for ΔpK_a at three different total ionic strengths are given in Appendix VI.

· Problem 1-41

What is the pH of a 0.05 M KCl solution containing 0.01 M K_2HPO_4 and 0.01 M KH_2PO_4?

Solution

The pH is *not* 7.2 even though $[HPO_4^{2-}] = [H_2PO_4^-]$.   If we correct for the activity coefficients of HPO_4^{2-} and $H_2PO_4^-$, we still will not obtain the correct pH.   We must correct for the total ionic strength, part of which results from the KCl.   First calculate the ionic strength.   The solution contains: 0.01 M $HPO_4^{2-} + 0.01\ M\ H_2PO_4^- + 0.05\ M\ Cl^- + 0.08\ M\ K^+$ (one K^+ for every $H_2PO_4^-$ and Cl^-; two K^+ for every HPO_4^{2-}).

$$\frac{\Gamma}{2} = \frac{[HPO_4^{2-}](2)^2 + [H_2PO_4^-](1)^2 + [Cl^-](1)^2 + [K^+](1)^2}{2}$$

$$= \frac{(0.01)(2)^2 + (0.01)(1) + (0.05)(1) + (0.08)(1)}{2}$$

$$= \frac{0.18}{2} = 0.09$$

From Appendix VI, we see that at ~ 0.10 ionic strength, ΔpK_a for a conjugate acid with a charge of -1 is -0.32.

$$pK_a' = pK_a + \Delta pK_a \qquad pK_a' = 7.2 + (-0.32)$$

$$pK_a' = 6.88$$

Since $[HPO_4^{2-}] = [H_2PO_4^-]$, $pH = pK_a'$.

$$\boxed{\textbf{pH} = \textbf{6.88}}$$

BUFFERS OF CONSTANT IONIC STRENGTH

In order to determine the effect of pH on a reaction, we must make certain that all the buffers used are of the same ionic strength (or else establish that changing ionic strength has no effect).   Facts not often appreciated are (a) a buffer of any given composition has a different ionic strength at different pH values, and (b) two buffers of different composition may have different ionic strengths at the same pH.   The simplest way to deal with a series of buffers of varying ionic strengths is to determine which buffer has the greatest ionic strength and then make all the others up to that maximum by adding a neutral, noninhibitory salt, such as KCl.

· Problem 1-42

(a)   Which buffer has the greater ionic strength: a 0.05 M Tris buffer, pH 7.5, or a 0.05 M phosphate buffer, pH 7.5?   (b) How can the ionic strengths be equalized?

Solution

(a)   The Tris buffer contains $Tris^0$, $Tris^+$, and a counter ion, for example, Cl^-. (The buffer might have been made by titrating $Tris^0$ with HCl.) $Tris^0$ is uncharged and, thus, has no effect on ionic strength. The concentrations of H^+ and OH^- are exceedingly low and, thus, can also be neglected. If we kept track of the amount of HCl added to attain pH 7.5, we would automatically know the concentration of $Tris^+$. If not, we can calculate $[Tris^+]$:

$$pH = pK_a + \log \frac{[Tris^0]}{[Tris^+]} \qquad 7.5 = 8.1 + \log \frac{[Tris^0]}{[Tris^+]}$$

$$-0.6 = \log \frac{[Tris^0]}{[Tris^+]} \qquad [Tris^+] = \frac{4}{5} \times 0.05 \, M = 0.04 \, M$$

$$\therefore \quad [Cl^-] = 0.04 \, M$$

A more accurate calculation would include γ_{Tris^+} (about 0.9 for monopositive ions in the region of 0.05 M).

The phosphate buffer contains $H_2PO_4^-$, HPO_4^{2-} and, for example, K^+. The pH is in the region of pK_{a_2}. Therefore, there is *approximately* 0.025 M of each phosphate species present. (If we prepared the buffer ourselves, we would know the exact amounts of HPO_4^{2-} and $H_2PO_4^-$ present.) The activity coefficients of HPO_4^{2-} and $H_2PO_4^-$ in the region of 0.025 M are 0.64 and 0.88, respectively (estimated from a semi-log plot of the values given in Appendix V).

$$pH = pK_{a_2} + \log \frac{\gamma_{HPO_4^{2-}}[HPO_4^{2-}]}{\gamma_{H_2PO_4^-}[H_2PO_4^-]} \qquad 7.5 = 7.2 + \log \frac{0.64[HPO_4^{2-}]}{0.88[H_2PO_4^-]}$$

$$0.3 = \log \frac{0.64[HPO_4^{2-}]}{0.88[H_2PO_4^-]} \qquad \therefore \quad \frac{0.64[HPO_4^{2-}]}{0.88[H_2PO_4^-]} = \frac{2}{1}$$

$$\frac{[HPO_4^{2-}]}{[H_2PO_4^-]} = \frac{(2)(0.88)}{(1)(0.64)} = \frac{2.75}{1}$$

$$\therefore \quad [HPO_4^{2-}] = \frac{2.75}{3.75} \times 0.05 = 0.037 \, M$$

$$[H_2PO_4^-] = \frac{1}{3.75} \times 0.05 = 0.013 \, M$$

The phosphate buffer contains: 0.037 M HPO_4^{2-}, 0.013 M $H_2PO_4^-$, and $(0.037)(2) + (0.013)(1) = 0.087 \, M$ K^+.

The ionic strengths of the two buffers are:

Tris:
$$\frac{\Gamma}{2} = \frac{[Tris^+](1)^2 + [Cl^-](1)^2}{2} = \frac{0.04 + 0.04}{2}$$

$$\boxed{\frac{\Gamma}{2} = 0.04}$$

Phosphate:
$$\frac{\Gamma}{2} = \frac{[HPO_4^{2-}](2)^2 + [H_2PO_4^-](1)^2 + [K^+](1)^2}{2}$$
$$= \frac{(0.037)(4) + (0.013)(1) + (0.087)(1)}{2} = \frac{0.248}{2}$$

$$\boxed{\frac{\Gamma}{2} = 0.124}$$

Thus, the 0.05 M phosphate buffer has about three times the ionic strength of the 0.05 M Tris buffer at pH 7.5.

(b)   The Tris buffer can be made up to 0.124 ionic strength by adding $0.124 - 0.04 = 0.084$ M KCl.   (The pH may change slightly when the ionic strength is adjusted from 0.04 to 0.124.)

E. AMINO ACIDS AND PEPTIDES

The common amino acids are simply weak polyprotic acids.   Calculations of pH, buffer preparation, and capacity, and so on, are done exactly as shown in the preceding sections.   Neutral amino acids (e.g., glycine, alanine, threonine) are treated as diprotic acids (Table 1-1).   Acidic amino acids (e.g., aspartic acid, glutamic acid) and basic amino acids (e.g., lysine, histidine, arginine) are treated as triprotic acids, exactly as shown earlier for phosphoric acid.

• Problem 1-43

"Glycine" can be obtained in three forms: (a) glycine hydrochloride, (b) isoelectric glycine (sometimes called glycine, free base), and (c) sodium glycinate.   Draw the structures of these three forms.

Solution

(a) $COOH$
Cl^- $H_3\overset{+}{N}$—$\overset{|}{\underset{|}{C}}$—H
H
glycine hydrochloride
(AA^{+1})

(b) COO^-
$H_3\overset{+}{N}$—$\overset{|}{\underset{|}{C}}$—H
H
isoelectric glycine
(AA^0)

(c) COO^- Na^+
H_2N—$\overset{|}{\underset{|}{C}}$—H
H
sodium glycinate
(AA^{-1})

• Problem 1-44

Calculate the pH of a 0.1 M solution of (a) glycine hydrochloride, (b) isoelectric glycine, and (c) sodium glycinate.

Solution

(a)   Glycine hydrochloride is essentially a diprotic acid.   Because the carboxyl group is so much stronger an acid ($K_{a_1} = 4.57 \times 10^{-3}$, $pK_{a_1} = 2.34$) than

Table 1-1 Predominant Species and pH at Key Points Along the Titration Curve of Weak Diprotic Acids

Start	pK_{a_1} (*halfway to 1st equiv. point*)	*First Equivalence Point*	pK_{a_2} (*halfway to 2nd equiv. point*)	*Second Equivalence Point*
H_2A	50% H_2A 50% HA^-	HA^-	50% HA^- 50% A^{2-}	A^{2-}
H_2CO_3 (CO_2)	50% $H_2CO_3 + CO_2$ 50% HCO_3^-	HCO_3^-	50% HCO_3^- 50% CO_3^{2-}	CO_3^{2-}
HOOC-$(CH_2)_2$-COOH	50% HOOC-$(CH_2)_2$-COOH 50% $^-$OOC-$(CH_2)_2$-COOH	$^-$OOC-$(CH_2)_2$-COOH	50% $^-$OOC-$(CH_2)_2$-COOH 50% $^-$OOC-$(CH_2)_2$-COO$^-$	$^-$OOC-$(CH_2)_2$-COO$^-$
H_3N^+-CHR-COOH	50% H_3N^+-CHR-COOH 50% H_3N^+-CHR-COO$^-$	H_3N^+-CHR-COO$^-$ (pH = pI)	50% H_3N^+-CHR-COO$^-$ 50% H_2N-CHR-COO$^-$	H_2N-CHR-COO$^-$
H_3N^+-R-NH_3^+	50% H_3N^+-R-NH_3^+ 50% H_3N^+-R-NH_2	H_3N^+-R-NH_2	50% H_3N^+-R-NH_2 50% H_2N-R-NH_2	H_2N-R-NH_2
$pH \simeq \dfrac{pK_{a_1} + p[Conc]}{2}$	$pH = pK_{a_1}$	$pH = \dfrac{pK_{a_1} + pK_{a_2}}{2}$	$pH = pK_{a_2}$	$pOH \simeq \dfrac{pK_{b_2} + p[Conc]}{2}$ $pH = 14 - pOH$

the charged amino group ($pK_{a_2} = 9.6$), the pH of the solution is established almost exclusively by the extent to which the carboxyl ionizes.

$$
\underset{\substack{+\\ AA^{+1}}}{H_3N-\overset{\overset{\textstyle COOH}{|}}{\underset{\underset{\textstyle H}{|}}{C}}-H} \;\rightleftharpoons\; \underset{\substack{+\\ AA^{0}}}{H_3N-\overset{\overset{\textstyle COO^-}{|}}{\underset{\underset{\textstyle H}{|}}{C}}-H} \;+\; H^+ \quad K_{eq}=K_{a_1}
$$

$$
K_{a_1} = \frac{[AA^0][H^+]}{[AA^{+1}]}
$$

Let
$$
y = M \text{ of } AA^{+1} \text{ that ionizes}
$$
$$
\therefore\; y = M \text{ of } H^+ \text{ produced}
$$
and
$$
y = M \text{ of } AA^0 \text{ produced}
$$
and
$$
0.1 - y = M \text{ of } AA^{+1} \text{ remaining at equilibrium}
$$
$$
K_{a_1} = \frac{(y)(y)}{(0.1 - y)} = 4.57 \times 10^{-3}
$$

Because of the proximity of the amino and carboxyl groups, the carboxyl group is a stronger acid than that of acetic acid. The y in the denominator cannot be ignored.

$$
4.57 \times 10^{-4} - 4.57 \times 10^{-3}y = y^2
$$
$$
y^2 + 4.57 \times 10^{-3}y - 4.57 \times 10^{-4} = 0
$$
$$
y = \frac{-b \pm \sqrt{b^2 - 4ac}}{2a}
$$

where $a = 1$ $b = 4.57 \times 10^{-3}$ $c = -4.57 \times 10^{-4}$.
Solving for y, we obtain:

$$
\boxed{[H^+] = 1.92 \times 10^{-2}\, M} \qquad \boxed{pH = 1.72}
$$

Thus, glycine hydrochloride is 19.2% ionized in a 0.1 M solution.

(b)   Isoelectric glycine is an intermediate ion of a polyprotic acid.

   The pH of a solution of an intermediate ion is essentially independent of the concentration of the ion. The pH may be calculated from the pK_a values on either side of the ion; that is, from the pK_a of the next acid group to ionize and the pK_a of the previous acid group ionized.

$$
pH = \frac{pK_{a_1} + pK_{a_2}}{2} = \frac{2.34 + 9.6}{2} = \frac{11.94}{2}
$$

$$
\boxed{pH = 5.97}
$$

At this point, it would be convenient to review *why* the pH is the average of pK_{a_2} and pK_{a_1}. The p*I* is defined as that pH where the predominant ionic form is AA^0, and any small concentration of AA^+ present is balanced by an equal concentration of AA^-. We know that:

$$K_{a_1} = \frac{[AA^0][H^+]}{[AA^+]} \quad \text{and} \quad K_{a_2} = \frac{[AA^-][H^+]}{[AA^0]}$$

$$\therefore \qquad [AA^0] = \frac{K_{a_1}[AA^+]}{[H^+]} \quad \text{and} \quad [H^+] = \frac{K_{a_2}[AA^0]}{[AA^-]}$$

Substituting for $[AA^0]$:

$$[H^+] = \frac{K_{a_2}K_{a_1}[AA^+]}{[AA^-][H^+]} \quad \text{or} \quad [H^+]^2 = \frac{K_{a_2}K_{a_1}[AA^+]}{[AA^-]}$$

But: $\qquad [AA^+] = [AA^-]$

$$\therefore \qquad \boxed{[H^+] = \sqrt{K_{a_2}K_{a_1}}} \quad \text{and} \quad \boxed{pH = \frac{pK_{a_2} + pK_{a_1}}{2}} \qquad (40)$$

(c) Sodium glycinate is a diprotic base. Both the un-ionized amino group and the carboxylate ion can accept a proton from water. However, because the amino group is a much stronger base than the carboxylate ion, the pH of the solution depends almost entirely on the extent to which the amino group ionizes. We can check the relative base strengths by calculating K_b for each group. For the amino group,

$$K_{b_1} = \frac{K_w}{K_{a_2}} = \frac{10^{-14}}{10^{-9.6}} = 10^{-4.4} = 3.98 \times 10^{-5}$$

For the carboxylate ion,

$$K_{b_2} = \frac{K_w}{K_{a_1}} = \frac{10^{-14}}{10^{-2.34}} = 10^{-11.66} = 2.19 \times 10^{-12}$$

$$K_{b_1} = \frac{[OH^-][AA^0]}{[AA^{-1}]} = \frac{(y)(y)}{(0.1 - y)}$$

Because the concentration of sodium glycinate is large compared to K_{b_1}, we can neglect the y in the denominator.

$$3.98 \times 10^{-5} = \frac{y^2}{0.1} \qquad y^2 = 3.98 \times 10^{-6}$$

$$y = \sqrt{3.98 \times 10^{-6}} = 1.995 \times 10^{-3}$$

$$[OH^-] = 1.995 \times 10^{-3}\ M \simeq 2 \times 10^{-3}\ M$$

$$[H^+] = \frac{K_w}{[OH^-]} = \frac{1 \times 10^{-14}}{2 \times 10^{-3}} = 5 \times 10^{-12}$$

$$pH = \log \frac{1}{[H^+]} = \log \frac{1}{5 \times 10^{-12}} = \log 2 \times 10^{11}$$

$$\boxed{pH = 11.3}$$

· Problem 1-45

(a) Draw the structures of the various forms of "aspartic acid" that may be obtained.   (b)–(e) Show how each form ionizes in water.

Solution

(a)   "Aspartic acid" may be obtained in four forms: aspartic hydrochloride (AA^{+1}), isoelectric aspartic acid (AA^0), monosodium aspartate (AA^{-1}), and disodium aspartate (AA^{-2}).   The structures are shown below.

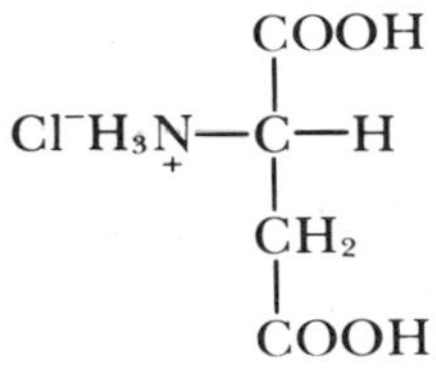

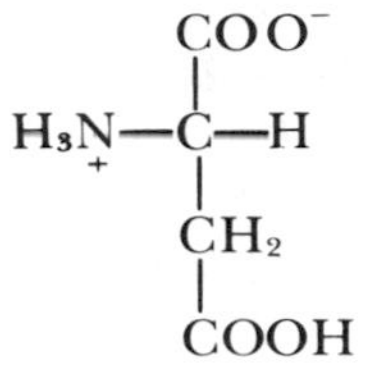

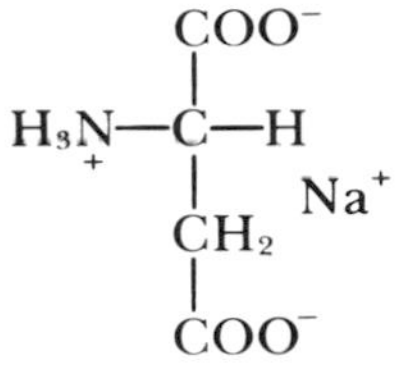

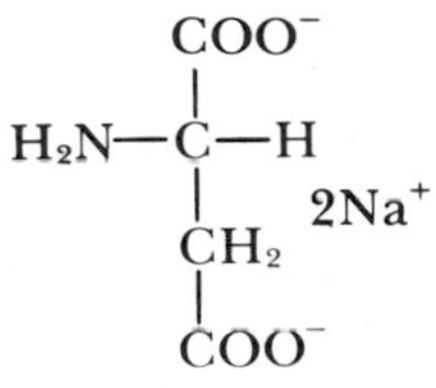

| aspartic hydro- | isoelectric aspartic | monosodium | disodium aspartate |
| chloride (AA^{+1}) | acid (AA^0) | aspartate (AA^{-1}) | (AA^{-2}) |

(b)   Aspartic hydrochloride ionizes as a typical polyprotic acid.   The pK_a values for the three acidic groups are 2.1 (α-COOH), 3.86 (β-COOH), and 9.82 (α-NH$_3^+$).   Because the α-COOH is so much stronger an acid than the other two groups, the pH of an aspartic hydrochloride solution depends almost exclusively on the concentration of aspartic hydrochloride and the extent to which the α-COOH ionizes.

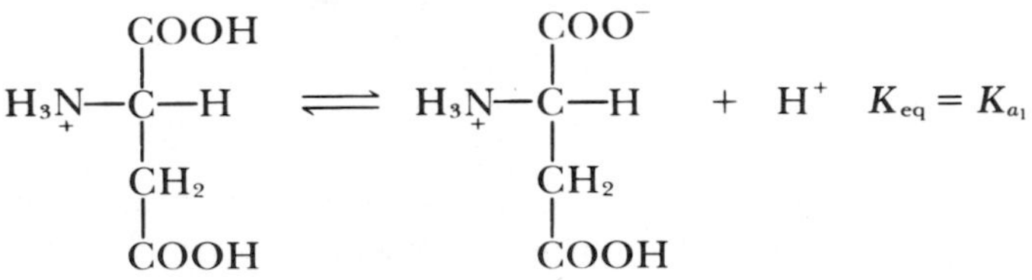

The pH calculations may be made exactly as described in the preceding problem for glycine hydrochloride.

$$K_{a_1} = \frac{[AA^0][H^+]}{[AA^{+1}]}$$

(c)   Disodium aspartate ionizes as a typical polyprotic base.   The pK_b values for the three basic groups can be calculated from their respective pK_a values as shown in Table 1-2.

$$pK_b = 14 - pK_a$$

Table 1-2

Conjugate Acid	pK_a	Conjugate Base	pK_b
α-COOH	2.1 (pK_{a_1})	α-COO$^-$	11.9 (pK_{b_3})
β-COOH	3.86 (pK_{a_2})	β-COO$^-$	10.14 (pK_{b_2})
α-NH$_3^+$	9.82 (pK_{a_3})	α-NH$_2$	4.18 (pK_{b_1})

Note that the pK_a values of the conjugate acids are numbered in decreasing order of acid strength. The pK_b values are numbered in decreasing order of base strength. Therefore, the α-NH$_3^+$ group is the weakest acid and its pK_a value is designated pK_{a_3}. The conjugate base of the α-NH$_3^+$ group is the α-NH$_2$ group which is the strongest of the basic groups. Hence, its pK_b value is designated pK_{b_1}. Because pK_{b_2} is almost 6 pH units less than pK_{b_1} (K_{b_2} is almost 10^6 times less than K_{b_1}), the pH of a disodium aspartate solution is established almost exclusively by the concentration of the salt and the extent of ionization of the α-NH$_2$ group.

Calculations of pH are done exactly as described in the previous problem for sodium glycinate.

$$K_{b_1} = \frac{[AA^{-1}][OH^-]}{[AA^{-2}]}$$

(d) The two remaining forms of aspartic acid are intermediate ions of polyprotic acids. For example, consider isoelectric aspartic acid. The compound ionizes both as an acid and a base.

$$\underset{\underset{\text{AA}^0}{}}{\overset{\text{COO}^-}{\underset{|}{\overset{|}{\underset{\overset{+}{H_3N}}{\overset{}{-}}}}}}\overset{}{\underset{\overset{|}{\underset{\overset{|}{\text{CH}_2}}{}}}{\text{C}}}\text{—H} \quad + \text{ HOH} \quad \rightleftharpoons \quad \underset{\underset{\text{AA}^{+1}}{}}{\overset{\text{COOH}}{\underset{|}{\overset{|}{\underset{\overset{+}{H_3N}}{}}}}}\text{—H} \quad + \text{ OH}^- \qquad K_{eq} = K_{b_3} = \frac{K_w}{K_{a_1}}$$

The pH of solutions of isoelectric aspartic acid may be calculated from the pK_{a_1} and pK_{a_2} values (the pK_a values on either side of isoelectric aspartic acid in its titration curve).

$$pH = \frac{pK_{a_1} + pK_{a_2}}{2} = \frac{2.1 + 3.86}{2} = \frac{5.96}{2} = 2.98$$

(e) The remaining form, monosodium aspartate, is also an intermediate ion of a polyprotic acid.   Its ionizations as an acid and as a base are shown below.

$$\underset{\underset{\text{AA}^{-1}}{}}{\overset{\text{COO}^-}{\underset{\overset{\overset{|}{\text{CH}_2}}{|}}{\underset{\overset{+}{H_3N}}{}}}}\text{—H} \quad \rightleftharpoons \quad \underset{\underset{\text{AA}^{-2}}{}}{\overset{\text{COO}^-}{\underset{\overset{\overset{|}{\text{CH}_2}}{|}}{H_2N}}}\text{—H} \quad + \text{ H}^+ \qquad K_{eq} = K_{a_3}$$

$$\underset{\underset{\text{AA}^{-1}}{}}{\overset{\text{COO}^-}{\underset{\overset{\overset{|}{\text{CH}_2}}{|}}{\underset{\overset{+}{H_3N}}{}}}}\text{—H} \quad + \text{ HOH} \quad \rightleftharpoons \quad \underset{\underset{\text{AA}^0}{}}{\overset{\text{COO}^-}{\underset{\overset{\overset{|}{\text{CH}_2}}{|}}{\underset{\overset{+}{H_3N}}{}}}}\text{—H} \quad + \text{ OH}^- \qquad K_{eq} = K_{b_2} = \frac{K_w}{K_{a_2}}$$

The group shown ionizing as an acid (the α-NH$_3^+$) is the only remaining acidic group.   The group shown ionizing as a base (the β-COO$^-$ group) is the stronger of the two basic groups present.   The ionization of the α-COO$^-$ group as a base contributes little toward establishing the pH of the solution because it is so much weaker a base than is the β-COO$^-$ group.   The pH of solutions of monosodium aspartate may be calculated from the pK_{a_2} and pK_{a_3} values.

$$pH = \frac{pK_{a_2} + pK_{a_3}}{2} = \frac{3.86 + 9.82}{2} = \frac{13.68}{2} = 6.84$$

· **Problem 1-46**

Sketch the pH curves for the titration of 100 ml of 0.1 M alanine hydrochloride with KOH in the (a) absence and (b) presence of excess formaldehyde.

Solution

(a) The titration curve resembles that of a typical diprotic acid with two buffering plateaus in the regions of the pK_a values. It takes 1 mole of OH^- per mole of amino acid to go from the starting point to the first end (equivalence) point and 0.5 mole of OH^- per mole of amino acid to get to the midway (the pK_{a_1}) position. To go from the first endpoint to the second end (equivalence) point, another mole of KOH per mole of amino acid is required; 1.5 moles of KOH per mole of amino acid hydrochloride bring the pH to the point midway between the first and second equivalence points (to the pK_{a_2} value).

(b) Formaldehyde reacts with amino groups to form methylol derivatives.

$$\underset{\substack{\\ R}}{\overset{\substack{COOH \\ |}}{H_2N-C-H}} \;\overset{HCHO}{\rightleftharpoons}\; \underset{\substack{HOH_2C \quad R}}{\overset{\substack{COOH \\ |}}{H-N-C-H}} \;\overset{HCHO}{\rightleftharpoons}\; \underset{\substack{HOH_2C \quad R}}{\overset{\substack{COOH \\ |}}{HOH_2C-N-C-H}}$$

The methylol derivatives are stronger acids (weaker bases) than are the original unsubstituted amino groups. In other words, the pK_{a_2} value for the substituted amino acid is lower than the pK_{a_2} value for the original amino acid. The titration curves are sketched in Figure 1-7. Note that formaldehyde has no effect on the amounts of KOH required to titrate the amino acid to pK_{a_1}, pK_{a_2} (or pK'_{a_2}), and the equivalence points. Also note that only the pK_{a_2} value is shifted; formaldehyde has no effect on the α-COOH group.

· **Problem 1-47**

Calculate the isoelectric point (pI) and the pH at which the maximum total number of charges are present (pH_m) for (a) glycine, (b) aspartic acid, and (c) lysine.

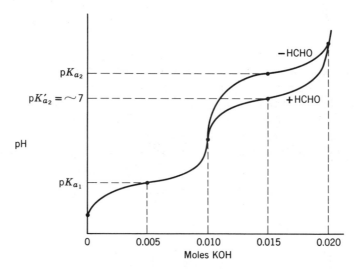

Figure 1-7· Titration curve of 0.01 moles of alanine hydro-chloride in the presence and in the absence of formaldehyde.

Solution

The isoelectric point, pI, is the pH at which the amino acid or peptide carries no net charge; that is, the predominant ionic form is the isoelectric species, AA^0, and (because the isoelectric form ionizes both as an acid and as a base) there are equal amounts of the ionic forms AA^{+1} and AA^{-1}. (The ionization of AA^0 to form AA^{+1} and AA^{-1} is a disproportionation reaction, as described earlier.) The pI may also be thought of as the pH of a solution of the isoelectric form of the amino acid.

The pI of amino acids is the pH at one equivalence point along the titration curve, specifically the equivalence point at which all the AA^{+1} is converted to AA^0. The pH at this point is, as usual, the average of the pK_a value to follow and the pK_a value just passed. Similarly, pH_m is the pH at one equivalence point and may be similarly calculated. To determine pI and pH_m simply sketch the titration curve and indicate the predominant ionic species present at each key point. Or, prepare a table showing the ionic form of each titratable group at key points. For simplicity, assume that you are starting with the maximally protonated amino acid or peptide.

(a) *Glycine.* As shown in Table 1-3, glycine exists as the "zwitterion" or isoelectric form at the first equivalence point. At this point, glycine also bears the greatest total number of charges. Therefore:

$$pI = pH_m = \frac{pK_{a_1} + pK_{a_2}}{2} = \frac{2.34 + 9.6}{2} = \frac{11.94}{2} = 5.97$$

(b) *Aspartic acid.* We can see from Table 1-3 and Figure 1-8 that aspartic acid carries no net charge at the first equivalence point.

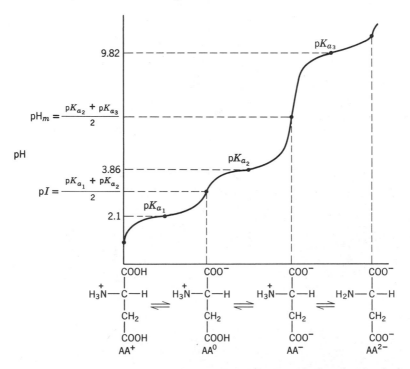

Figure 1-8 Titration curve of aspartic acid. For clarity, the vertical axis is not drawn to scale.

Table 1-3   Aspartic Acid

| Ionizable Group | Start | Predominant Ionic Form at Different Positions along the Titration Curve | | | | | |
		pK_{a_1}	First Equivalence	pK_{a_2}	Second Equivalence	pK_{a_3}	Third Equivalence
α-carboxyl	COOH	COOH$(\frac{1}{2})$ COO$^-(\frac{1}{2})$	COO$^-$	COO$^-$	COO$^-$	COO$^-$	COO$^-$
β-carboxyl	COOH	COOH	COOH	COOH$(\frac{1}{2})$ COO$^-(\frac{1}{2})$	COO$^-$	COO$^-$	COO$^-$
α-amino	NH$_3^+$	NH$_3^+$	NH$_3^+$	NH$_3^+$	NH$_3^+$	NH$_3^+(\frac{1}{2})$ NH$_2(\frac{1}{2})$	NH$_2$
Net charge	+1	$+\frac{1}{2}$	0	$-\frac{1}{2}$	-1	$-1\frac{1}{2}$	-2

Table 1-4 Lysine

Ionizable Group	Predominant Ionic Form at Different Positions along the Titration Curve						
	Start	pK_{a_1}	First Equivalence	pK_{a_2}	Second Equivalence	pK_{a_3}	Third Equivalence
α-carboxyl	COOH	$COOH(\frac{1}{2})$ $COO^{-}(\frac{1}{2})$	COO^-	COO^-	COO^-	COO^-	COO^-
α-amino	NH_3^+	NH_3^+	NH_3^+	$NH_3^{+}(\frac{1}{2})$ $NH_2(\frac{1}{2})$	NH_2	NH_2	NH_2
ε-amino	NH_3^+	NH_3^+	NH_3^+	NH_3^+	NH_3^+	$NH_3^{+}(\frac{1}{2})$ $NH_2(\frac{1}{2})$	NH_2
Net charge	$+2$	$+1\frac{1}{2}$	$+1$	$+\frac{1}{2}$	0	$-\frac{1}{2}$	-1

$$pI = \frac{pK_{a_1} + pK_{a_2}}{2} = \frac{2.09 + 3.86}{2} = \frac{5.95}{2} = 2.98$$

We can also see that aspartic acid carries the maximum total number of charges at the second equivalence point.

$$pH_m = \frac{pK_{a_2} + pK_{a_3}}{2} = \frac{3.86 + 9.82}{2} = \frac{13.68}{2} = 6.84$$

When constructing Table 1-3, we assumed that at the first equivalence point the α-carboxyl is completely ionized and that the β-carboxyl is completely un-ionized. These assumptions, of course, are not entirely true; the actual degree to which the α- and β-carboxyls are ionized can be calculated using the Henderson-Hasselbalch equation. If we carry out the calculation, we find that the proportion of α-carboxyl that is still in the COOH form exactly equals the proportion of β-carboxyl in the COO⁻ form. (At pH 2.98, we are just as far above the pK_{a_1} for the α-carboxyl as we are below the pK_{a_2} for the β-carboxyl.) Thus, to determine the net charge on the molecule, we are justified in tallying only the predominant ionic forms at each key point along the titration curve.

(c) *Lysine.* We see from Table 1-4 and Figure 1-9 that lysine carries no net charge at the second equivalence point.

$$pI = \frac{pK_{a_2} + pK_{a_3}}{2} = \frac{8.95 + 10.53}{2} = \frac{19.48}{2} = 9.74$$

We also see that lysine carries the maximum total number of charges at the first equivalence point (all ionizable groups charged).

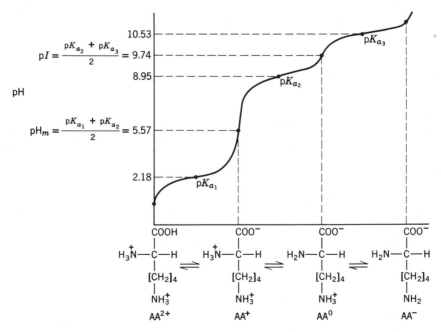

Figure 1-9 Titration curve of lysine. For clarity, the vertical axis is not drawn to scale.

$$\text{pH}_m = \frac{\text{p}K_{a_1} + \text{p}K_{a_2}}{2} = \frac{2.18 + 8.95}{2} = \frac{11.13}{2} = 5.57$$

We assumed that at the second equivalence point the α-amino group is completely uncharged and the ϵ-amino group is completely ionized. These assumptions are valid for calculating the net charge on lysine for the same reasons described earlier concerning aspartic acid.

· Problem 1-48

An enzyme catalyzing the decarboxylation of lysine accepts only the isoelectric form as a substrate. What is the actual concentration of isoelectric lysine in a $10^{-3}\ M$ solution of lysine in a buffer at pH 7.60?

Solution

At pH 7.60, the predominant ionic forms of lysine are AA^+ and AA^0 (Fig. 1-9). The equilibrium between these two species is described by K_{a_2}.

$$\text{pH} = \text{p}K_{a_2} + \log\frac{[AA^0]}{[AA^+]} \qquad 7.60 = 8.95 + \log\frac{[AA^0]}{[AA^+]}$$

$$1.35 = \log\frac{[AA^+]}{[AA^0]} \qquad \frac{[AA^+]}{[AA^0]} = \frac{22.4}{1}$$

$$\therefore\ [AA^0] = \frac{1}{23.4} \times 10^{-3} \qquad \boxed{[AA^0] = 4.27 \times 10^{-5}\ M}$$

· Problem 1-49

It is generally assumed that the completely uncharged form of a neutral amino acid shown below does not exist in solution. Instead, the major species of, for example, glycine at pH values around pI is the isoelectric form.

$$
\begin{array}{cc}
\text{COOH} & \text{COO}^- \\
| & | \\
\text{H}_2\text{N}-\text{C}-\text{H} & \text{H}_3\overset{+}{\text{N}}-\text{C}-\text{H} \\
| & | \\
\text{H} & \text{H} \\
\text{Uncharged form} & \text{Isoelectric form}
\end{array}
$$

Assuming p$K_{a_1} \simeq 2.5$ and p$K_{a_2} \simeq 9.5$, estimate the fraction of the total glycine present as the "rare" uncharged form at pH 5.5.

Solution

From the Henderson-Hasselbalch equation, we find that the ratio of R-COOH/R-COO$^-$ at pH 5.5 is about 10^{-3}. (The solution is 3 pH units above pK_{a_1}.) The ratio of R-NH$_2$/R-NH$_3^+$ is about 10^{-4}. (The solution is 4 pH units below pK_{a_2}.) Therefore, the combination, which gives the ratio of H$_2$N-R-COOH/H$_3$N$^+$-R-COO$^-$, is $10^{-3} \times 10^{-4} = 10^{-7}$.

$$\therefore\ \boxed{\text{one ten-millionth of the total glycine is present as the completely uncharged species}}$$

Additional problems on the charge properties of amino acids, peptides, and proteins will be found in Chapter 2.

· Problem 1-50

Calculate the ionic strength of a 0.05 M glycine buffer, pH 9.6.

Solution

The pH = pK_{a_2}; therefore $[AA^0] = [AA^-]$.   Of the glycine species present only the net charged AA^- contributes to the ionic strength.   If the buffer was prepared by titrating isoelectric glycine (AA^0) with NaOH, the solution contains $0.025 M$ AA^0, $0.025 M$ AA^-, and $0.025 M$ Na^+.   The ionic strength of a 1:1 salt is identical to the molarity of either ion.

$$\therefore \quad \boxed{\frac{\Gamma}{2} = 0.025}$$

If the buffer was prepared by titrating 0.05 moles of glycine hydrochloride (AA^+) with NaOH, then at pH 9.6 the buffer would contain $0.050 M$ Cl^-, $0.075 M$ Na^+, $0.025 M$ AA^-, and $0.025 M$ AA^0, of which the first three would contribute to the ionic strength:

$$\frac{\Gamma}{2} = \frac{(0.050)+(0.075)+(0.025)}{2} = \boxed{0.075}$$

Thus, if a low ionic strength is an important consideration, the first method of preparing the buffer is preferred.

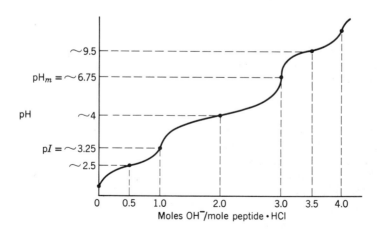

Figure 1-10   Titration curve of glutamylserylglutamylvaline hydrochloride.   For clarity, the vertical axis is not drawn to scale.

· **Problem 1-51**

Sketch the pH curve for the titration of 1 mole of glutamylserylglutamylvaline hydrochloride.

Solution

Titration and buffer calculations involving peptides are done exactly as shown earlier for polyprotic acids.   We must remember that the amino acid carboxyl groups used in forming the peptide bond are no longer available for titration.   The structure of the fully protonated glutamylserylglutamylvaline is shown below.   It is assumed that the α-COOH group retains a pK_a of ~ 2.5

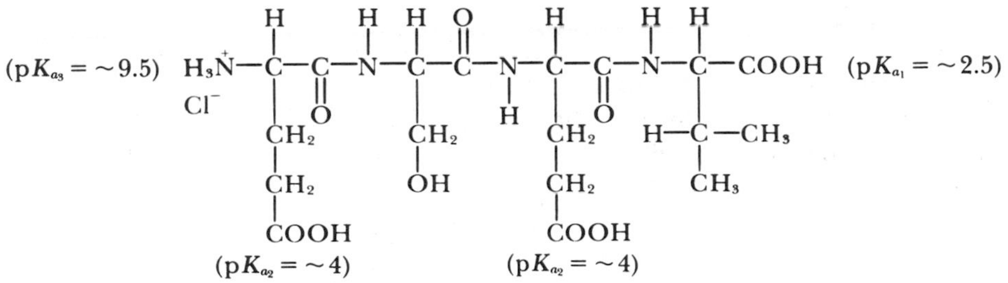

$$(pK_{a_3} = \sim 9.5) \quad H_3\overset{+}{N}-\underset{\underset{\underset{\underset{COOH}{(pK_{a_2} = \sim 4)}}{CH_2}}{CH_2}}{\overset{H}{\underset{|}{C}}}-\underset{O}{\overset{H}{\overset{||}{C}}}-\underset{|}{\overset{H}{\underset{|}{N}}}-\underset{\underset{OH}{CH_2}}{\overset{H}{\underset{|}{C}}}-\overset{H}{\overset{||}{\overset{O}{C}}}-\underset{H}{\overset{H}{\underset{|}{N}}}-\underset{\underset{\underset{COOH}{(pK_{a_2} = \sim 4)}}{CH_2}}{\overset{H}{\underset{|}{C}}}-\underset{CH_3}{\overset{H-\underset{|}{C}-CH_3}{}}-\overset{H}{\underset{|}{N}}-\overset{H}{\underset{|}{C}}-COOH \quad (pK_{a_1} = \sim 2.5)$$

$$Cl^-$$

and that both γ-COOH groups have identical pK_a values of ~ 4.0.   In reality, the α-COOH group would probably be a much weaker acid once the α-amino group of valine is tied up in a peptide bond.   The theoretical titration curve is shown in Figure 1-10.   In practice, two distinct buffering plateaus at pK_{a_1} and pK_{a_2} would not be as obvious as shown.

F. BLOOD BUFFERS

THE HCO$_3^-$/CO$_2$ SYSTEM

The HCO$_3^-$/CO$_2$ system is one of the two major blood buffers.   Carbonic acid ionizes as a typical weak diprotic acid:

$$H_2CO_3 \underset{pK_{a_1} = 3.8}{\overset{K_{a_1} = 1.58 \times 10^{-4}}{\rightleftharpoons}} H^+ + HCO_3^-$$

$$pK_{a_2} = 10.25 \; \Updownarrow \; K_{a_2} = 5.6 \times 10^{-11}$$

$$H^+$$
$$+$$
$$CO_3^{2-}$$

However, most of the conjugate acid dissolved in blood and cytoplasm is present as CO$_2$, not H$_2$CO$_3$.   The dissolved CO$_2$ is in equilibrium with CO$_2$ in the gas phase.   A more complete presentation of the CO$_2$ buffer system is shown below.

$$CO_2(gas)$$

$$CO_2 + H_2O \underset{K_{eq_1}}{\overset{}{\rightleftharpoons}} H_2CO_3 \underset{K_{a_1}}{\overset{}{\rightleftharpoons}} H^+ + HCO_3^-$$

(dissolved)

The equilibrium between CO_2 (gas) and CO_2 (dissolved) is given by:

$$[CO_2]_{dissolved} = k\,(P_{CO_2}) \tag{41}$$

That is, the concentration of dissolved CO_2 is directly proportional to the partial pressure of CO_2 in the gas phase. At 37°C and an ionic strength of 0.15, $k = 3.01 \times 10^{-5}$ when P_{CO_2} is expressed in terms of mm Hg. The equilibrium constant for the reaction: dissolved $CO_2 + H_2O \rightleftharpoons H_2CO_3$ is about 5×10^{-3}:

$$K_{eq_1} = \frac{[H_2CO_3]}{[CO_2]_{dis.}} = 5 \times 10^{-3}$$

Thus, the overall equilibrium constant between dissolved CO_2 and $H^+ + HCO_3^-$ is given by:

$$K_a' = \frac{[H^+][HCO_3^-]}{[CO_2]} = K_{eq_1} \times K_{a_1}$$

$$= (5 \times 10^{-3})(1.58 \times 10^{-4}) = 7.9 \times 10^{-7}$$

and $pK_a' = 6.1.$

The relationship can also be written as:

$$K_a' = \frac{[H^+][HCO_3^-]}{(3.01 \times 10^{-5})P_{CO_2}} = 7.9 \times 10^{-7}$$

At any pH:

$$pH = 6.1 + \log\frac{[HCO_3^-]}{[CO_2]} \quad \text{and} \quad pH = 6.1 + \log\frac{[HCO_3^-]}{(3.01 \times 10^{-5})P_{CO_2}}$$

For all practical purposes, a bicarbonate buffer can be considered to be composed of HCO_3^- (conjugate base) and dissolved CO_2 (conjugate acid).

The pH of blood is maintained at about 7.4.   If the pK_a' of CO_2 is 6.1, how can the HCO_3^-/CO_2 help buffer blood at pH 7.4?   Everything we have learned so far suggests that a buffer is effective only in the region of its pK_a.   The key here is that *in vivo* the HCO_3^-/CO_2 buffer is an *open system* in which the concentration of dissolved CO_2 is maintained constant.   Any excess CO_2 produced by the reaction $H^+ + HCO_3^- \rightarrow H_2O + CO_2$ is expelled by the lungs.   In contrast, the usual laboratory buffer is a *closed system*.   The concentration of conjugate acid increases when H^+ reacts with the conjugate base.   The effectiveness of the open system is illustrated below.

· Problem 1-52

Blood plasma contains a total carbonate pool (essentially $HCO_3^- + CO_2$) of 2.52×10^{-2} M.   (a) What is the HCO_3^-/CO_2 ratio and the concentration of each buffer component present at pH 7.4?   (b) What would the pH be if 10^{-2} M H^+ is added under conditions where the increased $[CO_2]$ cannot be released?   (c) What would the pH be if 10^{-2} M H^+ is added and the excess CO_2 eliminated (thereby maintaining the original $[CO_2]$)?

Solution

(a)
$$pH = pK'_a + \log\frac{[HCO_3^-]}{[CO_2]} \qquad 7.4 = 6.1 + \log\frac{[HCO_3^-]}{[CO_2]}$$

$$1.3 = \log\frac{[HCO_3^-]}{[CO_2]} \qquad \therefore \qquad \boxed{\frac{[HCO_3^-]}{[CO_2]} = \frac{20}{1}}$$

$$[HCO_3^-] = \frac{20}{21} \times 2.52 \times 10^{-2} = \boxed{\mathbf{2.40 \times 10^{-2}\ M}}$$

$$[CO_2] = \boxed{\mathbf{1.2 \times 10^{-3}\ M}}$$

(b)   If 0.01 M H^+ is added:

$$[HCO_3^-]_{final} = 0.024 - 0.010 = 0.014\ M$$
$$[CO_2]_{final} = 0.0012 + 0.010 = 0.0112\ M$$
$$pH = 6.1 + \log\frac{0.014}{0.0112} = 6.1 + \log 1.25$$

$$= 6.1 + 0.097 \qquad \boxed{\mathbf{pH = 6.2}}$$

Clearly, in a closed system, the HCO_3^-/CO_2 mixture has very little buffer capacity at pH 7.4.

(c)   In an open system:

$$0.024\ M\ HCO_3^- + 0.01\ M\ H^+ + 0.0012\ M\ CO_2$$
$$\downarrow$$
$$0.014\ M\ HCO_3^- + 0.0112\ M\ CO_2$$
$$\longrightarrow 0.01\ M\ CO_2\ \text{exhaled}$$
$$\downarrow$$
$$0.014\ M\ HCO_3^- + 0.0012\ M\ CO_2$$

$$pH = 6.1 + \log\frac{0.014}{0.0012} = 6.1 + \log 11.667$$

$$= 6.1 + \log 1.07 \qquad \boxed{\mathbf{pH = 7.16}}$$

In an open system, the pH decreases only 0.24 pH unit.   At first glance, it would seem that in an open system, the HCO_3^- reserve would be rapidly depleted.   However, *in vivo*, HCO_3^- is constantly replenished by the oxidative metabolic pathways, as described in the following section.

· **Problem 1-53**

The pH of a sample of arterial blood is 7.42.   Upon acidification of 10 ml of the blood, 5.91 ml of CO_2 (corrected for standard temperature and pressure) are produced.   Calculate (a) the total concentration of dissolved CO_2 in the blood $[CO_2 + HCO_3^-]$, (b) the concentrations of dissolved CO_2 and HCO_3^-, and (c) the partial pressure of the dissolved CO_2 in terms of mm Hg.

Solution

(a)   First calculate the number of moles of CO_2 represented by 5.91 ml at S.T.P.   One mole of a "perfect" gas occupies 22.4 liters at S.T.P.   The experimental value for CO_2 is 22.26 liters.

$$\therefore \quad \frac{5.91 \times 10^{-3} \text{ liters}}{22.26 \text{ liters/mole}} = 0.265 \times 10^{-3} \text{ moles}$$

This amount of CO_2 came from 10 ml of blood.

$$\therefore \quad \text{concentration of "total } CO_2\text{"} = \frac{26.5 \times 10^{-5} \text{ moles}}{10 \times 10^{-3} \text{ liters}} = \boxed{\textbf{2.65} \times \textbf{10}^{-2} \textbf{ \textit{M}}}$$

(b)
$$pH = pK_{a_1} + \log \frac{[HCO_3^-]}{[CO_2]}$$

$$7.42 = 6.1 + \log \frac{[HCO_3^-]}{[CO_2]} \qquad 1.32 = \log \frac{[HCO_3^-]}{[CO_2]}$$

$$\frac{[HCO_3^-]}{[CO_2]} = \text{antilog of } 1.32 = \frac{20.89}{1}$$

$$[HCO_3^-] = \frac{20.89}{21.89} \times 2.65 \times 10^{-2} M = \boxed{\textbf{2.53} \times \textbf{10}^{-2} \textbf{ \textit{M}}}$$

$$[CO_2] = \frac{1}{21.89} \times 2.65 \times 10^{-2} M = \boxed{\textbf{1.21} \times \textbf{10}^{-3} \textbf{ \textit{M}}}$$

(c)
$$[CO_2]_{\text{dissolved}} = kP_{CO_2}$$

$$P_{CO_2} = \frac{[CO_2]}{k} = \frac{1.21 \times 10^{-3}}{3.01 \times 10^{-5}} = \boxed{\textbf{40.22 mm Hg}}$$

HEMOGLOBIN

Aside from its well-known function as an oxygen carrier, hemoglobin plays an important role as a blood buffer.   In order to understand the interrelationship between oxygen uptake and release, and the buffering action of hemoglobin, we must consider the interaction of several simultaneous equilibria.   A greatly simplified version* of the equilibria is developed below.

---

* The simplified version considers only the binding of a single molecule of O_2 to a hemoglobin monomer.   In this way, O_2 binding to HHgb or Hgb can be described by a single oxygen

At any time, hemoglobin is present as a mixture of deoxygenated and oxygenated forms. The proportion of each depends on the concentration (partial pressure) of O_2.

$$\text{``deoxyhemoglobin''} + O_2 \rightleftharpoons \text{``oxyhemoglobin''}$$

Hemoglobin contains many ionizable groups. One in particular is a histidine residue that has a pK_a around neutrality. Thus, at any time, hemoglobin is also present as a mixture of protonated and unprotonated forms. The proportion of each depends on the pH of the blood.

$$\text{``H Hemoglobin''}$$
$$\Updownarrow$$
$$\text{``Hemoglobin''}$$
$$+$$
$$H^+$$

In order to combine the two simultaneous equilibria, we must recognize that "deoxyhemoglobin" represents a mixture of protonated deoxyhemoglobin (conjugate acid) plus nonprotonated deoxyhemoglobin (conjugate base). Similarly, "oxyhemoglobin" represents a mixture of H oxyhemoglobin (conjugate acid) and oxyhemoglobin (conjugate base). To state it in another way: the conjugate acid of hemoglobin exists in oxygenated and deoxygenated forms. So does the conjugate base of hemoglobin. Thus, there are really four species of hemoglobin present at any time. The proportion of each depends on the concentration of O_2 and the pH. The combined equilibria are shown in Figure 1-11.

Hemoglobin in the red blood cells arrives in the lungs mainly as a mixture of deoxy forms, HHgb + Hgb. The proportion of each is governed by the pH and the pK_a of deoxyhemoglobin. Since the pH is about 7.4, and pK_a is about 7.7, approximately two thirds of the total deoxyhemoglobin is present as the conjugate acid form. In the lungs, hemoglobin picks up oxygen. The horizontal equilibrium shifts to the right (reaction 1). $HHgbO_2$ is a stronger acid than HHgb. (A conformational change in the molecule upon oxygenation decreases the pK_a of the histidine group in the region of the heme to about 6.2.) As a result, the vertical equilibrium shifts downward (reaction 2) and H^+ is released. The increased $[H^+]$ forces the $H^+ + HCO_3^-$ equilibrium to the right (reaction 3). This results in the removal of H^+ and the release of CO_2 to the atmosphere. The oxygenated hemoglobin (mostly as the conjugate base $HgbO_2$ at pH 7.4) is transported to the tissues where the low O_2 partial pressure causes the horizontal equilibrium to shift to the left. O_2 is given off (reaction 4). Hgb is a stronger base than $HgbO_2$ (which follows, if $HHgbO_2$ is a stronger acid than HHgb). Thus, after O_2 is released, the vertical equilibrium shifts upward. As a result, H^+ ions (produced from the oxidation of food—reaction 6) are taken up by Hgb (reaction 7). Stated differently (but equivalently): in the tissues, the higher

---

dissociation constant. The "percent saturation versus $[O_2]$" curve for this model would be a hyperbola. Hemoglobin is actually a tetramer which displays "cooperative" binding of four molecules of O_2. The O_2 binding curve is sigmoidal (Chapter 4). The simplified scheme used in the present discussion is incorrect, but it conveys the essential features of the interaction between O_2, H^+, and CO_2.

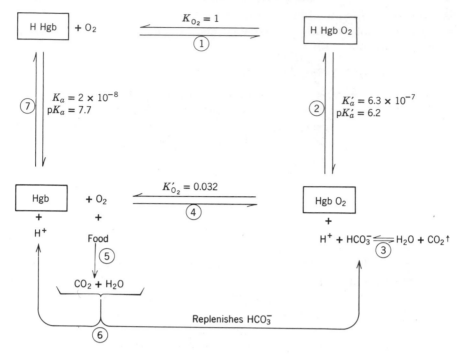

Figure 1-11   A simplified model for the oxygen/H$^+$ equilibria of hemo-globin. K_{O_2} is arbitrarily taken as unity.   All constants are dissociation constants.   pK_a is taken as 7.7, but values of 7.71 to 8.18 have been reported.   pK_a' is taken as 6.2, but values of 6.17 to 6.68 have been reported.

[H$^+$] forces the vertical equilibrium upward.   The conjugate acid of hemo-globin has a lower O_2 affinity (higher O_2 dissociation constant) than the conjugate base of hemoglobin.   Consequently, O_2 is released (i.e., the hori-zontal equilibrium shifts to the left).   Of course, both the H$^+$ and O_2 shifts occur simultaneously.   The sequence of events can be summarized as shown below.

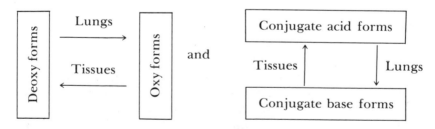

or, the overall equilibrium shifts as follows:

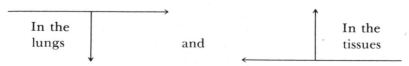

As an overall result, the oxidation of food yields CO_2, H$^+$, and HCO_3^- yet the concentrations of all three (hence, the pH of the tissues and blood) remain essentially constant.

· **Problem 1-54**

Consult Figure 1-11.   If $K_a = 2 \times 10^{-8}$, $K'_a = 6.3 \times 10^{-7}$, and $K_{O_2} = 1$, why must K'_{O_2} equal 0.032?

Solution

The overall equilibrium constant between any two points is the same regardless of the path taken. Thus, K_{eq} for the sequence $HHgbO_2 \rightleftharpoons O_2 + HHgb \rightleftharpoons H^+ + Hgb$ equals K_{eq} for the sequence $HHgbO_2 \rightleftharpoons H^+ + HgbO_2 \rightleftharpoons O_2 + Hgb$. The overall K_{eq} of a sequence of reactions is the product of the K_{eq}'s of each step.

$$\therefore \quad K_{O_2} \times K_a = K'_a \times K'_{O_2}$$

$$K'_{O_2} = \frac{K_{O_2} \times K_a}{K'_a} = \frac{(1)(2 \times 10^{-8})}{(6.3 \times 10^{-7})}$$

$$\boxed{K'_{O_2} = 0.032}$$

Thus, $HgbO_2$ has a lower oxygen dissociation constant (higher O_2 binding constant) than $HHgbO_2$.

· **Problem 1-55**

How many moles of H^+ can be taken up by hemoglobin at pH 7.4 as a consequence of the release of one mole of O_2?   Assume $pK_a = 7.7$ and $pK'_a = 6.2$.

Solution

Using the Hendersen-Hasselbalch equation, we can calculate that, at pH 7.4, $15.85/16.85 = 0.941$ of the total oxyhemoglobin is in the $HgbO_2$ form, and $1/16.84 = 0.059$ is in the $HHgbO_2$ form.   Similarly, we can calculate that, at pH 7.4, one third $= 0.333$ of the total deoxyhemoglobin is in the Hgb form and two thirds $= 0.667$ is in the HHgb form.   Thus after the release of one mole of O_2, the conjugate base/conjugate acid ratio must decrease to maintain equilibrium.   That is, the conjugate base/conjugate acid equilibrium ratio of $15.85:1$ for oxyhemoglobin is much higher than the corresponding equilibrium ratio of $0.5:1$ for deoxyhemoglobin.   Therefore, when oxyhemoglobin is converted to deoxyhemoglobin, the amount of the conjugate base must decrease (by picking up an H^+ to become the conjugate acid).   Specifically, the release of one mole of O_2 forces the uptake of $(0.667 - 0.059)$ or $(0.941 - 0.333) = 0.608$ moles of H^+.   Under physiological conditions, slightly more than 0.6 moles of H^+ are formed (from CO_2 entering the blood) for every mole of O_2 released.   This extra H^+ is partially absorbed by the noncarbonate blood buffers (phosphate and plasma proteins).   As a result, the pH of venous blood is actually 7.38 in normal resting individuals, compared to pH 7.41 for arterial blood.

· **Problem 1-56**

The oxygen binding curve for hemoglobin is shown in Figure 1-12.   In the presence of CO_2, the partial pressure of O_2 required for any given fraction of

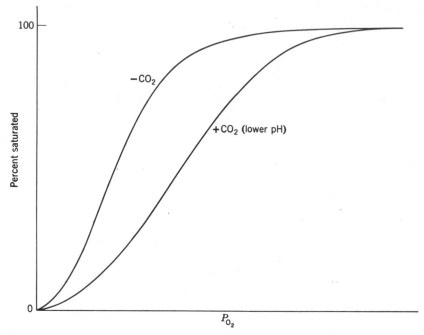

Figure 1-12 Oxygen binding by hemoglobin in the absence and in the presence of CO_2.

saturation is increased. Suggest an explanation for this "Bohr effect," as it is known.

Solution

As shown in Figure 1-11, $HHgbO_2$ has a greater oxygen dissociation constant than $HgbO_2$ or, in other words, $HHgb$ has a lower O_2 affinity than does Hgb. In effect, CO_2 shifts the equilibrium between the conjugate acid and conjugate base forms of hemoglobin. *In vivo*, CO_2 rapidly diffuses into the red blood cells, and reacts with water to produce $H^+ + HCO_3^-$. (In fact, the red blood cells contain an enzyme called carbonic anhydrase that accelerates the reaction $CO_2 + H_2O \rightleftharpoons H^+ + HCO_3^-$.) The decreased pH shifts the vertical equilibria of Figure 1-11 upward. Thus, hemoglobin is forced to a form with a lower O_2 affinity and O_2 is given off. In other words, at any given P_{O_2}, less O_2 is bound to hemoglobin in the presence of CO_2 because CO_2 causes the pH to decrease. As a result, a higher P_{O_2} is required to attain any given percent saturation.

GENERAL REFERENCES

Christensen, H. N., *Body Fluids and the Acid-Base Balance.* W. B. Saunders (1965).

Gomori, G., "Preparation of Buffers for Use in Enzyme Studies" in *Methods in Enzymology.* S. P. Colowick and N. O. Kaplan, eds., Vol. 1, p. 138, Academic Press (1955).

Good, N. E. and S. Izawa, "Hydrogen Ion Buffers" in *Methods in Enzymology.* A. San Pietro, ed., Vol. 24, p. 53, Academic Press (1972).

Masoro, E. J. and P. D. Siegel, *Acid-Base Regulation: Its Physiology and Pathophysiology.* W. B. Saunders (1971).

PRACTICE PROBLEMS

Consult Appendices IV and VII for K_a and pK_a values.   Answers to the Practice Problems are given on pages 421–423.

Concentrations of Solutions

1. A solution contains 15 g of $CaCl_2$ in a total volume of 190 ml.   Express the concentration of this solution in terms of (a) g/liter, (b) % (w/v), (c) mg %, (d) M, and (e) osmolarity.   (f) What is the ionic strength of the solution?

2. A solution was prepared by dissolving 8.0 g of solid ammonium sulfate (MW = 132.14, specific volume = 0.565 ml/g) in 35 ml (i.e., 35 g) of water.   (a) What is the final volume of the solution?   Express the concentration of ammonium sulfate in terms of (b) % (w/w), (c) % (w/v), (d) m, (e) M, (f) osmolarity, and (g) mole fraction ammonium sulfate.   (h) What is the ionic strength of the solution?

3. Starting with 150 ml of a 40% saturated ammonium sulfate solution at 0°C calculate (a) the weight of solid ammonium sulfate that must be added to bring the solution to 60% saturation, and (b) the volume of saturated solution that must be added to attain 60% saturation.   The specific volume of ammonium sulfate is 0.565 ml/g.

4. What is the molarity of pure ethanol— that is, how many moles are present in 1 liter of pure ethanol?   The density of ethanol is 0.789 g/ml.   The MW of ethanol is 46.07.

Strong Acids and Bases—pH

5. Calculate the pH, pOH, and number of H^+ and OH^- ions per liter in each of the following solutions: (a) 0.01 M HCl, (b) 10^{-4} M HNO_3, (c) 0.0025 M H_2SO_4, (d) 3.7×10^{-5} M KOH, (e) 5×10^{-8} M HCl, (f) 2.9×10^{-3} M NaOH, (g) 1 M HCl, (h) 10 M HNO_3, and (i) 3×10^{-5} N H_2SO_4.

6. Calculate the H^+ ion concentration (M), the OH^- ion concentration, and the number of H^+ and OH^- ions per liter in solutions having pH values of (a) 2.73, (b) 5.29, (c) 6.78, (d) 8.65, (e) 9.52, (f) 11.41, and (g) 0.

7. Calculate the (a) $[H^+]$, (b) $[OH^-]$, (c) pH, and (d) pOH of the final solution obtained after 100 ml of 0.2 M NaOH are added to 150 ml of 0.4 M H_2SO_4.

8. Calculate the pH and pOH of a solution obtained by adding 0.2 g of solid KOH to 1.5 liters of 0.002 M HCl.

9. The pH of a 0.10 M HCl solution was found to be 1.15.   Calculate the (a) a_{H^+} and (b) γ_{H^+} in this solution.

10. The activity coefficient of the hydroxyl ion (γ_{OH^-}) is 0.72 in a 0.1 M solution of KOH.   Calculate the pH and pOH of this solution.

11. Concentrated H_2SO_4 is 96% H_2SO_4 by weight and has a density of 1.84 g/ml.   Calculate the volume of concentrated acid required to make (a) 750 ml of 1 N H_2SO_4, (b) 600 ml of 1 M H_2SO_4, (c) 1000 g of a dilute H_2SO_4 solution containing 12% H_2SO_4 by weight, (d) an H_2SO_4 solution containing 6.5 equivalents per liter, and (e) a dilute H_2SO_4 solution of pH 3.8.

12. Concentrated HCl is 37.5% HCl by weight and has a density of 1.19.   (a) Calculate the molarity of the concentrated acid.   (b) Describe the preparation of 500 ml of 0.2 M HCl.   (c) Describe the preparation of 350 ml of 0.5 N HCl.   (d) Describe the preparation of an HCl solution containing 25% HCl by weight.   (e) Describe the preparation of a dilute HCl solution having a pH of 4.7.

13. Calculate the weight of solid NaOH required to prepare (a) 5 liters of a 2 M solution, (b) 2 liters of a solution of pH 11.5, and (c) 500 ml of 62% w/w solution.   The density of 62% NaOH solution is 1.15 g/ml.

14. How many milliliters of 0.12 M H_2SO_4 are required to neutralize exactly *half* of the OH^- ions present in 540 ml of 0.18 N NaOH?

15. How many grams of solid Na_2CO_3 are required to neutralize exactly 2 liters of an HCl solution of pH 2.0?

16. How many milliliters of 0.15 M KOH are required to neutralize exactly 180 g of pure H_2SO_4?

Weak Acids and Bases—Buffers

When solving the problems below, assume that $\gamma = 1$ for all substances unless otherwise indicated.

17. The weak acid, HA, is 2.4% dissociated in a 0.22 M solution. Calculate (a) the K_a, (b) the pH of the solution, (c) the amount of 0.1 N KOH required to neutralize 550 ml of the weak acid solution, and (d) the number of H^+ ions in 550 ml of the weak acid solution.

18. The pH of a 0.27 M solution of a weak acid, HA, is 4.3. (a) What is the H^+ ion concentration in the solution? (b) What is the degree of ionization of the acid? (c) What is the K_a?

19. The K_a of a weak acid, HA, is 3×10^{-4}. Calculate (a) the OH^- ion concentration in the solution and (b) the degree of dissociation of the acid in a 0.15 M solution.

20. At what concentration of weak acid, HA (in terms of its K_a), will the acid be 25% dissociated?

21. (a) Calculate the pH of a 0.05 M solution of ethanolamine, $K_b = 2.8 \times 10^{-5}$. (b) What is the degree of ionization of the amine?

22. Calculate the pK_a and pK_b of weak acids with K_a values of (a) 6.23×10^{-4}, (b) 2.9×10^{-5}, (c) 3.4×10^{-5}, and (d) 7.2×10^{-6}.

23. Calculate the pK_b and pK_a of weak bases with K_b values of (a) 2.1×10^{-5}, (b) 3.1×10^{-6}, (c) 7.8×10^{-5}, and (d) 9.2×10^{-4}.

24. Calculate the pH of a 0.2 M solution of an amine that has a pK_a of 9.5.

25. What is the pH of a 0.20 M solution of (a) H_3PO_4, (b) KH_2PO_4, (c) K_2HPO_4, (d) K_3PO_4, (e) potassium acetate, (f) NH_4Br, (g) sodium phenolate, (h) trisodium citrate, (i) disodium citrate, and (j) ethanolamine hydrochloride?

26. How many milliliters of 0.1 M KOH are required to titrate completely 270 ml of 0.4 M propionic acid?

27. How many milliliters of 0.2 M KOH are required to titrate completely 650 ml of 0.05 M citric acid?

28. What are the final hydrogen ion concentration and pH of a solution obtained by mixing 100 ml of 0.2 M KOH with 150 ml of 0.1 M HOAc? pK_a of HOAc = 4.77.

29. What are the final hydrogen ion concentration and pH of a solution obtained by mixing 200 ml of 0.4 M aqueous NH_3 with 300 ml of 0.2 M HCl? ($K_b = 1.8 \times 10^{-5}$)

30. What are the final hydrogen ion concentration and pH of a solution obtained by mixing 250 ml of 0.1 M citric acid with 300 ml of 0.1 M KOH? pK_a's are 3.06, 4.74, and 5.40.

31. What are the final hydrogen ion concentration and pH of a solution obtained by mixing 400 ml of 0.2 M NaOH with 150 ml of 0.1 M H_3PO_4? pK_a's are 2.12, 7.21, and 12.32.

32. What are the concentrations of NH_3 and NH_4Cl in a 0.15 M "ammonia" buffer, pH 9.6 ($K_b = 1.8 \times 10^{-5}$)?

33. (a) What is the pH of a solution containing 0.01 M HPO_4^{2-} and 0.01 M PO_4^{3-} (assume $\gamma = 1$)? (b) Calculate the actual pH by using the activity coefficients listed in Appendix V. (c) What is pK'_{a_3} at a total ionic strength of 0.1? (Consult Appendix VI.)

34. What is the pH of a solution containing 0.3 M Tris(hydroxymethyl)aminomethane (free base) and 0.2 M Tris hydrochloride? $pK_a = 8.1$.

35. What is the pH of a solution containing 0.2 g/liter Na_2CO_3 and 0.2 g/liter $NaHCO_3$? ($pK_{a_2} = 10.25$)

36. What is the pH of a solution prepared by dissolving 5.35 g of NH_4Cl in a liter of 0.2 M NH_3? (K_b of NH_3 is 1.8×10^{-5})

37. Describe the preparation of 2 liters of 0.25 M formate buffer, pH 4.5, starting from 1 M formic acid and solid sodium formate (HCOONa). pK_a of HCOOH is 3.75.

38. Describe the preparation of 40 liters of 0.02 M phosphate buffer, pH 6.9, starting from (a) a 2 M H_3PO_4 solution and a 1 M KOH solution, (b) a 0.8 M H_3PO_4 solution and solid NaOH, (c) a commercial concentrated H_3PO_4 solution and 1 M KOH, (d) 1 M solutions of KH_2PO_4 and Na_2HPO_4, (e) solid KH_2PO_4 and K_2HPO_4, (f) solid K_2HPO_4 and 1.5 M HCl, (g) 1.2 M K_2HPO_4 and 2 M H_2SO_4, (h) solid KH_2PO_4 and 2 M KOH, (i) 1.5 M KH_2PO_4 and 1 M NaOH, and (j) solid Na_3PO_4 and 1 M HCl.

39. What volume of glacial acetic acid (density 1.06 g/ml) and what weight of solid potassium acetate are required to prepare 5 liters of 0.2 M acetate buffer, pH 5.0?

40. An enzyme-catalyzed reaction was carried out in a solution buffered with 0.03 M phosphate, pH 7.2. As a result of the reaction, 0.004 mole/liter of acid was formed. (a) What was the pH at the end of the reaction? (b) What would the pH be if no buffer were present? (c) Write the chemical equation showing how the phosphate buffer resisted a large change in pH.

41. An enzyme-catalyzed reaction was carried out in a solution containing 0.2 M Tris buffer. The pH of the reaction mixture at the start was 7.8. As a result of the reaction, 0.033 mole/liter of H^+ was consumed. (a) What was the ratio of $Tris^0$ (free base) to $Tris^+$ Cl^- at the start of the reaction? (b) What was the $Tris^0/Tris^+$ ratio at the end of the reaction? (c) What was the final pH of the reaction mixture? (d) What would the final pH be if no buffer were present? (e) Write the chemical equations showing how the Tris buffer maintained a near constant pH during the reaction. The pK_a of Tris is 8.1.

42. When a sulfate ester is hydrolyzed, an H^+ ion is produced:

$$R\text{-}O\text{-}SO_3^- + H_2O \longrightarrow ROH + SO_4^{2-} + H^+$$

The above reaction was carried out in 1.0 ml of 0.02 M Tris buffer, pH 8.10, containing 0.01 M R-O-SO$_3^-$ and an enzyme called a sulfatase that catalyzes the reaction. At the end of 10 minutes, the pH of the reaction mixture decreased to 7.97. How many μmoles of R-O-SO$_3^-$ were hydrolyzed during the 10 minute incubation period?

43. Calculate the (a) instantaneous and (b) practical buffer capacity in the acid and alkaline directions of a 0.01 M phosphate buffer, pH 6.8.

44. The pK_a values of malic acid are 3.40 and 5.05. (a) What is the pH of a solution of 0.05 M monosodium malate? (b) What is the pH at the first equivalence point when malic acid is titrated with KOH? (c) What are the ratios and concentrations of all malate species in a 0.05 M solution at pH 4.70?

Amino Acids and Peptides

45. Calculate the pH of a 1 mM solution of (a) alanine hydrochloride, (b) isoelectric alanine, and (c) the sodium salt of alanine.

46. Calculate the volume of 0.1 M KOH required to titrate completely (a) 450 ml of 0.25 M alanine hydrochloride, (b) 200 ml of 0.10 M isoelectric serine, (c) 400 ml of 0.15 M monosodium glutamate, and (d) 400 ml of 0.15 M isoelectric glutamic acid.

47. Calculate the volume of 0.2 M HCl required to titrate completely (a) 200 ml of 0.25 M isoelectric leucine, (b) 375 ml of 0.25 M isoelectric glutamic acid, (c) 490 ml of 0.25 M isoelectric lysine, and (d) 125 ml of 0.25 M sodium salt of lysine.

48. Calculate the pH of a solution obtained by adding 20 ml of 0.20 M KOH to 480 ml of 0.02 M isoelectric glycine.

49. What is the major ionic species present at pH 7.5 in 0.15 M solutions of (a) leucine, (b) aspartic acid, and (c) lysine.

50. Describe the preparation of 1 liter of 0.2 M histidine buffer, pH 6.5, starting from solid histidine hydrochloride monohydrate (MW = 209.6) and 1 M KOH. pK_{a_2} of histidine = 6.0. (His. HCl is AA^+).

Blood Buffers

51. What is the concentration of dissolved CO_2 in equilibrium with an atmosphere containing a partial pressure of CO_2 of 40 mm Hg?

52. What are the ratios of $CO_2/HCO_3^-/CO_3^{2-}$ in blood plasma at pH 7.4? ($pK_{a_1} = 6.1$, $pK_{a_2} = 10.25$)

53. Blood plasma at pH 7.4 contains 2.4×10^{-2} M HCO_3^- and 1.2×10^{-3} M CO_2. Calculate the pH after the addition of 3.2×10^{-3} M H^+. Assume that the concentration of dissolved CO_2 is maintained constant at 1.2×10^{-3} M by the release of excess CO_2.

54. Consider the hemoglobin reaction scheme shown in Figure 1-11. If $K_a = 6.6 \times 10^{-9}$, $K_a' = 2.4 \times 10^{-7}$, and $K_{O_2} = 1$, what must K'_{O_2} be?

55. The pK_a's of $HHgbO_2$ and $HHgb$ from an aquatic mammal are 6.62 and 8.18, respectively. Calculate the number of moles of H^+ taken up by the hemoglobin per mole of O_2 released at pH 7.4.

2

CHEMISTRY OF BIOLOGICAL MOLECULES

A. AMINO ACIDS, PEPTIDES, AND PROTEINS

The following two sections illustrate how a knowledge of the acid-base properties of amino acids and peptides can be used to advantage in designing separation procedures, or predicting separation patterns.

ION-EXCHANGE CHROMATOGRAPHY

Ion-exchange chromatography is widely used to separate and analyze mixtures of amino acids. The most common ion-exchange resin used for this purpose is a sulfonic acid cation exchanger of the Dowex-50 (polystyrene) type. The structure of the resin is:

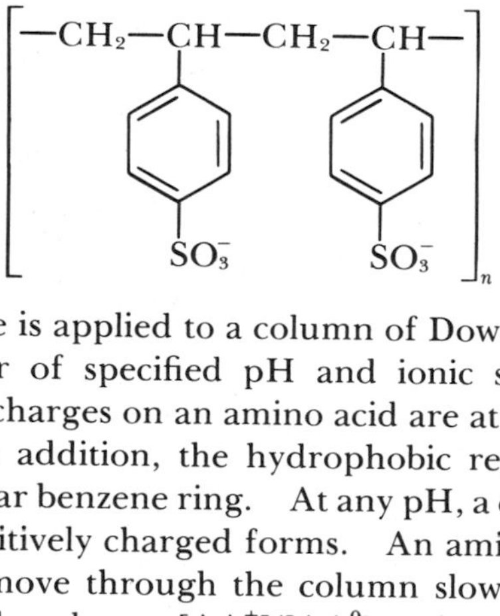

The amino acid mixture is applied to a column of Dowex-50 and then eluted by percolating a buffer of specified pH and ionic strength through the column. The positive charges on an amino acid are attracted to the resin by electrostatic forces. In addition, the hydrophobic regions of amino acids interact with the nonpolar benzene ring. At any pH, a certain fraction of any amino acid exists in positively charged forms. An amino acid with a higher $[AA^+]/[AA^0]$ ratio will move through the column slower than one of equal nonpolar character with a lower $[AA^+]/[AA^0]$ ratio. In other words, the amino acid with the lower $[AA^+]/[AA^0]$ ratio will elute before the one with a higher ratio (provided they have equal nonpolar attractions for the resin).

A rapid estimate of the effective charge on an amino acid can be made by comparing its pI with the pH of the buffer used:

$$\text{Let } \Delta p = pI - pH$$

If Δp is positive, the amino acid carries a net positive charge. An amino acid with a greater Δp will stick more tightly to a cation-exchange resin than an equally hydrophobic amino acid with a lower Δp. If Δp is negative, the amino acid carries a net negative charge and, consequently, will have very little attraction for the resin.

· Problem 2-1

A solution containing aspartic acid ($pI = 2.98$), glycine ($pI = 5.97$), threonine ($pI = 6.53$), leucine ($pI = 5.98$), and lysine ($pI = 9.74$) in a pH 3.0 citrate buffer was applied to a Dowex-50 cation-exchange column equilibrated with the same buffer. The column was then eluted with the buffer and fractions collected. In what order will the five amino acids elute from the column?

Solution

Aspartic acid, with two partially ionized (negatively charged) carboxyl groups (and the lowest Δp) will elute first. Of the three neutral amino acids, threonine has the greatest $[AA^+]/[AA^0]$ ratio (i.e., the highest Δp), but it is highly polar because of the OH group. Consequently, threonine elutes before glycine and leucine. Glycine and leucine have about the same Δp, but leucine is decidedly more nonpolar than glycine. Therefore, glycine elutes third, followed by leucine. Lysine has a high effective positive charge because of its additional amino group ($pI = 9.74$, $\Delta p = 6.74$). Therefore lysine elutes last, or not at all, unless the pH and/or the ionic strength of the eluting buffer are increased.

ELECTROPHORESIS

Charged compounds such as amino acids may be separated by taking advantage of their different mobilities in an electric field. Electrophoretic mobility of a compound on a buffered solid support depends approximately on the charge/mass ratio. This can be expressed mathematically as:

$$\text{Mobility} = \frac{-k\,\Delta p}{\text{MW}} = \frac{k\,(\text{pH} - \text{p}I)}{\text{MW}} \qquad (1)$$

where k is a constant related to the voltage, electrophoretic medium, and the like. A positive mobility value as defined by the above equation indicates movement toward the positive pole; a negative value indicates movement toward the negative pole. The above equation is not exact. Anomalies occur because of different degrees of hydration of different ions, differential binding of ions to the support, and the like.

Electrophoretic mobility at any given pH on a given medium is usually expressed in terms of $cm^2\ volt^{-1}\ sec^{-1}$, as shown below:

$$\text{mobility} = \frac{\text{distance per unit time}}{\text{electrical field strength}} = \frac{\text{cm/sec}}{\text{volts/cm}} = cm^2 \times volt^{-1} \times sec^{-1}$$

· Problem 2-2

What are the relative electrophoretic mobilities of glycine, leucine, aspartic acid, glutamic acid, and lysine at pH 4.70?

Solution

Using the approximate equation we would predict that lysine moves the fastest toward the negative pole, followed by glycine and then leucine. Aspartate moves faster than glutamate toward the positive pole. The calculations are summarized below:

Amino Acid	MW	pI	Mobility $\propto \dfrac{pH - pI}{MW}$ at pH 4.70
Lysine	146.2	9.74	−0.0345
Glycine	75.1	5.97	−0.0169
Leucine	131.2	5.98	−0.0098
Glutamic acid	147.1	3.22	+0.0100
Aspartic acid	133.1	2.98	+0.0129

PRIMARY STRUCTURE—SEQUENCING

The sequence of amino acids in the polypeptide chain (i.e., the *primary structure* of a polypeptide or protein) can be established by selective chemical and enzymatic cleavage of the protein followed by separation, amino acid analysis, and sequence determination of all peptide fragments. The entire amino acid sequence is established by overlapping identical regions of the individual fragments. The following problem illustrates the procedure.

· **Problem 2-3**

Partial hydrolysis of a protein yielded a number of polypeptides. One of them was purified. Deduce the sequence of amino acids in this polypeptide from the following information:

(a) Complete acid hydrolysis yielded ala + arg + 2 ser + lys + phe + met + trp + pro.

(b) Treatment with fluorodinitrobenzene (FDNB, the Sanger reagent) followed by complete acid hydrolysis yielded dinitrophenylalanine (DNP-ala) and ε-dinitrophenyllysine (ε-DNP-lys) as the only DNP derivatives.

(c) Neither carboxypeptidase A nor carboxypeptidase B released a C-terminal amino acid.

(d) Treatment with cyanogen bromide (CNBr) yielded two peptides. One contained ser + trp + pro. The other contained all the remaining amino acids (including the second ser).

(e) Treatment with chymotrypsin yielded three peptides. One contained only ser + pro. Another contained only met + trp. The third contained phe + lys + ser + arg + ala.

(f) Treatment with trypsin yielded three peptides. One contained only ala + arg. Another contained only lys + ser. The third contained phe + trp + met + ser + pro.

Solution

(a) FDNB reacts with free amino groups yielding the DNP-amino acid derivative upon hydrolysis. Thus the N-terminal amino acid is alanine. Lysine is in the interior of the chain and has its ϵ-amino group free. Thus, the peptide is linear, not circular.

(b) Carboxypeptidase A will cleave all C-terminal amino acids *except* arginine, lysine, or proline. Carboxypeptidase B will cleave *only* C-terminal arginine or lysine. Neither will act on any C-terminal amino acid if the next-to-last (penultimate) amino acid is proline. The lack of product with both enzymes suggest that proline is the last or penultimate residue.

Note. Carboxypeptidases will continue to cleave susceptible C-terminal amino acids. So what we really determine is the identity of the amino acid that is released most rapidly.

(c) Cyanogen bromide cleaves specifically on the carboxyl side of methionine residues. (The methionine is converted to homoserine.) The data so far suggest that the tripeptide released by CNBr is C-terminal. (It contains the pro.) Thus, the last four residues include met, followed by trp, ser, and pro (but the sequence of the last three is still unknown).

(d) Chymotrypsin cleaves on the carboxyl side of phenylalanine, tyrosine, tryptophan, and leucine provided the next residue (donating the amino group) is not proline. The composition of one of the chymotrypsin dipeptides confirms that the C-terminal end of the original peptide is met-trp-ser-pro. The amino acid preceding the methionine must be phenylalanine (the only remaining residue susceptible to chymotrypsin). Thus, the terminal sequence is phe-met-trp-ser-pro.

(e) Trypsin cleaves on the carboxyl side of lysine and arginine provided the next amino acid (donating the amino group) is not proline. Since alanine is N-terminal, the beginning sequence must be ala-arg-ser-lys.

The overall sequence is shown below:

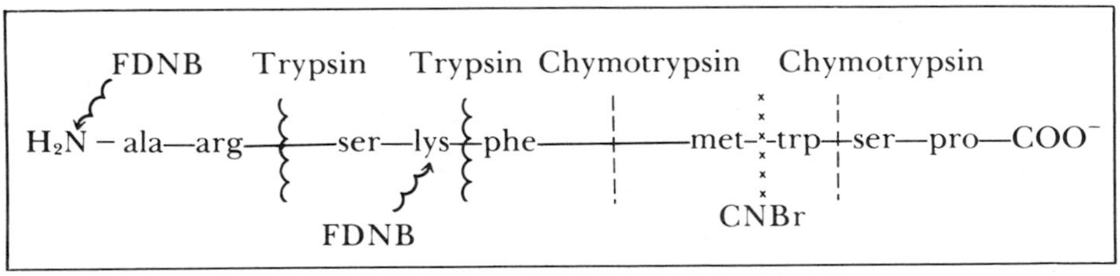

If another major polypeptide fragment was shown to contain met-trp-ser-pro-glu-glu-thr-leu-val-gly, then the met-trp-ser-pro overlap suggests a sequence of ala-arg-ser-lys-phe-met-trp-ser-pro-glu-glu-thr-leu-val-gly.

· Problem 2-4

Upon complete acid hydrolysis, a peptide yielded gly + ala + 2 cys + arg + glu + ile + thr + phe + val + NH_4^+. Reduction of the original peptide with mercaptoethanol, followed by alkylation of the cysteine residues with iodoacetate yielded two smaller peptides (*A* and *B*). Suggest a likely structure of the original peptide from the following data:

Peptide A: (a) Contained ala + gly + cys + glu + arg + ile + NH_4^+
 (b) Carboxypeptidase A liberated isoleucine
 (c) Treatment with phenylisothiocyanate (PITC, the Edman reagent) yielded the phenylthiohydantoin derivative of glycine (PTH-glycine)
 (d) Treatment with trypsin yielded two peptides. One contained glutamate + isoleucine + NH_4^+. The other contained gly + ala + cys + arg

Peptide B: (e) Contained thr + val + cys + phe
 (f) Carboxypeptidase A liberated valine
 (g) Chymotrypsin liberated valine and a tripeptide containing cys + thr + phe
 (h) The Edman degradation yielded PTH-threonine

Solution

The Edman reagent attacks free amino groups releasing the PTH-amino acid derivative. Mercaptoethanol reduces disulfide bonds. Iodoacetate alkylates the –SH groups, thereby preventing reoxidation. The NH_4^+ produced by acid hydrolysis must have come from an amide. Thus, the glutamate probably was present as glutamine. The original peptide probably has the structure shown below:

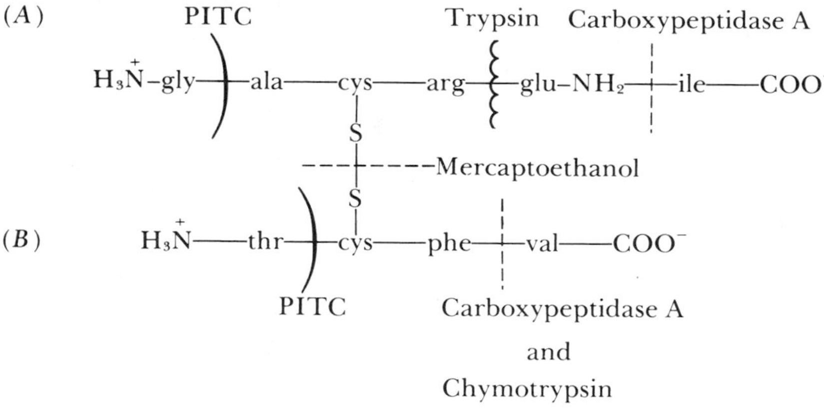

The ala and cys of peptide *A* could be reversed.

VARIETY OF PEPTIDES AND PROTEINS

The following problems illustrate the tremendous variety of polypeptides and proteins that can be made from 20 amino acids, and the variety of amino acid analysis patterns they would yield.

· Problem 2-5

How many different linear tripeptides can be made from three different L-α-neutral amino acids (a) using any of the three amino acids for any position (repetition allowed), (b) using each amino acid only once in the chain? (c) How many qualitatively different amino acid analysis patterns containing *all three* amino acids are possible? (d) How many *qualitatively* unique amino acid analysis patterns are possible among the total number of possible tripeptides calculated in a? (e) How many *quantitatively* unique amino acid analysis patterns are possible among the total number of possible tripeptides calculated in a?

Solution

(a) There are three possibilities for the first position and, because any of the three amino acids can be used for any position, there are also three possibilities for the second position and three possibilities for the third position. Thus, the total number of tripeptides possible is $3 \times 3 \times 3 = 27$. These are shown below calling the three amino acids a, b, and c.

	$a = 2^{nd}$	$b = 2^{nd}$	$c = 2^{nd}$
a is the first amino acid	aaa aab aac	aba abb abc	aca acb acc
b is the first amino acid	baa bab bac	bba bbb bbc	bca bcb bcc
c is the first amino acid	caa cab cac	cba cbb cbc	cca ccb ccc

In general, the total number of different linear arrangements of n objects taken in groups of r at a time is given by:

$$_nN_r^{tot} = n^r \tag{2}$$

where any of the n objects can be used as many as r times. In this case, $n = 3$ and $r = 3$. Therefore $_nN_r^{tot} = 3^3 = 27$.

(b) If each amino acid is used only once in the chain, then there are three possibilities for the first position, two possibilities for the second position, and only one possibility for the third position. Thus, the total number of different tripeptides containing all three amino acids is $3 \times 2 \times 1 = 6$. These are shown below.

$a = $ first	$b = $ first	$c = $ first
abc acb	bac bca	cab cba

In general, the number of *permutations*, or ordered arrangements of a set of n objects taken in groups of r at a time, is given by

$$_nP_r = n(n-1)(n-2)\ldots\ldots(n-r+1) \tag{3}$$

where n is equal to or greater than r (each of the r objects in a permutation is different). If all n objects are used:

$$\boxed{{_nP_n = n!}} \tag{4}$$

In this problem three amino acids are taken three at a time.

$$\therefore \quad {_nP_n} = 3! = 3 \times 2 \times 1 = \boxed{6}$$

(c) Obviously, only one pattern is possible, obtained from any of the six permutations shown in b. In general however, the number of *combinations* of n objects taken in groups of r at a time is given by:

$$\boxed{{_nC_r} = \frac{n!}{r!(n-r)!}} \tag{5}$$

where n is equal to or greater than r. (Each of the r objects in a combination is different.) All six permutations shown in b represent the same combination. When $n = 3$ and $r = 3$:

$$_nC_r = \frac{3!}{3!(3-3)!} = \frac{3!}{(3!)(0!)} = \frac{3 \times 2 \times 1}{(3 \times 2 \times 1)(1)} = \boxed{1}$$

(Remember, $0! = 1$.)

(d) If each of the 27 tripeptides shown in part a are hydrolyzed and passed through an amino acid analyzer, we would observe only seven unique qualitative composition patterns:

(1) *Only a* (from aaa)
(2) *Only b* (from bbb)
(3) *Only c* (from ccc)
(4) *a + b* (from aab, aba, abb, baa, bab, and bba)
(5) *a + c* (from aac, aca, acc, caa, cac, and cca)
(6) *b + c* (from bbc, bcb, bcc, cbb, cbc, and ccb)
(7) *a + b + c* (from abc, acb, bac, bca, cab, and cba)

In general, the total number of unique qualitative patterns equals the sum of all the combinations of n objects taken 1, 2, 3, ... up to r at a time, where n is equal to or greater than r. When n is equal to or less than r (as in this problem), the total number of unique qualitative patterns is the same as the total number of combinations of n objects taken in groups of 1, 2, ... up to n at a time:

$$\boxed{{_nC_n^{tot}} = 2^n - 1} \tag{6}$$

With $n = 3$, Equation 5 used three times (for $r = 1$, 2, and 3) yields $3 + 3 + 1 = 7$. Equation 6 yields $2^3 - 1 = 8 - 1 = 7$.

(e) Of the 27 tripeptides shown in a, some yield the same quantitative patterns. For example, aab, aba, and baa all yield 2 moles of a for every mole of b, and no c. All together there are 10 different quantitative patterns.

	Molar Ratio a:	b:	c:
(1)	3	0	0
(2)	0	3	0
(3)	0	0	3
(4)	2	1	0
(5)	0	2	1
(6)	2	0	1
(7)	1	2	0
(8)	1	0	2
(9)	0	1	2
(10)	1	1	1

In general, the total number of *quantitatively different* patterns obtained from n objects taken in groups of r at a time is given by:

$$ {}_nN_r = \frac{(n + r - 1)!}{r!(n - 1)!} \tag{7} $$

For $n = 3$ and $r = 3$:

$$ {}_nN_r = \frac{(3 + 3 - 1)!}{3!(3 - 1)!} = \frac{5!}{3!2!} = \frac{5 \times 4 \times 3 \times 2}{3 \times 2 \times 2} = 10 $$

Applying Equations 2 to 6 to a linear dodecapeptide (12 amino acids), we would find that with 20 different amino acids it is possible to construct 4.09×10^{15} different linear dodecapeptides, using each amino acid as many as 12 times (Equation 2). Only 6.03×10^{13} dodecapeptides can be made if each amino acid is used once (Equation 3). Among the 6.03×10^{13} dodecapeptides, there are 125,970 unique combinations (Equation 5). The rest represent the same combinations in a different order. The 4.096×10^{15} different dodecapeptides yield 9.106×10^5 qualitatively different amino acid analyses (Equation 5 used 12 times). Quantitatively, 1.41×10^8 different amino acid analyses are possible (Equation 6).

PROTEIN CONFORMATION

The four atoms of the peptide bond and the two consecutive α-carbons lie in a single plane. The H and the O atoms are *trans* to each other (Fig. 2-1). The peptide bond is rigid, but the planes can rotate about the α-carbon atoms. The distance between consecutive α-carbon atoms remains

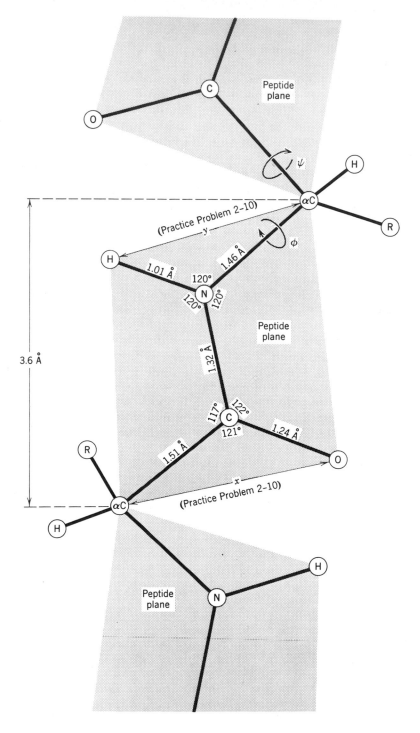

Figure 2-1 The peptide bond is rigid and fixed in a plane. The planes can rotate about the α-carbons. The rotation is described by the angles φ and ψ.

3.6 Å.   Thus, a fully extended polypeptide chain of 100 amino acid residues is 3600 Å long.

The linear polypeptide chain can assume a number of different secondary structures, depending on the nature of the R-groups present.   One common structure is called the α-helix (Fig. 2-2).   This is a right-handed helix with 3.6 amino acid residues per turn.   (A right-handed helix can be visualized by making a fist with your right hand, with the thumb extended.   Your thumb points in the direction that the helix progresses, while your other fingers indicate the rotation of the helix.   Thus, a right-handed helix progresses upward while coiling counter-clockwise.)   The α-helix has a pitch of 5.4 Å.   That is, for each turn, the helix rises 5.4 Å along the axis.   The α-helix is stabilized by intrachain hydrogen bonds between the —C=O of each peptide bond and the —NH of the peptide bond four residues away.   (If the H is called atom number 1, the hydrogen-bonded oxygen is the 13th atom along the chain.   Thus, the coil is designated a 3.6_{13} helix.)   The α-helix is prevented from forming by two or more consecutive residues with like charges (e.g., lysine, glutamate) or by two or more consecutive residues with bulky R-groups that branch at the β-carbon (e.g., isoleucine, threonine, valine).   In these cases, the polypeptide chain may assume a random coil structure.   Proline cannot participate in forming an α-helix because the nitrogen atom is in a rigid ring.   Thus, no rotation about the α-carbon is possible.   Also, there are no hydrogen atoms on the nitrogen of a proline residue, so no intrachain hydrogen bonds can form.   Successive serine residues disrupt the α-helix because of the tendency of the OH groups to hydrogen bond strongly to water.   Stretches of proline and serine coil into helical arrangements other than an α-helix.

Repeating sequences of amino acids with small, compact R-groups (e.g., glycine, alanine) tend to form the β, or pleated sheet, structure, which consists of parallel (Fig. 2-3a) or antiparallel (Fig. 2-3b) polypeptide chains linked by interchain hydrogen bonds.   Silk is an example of the antiparallel sheet.

Most nonfibrous proteins have a very precise and compact three-dimensional or tertiary structure formed when the α-helix and random coil of the polypeptide chain bends, twists, and folds over and back upon itself.   The tertiary structure is stabilized by interactions of amino acid R-groups (Fig. 2-4a), and thus, is dictated by the primary structure.   The biochemical function of a protein is intimately tied to its tertiary structure.   That is, to function in a certain way, a protein must have the correct tertiary structure.   Stated conversely: only one specific tertiary structure will permit a protein to serve optimally a specific function (see also Figs. 4-3 and 4-4).

Many proteins have still another order of structural complexity—a quaternary structure formed by the noncovalent association of tertiary-structured subunits (Fig. 2-4b).   Often, only the quaternary structured protein (dimer, tetramer, and so on) shows full activity.

· Problem 2-6

(a) Calculate the axial length of an α-helix containing 78 amino acids.   (b) How long would the polypeptide chain be if it were fully extended?

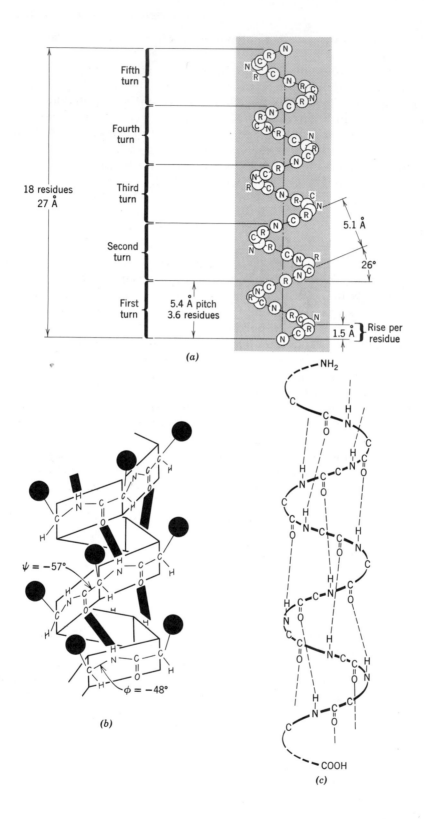

Fifth
turn

Fourth
turn

Third
turn

Second
turn

First
turn

18 residues
27 Å

5.1 Å

26°

5.4 Å pitch
3.6 residues

1.5 Å   Rise per
        residue

(a)

$\psi = -57°$

$\phi = -48°$

(b)

NH₂

COOH

(c)

Solution

(a) The α-helix rises 5.4 Å for every 3.6 residues (or 1.5 Å per residue).

$$\therefore \quad \frac{5.4 \text{ Å}}{3.6 \text{ residues}} = \frac{1.5 \text{ Å}}{1 \text{ residue}} = \frac{\text{length}}{78 \text{ residues}}$$

$$\text{length} = (78)(1.5 \text{ Å}) = \boxed{\textbf{117 Å}}$$

(b) In the fully extended chain, the distance between residues is 3.6 Å.

$$\therefore \quad \text{length} = (78)(3.6 \text{ Å}) = \boxed{\textbf{280.8 Å}}$$

· Problem 2-7

The average molecular weight of an amino acid residue is 120.   The average density of a soluble protein is 1.33 g/cm³.   Calculate (a) the specific volume of an average soluble protein, (b) the weight of a single molecule of a protein containing 270 amino acids, and (c) the volume occupied by a single molecule of this protein.   (d) Will a molecule of this protein fit completely within a cell membrane 100 Å thick?   Assume that the molecule is spherical.

Solution

(a) Specific volume, $\bar{v}$, is the reciprocal of density (milliliters occupied by 1.0 g of the substance).

$$\bar{v} = \frac{1}{\rho} = \frac{1}{1.33} = \boxed{\textbf{0.75 ml/g}}$$

(b)   MW $= (270)(120) = 32{,}400$

$\therefore$ 1 mole weighs 32,400 g.   One mole contains 6.023×10^{23} molecules. Therefore, the weight of a single molecule is given by:

$$\text{Wt}_{\text{g/molecule}} = \frac{\text{MW}}{N} = \frac{32.4 \times 10^{3}}{6.023 \times 10^{23}} = \boxed{\textbf{5.38} \times \textbf{10}^{-20} \textbf{ g/molecule}}$$

Figure 2-2   Three representations of the α-helix.   The helix rises 1.5 Å per residue and completes a turn every 3.6 residues to yield a pitch of 5.4 Å.   When the NH₂ terminus is placed at the top, each —C=O is hydrogen-bonded to an —NH four residues (but three peptide planes) below.   (*a*) Redrawn from from E. E. Conn and P. K. Stumpf, *Outlines of Biochemistry.* Wiley (1972).   R represents the α-carbon.   (*b*) Redrawn from R. Barker, *Organic Chemistry of Biological Molecules.* Prentice-Hall (1971).   (*c*) Redrawn from K. D. Kopple, *Peptides and Amino Acids.*   Benjamin (1966).   For clarity, the R-groups on the α-carbons are not shown.

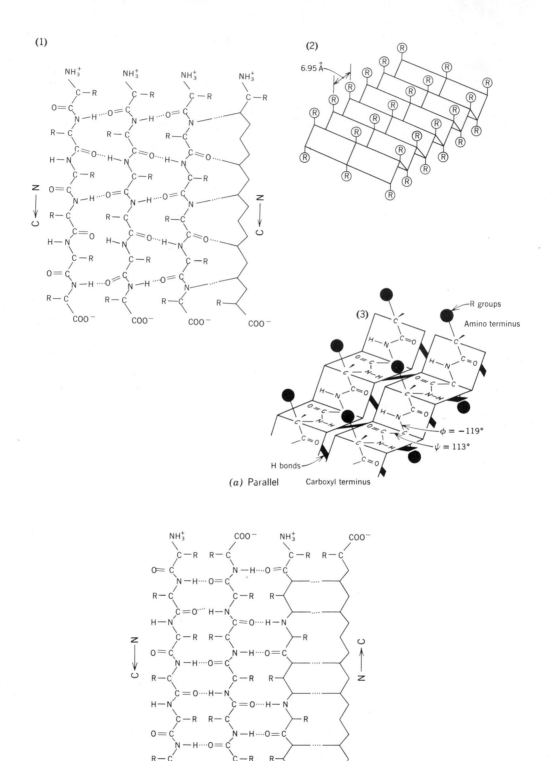

Figure 2-3 (a) Three representations of the parallel β-structure. Representation 3 is redrawn from Barker (1971). (b) Antiparallel β-structure.

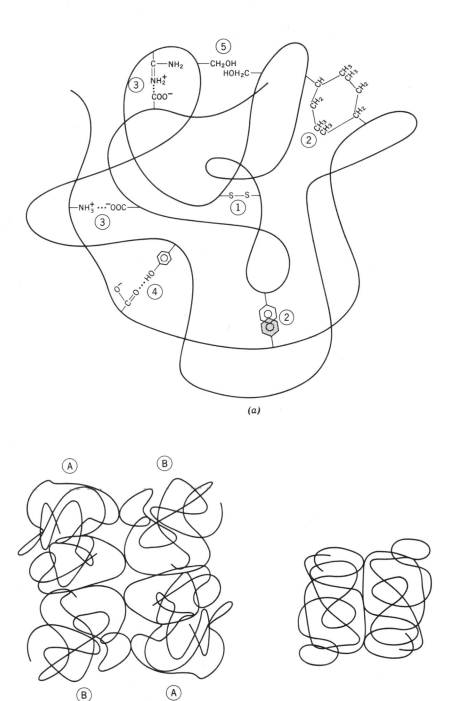

(a)

Tetramer
(composed of two different subunits)　　(b)　　Dimer
(composed of identical subunits)

Figure 2-4　(*a*) The tertiary structure of a protein is stabilized by (1) covalent disulfide bonds formed by the oxidation of two cysteine residues; (2) hydrophobic interactions; (3) ionic interactions (salt linkages); (4) hydrogen bonds; and (5) dipole–dipole interactions. (*b*) The association of tertiary-structured subunits to form a dimer and a tetramer (quaternary structures). The subunits of an oligomer are not always identical.

(c) $\text{vol}_{ml/molecule} = \bar{v}_{ml/g} \times \text{wt}_{g/molecule} = \dfrac{\text{wt}_{g/molecule}}{\rho_{g/ml}}$

$$= (0.75)(5.38 \times 10^{-20}) = \boxed{\mathbf{4.035 \times 10^{-20} \ ml/molecule}}$$

In general, the volume of a molecule is given by:

$$\boxed{\mathbf{vol = \dfrac{MW\bar{v}}{N} = \dfrac{MW}{\rho N}}} \qquad (8)$$

(d)   The volume of a sphere is $\frac{4}{3}\pi r^3$ where r is the radius.

$$\text{vol} = 4.035 \times 10^{-20} \ ml = 4.035 \times 10^{-20} \ cm^3$$

$$4.035 \times 10^{-20} = \tfrac{4}{3}\pi r^3$$

$$r^3 = \frac{(3)(4.035 \times 10^{-20})}{(4)(3.142)} = 9.63 \times 10^{-21}$$

$$r = \sqrt[3]{9.63 \times 10^{-21}} = \sqrt[3]{9.63} \times \sqrt[3]{10^{-21}}$$

$$\boxed{\boldsymbol{r = 2.13 \times 10^{-7} \ cm}} \quad \text{or} \quad \boxed{\boldsymbol{r = 21.3 \ \text{Å}}}$$

The diameter of the molecule is 42.6 Å and will fit into a 100 Å-thick membrane.

· Problem 2-8

E. coli is a rod-shaped bacterium about 2 μ long and 1 μ in diameter. The average density of a cell is 1.28 g/ml. Approximately 13.5% of the wet weight of E. coli is soluble protein. Estimate the number of molecules of a particular enzyme per cell if the enzyme has a MW of 100,000 and represents 0.1% of the total soluble protein.

Solution

The volume of an E. coli cell (assuming it is a cylinder) is given by:

$$\text{vol} = \pi r^2 l$$

where $r = 0.5 \ \mu = 5 \times 10^{-5} \ cm$

and $l = 2 \ \mu = 2 \times 10^{-4} \ cm$

$$\text{vol} = (3.14)(5 \times 10^{-5})^2(2 \times 10^{-4}) = (3.14)(25 \times 10^{-10})(2 \times 10^{-4})$$

$$\boxed{\mathbf{vol = 1.57 \times 10^{-12} \ cm^3}}$$

The weight of a single *E. coli* cell is given by:

$$wt = vol \times \rho = (1.57 \times 10^{-12})(1.28)$$

$$\boxed{\textbf{wt} = \textbf{2} \times \textbf{10}^{-12}\,\textbf{g}}$$

In a single cell, the amount of the enzyme is:

$$13.5\% \times 0.1\% \times 2 \times 10^{-12} = (0.135)(10^{-3})(2 \times 10^{-12})$$

$$= 2.7 \times 10^{-16}\,g$$

$$\text{number of molecules} = \frac{wt_g \times N}{MW}$$

where N = Avogadro's number

$$\text{number of molecules} = \frac{(2.7 \times 10^{-16})(6.023 \times 10^{23})}{10^5}$$

$$= \boxed{\textbf{1626 molecules per cell}}$$

In general:

$$\boxed{\textbf{number of molecules per cell} = \frac{N\rho V f}{MW}} \tag{9}$$

where N = Avogadro's number = 6.023×10^{23}
ρ = the density of the cell (g/cm³ or g/ml)
V = the volume of the cell (cm³ or ml)
f = the weight fraction of the cell represented by the compound (as a decimal)
MW = the molecular weight of the compound

· Problem 2-9

The specific volume of a leucine residue is 0.90 ml/g.   The specific volume of a glycine residue is 0.64.   Calculate the specific volume and density of a synthetic polypeptide containing 60% leucine residues and 40% glycine residues by weight.

Solution

The specific volume of a peptide or protein is given by:

$$\boxed{\bar{v}_P = \Sigma \textbf{ (weight fraction of residue} \times \textbf{specific volume of residue)}} \tag{10}$$

Note that weight fraction is used so that the answer comes out in milliliters per 1 g of protein. Equation 10 can also be written as:

$$\bar{v}_P = \frac{\Sigma(\bar{v}_{\text{residue}} \times \text{MW}_{\text{residue}})}{\text{MW}_P} \tag{11}$$

Similarly, the density of a protein is given by:

$$\rho_P = \Sigma \text{ (volume fraction of residue} \times \text{density of residue)} \tag{12}$$

In this case volume fraction is used to obtain an answer in terms of grams per 1 cm³ (ml) of protein. We know the weight fractions, so using Equation 10:

$$\bar{v}_P = (0.60)(0.90) + (0.40)(0.64) = 0.54 + 0.256$$

$$\bar{v}_P = 0.796 \text{ ml/g} \qquad \rho_P = \frac{1}{\bar{v}} = \frac{1}{0.796}$$

$$\rho_P = 1.256 \text{ g/ml}$$

· **Problem 2-10**

Lactic dehydrogenase (LDH) is a tetramer composed of two kinds of subunits called M and H. A large fraction of the skeletal muscle LDH is MMMM. A large fraction of the heart LDH is HHHH. Other tissues contain hybrid *isozymes* containing both M and H subunits. How many different LDH isozymes are possible? Assume that the subunits are arranged in a "square" or "tetrahedral" fashion so that there is no "sequence" (i.e., HMMM is the same as MMMH).

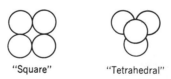

"Square"   "Tetrahedral"

Solution

There are five possible isozymes of LDH: HHHH, HHHM, HHMM, HMMM, and MMMM. We could have calculated the number from Equation 7, which gives the total number of quantitatively different ways of arranging n objects in groups of r:

$$_nN_r = \frac{(n+r-1)!}{r!(n-1)!} = \frac{(2+4-1)!}{4!(2-1)!} = \frac{5!}{4!} = 5$$

MOLECULAR WEIGHT FROM COMPOSITION

Molecular weight calculations from amino acid or prosthetic group analysis is based on the simple fact that there must be *at least* one mole of any residue present per mole of protein.   There may be more than one mole of any given residue present.   Thus, this method yields a minimum molecular weight.

· Problem 2-11

Hemoglobin contains 0.335% iron by weight. Calculate the minimum molecular weight of hemoglobin.

Solution

At least one atom of Fe must be present per molecule of hemoglobin.   A gram-atom of Fe weighs 55.85 g.   Therefore, the *minimum* molecular weight of hemoglobin is the weight that contains 55.85 g of Fe.   Another way to look at it is that 55.85 g represents 0.335% of the minimum molecular weight.

or

$$\frac{100 \text{ g protein}}{0.335 \text{ g Fe}} = \frac{MW_{min}}{55.85 \text{ g Fe}}$$

$$0.335\% \times MW_{min} = 55.85$$

$$3.35 \times 10^{-3} \, MW_{min} = 55.85$$

$$MW_{min} = \frac{(100)(55.85)}{(0.335)}$$

$$MW_{min} = \frac{55.85}{3.35 \times 10^{-3}}$$

$$\boxed{MW_{min} = 16{,}672}$$

$$\boxed{MW_{min} = 16{,}672}$$

In general:

$$\boxed{MW_{min} = \frac{MW_{constituent} \times 100}{\% \text{ of constituent}}}$$

(13)

Physical measurements suggest a molecular weight of about 65,000.   Thus, hemoglobin is a tetramer, containing one Fe per monomer.

· Problem 2-12

Amino acid analysis of 1.0 mg of a pure enzyme yielded 58.1 μg of leucine (MW = 131.2) and 36.2 μg of tryptophan (MW = 204.2). What is the minimum MW of the enzyme?

Solution

The minimum MW based on leucine content is calculated as follows:

$$\frac{10^{-3} \text{ g enzyme}}{58.1 \times 10^{-6} \text{ g leucine}} = \frac{MW}{131.2}$$

$$MW_{min} = 2258$$

The minimum MW based on tryptophan content is calculated as follows:

$$\frac{10^{-3}\,\text{g enzyme}}{36.2 \times 10^{-6}\,\text{g trp}} = \frac{\text{MW}}{204.2}$$

$$\text{MW}_{min} = 5641$$

Each calculation assumes that only one residue of each amino acid is present in a molecule of the enzyme. The molar ratio of leucine/tryptophan is:

$$\frac{58.1/131.2}{36.2/204.2} = \frac{0.443}{0.177} = \frac{2.5}{1}$$

There must be a whole number of residues of each amino acid present in a molecule of protein. Therefore, the actual ratio must be 5:2. The actual minimum MW then is the weight that contains 5 g-residues of leucine and 2 g-residues of tryptophan.

$$5 \times 2258 = \boxed{\textbf{11,290}}$$

or

$$2 \times 5641 = \boxed{\textbf{11,282}}$$

Of course, if a molecule of enzyme contained (for example) 25 leucine residues and 10 tryptophan residues, the true MW would be about 56,450.

MOLECULAR WEIGHT FROM GEL FILTRATION

Gel filtration is a molecular sieving process that can be used to fractionate molecules according to size. The sieving medium is a porous gel such as *Sephadex* (cross-linked dextran), *Biogel* (polyacrylamide), or controlled-pore glass beads. The gel can be considered to have pores of a fixed diameter range (Fig. 2-5a). Thus, molecules much smaller than the pores can freely penetrate the gel particles. Molecules with diameters much larger than the largest pore are completely excluded from the gel. Intermediate size molecules can pass into some of the gel particles but, compared to the very small molecules, a greater proportion of the intermediate size molecules will be outside the gel at any time. The gel filtration process is illustrated in Figure 2-5b. A small amount of solution containing molecules of different sizes is placed on a column containing the gel. The solution is then washed through the column with an appropriate buffer. The total liquid volume of the column is the sum of the liquid volume outside the gel particles plus the liquid volume inside the particles: $V_{tot\,liq} = V_o + V_i$. Very large molecules, which have only V_o to pass through, will elute from the column first (Fig. 2-5c). The intermediate size molecules elute next. Very small molecules, which must pass through $V_o + V_i$, elute last. Gel filtration is the most convenient method for estimating the molecular weight of a protein. It does not require a pure, homogeneous protein. All that is needed is a method for detecting the protein as it comes off the column.

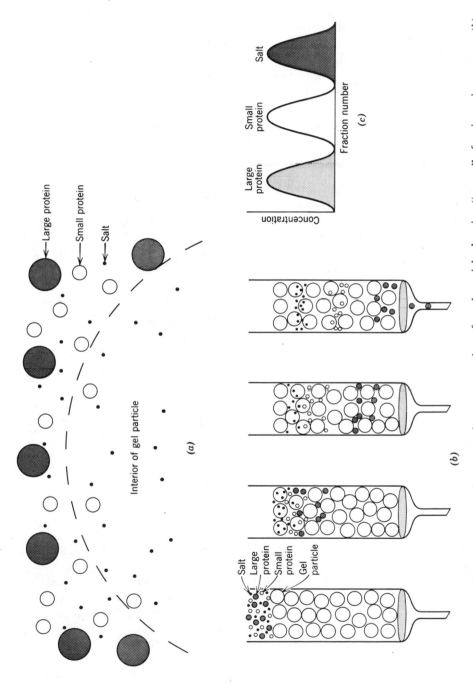

Figure 2-5 Gel filtration. (*a*) Diagrammatic representation of a gel particle showing "pores" of a given size range. (*b*) Progressive separation of three kinds of molecules of different size. (*c*) Elution pattern.

· Problem 2-13

(a) Predict the order of elution when a mixture containing the following compounds is passed through a column containing a gel that excludes all proteins of MW 200,000 and higher: cytochrome c (MW = 13,000), tryptophan synthetase (MW = 117,000), hexokinase (MW = 96,000), ATP sulfurylase (MW = 440,000), glucose oxidase (MW = 154,000), and xanthine oxidase (MW = 300,000).   (b) What factors other than molecular weight will influence the elution volume, V_e, of a protein from a Sephadex column?

Solution

(a)   The proteins will elute in the order largest to smallest.   ATP sulfurylase and xanthine oxidase are both completely excluded and, consequently, will not separate from each other.   Both enzymes will elute in a volume equal to V_o.   The elution order is ATP-sulfurylase + xanthine oxidase, glucose oxidase, tryptophan synthetase, hexokinase, cytochrome c.

(b)   Two important factors that will influence the V_e of a protein are shape and amino acid content.   A nonspherical molecule will appear to be larger than a spherical molecule of the same MW, and elute earlier than expected.   The hydrophobic amino acid residues of a protein may interact with the dextran of Sephadex.   Thus, a protein with an above-average content of hydrophobic amino acids will elute later than expected and, hence, appear to be smaller than it really is.   Sephadex contains a small amount of —COO$^-$ groups that might retard proteins with a high content of basic amino acids.

· Problem 2-14

Estimate the MW of the diphtheria toxin protein if it elutes from the calibrated Sephadex G-200 column (Fig. 2-6) just before hemoglobin at a V_e/V_o of 2.1.

Solution

A V_e/V_o of 2.1 corresponds to a MW of about 71,000.

MOLECULAR WEIGHT FROM SDS GEL ELECTROPHORESIS

Sodium dodecyl sulfate (SDS) is an anionic detergent that very effectively disrupts the quaternary structure of most multimeric proteins.   Many SDS molecules bind tightly to the subunits thereby obscuring the original charge on the protein (Fig. 2-7a).   Thus, during electrophoresis, all SDS-protein complexes migrate toward the positive pole with about the same charge/mass ratio.   If electrophoresis is carried out on polyacrylamide gels, the mobility of an SDS-protein complex will depend almost exclusively on its size.   In effect, SDS gel electrophoresis is gel filtration with an electric field as the driving force, instead of bulk solution flow.   Thus, a standard curve can be prepared using proteins with known monomer molecular weights (Fig. 2-7b).   The MW of an unknown can then be easily determined.   SDS gel electrophoresis can be performed on nonhomogeneous preparations if it is possible to locate

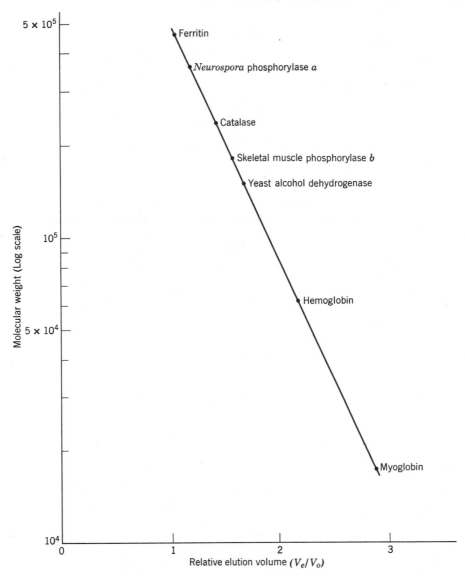

Figure 2-6   Calibration of a Sephadex G-200 gel filtration column. V_e/V_o represents the elution volume of a given protein relative to the elution volume of a very
large molecule that is completely excluded from the gel.   The standard proteins
are more or less spherical and have average amino acid contents.

the position of the protein of interest (e.g., by some specific stain for that
protein, or one of its reaction products if the protein is an enzyme).

· **Problem 2-15**

Muscle glycogen phosphorylase *a* elutes from a calibrated Biogel P-300
column at a position corresponding to a MW of 360,000.   SDS gel electrophoresis suggests a MW of 90,000.   A microbiological assay on an enzymatically hydrolyzed sample of phosphorylase *a* disclosed the presence of
1.86 μg of pyridoxal (MW = 167.2) per milligram of protein.   What conclusions can be drawn about the structure of phosphorylase *a*?

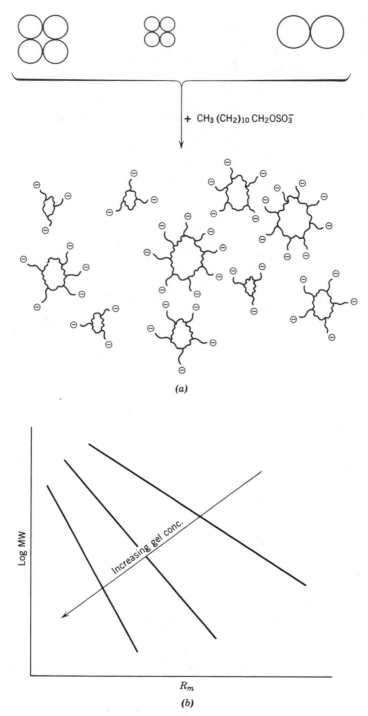

$+$ CH$_3$ (CH$_2$)$_{10}$ CH$_2$OSO$_3^-$

(a)

Log MW

Increasing gel conc.

R_m

(b)

Figure 2-7 (a) Disruption of quaternary protein structure by SDS. The SDS binds tightly to the subunits obscuring the original charge on the protein. In effect, all the SDS-protein complexes that form have the same charge/mass ratio. (b) Standard curve for SDS-gel electrophoresis. R_m represents the mobility (mm) relative to the "front" which is usually a negatively charged dye, such as bromphenol blue.

Solution

$$\frac{1.86 \times 10^{-6} \, \mu\text{g pyridoxal}}{1.0 \times 10^{-3} \, \text{g phosphorylase}} = \frac{167.2}{\text{MW}_{min}}$$

$$\boxed{\text{MW}_{min} = 89{,}892}$$

The cumulative results suggest that phosphorylase *a* is a tetramer (MW = 360,000) composed of four subunits of the same MW (90,000), and that each subunit contains one pyridoxal group.

MOLECULAR WEIGHT FROM OSMOTIC PRESSURE

When a solution is separated from the pure solvent by a membrane that is permeable to the solvent but not to the solute, molecules of solvent migrate through the membrane into the solution compartment (Fig. 2-8*a*). The pressure that must be exerted to prevent the passage of solvent molecules is known as the osmotic pressure (π). The osmotic pressure of a solution depends on the concentration of solute and the temperature of the solution.   The relationship (shown below) is identical to the *PVT* relationship of gases.

$$\boxed{\pi V = nRT} \tag{14}$$

where π = osmotic pressure in atm
         V = volume of the solution in liters
         n = number of moles of solute
         R = gas constant, 0.0821 liter-atm/mole-°K
         T = the absolute temperature

$$\pi = \frac{n}{V} RT$$

$$\therefore \quad \boxed{\pi = MRT} \tag{15}$$

where M = molarity of the solution.
    The molecular weight of the solute may be determined from measurement of π.

$$\pi V = \frac{\text{wt}_g}{\text{MW}} RT \qquad \text{MW} = \frac{\text{wt}_g}{V} \frac{RT}{\pi} \qquad \therefore \quad \boxed{\text{MW} = \frac{CRT}{\pi}} \tag{16}$$

where C = concentration in g/liter.

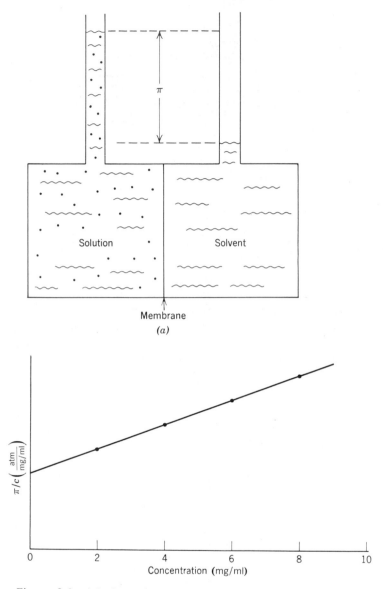

Figure 2-8 (*a*) A static osmometer. (*b*) Correction for the nonideality of the solution.

Only extremely dilute solutions (in which there are no interactions between particles or particles and solvent) obey the relationship $\pi V = nRT$. Yet, relatively concentrated solutions must be used to obtain good measurements of π. Usually, the observed value of π must be corrected for nonideality. This is accomplished by plotting π/C (the *reduced* or *specific* osmotic strength) against C, and extrapolating to zero C. The MW is then calculated from:

$$MW = \frac{RT}{(\pi/C)_{C \to 0}}$$

(17)

· **Problem 2-16**

At 10°C, the osmotic pressure of a protein solution was 3.05×10^{-3} atm at 4 mg/ml, and 1.40×10^{-3} atm at 2 mg/ml.   Calculate MW.

Solution

The data and the values of π/C are tabulated below:

C	π	π/C
4 mg/ml	3.05×10^{-3}	7.625×10^{-4}
2 mg/ml	1.40×10^{-3}	7.000×10^{-4}
$\therefore$ at 0 mg/ml	—	6.375×10^{-4}

The value of π/C decreases by 0.625×10^{-4} for a decrease in C of 2 mg/ml.   Therefore, at infinite dilution, π/C would be 6.375×10^{-4}.   (In practice, many values of π would be obtained and the values of π/C plotted as shown in Figure 2-8*b*.)

$$\text{MW} = \frac{RT}{(\pi/C)_{C \to 0}} = \frac{(0.0821)(283)}{6.375 \times 10^{-4}}$$

$$\boxed{\text{MW} = 36{,}446}$$

MOLECULAR WEIGHT FROM SEDIMENTATION VELOCITY

The molecular weight of a protein or other macromolecule can be obtained from ultracentrifugation studies. The principles involved are rather simple: a high-molecular-weight molecule sediments faster and diffuses slower than a lower-molecular-weight molecule of the same density. This is expressed in the Svedberg equation:

$$\boxed{\text{MW} = \frac{RTs}{D(1 - \bar{v}\rho_s)}} \tag{18}$$

where R = the gas constant (8.314×10^{7} ergs mole^{-1} degree^{-1})
   T = the absolute temperature (°K)
   D = the diffusion coefficient (cm^2/sec)
      = the amount of the compound diffusing per second across an area of 1 cm^2 at a unit concentration gradient (i.e., 1 M higher on one side of a membrane)
   $\bar{v}$ = the specific volume of the macromolecule (the inverse of its density)*

---

* Actually, $\bar{v}$ represents the partial specific volume of the solute, that is, the increase in the volume of the solution caused by the addition of 1 g of solute.   If there are no significant structural changes in the solute as a result of interaction with the solvent, $\bar{v}$ can be taken as the sum of the fractional specific volumes of the subunits or components of the solute as described in Problem 2-9.

ρ_s = the density of the solvent

s = the sedimentation coefficient (sec)

$$s = \frac{dx/dt}{\omega^2 x} \qquad \text{or} \qquad s = \frac{2.3 \log x}{t\omega^2} \qquad (19)$$

where dx/dt = the rate at which the moving boundary sediments (Fig. 2-9a,b)

x = the distance of the boundary from the center of rotation at any time, t

ω = the angular velocity (radians/sec)

The sedimentation coefficient has units of seconds. The expression for s is, in fact, another version of:

$$\text{velocity} = \text{acceleration} \times \text{time} \qquad \text{or} \qquad \text{time} = \frac{\text{velocity}}{\text{acceleration}}$$

where velocity = dx/dt

and acceleration in a centrifugal field = $\omega^2 x$

$$= (\text{radians/sec})^2 \times \text{cm}$$

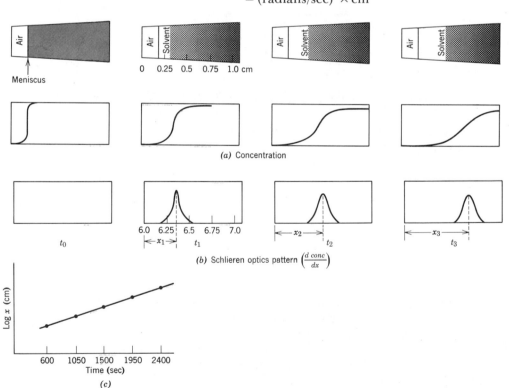

(a) Concentration

(b) Schlieren optics pattern $\left(\frac{d\ conc}{dx}\right)$

(c)

Figure 2-9 Sedimentation velocity measurement. (*a*) The distribution of protein in the cell as a function of centrifugation time. The protein is sedimenting toward the right. (*b*) Schlieren optics pattern. The optical system measures the change in refractive index of the solution. Thus, the pattern gives the concentration gradient of protein along the sedimentation path. (*c*) Plot of log x versus time where x is the distance the boundary has moved (i.e., the distance from the meniscus to the peak of the Schlieren pattern).

The sedimentation constant is obtained by measuring the position of the boundary at different times and plotting $\log x$ versus time (Fig. 2-9c). The slope of the plot is $\log x/t$. Therefore:

$$s = \frac{2.3(\text{slope})}{\omega^2} \tag{20}$$

If the rotation of the centrifuge is given in terms of revolutions per minute (RPM), then ω in terms of radians per second is obtained from:

$$\omega = \frac{\text{RPM}}{60}(2\pi) \tag{21}$$

because one revolution (the circumference of a circle) is 2π radians. For convenience, s is usually expressed in terms of Svedberg units, where $1\,s = 10^{-13}$ sec. If the s value is determined at 20°C in aqueous solution, the symbol $s_{20,w}$ is used. The diffusion coefficient, D, must be obtained from an independent experiment.

The derivation of the Svedberg equation is based on the fact that a molecule in a centrifugal field will sediment at a rate so that the centrifugal force, F_c, exactly equals the sum of the buoyant force, F_b, plus the frictional force (drag), F_f.

$$F_c = F_b + F_f \qquad \text{or} \qquad F_c - F_b = F_f$$

where $F_c - F_b$ can be considered to be the net centrifugal force:

$$\text{net } F_c = m_{\text{eff}}a = m_{\text{eff}}\omega^2 x$$

where m_{eff} is the effective mass of the molecule, that is, its real mass minus the mass of an equal volume of solvent (which the molecule displaces).

$$m_{\text{eff}} = (\text{volume}_P \times \text{density}_P) - (\text{volume}_P \times \text{density}_S)$$

$$= \text{volume}_P(\rho_P - \rho_S) = \frac{(\text{MW})\bar{v}}{N}(\rho_P - \rho_S)$$

$$\therefore \quad \text{net } F_c = \frac{(\text{MW})\bar{v}\omega^2 x}{N}(\rho_P - \rho_S)$$

$$\bar{v} = 1/\rho_P$$

$$\therefore \quad \text{net } F_c = \frac{(\text{MW})\omega^2 x}{N}(1 - \bar{v}\rho_S) \tag{22}$$

The frictional force is given by:

$$F_f = \frac{RT}{DN}\frac{dx}{dt} \tag{23}$$

where N is Avogadro's number, dx/dt is the rate of sedimentation of the molecule, and D is the diffusion constant, which must be determined independently. D is simply the proportionality constant of Fick's first law of diffusion:

$$\frac{dc}{dt} = DA\frac{\delta c}{\delta x}$$

(24)

Fick's first law states that the rate at which a substance diffuses across a given area, A, is proportional to the concentration gradient across that area.
   Equating 22 and 23:

$$\frac{(MW)\omega^2 x}{N}(1 - \bar{v}\rho_s) = \frac{RT\frac{dx}{dt}}{DN}$$

or

$$MW = \frac{RT\frac{dx}{dt}}{D\omega^2 x(1 - \bar{v}\rho_s)} = \frac{RTs}{D(1 - \bar{v}\rho_s)}$$

(25)

In practice, s values would be determined at several different concentrations of protein and extrapolated to zero concentration.

· **Problem 2-17**

An ultracentrifuge is operating at 58,000 RPM.   (a) Calculate ω in radians per second.   (b) Calculate the centrifugal force at a point 6.2 cm from the center of rotation.   (c) How many "g's" is this equivalent to?

Solution

(a)
$$\omega = \frac{RPM}{60}(2\pi) = \frac{(58,000)(2)(3.14)}{60}$$

$$\boxed{\omega = 6070.7 \text{ radians/sec}}$$

(b)
$$a = \omega^2 x = (6070.7)^2(6.2) = (36.85 \times 10^6)(6.2)$$

$$\boxed{a = 2.284 \times 10^8 \text{ cm/sec}^2}$$

Note that a radian has no units since it is simply the ratio of arc length to radius length.

(c)   The earth's gravitational field $= g = 980$ cm/sec^2.

$$a = \frac{2.284 \times 10^8}{980} = \boxed{233,061 \times g}$$

· **Problem 2-18**

At 20°C, human serum albumin has a diffusion coefficient of 6.1×10^{-7} cm^2/sec and a sedimentation coefficient of 4.6 s. The density of water at 20°C is 0.998. Calculate the MW of the albumin, assuming a specific volume of 0.74 at 20°C.

Solution

$$MW = \frac{RTs}{D(1 - \bar{v}\rho_s)} = \frac{(8.314 \times 10^7)(293)(4.6 \times 10^{-13})}{(6.1 \times 10^{-7})[1 - (0.74)(0.998)]}$$

$$= \frac{11.205 \times 10^{-3}}{(6.1 \times 10^{-7})(0.2615)} = \frac{11.205 \times 10^{-3}}{1.595 \times 10^{-7}}$$

$$\boxed{MW = 70,253}$$

See also Problem 6-28 for a description of molecular weight determination by affinity labeling.

B. CARBOHYDRATES

· **Problem 2-19**

How many different aldohexose stereoisomers are possible (excluding anomers)?

Solution

The number of stereoisomers depends on the number of asymmetric carbon atoms present. For nonlike-ended molecules:

$$\boxed{\text{number of stereoisomers} = 2^n} \tag{26}$$

where n = the number of asymmetric carbon atoms. For an aldohexose in the straight chain form there are four asymmetric carbon atoms: numbers 2, 3, 4, and 5.

$$\text{number of stereoisomers} = 2^4 = \boxed{16}$$

Half are D-sugars (OH on carbon 5 to the right) and half are L-sugars (OH on carbon 5 to the left). A fifth asymmetric carbon atom is produced upon ring formation yielding the α and β anomers of each of the 16 sugars.

· **Problem 2-20**

How many different disaccharides containing D-galactopyranose plus D-glucopyranose are possible?

Solution

There are 20 possible disaccharides containing galactose plus glucose in the pyranose ring forms:

Galactosides:	1–2, 1–3, 1–4, and 1–6	= 4
	linked α or β	∴ ×2
		= 8
Glucosides:	1–2, 1–3, 1–4, and 1–6	= 4
	linked α or β	∴ ×2
		= 8
Nonreducing disaccharides:	α–α, α–β, β–α, and β–β	= 4
	(1–1 linked)	

Total: **20**

In contrast, two neutral amino acids can make only two different dipeptides; an acidic and a basic amino acid can make only four different dipeptides. There are about 20 different common monosaccharides (including the amino sugars, uronic acids, and other derivatives). Thus, the potential for variability among polysaccharides is even greater than that of proteins. However, almost all polysaccharides found in nature are composed of *repeating units* of fewer than 5 different monosaccharides.

· **Problem 2-21**

A 25 mg sample of *E. coli* glycogen was hydrolyzed in 2 ml of 2 N H_2SO_4. The hydrolysate was neutralized and diluted to 10 ml. The glucose content of the final solution was found to be 2.35 mg/ml. What is the purity of the isolated glycogen?

Solution

The amount of glucose obtained upon hydrolysis is:

$$2.35 \text{ mg/ml} \times 10 \text{ ml} = 23.5 \text{ mg}$$

$$\frac{23.5 \text{ mg}}{180 \text{ mg/mmole}} = 0.1306 \text{ mmole}$$

The glycogen sample contains 0.1306 mmole of glucose. However, when glucose (or any other monosaccharide) is polymerized, an average of one mole of H_2O is removed per glycosidic bond formed. Thus, the MW of a glucose residue is $180 - 18 = 162$. The amount of glucose obtained corresponds to:

$$0.1306 \text{ mmole} \times 162 \text{ mg/mmole} = 21.16 \text{ mg glycogen}$$

The purity of the glycogen is:

$$\frac{(21.16)(100)}{25} = \boxed{\textbf{84.6\%}}$$

(Or, the sample was not completely hydrolyzed.)

· **Problem 2-22**

Exactly 81.0 mg of glycogen were exhaustively methylated and then acid hydrolyzed. The methylated products were separated and identified by thin layer chromatography. Exactly 62.5 μmoles of 2,3-dimethylglucose were obtained. (a) What percent of the total glucose residues are branch points? (b) What were the other products of the methylation and hydrolysis and how much of each were formed?

Solution

(a)   The total amount of glucose present is:

$$\frac{81 \times 10^{-3}\,\text{g}}{162\,\text{g/mole}} = 5 \times 10^{-4}\,\text{moles} = 500\,\mu\text{moles}$$

The 2,3-dimethylglucose comes from the branch points (Fig. 2-10). Therefore,

$$\% \text{ branch points} = \frac{62.5}{500} \times 100 = \boxed{\textbf{12.5\%}}$$

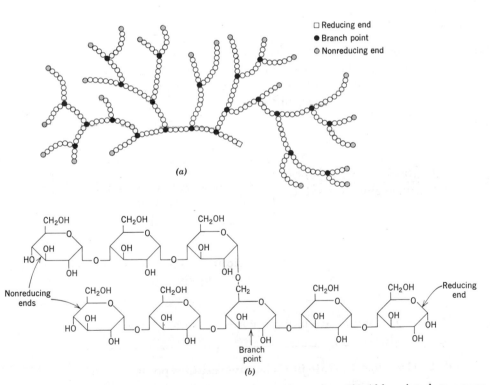

(a)

(b)

Figure 2-10   (a) Branched structure of glycogen or amylopectin. (b) Abbreviated structure of glycogen or amylopectin. Each free OH group can be methylated with dimethyl sulfate. The branch points yield 2,3-dimethylglucose upon hydrolysis. Nonreducing end groups (NRE) yield 2,3,4,6-tetramethylglucose upon hydrolysis. The reducing end of the molecule (RE) is converted to 1,2,3,6-tetramethylglucose, but upon hydrolysis of the glycosidic bonds, the methyl group on carbon-1 is removed as methanol. (The other methyl groups are present as stable ethers.) Thus the RE and all the remaining 1–4 linked residues yield 2,3,6-trimethylglucose. Free OH groups on the hydrolysis products indicate that the OH was used in ring formation or in the glycosidic linkage between monosaccharide residues.

(b) For every mole of branch point, there is a mole of nonreducing end groups (NRE).

$$\therefore \quad \boxed{\mu\text{moles of 2,3,4,6-tetramethylglucose} = 62.5}$$

There are actually $n + 1$ residues (molecules) of NRE for n residues of branch points. However, in a high MW molecule such as glycogen, the molar difference between $n + 1$ and n residues is insignificant. (This point is illustrated in more detail in the following problem.)

The glucose that does not appear as dimethylglucose or tetramethylglucose appears as trimethylglucose.

$$\therefore \quad \mu\text{moles of 2,3,6-trimethylglucose} = 500 - (2 \times 62.5)$$
$$= 500 - 125$$

$$\boxed{\mu\text{moles of 2,3,6-trimethylglucose} = 375}$$

· **Problem 2-23**

The glycogen described in the previous problem has a MW of 3×10^6. (a) How many glucose residues does a molecule of this glycogen contain? (b) How many of the residues are at branch points? (c) How many of the residues are at the nonreducing ends?

Solution

(a) $\dfrac{3 \times 10^6 \text{ g/mole glycogen}}{162 \text{ g/mole glucose residue}} = \boxed{\begin{array}{c}\textbf{18,519 moles glucose per mole} \\ \textbf{glycogen}\end{array}}$

$= \boxed{\begin{array}{c}\textbf{18,519 residues of glucose per molecule} \\ \textbf{of glycogen}\end{array}}$

(b) 12.5% of the residues are branch points.

$(0.125)(18,519) = \boxed{\textbf{2315 branch point residues per molecule}}$

(c) There are $n + 1$ NRE residues per n branch point residues.

$$\therefore \quad n + 1 = \boxed{\textbf{2316 NRE residues per molecule}}$$

Thus, the 81 mg of glycogen represents

$$\frac{81 \times 10^{-3}\,g}{3 \times 10^{6}\,g/mole} = 2.7 \times 10^{-8}\,moles = 0.027\,\mu moles \text{ of glycogen}$$

and $(0.027)(18{,}519) = 500\,\mu$moles of glucose (total)

including $(0.027)(2315) = 62.505\,\mu$moles of branch points

and $(0.027)(2316) = 62.532\,\mu$moles of NRE

Thus, the assumption that the number of moles of branch points = the number of moles of NRE made in the previous problem is quite valid.

· Problem 2-24

A trisaccharide isolated from milk was completely hydrolyzed by β-galactosidase to galactose and glucose in a 2:1 ratio. Reduction of the original trisaccharide with $NaBH_4$, followed by exhaustive methylation, acid hydrolysis, then another $NaBH_4$ reduction step, and finally acetylation with acetic anhydride yielded three products: (a) 2,3,4,6-tetramethyl-1,5-diacetyl-galactitol, (b) 2,3,4-trimethyl-1,5,6-triacetylgalactitol, and (c) 1,2,3,5,6-pentamethyl-4-acetylsorbitol. What is the structure of the trisaccharide?

Solution

The reduction prior to methylation converts the reducing end residue to an open-chain sugar alcohol. This permits methylation at positions 1 and 5. Thus, glucose (which yields the sorbitol derivative) is at the reducing end. Reduction after methylation and hydrolysis opens the rings of the other monosaccharides. The new OH groups (as well as the OH groups that were used in linkage to other sugars) can then be acetylated. Thus, assuming that the three residues are all in the pyranose ring form, the most likely structure of the original trisaccharide is gal $(\beta 1 \rightarrow 6)$ gal $(\beta 1 \rightarrow 4)$ glu as shown below:

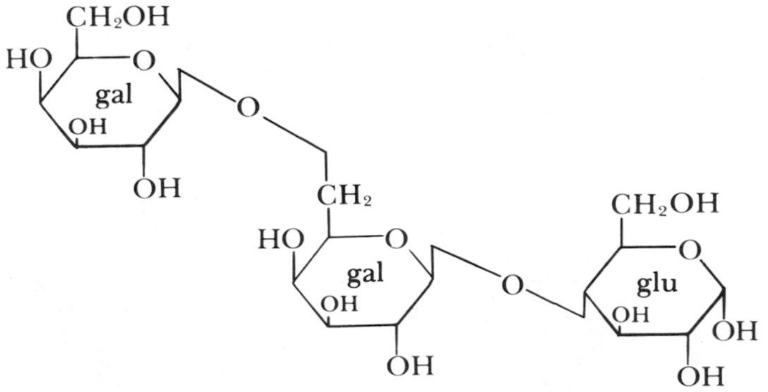

· Problem 2-25

Upon treatment with periodate, a 200 mg sample of cellulose released 4.12 μmoles of formic acid. (a) What is the average MW of the cellulose molecules? (b) What is the average chain length of a cellulose molecule?

Solution

(a)   Periodate (IO_4^-) cleaves the bonds between carbon atoms containing oxidizable functions.   In the process, the carbons are raised one oxidation state.   Thus, a glycol is converted to two aldehyde groups, an aldehyde adjacent to an alcohol group is converted to formic acid, while the alcohol group is oxidized to an aldehyde.   A chain containing three consecutive alcohol groups yields two aldehyde groups and one formic acid.   In effect, the middle alcohol group is oxidized in two steps.   Cellulose is a linear (unbranched) polymer composed of $\beta 1 \to 4$ linked glucose residues.   Each chain yields three molecules of formic acid, as shown in Figure 2-11.   Thus, 200 mg of cellulose corresponds to:

$$\frac{4.12}{3} = 1.373 \ \mu\text{moles of cellulose}$$

$$MW = \frac{wt_g}{\text{number of moles}}$$

$$MW = \frac{200 \times 10^{-3}}{1.373 \times 10^{-6}} \qquad \boxed{\textbf{MW = 145,666}}$$

(b)   Number of glucose residues per chain $= \dfrac{\text{MW of cellulose}}{\text{MW of glucose residue}}$

$$= \frac{145,666}{162} = \boxed{\textbf{899}}$$

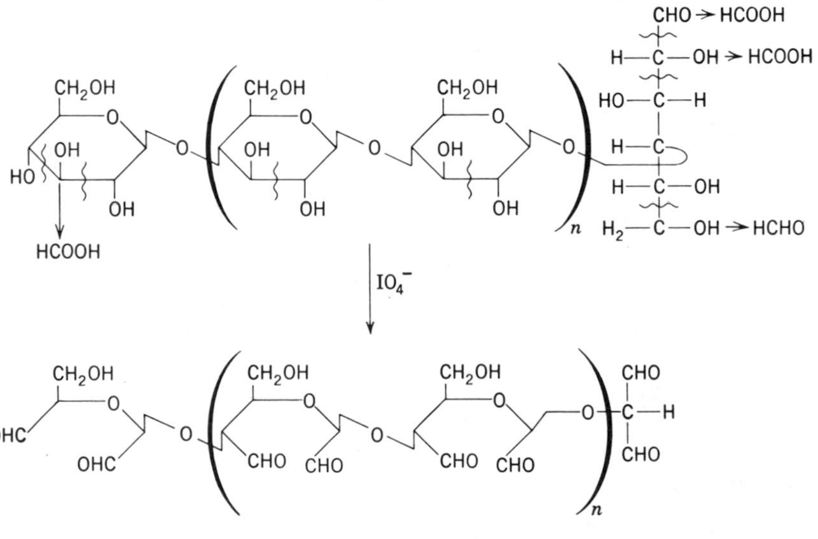

Figure 2-11   Periodate oxidation of cellulose.   Each linear chain yields three molecules of formic acid.   For clarity, the reducing end residue is shown in the open chain form.

· **Problem 2-26**

A 500 mg sample of glycogen was treated with radioactive cyanide ($KC^{14}N$). Exactly 0.193 μmoles of $C^{14}N^-$ were incorporated. Another 500 mg sample of the glycogen was treated with methanol containing 3% HCl in order to form the methyl glucoside of the reducing end. (Methanol will not methylate the other OH groups.) The methyl glucoside was then treated with periodate. Exactly 347 μmoles of formic acid were produced. Calculate (a) the MW of the glycogen and (b) the degree of branching.

Solution

(a) Cyanide adds to the "aldehyde" group of carbon number one of the reducing end. Since glycogen contains only one reducing end residue per molecule, the number of moles of CN^- added equals the number of moles of glycogen present. Therefore the 500 mg of glycogen represents 0.193 μmoles.

$$MW = \frac{wt_g}{\text{number of moles}} = \frac{500 \times 10^{-3}}{0.193 \times 10^{-6}}$$

$$\boxed{MW = 2.59 \times 10^6}$$

(b) Methylation renders carbon number one of the reducing end residue resistant to periodate oxidation. All the formic acid produced now comes from carbon number three of the NRE residues. Thus, 347 μmoles of formic acid indicate that there are 347 μmoles of NRE residues (and 347 μmoles of branch point residues) present.

$$\text{total glucose present} = \frac{500 \times 10^{-3} \text{ g}}{162 \text{ g/mole residue}} = 3.086 \times 10^{-3} \text{ moles}$$

$$= 3086 \text{ } \mu\text{moles}$$

$$\% \text{ branch points} = \frac{347}{3086} \times 100 = \boxed{11.24\%}$$

See also the problems in Chapter 5, Section C on Optical Rotation.

C. LIPIDS

· **Problem 2-27**

(a) How many "different" triglycerides (triacylglycerols) can be made from glycerol and three different fatty acids (e.g., lauric, palmitic, and stearic)? (b) How many triglycerides of *quantitatively* different composition can be made from glycerol and three different fatty acids?

Solution

(a) Proceeding as we did earlier for calculating the total number of different tripeptides that can be made from three amino acids (Problem 2-5), we would conclude that:

$$ {}_{n}N_{r}^{\text{tot}} = n^{r} = 3^{3} = \boxed{\mathbf{27}} $$

However, the middle carbon atom of the symmetrical glycerol molecule becomes assymmetric when two different fatty acids are esterified on the first and third positions. So we might ask ourselves: is L-stearodipalmitin, for example, really different from L-dipalmitostearin? The structures are shown below.

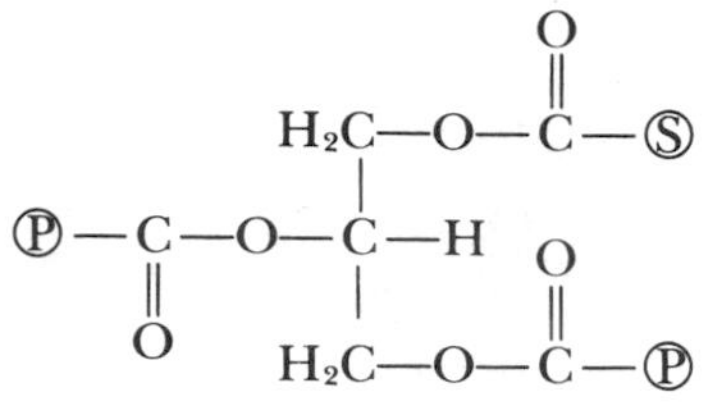

L-stearodipalmitin (L-SPP)
↓ rotate 180° in the plane of the page
D-dipalmitostearin (D-PPS)

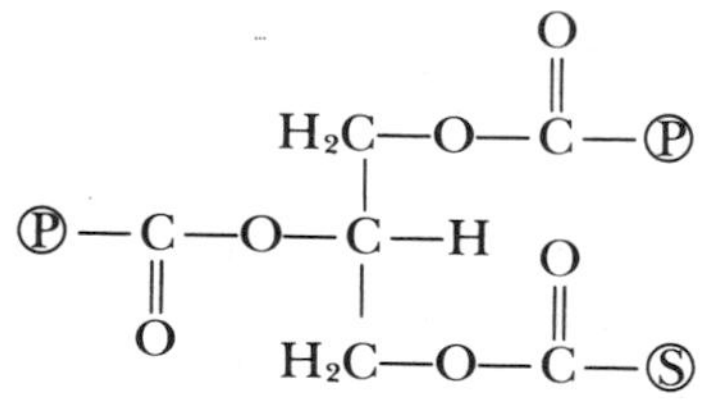

L-dipalmitostearin (L-PPS)
↓ rotate 180° in the plane of the page
D-stearodipalmitin (D-SPP)

It is clear that L-SPP is indeed different from L-PPS. However, L-SPP is identical to D-PPS, while L-PPS is identical to D-SPP. Thus, there really are 27 different triglycerides possible. These can be considered 27 different L-triglycerides *or* 27 different D-triglycerides. If we count only one member of a D-L pair, then there are fewer than 27 different possibilities.* In this case, SPP is considered to be the same as PPS. In other words, some of the 27 different L-forms are equivalent to the enantiomers (D-forms) of other L-forms. Specifically, there are 18 different triglycerides with the six different combinations of end substituents shown below.

P	S	L	P	P	S	
┼	┼	┼	┼	┼	┼	= 6
P	S	L	S	L	L	

where P = palmitic, S = stearic, and L = lauric. Each of the above six quantitatively different structures can have any of the three fatty acids at the 2, or β, position ($6 \times 3 = 18$). In general, the total number of different triglycerides (excluding enantiomers) can be derived from Equation 7:

*(Including nine like-ended molecules which are really neither D nor L.

$$N = \left(\frac{(n+r-1)!}{r!(n-1)!}\right)n = \frac{n(n+1)!}{2(n-1)!} = \frac{n^3 + n^2}{2} \qquad (27)$$

where n = the number of fatty acids available
        r = 2 (the two ends)

(b)   The total number of quantitatively different compositions is given by Equation 7 with $r = 3$:

$$_nN_r = \frac{(n+r-1)!}{r!(n-1)} = \frac{(n+2)!}{3!(n-1)!} = \frac{5 \times 4 \times 3 \times 2}{3 \times 2 \times 2}$$

$$\boxed{N = 10}$$

There are fewer triglycerides than calculated in part a because some of the 18 different triglycerides have the same quantitative composition (e.g., PSS and SPS).

· Problem 2-28

Calculate the saponification number of palmitodistearin.

Solution

The saponification of palmitodistearin is shown below.

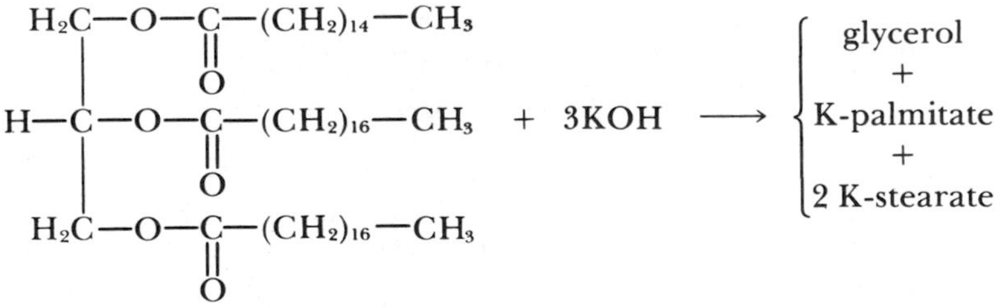

$$
\begin{array}{l}
\text{H}_2\text{C}-\text{O}-\underset{\underset{\text{O}}{\|}}{\text{C}}-(\text{CH}_2)_{14}-\text{CH}_3 \\[2mm]
\text{H}-\overset{|}{\underset{|}{\text{C}}}-\text{O}-\underset{\underset{\text{O}}{\|}}{\text{C}}-(\text{CH}_2)_{16}-\text{CH}_3 \quad + \quad 3\text{KOH} \quad \longrightarrow \\[2mm]
\text{H}_2\text{C}-\text{O}-\underset{\underset{\text{O}}{\|}}{\text{C}}-(\text{CH}_2)_{16}-\text{CH}_3
\end{array}
\left\{
\begin{array}{l}
\text{glycerol} \\
+ \\
\text{K-palmitate} \\
+ \\
2\ \text{K-stearate}
\end{array}
\right.
$$

Three moles of KOH are required to saponify one mole of triglyceride. The saponification number is defined as the number of milligrams of KOH required to saponify 1.0 g of triglyceride.

MW of KOH = 56   MW of palmitodistearin = 862

$\therefore$ (3)(56) = 168 g of KOH is required to saponify 862 g of the triglyceride.

$$\frac{168 \times 10^3 \text{ mg KOH}}{862 \text{ g triglyceride}} = 194.9 \text{ mg KOH/g triglyceride}$$

or

$$\boxed{\text{saponification number} = 194.9}$$

· **Problem 2-29**

A 250 mg sample of pure olive oil required 47.5 mg of KOH for complete saponification.   Calculate the average MW of the triglycerides in the olive oil.

Solution

$$\text{amount of KOH required} = \frac{47.5 \times 10^{-3} \text{ g}}{56 \text{ g/mole}} = 8.482 \times 10^{-4} \text{ moles}$$

It takes 3 moles of KOH for each mole of triglyceride.

$$\therefore \quad \text{amount of triglyceride present} = \frac{8.482 \times 10^{-4}}{3} = 2.827 \times 10^{-4} \text{ moles}$$

$$\text{number of moles} = \frac{\text{wt}_g}{\text{MW}} \quad \text{or} \quad \text{MW} = \frac{\text{wt}_g}{\text{number of moles}}$$

$$\text{MW} = \frac{250 \times 10^{-3}}{2.827 \times 10^{-4}} \qquad \boxed{\text{MW} = 884}$$

Or, in general:

$$\boxed{\text{MW}_\text{aver} = \frac{3 \times 56 \times 1000}{\text{saponification number}} = \frac{168{,}000}{\text{saponification number}}} \tag{28}$$

The saponification number of the olive oil is $\dfrac{47.5 \text{ mg KOH}}{0.250 \text{ g oil}} = 190$

$$\text{MW}_\text{aver} = \frac{168{,}000}{190} = \boxed{884}$$

· **Problem 2-30**

The olive oil described in the previous problem was reacted with iodine.   Exactly 578 mg of I_2 were absorbed by 680 mg of the oil.   (a) On the average, how many double bonds are present in a molecule of triglyceride?   (b) What is the iodine number of the oil?

Solution

(a)   Calculate the number of moles of I_2 absorbed by a mole of the oil.   Each mole of I_2 adds across a double bond.

$$\frac{0.578 \text{ g } I_2}{0.680 \text{ g oil}} = \frac{\text{wt}_g \text{ } I_2}{884 \text{ g oil}}$$

$$\text{wt}_g \text{ } I_2 \text{ absorbed} = \frac{(884)(0.578)}{(0.680)} = 751.4 \text{ g } I_2 \text{ per mole of oil}$$

$$\text{MW of } I_2 = (2)(126.9) = 253.8$$

$$\frac{751.4 \text{ g } I_2}{253.8 \text{ g/mole}} = 2.96 \text{ moles } I_2/\text{mole of oil}$$

Thus, on the average, there are three double bonds per molecule of triglyceride.

(b)   Iodine number is defined as g iodine absorbed per 100 g of oil or fat.

$$\text{iodine number} = \frac{751.4}{884} \times 100 = \boxed{85}$$

· Problem 2-31

An acidic, lipid-soluble organic compound with prostaglandin activity was isolated from a Pacific coral.   Elemental analysis gave 67.80% C, 9.60% H, and 22.60% O.   What is (a) the simplest empirical formula and (b) the minimum MW of the compound?

Solution

The simplest empirical formula of a compound may be calculated from elemental composition data.   The calculations are based on the fact that a molecule of the compound must contain at least 1 atom of every element shown to be present.   Once the simplest empirical formula is known, the minimal molecular weight can be calculated as the sum of the atomic weights of all elements present.

(a)   First, divide the percent composition values by the atomic weights of each element present.   This yields the number of gram-atoms of each element per 100 g of compound.

$$C = \frac{67.80}{12} \qquad H = \frac{9.60}{1} \qquad O = \frac{22.60}{16}$$
$$C = 5.65 \qquad H = 9.60 \qquad O = 1.41$$

Next, divide each of the relative molar amounts shown above by the smallest relative value.

$$C = \frac{5.65}{1.41} \qquad H = \frac{9.6}{1.41} \qquad O = \frac{1.41}{1.41}$$
$$C = 4 \qquad H = 6.8 \qquad O = 1$$

There must be a whole number of atoms of each element present.   Thus, the simplest empirical formula might be C_4H_7O.   However, if the elemental analysis is quite accurate, then we can multiply the $4:6.8:1$ ratio by integers until whole numbers result.   In this case, multiplying by 5 yields:

$$\boxed{C_{20}H_{34}O_5}$$

(b)   The minimum MW is $(20)(12) + (34)(1) + (5)(16) = \boxed{354}$

· Problem 2-32

A 12 mg sample of the unknown compound described in the previous problem was dissolved in 0.8 g of pure camphor.   The freezing point of the solution was 176.8°C. The freezing point of pure camphor is 178.4°C. The molal freezing point depression constant for camphor is 37.7°. Calculate the apparent MW of the unknown compound.

Solution

One mole of solute dissolved in 1000 g of solvent (to produce a 1 molal solution) depresses the freezing point by the molal freezing point depression constant (K_f). The degree to which the freezing point of any solution is depressed (compared to that of the pure solvent) is directly proportional to the molality of the solute present.

$$\frac{MW_g \text{ solute}/1000 \text{ g solvent}}{K_f} = \frac{(1000) \text{ wt}_g \text{ solute}/\text{wt}_g \text{ solvent}}{\Delta T_{f.p.}}$$

or

$$\boxed{MW_{solute} = \frac{(1000)(K_f) \text{ wt}_g \text{ solute}}{(\Delta T_{f.p.}) \text{ wt}_g \text{ solvent}}} \tag{29}$$

$$MW = \frac{(1000)(37.7)(12 \times 10^{-3})}{(178.4 - 176.8)(0.8)} = \frac{452.4}{(1.6)(0.8)}$$

$$\boxed{MW = 353.4}$$

Thus, the assigned formula $C_{20}H_{34}O_5$ is correct. The freezing point depression is a colligative property—it depends on the number of particles of solute present. If the solute dissociates in the solvent, the calculation yields a low value for the molecular weight.

· **Problem 2-33**

The density of a homogeneous membrane preparation is 1.15. The average density of the membrane proteins is 1.30 g/cm³. The average density of the membrane lipids is 0.92 g/cm³. What are the weight percentages of protein and lipid in the membrane?

Solution

As shown by Equation 10:

$$\bar{v}_{membrane} = (\text{weight fraction protein} \times \bar{v}_{protein}) + (\text{weight fraction lipid} \times \bar{v}_{lipid})$$

$$\bar{v}_{membrane} = \frac{1}{1.15} = 0.87 \qquad \bar{v}_{protein} = \frac{1}{1.30} = 0.77 \qquad \bar{v}_{lipid} = \frac{1}{0.92} = 1.09$$

Let X = weight fraction of protein $\quad \therefore \quad (1 - X)$ = weight fraction of lipid

$$0.87 = (0.77 \, X) + 1.09 \, (1 - X) = 0.77 \, X + 1.09 - 1.09 \, X$$

$$0.22 = 0.32 \, X \qquad X = \frac{0.22}{0.32}$$

$$\text{weight fraction protein} = 0.688 \quad \text{or} \quad \boxed{\textbf{68.8\% protein}}$$

$$\text{weight fraction lipid} = 1.000 - 0.688$$

$$= 0.312 \quad \text{or} \quad \boxed{\textbf{31.2\% lipid}}$$

D. NUCLEOTIDES AND NUCLEIC ACIDS

· Problem 2-34

A sample of DNA purified from *Mycobacterium tuberculosis* contains 15.1% adenine on a molar basis. What are the percentages of the other bases present?

Solution

Double-stranded DNA from most sources contain equimolar amounts of adenine and thymine, and equimolar amounts of guanine and cytosine, that is, A = T and G = C.

$$\therefore \quad \text{if A} = 15.1\%, \text{ then T} = 15.1\%$$

and
$$G + C = (100 - 30.2)\% = 69.8\%$$
$$G = 34.9\% \text{ and } C = 34.9\%$$

In some organisms, 5-methylcytosine or 5-hydroxymethylcytosine replaces some of the cytosine. The fact that A = T and G = C, and that A + C = G + T (sum of the amino bases equals the sum of the keto bases) together with X-ray diffraction studies led to the idea that DNA is a double helix stabilized by hydrogen bonding (Fig. 2-12).

· Problem 2-35

(a) Calculate the length of a double-stranded DNA molecule of MW 3×10^7. (b) What is the volume occupied by one molecule of this DNA? (c) How many helical turns does a molecule of this DNA contain? Consult Figure 2-12.

Solution

(a) The average MW of a complementary pair of deoxy nucleotide residues is about 618. Therefore, the DNA contains:

$$\frac{3 \times 10^7 \text{ g/mole DNA}}{618 \text{ g/mole of nucleotide pair}} = 48,544 \text{ nucleotide pairs}$$

As shown in Figure 2-12a, the double helix rises 3.4 Å per nucleotide pair.

$$\therefore \quad \text{length} = (48,544)(3.4) = \boxed{\textbf{165,049 Å} = \textbf{16.50 } \boldsymbol{\mu}\textbf{m} = \textbf{16.50} \times \textbf{10}^{-4} \textbf{ cm}}$$

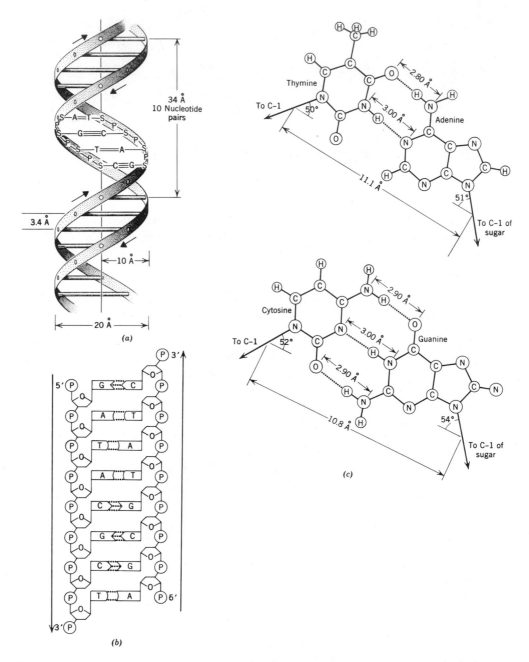

Figure 2-12 (*a*) The DNA of most organisms exists as a double-stranded, right-handed helix wound around a common axis. (*b*) The double helix is composed of two polynucleotide chains running in opposite directions (antiparallel). The bases project into the interior of the double helix. (*c*) The antiparallel structure is stabilized by hydrogen bonding between A-T and G-C pairs, and by nonpolar vertical interactions between the bases ("base stacking"). [(*a*) Redrawn from E. E. Conn and P. K. Stumpf, *Outlines of Biochemistry.* Wiley (1972); (*b*) and (*c*) Redrawn from J. R. Bronk, *Chemical Biology.* Macmillan (1973).]

(b)   The molecule can be considered to be a cylinder 16.50×10^{-4} cm long and 20×10^{-8} cm in diameter.

$$\text{vol} = \pi r^2 l$$

$$\text{vol} = (3.14)(10 \times 10^{-8})^2 (16.50 \times 10^{-4}) = \boxed{\textbf{5.18} \times \textbf{10}^{-17} \textbf{ cm}^3}$$

(c)   As shown in Figure 2-12, there are 10 nucleotide pairs per helical turn.

$$\therefore \quad 48{,}544 \text{ nucleotide pairs} = \boxed{\textbf{4854 helical turns}}$$

· Problem 2-36

The MW of bacteriophage T4 DNA is 1.3×10^8 (double stranded).   (a) How many amino acids can be coded for by T4 DNA?   (b) How many different proteins of MW 55,000 could be coded for by T4 DNA?

Solution

(a)   The genetic code is a triplet code.   That is, it takes a sequence of three nucleotides on the coding strand of DNA to specify one amino acid.   The DNA of T4 contains:

$$\frac{1.3 \times 10^8}{618} = 2.1 \times 10^5 \text{ nucleotide pairs}$$

$$= 2.1 \times 10^5 \text{ nucleotides in the coding strand}$$

$$\frac{2.1 \times 10^5}{3} = \boxed{\textbf{7} \times \textbf{10}^4 \textbf{ codons}}$$

(b)   The average MW of an amino acid residue is 120.   A protein of MW 55,000 contains:

$$\frac{55{,}000}{120} = 458 \text{ amino acids}$$

$$\therefore \quad 7 \times 10^4 \text{ codons can yield:}$$

$$\frac{7 \times 10^4}{458} = \boxed{\textbf{153 proteins of MW 55,000}}$$

(The actual number will be somewhat less because not all the DNA codes for specific proteins.)

· Problem 2-37

The 23 *s* ribosomal RNA of *E. coli* has a MW of 1.1×10^6.   Approximately 0.3% of the total *E. coli* DNA hybridizes with the 23 *s r* RNA.   The MW of the

E. coli DNA (chromosome) is 2.2×10^9.   How many copies of the 23 *s* *r*RNA gene does the *E. coli* chromosome have?

Solution

Approximately 0.3% of the *E. coli* chromosome (actually, 0.6% of the coding strand) codes for 23 *s* *r*RNA.   This corresponds to a segment of MW:

$$(0.3 \times 10^{-2})(2.2 \times 10^9) = 6.6 \times 10^6$$

The molecular weights of a ribonucleotide residue and a deoxyribonucleotide residue are about the same (320 and 309, respectively).   Since there is a 1:1 coding ratio between DNA and RNA, the *E. coli* chromosome must contain:

$$\frac{6.6 \times 10^6}{1.1 \times 10^6} = \boxed{\textbf{6 copies of the } r\textbf{RNA gene}}$$

· **Problem 2-38**

What is the MW of an *m*RNA that codes for a protein of MW 75,000?

Solution

A protein of MW 75,000 contains:

$$\frac{75,000}{120} = 625 \text{ amino acids}$$

It takes a *m*RNA containing $(3)(625) = 1875$ nucleotides to code for the protein (perhaps slightly more allowing for "start" and "stop" regions).   The average MW of a ribonucleotide residue is about 320.   Therefore, the MW of the *m*RNA is approximately:

$$(1875)(320) = \boxed{\textbf{600,000}}$$

In general, the ratio $MW_{mRNA}/MW_{protein}$ is 8 to 10, depending on amino acid composition.

· **Problem 2-39**

A sample of calf thymus DNA had a T_m of 86.0°C in 0.15 *M* NaCl + 0.015 *M* sodium citrate.   Under the same conditions, DNA samples of known base composition gave T_m values shown in Figure 2-13b.   (a) Calculate the G + C content of the calf thymus DNA.   (b) Derive a simple equation relating the % G + C content to the T_m for the conditions shown in Figure 2-13b.

Solution

When a solution of double-stranded DNA is heated, an increase in ultraviolet light absorption ($A_{260\,nm}$) is observed (Fig. 2-13a).   The increase results from an unwinding and separation of the two strands and is accompanied by a decrease in the viscosity of the solution.   The midpoint of the $A_{260\,nm}$ versus T curve ("melting curve") is called the "melting temperature," T_m.   Samples of DNA from different sources have different T_m values in a given ionic

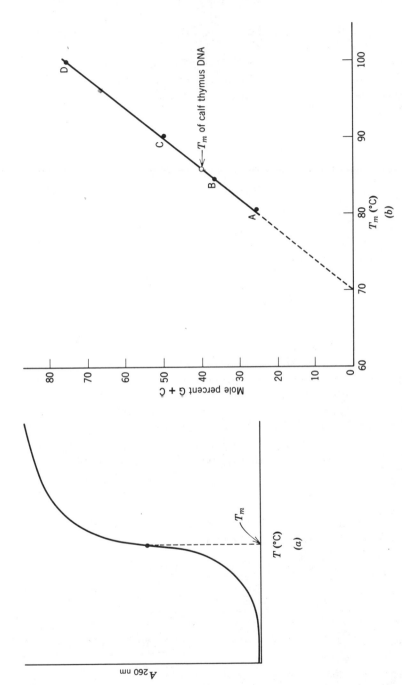

Figure 2-13 (*a*) "Melting curve" of double-stranded DNA. (*b*) Linear relationship between %G+C content and T_m. The linear relationship holds only between about 25 to 85% G+C.

medium. The T_m value varies almost linearly with the G + C content of the DNA (Fig. 2-13b). This is not unexpected since G and C form three hydrogen bonds, while A and T form only two hydrogen bonds, as shown in Figure 2-12c.

(a) From Figure 2-13b, we would predict that calf thymus DNA contains 40% G + C.

(b) The relationship between % G + C and T_m is a linear one. However, the plot shown in Figure 2-13b does not start at 0°C. To simplify matters, we can imagine that the vertical axis is shifted to the right so that the plot extrapolates to the origin (0% G + C at 70°C). The equation for the plot is now $y = mx$, where $y = $ % G + C, $m = $ the slope, and $x = T_m - 70$. The slope (from standard samples A and C) is $(50 - 25)/(90 - 80) = 25/10 = 2.5$.

$$\therefore \quad \boxed{\textbf{% G + C} = \textbf{(2.5)}(\textbf{\textit{T}}_m - \textbf{70})} \tag{30}$$

For a T_m of 86.0°C:

$$\% \text{ G + C} = 2.5(86.0 - 70.0) = 2.5(16) = 40\%$$

The above equation is valid only for the given ionic conditions.

· Problem 2-40

Samples of high MW, double-stranded DNA from three different organisms were mechanically sheared to short segments, and then melted. Upon cooling the solutions, the short segments reformed into short double helices. The rates of renaturation were measured. Sample B, which had a MW of 2×10^9 before shearing, renatured 3 times faster than sample A. The original MW of sample A was 6×10^9. Sample C renatured at a rate 5 times faster than sample B, and 15 times faster than sample A. All three solutions contained 1 mg DNA/ml. What was the original MW of the DNA in sample C?

Solution

A schematic representation of samples A and B are shown below. The vertical lines represent the positions of shearing.

Sample A: 1 mg/ml   *Sample B: 1 mg/ml*

The letters a, b, c, etc. represent polynucleotide sequences in one strand. The letters a′, b′, c′, etc. represent the complementary sequences in the other strand. We see that at equal weight concentrations (mg/ml), the molar concentration of DNA B (and of any given sequence of DNA B) is three times that of DNA A. Upon cooling the solution of DNA A, 1 collision out of 12 results in a successful renaturation. For example, segment a′ will form a short double helix with segment a, but not with any of the other 11 segments.   Upon cooling the solution of DNA B, 1 collision out of 4 results in a successful renaturation (e.g., g with g′, but not with h, h′, or another g).   Thus, the rate of renaturation is inversely proportional to the original MW of the DNA, or directly proportional to the concentration of complementary sequences.   Sample C at 1 mg/ml renatured 5 times faster than sample B and 15 times faster than sample A.   Therefore:

$$MW_C = \frac{MW_B}{5} = \frac{MW_A}{15}$$

$$= \frac{2 \times 10^9}{5} = \frac{6 \times 10^9}{15} = \boxed{\mathbf{4 \times 10^8}}$$

See also the problems on spectrophotometric determination of nucleotides in Section A of Chapter 5.

GENERAL REFERENCES

Barker, R., *Organic Chemistry of Biological Compounds*. Prentice-Hall (1971).

Barry, J. M. and E. M. Barry, *An Introduction to the Structure of Biological Molecules*. Prentice-Hall (1969).

Davidson, E. A., *Carbohydrate Chemistry*. Holt, Rinehart and Winston (1967).

Davidson, J. N., *The Biochemistry of Nucleic Acids*. 7th ed. Academic Press (1972).

Dickerson, R. E. and I. Geis, *The Structure and Action of Proteins*. Harper & Row (1969).

Guthrie, R. D. and J. Honeyman, *An Introduction to the Chemistry of Carbohydrates*. 3rd ed. Clarendon Press (1968).

Haschemeyer, R. H. and A. E. U. Haschemeyer, *Proteins*. Wiley-Interscience (1973).

PRACTICE PROBLEMS

Answers to Practice Problems are given on pages 423–424.

Amino Acids, Peptides, and Proteins

1.   In what order will the following amino acids elute from a Dowex-50 column at pH 3.2:   alanine   (pI = 6.02),   arginine   (pI = 10.76), glutamic acid (pI = 3.22), serine (pI = 5.68), and tryptophan (pI = 5.88)?

2.   What are the relative electrophoretic mobilities at pH 5.68 of the five amino acids given in problem 1?

3.   Deduce the sequence of amino acids in a peptide from the following information: (a)   composition = met + tyr + ser + phe + gly + lys + ala;   (b) Sanger's reagent yielded α,ε-diDNP-lysine as the sole DNP derivative; (c) CNBr yielded a dipeptide containing lys + met (i.e., homoserine) and a second peptide containing all the other amino acids; (d) glycine was released rapidly upon treatment of the original peptide with carboxypeptidase A; (e) chymotrypsin released three peptides.   One contained tyr + lys + met.   A second contained ala + gly.   The third contained ser + phe.

4.   Deduce the sequence of amino acids in a peptide from the following information: (a)   composition = phe + pro + glu + 2 lys; (b) treatment with the Edman reagent yielded PTH-glutamate; and (c) trypsin, carboxypeptidase A, and carboxypeptidase B did not release any smaller peptides or amino acids.

5. A peptide containing equimolar amounts of met + phe + asp + ser + thr was treated with CNBr. A peptide and a single amino acid (identified as homoserine) were released. Treatment of the original pentapeptide with chymotrypsin produced two fragments, one of which was significantly more acidic than the other. The acidic fragment contained methionine. Treatment of the original pentapeptide with carboxypeptidase A yielded serine very rapidly, followed by threonine. Deduce the amino acid sequence of the pentapeptide.

6. How many different linear α-linked hexapeptides can be made from 20 different L-α-amino acids (a) using any amino acid for any of the six different positions (repetition allowed), (b) using each amino acid only once in the chain? (c) How many amino acid analysis patterns containing six differnt amino acids are possible? (d) How many qualitatively different amino acid analysis patterns are possible? (e) How many quantitatively different amino acid analysis patterns are possible?

7. (a) Calculate the axial length of an α-helix containing 122 amino acid residues. (b) How long would the polypeptide chain be if it were fully extended? (c) Approximately what is the MW of the protein?

8. The specific volumes of serine, proline, alanine, and glycine residues are 0.64, 0.76, 0.74, and 0.64, respectively. (a) Calculate the specific volume and density of a synthetic polypeptide containing equimolar ratios of all four amino acids. (b) What is the diameter of a spherical molecule of the synthetic polypeptide if MW = 7810?

9. The ribosomes in *E. coli* account for about 5% of the cell volume. Assuming that each ribosome is approximately a sphere of 180 Å in diameter, calculate the number of ribosomes in an *E. coli* cell. Assume that *E. coli* is a cylinder 1 μ in diameter and 2 μ long.

10. Calculate the width of the peptide plane, that is, (a) the distance x, between the α-carbon and the oxygen, and (b) the distance, y, between the H attached to the nitrogen atom and the α-carbon. Consult Figure 2-1. (Hint: use plane trigonometry.)

11. Glutamine synthetase is a dodecamer composed of 12 identical subunits. Each subunit can be adenylated. How many quantita-

tively different forms of glutamine synthetase are possible?

12. A protein believed to be involved in membrane transport was obtained from a bacterium by osmotic shock. Amino acid analysis of 10 mg of the purified protein yielded 61 μg of tryptophan. What is the *minimum* molecular weight of the protein? (The MW of tryptophan is 204.1.)

13. The enzyme glutathione peroxidase contains 0.34% selenium by weight. (The atomic weight of Se is 78.96.) The MW of the enzyme determined from gel filtration is about 88,000. What is the likely quaternary structure of this enzyme?

14. (a) At 5°C, the osmotic pressure of a protein solution (5.0 mg/ml) was 2.86×10^{-3} atm. Estimate MW. (b) Another measurement made at 2.5 mg/ml yielded an osmotic pressure of 1.37×10^{-3} atm. Calculate a more accurate value for MW. (Extrapolate to zero concentration.) $R = 0.0821$ liter $\times$ atm $\times$ mole^{-1} $\times$ degree^{-1}.

15. The $s_{20,w}$ of lysozyme is 1.91×10^{-13} sec. D is 11.2×10^{-7} cm^2/sec, and $\bar{v} = 0.703$. Calculate MW.

Carbohydrates

16. (a) How many 2-ketopentose stereoisomers are possible (excluding anomers)? How many aldopentose stereoisomers are possible (excluding anomers)?

17. How many different disaccharides can be made from two molecules of D-glucopyranose?

18. A 75 mg sample of cellulose was acid hydrolyzed. The hydrolysate was found to contain 75 mg of glucose. What is the purity of the cellulose sample?

19. A disaccharide containing only glucose was exhaustively methylated and then acid hydrolyzed. The only products found were 3,4,6-trimethylglucose and 2,3,4,6-tetramethylglucose. What is the most likely structure of the disaccharide?

20. (a) A glucan isolated from *Aspergillus aculeatus* was completely hydrolyzed to a single disaccharide with an enzyme specific for α, 1–4 linkages. The disaccharide was

reduced with $NaBH_4$ and then treated with periodate. For each mole of reduced disaccharide, five moles of periodate were utilized yielding two moles of formic acid plus two moles of formaldehyde. What is a likely structure of the unreduced disaccharide? (b) Exhaustive methylation of the original polysaccharide followed by acid hydrolysis yielded approximately equal molar amounts of 2,4,6-trimethylglucose and 2,3,6-trimethylglucose plus a trace of 2,3,4,6-tetramethylglucose. What is a possible structure of the polysaccharide?

21. Exhaustive methylation of 32.4 mg of amylopectin followed by acid hydrolysis yielded 10 μ moles of 2,3,4,6-tetramethylglucose. (a) What are the other products? How much of each were obtained? (b) What percent of the glucose residues are linked via 1–6 bonds? (c) If the MW of the amylopectin is 1.2×10^6, how many branch point residues does a molecule of amylopectin contain?

22. Raffinose is a trisaccharide found in plants. It has the structure gal($\alpha 1 \rightarrow$ 6)glu($\alpha 1 \rightarrow 2\beta$)fru (i.e., galactose linked to carbon number six of the glucose residue of sucrose). (a) How many moles of periodate will one mole of raffinose consume? (b) How many moles of formic acid will be produced? (c) What are the products of exhaustive methylation of raffinose followed by hydrolysis?

23. Amylose is a linear α,1–4 glucan. A 648 mg sample of amylose was treated with periodate. Exactly 4.43 μ moles of formic acid were produced. (a) What is the average MW of the amylose in the sample? (b) How many μ moles of $^{14}CN^-$ will add to 648 mg of the amylose?

Lipids

24. (a) How many different L-triglycerides can be made from glycerol and four different fatty acids? (b) How many different triglycerides does this correspond to if we count only one member of a D-L pair? (c) How many triglycerides of quantitatively different composition can be made from glycerol and four different fatty acids?

25. Calculate the saponification number of tributyrin.

26. The saponification number of a sample of butter fat is 230. Calculate the average MW of the triglycerides present.

27. The iodine number of a sample of butter fat is 68. If the saponification number of the sample is 210, how many double bonds, on the average, are present in a molecule of triglyceride?

28. Progesterone, a female sex hormone, was found to contain 80.3% C, 9.5% H, and 10.2% O. A solution of 60 mg of progesterone in 750 mg of benzene freezes at 4.20°C. The freezing point of pure benzene is 5.50°C. The molal freezing point depression constant is 5.10°C. What is the empirical formula and MW of progesterone?

29. Calculate the density and specific volume of a lipoprotein containing 80% protein and 20% lipid by weight. The average densities of protein and lipid are 1.30 g/cm^3 and 0.92 g/cm^3, respectively.

Nucleotides and Nucleic Acids

30. Yeast DNA contains 32.8% thymine on a molar basis. Calculate the molar percentages of the other bases.

31. E. coli DNA (chromosome) has a MW of 2.2×10^9. (a) How many nucleotide pairs does the DNA contain? (b) How long is the DNA molecule? (c) An average E. coli cell has a volume of 1.57×10^{-12} cm^3. What fraction of the cell volume is occupied by the DNA?

32. E. coli can divide every 40 minutes. Thus, its DNA (MW $= 2.2 \times 10^9$) can be duplicated in 40 minutes (or less). Calculate (a) the number of internucleotide bonds made per minute, (b) the rate of chromosome duplication in terms of mm/min and μ m/min (assuming only one growing point), and (c) the rate at which the double helix unwinds (turns/min) during duplication.

33. What is the MW of a protein coded for by an m RNA containing 1000 nucleotides?

34. If 75% of the E. coli chromosome codes for specific proteins, how many different proteins of average MW 60,000 can be made?

35. What is the MW of the gene that codes for a t RNA containing 80 nucleotide residues?

36.   The following T_m data were obtained for double-stranded DNA in 10 mM phosphate buffer containing 1 mM EDTA.

Sample	% G + C	T_m (°C)
A	70	78.5
B	52.5	71.2
C	37.5	65.0
Z	?	73.3

(a) Calculate the % G + C content of sample Z.   (b) Derive an equation relating the % G + C content to the T_m for the above conditions.

37.   Sheared and melted segments of bacteriophage T2 DNA (MW = 1.3×10^8) renatured 19.2% as fast as sheared and melted DNA from T7 bacteriophage under the same conditions of temperature, ionic strength, and concentration.   What is the MW of T7 DNA?

3

BIOCHEMICAL ENERGETICS

A. ENERGY-YIELDING AND ENERGY-REQUIRING REACTIONS

THE LAWS OF THERMODYNAMICS

All processes that occur in the universe are subject to the basic laws of thermodynamics. The reactions that occur in living cells are no exception. The *first law of thermodynamics* states that energy can neither be created nor destroyed. In any given process, one form of energy may be converted into another but the total energy of the system plus its surroundings remains constant. The first law is simply a law of conservation of energy. Nothing is said about the relative usefulness of different forms of energy or the direction of a process or reaction.

The *second law of thermodynamics* states that all naturally occurring processes proceed in a direction that leads to a minimum potential energy level, that is, toward equilibrium. Such "spontaneous" reactions (as they are called) release energy as they progress toward equilibrium and, theoretically, the energy can be harnessed and made to do work. We are all familiar with "spontaneous" reactions. For example: heat flows from a warm body into a cooler body (never in the opposite direction); a wound spring spontaneously unwinds (an unwound spring never winds itself up); water flows downhill (never uphill); gases diffuse from a region of high pressure and concentration to a region of lower pressure and concentration (never in the opposite direction); the great pyramids will someday crumble away to sand (but the sand grains will never spontaneously assemble into a pyramid). In all of these spontaneous reactions, energy is conserved. For example, the heat lost by the warm body is gained by the cooler body. But certainly, *something* has been lost. That something is the *capacity* or *potential* to do more work (to transfer still more energy). While the total energy of a system and its surroundings remain constant, the energy is distributed in a qualitatively different way after a spontaneous reaction. A more complete statement of the second law that takes into account the unidirectionality of spontaneous processes and the decreased potential to do further work is this: the *entropy* of the universe is constantly increasing. Entropy, given the symbol S, is a measure of the *randomness* or *orderliness* of the energy and matter in a system. The more random, disordered, disorganized, or chaotic the system, the higher its entropy. The more organized, orderly, constrained, or highly structured the system, the lower its entropy. Only organized, nonrandom energy is useful (can be made to do work). An increase in entropy represents a loss of organization and, hence, a decrease in the potential to do further work. The second law leads directly to the *third law of thermodynamics*, which states that at a temperature of absolute zero (0°K), where all random motion ceases, the entropy of a perfect crystal of every substance is zero, that is, all the atoms are maximally ordered.

If the spontaneous direction of order is downhill, how then can we explain the biosynthesis of complex, highly organized macromolecules or, for that matter, the very existence of living cells? No laws of thermodynamics have been violated, just as no laws are violated when pyramids are built from sand. The natural tendency of matter and energy *in a given system* to run downhill can be counteracted by putting energy into that system, that is, by doing work on the system. The *total* energy—that of the system receiving the energy plus that of the system providing the energy—remains constant. The total entropy—that of the system receiving the energy plus its surroundings—increases. However, the entropy of just the system receiving the energy may increase, decrease, or remain constant. In general, a living cell takes in raw materials at a high entropy state from its environment and orders these materials into a lower entropy state at the expense of the environment, which also supplies chemical energy at a low entropy state. Photosynthetic organisms capture light energy.

While the entropy change that takes place during a process or reaction is of great interest to biologists, there are two related thermodynamic *functions of state* (as they are known) that are more easily measured or calculated. These are (a) the change in *free energy* and (b) the change in *enthalpy* or *heat content*. The free energy change is a measure of the maximum useful work that a reaction could perform at constant temperature and pressure, and depends on the displacement of the system from equilibrium. The enthalpy change is a measure of the heat flow that accompanies a reaction as it proceeds toward equilibrium at constant temperature, pressure, and volume. These concepts are examined in more detail below.

COUPLED REACTIONS

Chemical reactions may be classified as (a) "exergonic," those that yield energy (i.e., are capable of doing work) and (b) "endergonic," those that utilize (require) energy (i.e., work must be done to make them go). While it may not be immediately obvious why certain reactions are more exergonic than others, the student intuitively recognizes that biosynthetic processes (i.e., the formation of large macromolecules from their constitutive subunits) require energy. Work must be done to build complex structures from simple building blocks. Living cells are exceedingly complex and delicate. Yet they not only maintain their integrity over long periods of time but also grow and multiply. In terms of energetics, this is accomplished by catalyzing certain exergonic reactions and trapping some of the energy released in "energy-rich" compounds. Biosynthetic (endergonic) reactions then are driven by this trapped energy. For example, suppose that the reaction by which A is converted to B is exergonic, releasing 15 kcal of energy:

$$A \rightarrow B + 15 \text{ kcal} \tag{1}$$

All of this energy would be wasted if the reaction proceeded as written. In a living cell, a portion of the total energy is trapped by coupling reaction 1 to the endergonic synthesis of an "energy-rich" compound $X \sim Y$. If, for example, the synthesis of $X \sim Y$ requires 8 kcal, then the overall coupled reaction releases only 7 kcal.

$$A \xrightarrow[\ X+Y \quad\quad X\sim Y\ ]{} B + 7 \text{ kcal} \tag{2}$$

Now suppose that a complex molecule, C-D, is to be synthesized from its components $C + D$. The reaction is endergonic and will go only if 5 kcal of energy are supplied.

$$5 \text{ kcal} + C + D \rightarrow C\text{-}D \tag{3}$$

The 5 kcal of energy can be supplied by the energy-rich compound, $X \sim Y$. Because $X \sim Y$ has stored 8 kcal and only 5 kcal are required, 3 kcal are released by the overall coupled reaction.

$$C + D \xrightarrow[\ X\sim Y \quad\quad X+Y\ ]{} C\text{-}D + 3 \text{ kcal} \tag{4}$$

It is apparent that if $X \sim Y$ were degraded into its components without coupling its breakdown to an endergonic reaction, 8 kcal of energy would be wasted.

$$X \sim Y \rightarrow X + Y + 8 \text{ kcal} \tag{5}$$

These energy relationships are illustrated in Figure 3-1. Note that none of the reactions proceeds with 100% efficiency. Of the original 15 kcal made available, only 8 kcal are conserved in $X \sim Y$; of the 8 kcal conserved, only 5 kcal are trapped in C-D.

The coupled reaction concept adequately illustrates the principle of energy conservation and utilization in living cells. However, the actual mechanisms of energy coupling in living cells seldom involve the simultaneous catalysis of two reactions. Instead, the net effect is generally obtained by catalyzing two

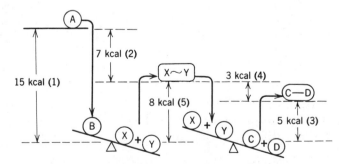

Figure 3-1 Energy relationships in the synthesis of C–D from C + D at the expense of the energy released when A is converted to B.

consecutive reactions involving a common intermediate. As an example of the formation of an energy-rich compound, consider the oxidation of glyceraldehyde-3-phosphate to 3-phosphoglyceric acid.

$$
\begin{array}{c}
\text{CHO} \\
| \\
\text{H--C--OH} \\
| \\
\text{H--C--H} \\
| \\
\text{OPO}_3^{2-}
\end{array}
\quad
\xrightarrow[\quad]{\text{NAD}^+ \quad \text{NADH+H}^+}
\quad
\begin{array}{c}
\text{COOH} \\
| \\
\text{H--C--OH} \\
| \\
\text{H--C--H} \\
| \\
\text{OPO}_3^{2-}
\end{array}
\;+\; 12\,\text{kcal}
\qquad (6)
$$

This oxidation-reduction reaction yields sufficient energy to drive the synthesis of ATP from ADP and P_i in a hypothetical coupled reaction.

$$
\begin{array}{c}
\text{CHO} \\
| \\
\text{H--C--OH} \\
| \\
\text{H--C--H} \\
| \\
\text{OPO}_3^{2-}
\end{array}
\quad
\xrightarrow[\substack{\text{ADP+P}_i \qquad \text{ATP} \\ \text{(8 kcal conserved)}}]{\text{NAD}^+ \quad \text{NADH+H}^+}
\quad
\begin{array}{c}
\text{COOH} \\
| \\
\text{H--C--OH} \\
| \\
\text{H--C--H} \\
| \\
\text{OPO}_3^{2-}
\end{array}
\;+\; 4\,\text{kcal}
\qquad (7)
$$

In the living cell the two reactions are not actually coupled as shown. Instead, an energy-rich acyl phosphate is formed simultaneously with the oxidation.

$$
\begin{array}{c}
\text{CHO} \\
| \\
\text{H--C--OH} \\
| \\
\text{H--C--H} \\
| \\
\text{OPO}_3^{2-}
\end{array}
\quad
\xrightarrow[\substack{\\ \text{P}_i \quad \text{dehydrogenase}}]{\substack{\text{NAD}^+ \quad \text{NADH+H}^+ \\ \text{glyceraldehyde-3-phosphate}}}
\quad
\begin{array}{c}
\text{O} \\
\| \\
\text{C--OPO}_3^{2-} \\
| \\
\text{H--C--OH} \\
| \\
\text{H--C--H} \\
| \\
\text{OPO}_3^{2-}
\end{array}
\;+\; 0\,\text{kcal}
\qquad (8)
$$

The energy is conserved by transferring the phosphate group to ADP in a subsequent reaction.

$$
\begin{array}{c}
\text{O} \\
\| \\
\text{C--OPO}_3^{2-} \\
| \\
\text{H--C--OH} \\
| \\
\text{H--C--H} \\
| \\
\text{OPO}_3^{2-}
\end{array}
\;+\; \text{ADP}
\quad
\xrightarrow[\text{Mg}^{++}]{\text{phosphoglycerate kinase}}
\quad
\begin{array}{c}
\text{COOH} \\
| \\
\text{H--C--OH} \\
| \\
\text{H--C--H} \\
| \\
\text{OPO}_3^{2-}
\end{array}
\;+\; \text{ATP} \;+\; 4\,\text{kcal}
$$
$$(9)$$

The sum of reactions 8 and 9 effectively equals the coupled reaction 7.

The formation of phosphoenolpyruvate (PEP) from 2-phosphoglyceric acid (2-PGA) illustrates a different principle.

$$
\begin{array}{c}
\text{COOH} \\
| \\
\text{H--C--OPO}_3 \\
| \\
\text{CH}_2\text{OH}
\end{array}
\quad
\xrightarrow{\text{enolase}}
\quad
\begin{array}{c}
\text{COOH} \\
| \\
\text{C--OPO}_3^{2-} \\
\| \\
\text{CH}_2
\end{array}
\;+\; \text{H}_2\text{O}
\qquad (10)
$$

In this reaction an energy-rich compound is produced from an energy-poor precursor by a simple dehydration reaction. The student is tempted to ask "where did the energy come from?" The answer lies in the restricted definition of "energy-rich" as used by biochemists. In spite of the fact that 2-PGA is considered relatively "energy-poor," it yields about the same amount of energy as PEP if both are *burned* to CO_2, H_2O, and P_i. The dehydration reaction results in a rearrangement of electrons so that a much larger portion of the total potential energy becomes available upon *hydrolysis*. An "energy-rich" compound (to the biochemist) is one that releases a relatively large amount of energy (7 to 18 kcal per mole under standard-state conditions) upon hydrolysis. (We can also consider the enolase-catalyzed reaction as an internal oxidation-reduction: the carbon carrying the OH is reduced; the carbon carrying the phosphate is oxidized.)

As an example of the utilization of an energy-rich compound, consider the synthesis of glucose-6-phosphate from glucose and inorganic phosphate.

$$glucose + P_i + 3 \text{ kcal} \rightarrow G\text{-}6\text{-}P \tag{11}$$

The reaction is endergonic and requires 3 kcal/mole, which can be supplied by the hydrolysis of ATP. The hypothetical coupled reaction is shown below.

$$glucose + P_i \longrightarrow G\text{-}6\text{-}P + HOH + 5 \text{ kcal} \tag{12}$$

$$ATP \qquad ADP + P_i$$
$$+ HOH$$

The reaction actually catalyzed by hexokinase is the sum of the two individual reactions.

$$glucose + ATP \xrightarrow[Mg^{++}]{\text{hexokinase}} G\text{-}6\text{-}P + ADP + 5 \text{ kcal} \tag{13}$$

Not only does ATP supply the energy, it supplies the phosphate group as well. In fact, the energy in energy-rich compounds is seldom released by hydrolysis in living cells. Instead, the potential energy is used as "group-transfer potential,"—the potential energy is used to transfer a portion of the energy-rich molecule to an acceptor that, in effect, "activates" the acceptor. ATP is used in cells as (a) a phosphate donor, (b) a pyrophosphate donor, (c) an AMP donor, and (d) an adenosine donor.

FREE ENERGY CHANGE (ΔG)

The energy released or utilized in a chemical reaction represents the *difference* between the energy contents of the products and the reactants. At constant temperature and pressure, the energy difference is called the "free energy difference" (or "Gibbs free energy change"), ΔG, and is the maximum potential of a reaction for performing useful work. By definition, ΔG is the free energy content of the products minus the free energy content of the reactants. Thus, for the exergonic reaction 1, we obtain the following:

$$A \rightarrow B + 15 \text{ kcal} \tag{1}$$
$$\Delta G = G_B - G_A$$

In order for A to yield B plus energy, the free energy content of A must be greater than the free energy content of B.

$$\Delta G = \text{(some value)} - \text{(some larger value)}$$

$$\Delta G = \text{a negative value}$$

$$\Delta G = -15 \text{ kcal}$$

Thus, because of the definition of ΔG, exergonic reactions have negative ΔG values while endergonic reactions have positive values. When a reaction is written in reverse, the sign of ΔG changes. For example, the synthesis of $X \sim Y$ *requires* 8 kcal per mole. Therefore, ΔG for the synthetic (endergonic) reaction is +8 kcal/mole.

$$X + Y \rightarrow X \sim Y \qquad \Delta G = +8 \text{ kcal/mole}$$

The degradation of $X \sim Y$ *releases* 8 kcal/mole.

$$X \sim Y \rightarrow X + Y \qquad \Delta G = -8 \text{ kcal/mole}$$

RELATIONSHIP BETWEEN ΔG AND THE [P]/[S] RATIO

Consider the reaction $S \rightarrow P$ where S is the reactant or "substrate" (of an enzyme) and P is the product. What is the relationship between the amount of energy released and the concentrations of S and P? By analogy, let us assume that the concentrations of S and P are liquid volumes separated into two arms of a U-tube as shown in Figure 3-2. Let us also place a waterwheel in the arm containing S. When the stopcock separating S and P is opened (i.e., when an enzyme catalyzing the $S \rightarrow P$ reaction is added), S is converted to P. As the level of S drops, the waterwheel turns (energy is released and work is done). S is converted to P until equilibrium is attained. For the particular reaction shown, the *volumes* of S and P at equilibrium are not equal, indicating that the equilibrium lies in favor of P. We can see from this analogy that the energy released depends on how far from equilibrium the original S/P ratio is. The greater the S/P ratio, the more the work that can be done in converting S to P. We can also see that if we start with an S/P ratio identical to the equilibrium S/P ratio (a P/S ratio equal to K_{eq}), no net transformation of S to P occurs; therefore, no work can be done (i.e., $\Delta G = 0$). If we start with a P/S ratio greater than K_{eq} (the system is displaced from equilibrium in favor of P), then work must be done (energy is required) to convert still more S to P (i.e., ΔG is positive). However, the conversion of P to S now occurs with the production of energy (a minus ΔG value) until equilibrium (the minimum energy level of the system) is attained.

The U-tube analogy is not perfect because most reactions that occur in a living cell never attain equilibrium. Instead, the reactants and products are maintained (within narrow limits) at *steady-state* levels that may be quite different from the equilibrium levels. Nevertheless, the analogy illustrates how the energy released or utilized in a reaction depends on the displacement of the system from equilibrium. A mathematical statement of the ΔG of a reaction then must contain two terms: one that indicates the actual concentrations of the substrates and products and one that states the equilibrium concentrations. Such an expression for the reaction $cS_1 + dS_2 + \cdots \rightarrow aP_1 +$

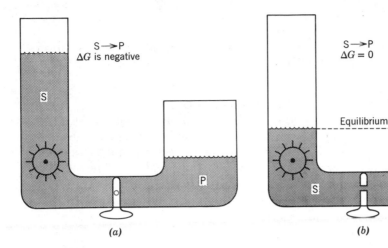

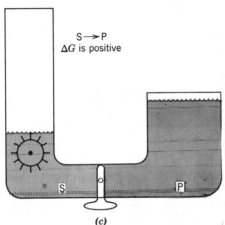

Figure 3-2 (*a*) The [S]/[P] ratio at the start of the reaction is greater than the ratio at equilibrium. The reaction will proceed "spontaneously" from S to P (if the stopcock is opened). (*b*) The [S]/[P] ratio at the end of the reaction equals the equilibrium ratio. [P]/[S] = K_{eq}. No further net reaction occurs. (*c*) The [P]/[S] ratio is greater than that at equilibrium. The reaction will proceed "spontaneously" in the direction [P] to [S] (if the stopcock is opened).

bP$_2$... is shown below:

$$\Delta G \qquad = \qquad RT \ln \frac{[P_1]^a \, [P_2]^b \ldots}{[S_1]^c \, [S_2]^d \ldots} \qquad - \qquad RT \ln K_{eq}$$

that is, ΔG...depends on $\underbrace{}$ the difference $\underbrace{}$
                                                        between

$\left\{ \begin{array}{l} \text{the actual product/} \\ \text{substrate ratios} \end{array} \right\}$ and $\left\{ \begin{array}{l} \text{the product/} \\ \text{substrate ratios} \\ \text{at equilibrium} \end{array} \right\}$

or

$$\Delta G = - RT \ln K_{eq} + RT \ln \frac{[P_1]^a \, [P_2]^b \ldots}{[S_1]^c \, [S_2]^d \ldots} \qquad (14)$$

or

$$\Delta G = - 2.3 \, RT \log K_{eq} + 2.3 \, RT \log \frac{[P_1]^a \, [P_2]^b \ldots}{[S_1]^c \, [S_2]^d \ldots} \qquad (15)$$

where R = the gas constant = 1.987 cal × mole^{-1} × °K^{-1}
$\quad\quad T$ = the absolute temperature, °K
$\quad\quad$ [P], [S], etc. = the actual concentrations of products and substrates
$\quad\quad a$, b, c, etc. = the coefficients of P, S, etc., in the balanced chemical
$\quad\quad\quad\quad\quad\quad\quad$ equation

At 25°C, 2.303 RT = 1364; at 37°C, 2.303 RT = 1419.

We can see that at equilibrium the [P]/[S] ratio equals K_{eq}; hence, $\Delta G = 0$.

To catalog and compare ΔG values for various reactions, chemists have agreed upon a "standard-state" where all reactants and products are considered to be maintained at steady-state concentrations of 1 M. The standard-state for gases is considered to be 1 atm partial pressure. Under this condition the log [P]/[S] term (regardless of the exponents) is zero. The ΔG under standard-state conditions is designated $\Delta G°$.

$$\Delta G = -1364 \log K_{eq} + 1364 \log 1$$
$$= -1364 \log K_{eq} + 0$$

$$\boxed{\Delta G° = -1364 \log K_{eq}} \quad \text{at 25°C} \quad (16)$$

Thus the $\Delta G°$ value of a reaction is related to the K_{eq}. In fact, both the $\Delta G°$ and K_{eq} values impart the same information, that is, in which direction and how far a reaction will proceed when all substrates and products are 1 M. The correspondence between $\Delta G°$ and K_{eq} is shown in Table 3-1.

The actual ΔG under any particular set of concentration conditions can be calculated from:

$$\Delta G = -1364 \log K_{eq} + 1364 \log \frac{[P_1]^a [P_2]^b \cdots}{[S_1]^c [S_2]^d \cdots} \quad (17)$$

or $\quad\quad$ $$\boxed{\Delta G = \Delta G° + 1364 \log \frac{[P_1]^a [P_2]^b \cdots}{[S_1]^c [S_2]^d \cdots}} \quad \text{at 25°C} \quad (18)$$

Table 3-1   Correspondence Between K_{eq} and $\Delta G°$ at 25°C

K_{eq}	$\log K_{eq}$	$\Delta G°$
0.0001	−4	+5456 cal
0.001	−3	+4092 cal
0.01	−2	+2728 cal
0.1	−1	+1364 cal
1	0	0
10	1	−1364 cal
100	2	−2728 cal
1000	3	−4092 cal
10,000	4	−5456 cal

A negative ΔG means that the reaction as written will proceed from left to right toward a state of minimum energy. Such reactions are said to be "spontaneous"—an unfortunate term that suggests a high velocity. A "spontaneous" reaction is indeed one that will go to the right as written *if it goes at all.* The magnitude of the negative ΔG even tells us how far to the right the reaction will go (because the magnitude of ΔG tells us how far the system is from equilibrium). However, a ΔG value says absolutely nothing about the *rate* at which the reaction will approach equilibrium. Many reactions with very large negative ΔG values do not proceed at a detectable rate (at life temperatures) in the absence of an appropriate catalyst (enzyme). The situation is analogous to that shown in Figure 3-2a. The system may be markedly displaced from equilibrium, yet nothing happens unless the stopcock is opened.

The ΔG of a reaction tells us only the *difference* between the free energy contents of the products and the original substrates. This difference must be the same regardless of the path taken or the number of steps involved. For example, suppose S can be connected to P by any of the three reaction sequences shown below:

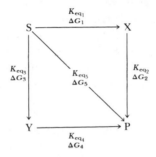

The overall ΔG is the same whether the reaction sequence is $S \to P$, $S \to X \to P$ or $S \to Y \to P$. If the overall ΔG is the same, then $\Delta G_1 + \Delta G_2 = \Delta G_3 + \Delta G_4 = \Delta G_5$. Similarly, the overall equilibrium constant is the same:

$$K_{eq_1} \times K_{eq_2} = K_{eq_3} \times K_{eq_4} = K_{eq_5}$$

EFFECT OF NONSTANDARD [H$^+$]

When the H$^+$ ion appears as a substrate or product, its standard-state concentration is also taken as $1\ M$ (i.e., pH = 0). However, almost all enzymes are denatured at pH 0 and, consequently, there is no reaction to study. Because of this, biochemists have adopted a modified standard-state in which all substrates and products *except* H$^+$ are considered to be $1\ M$. The H$^+$ ion concentration is taken to be some physiological value (e.g., $10^{-7}\ M$). The relationship between $\Delta G°$ and the modified standard-state free energy change, designated $\Delta G'$, can be easily calculated. For example, consider a reaction that yields an H$^+$ ion as a product.

$$S \rightarrow P + H^+$$

$$\Delta G = \Delta G^\circ + 1364 \log \frac{[P][H^+]}{[S]}$$

$$\Delta G' = \Delta G^\circ + 1364 \log [H^+] \qquad \text{when [S] and [P] = 1 } M$$

$$= \Delta G^\circ - 1364 \log \frac{1}{[H^+]}$$

$$\boxed{\Delta G' = \Delta G^\circ - 1364 \text{ pH}} \qquad \text{at } 25^\circ C \qquad\qquad (19)$$

For a reaction involving H^+ ion as a substrate, the relationship is:

$$\boxed{\Delta G' = \Delta G^\circ + 1364 \text{ pH}} \qquad \text{at } 25^\circ C \qquad\qquad (20)$$

Equations 19 and 20 assume that there is no change in the ionization of a group as a result of the reaction. That is, ionizable compounds are not involved in the reaction or, if a group does ionize, it has the same pK_a in the substrate and product. (The H^+ that appears as a product might come from the oxidation of a —CHOH— group to a —C=O group.) Equation 19 can be used to predict $\Delta G'$ at some pH other than zero even when a product ionizes *provided* the standard-state is taken as 1 M of the fully ionized product. This is the same as writing the equation as $S \rightarrow P + H^+$ where P would be, for example, A^-. As we shall see however, biochemists prefer to take 1 M *total* product (ionized P plus un-ionized P) as the standard-state. In this case, a different equation (derived later) must be used to obtain $\Delta G'$ in relation to ΔG°.

The $\Delta G'$ of a reaction under physiological standard-state conditions (all substrates and products at 1 M *except* H^+) is given by:

$$\boxed{\Delta G' = -2.3 \, RT \log K'_{eq}} \qquad\qquad (21)$$

or $\qquad$ $$\boxed{\Delta G' = -1364 \log K'_{eq}} \qquad \text{at } 25^\circ C \qquad\qquad (22)$$

where K'_{eq} is the equilibrium constant of the reaction at the specified pH (usually around 7). The ΔG of a reaction when all substrates and products are not 1 M (but pH = 7) is given by:

$$\Delta G = -2.3 \, RT \log K'_{eq} + 2.3 \, RT \log \frac{[P_1]^a \, [P_2]^b \ldots}{[S_1]^c \, [S_2]^d \ldots}$$

or $\qquad$ $$\boxed{\Delta G = \Delta G' + 2.3 \, RT \log \frac{[P_1]^a \, [P_2]^b \ldots}{[S_1]^c \, [S_2]^d \ldots}} \qquad\qquad (23)$$

or

$$\Delta G = \Delta G' + 1364 \log \frac{[P_1]^a \ [P_2]^b \dots}{[S_1]^c \ [S_2]^d \dots}$$

at 25°C   (24)

Thus, if K'_{eq} or $\Delta G'$ is specified, there is no further need to include H^+ or pH in the calculations unless we are interested in calculating $\Delta G'$ at another pH.

DIFFERENT CONVENTIONS FOR DEFINING "K_{eq}" AND "ΔG"

We would expect the "K_{eq}" of a reaction to be a fixed value for a given set of conditions (e.g., temperature, pressure, total ionic strength).   However, the value of "K_{eq}" depends on *how we define it*.   For example, consider the acid-catalyzed hydrolysis of ethyl acetate.   The reaction can be written:

$$EtOAc + H_2O \ \underset{}{\overset{H^+}{\rightleftharpoons}} \ EtOH + HOAc$$

We can define the K_{eq} as follows:

$$K_{eq_1} = \frac{[EtOH][HOAc]}{[EtOAc][H_2O]}$$ (Convention I)

Suppose we measure the equilibrium concentrations and find $[EtOH] = 2 \times 10^{-1} \ M$, $[HOAc] = 2 \times 10^{-1} \ M$, and $[EtOAc] = 2.18 \times 10^{-3} \ M$.   As defined, K_{eq_1} contains $[H_2O]$, which is present at very close to $(1000 \ g/liter)/(18 \ g/mole) = 55.6 \ M$.

$$\therefore \quad K_{eq_1} = \frac{(0.2)(0.2)}{(2.18 \times 10^{-3})(55.6)} = 0.33$$

and

$$\Delta G_1^{\circ} = -1364 \log 0.33 = +656.7 \ cal/mole$$

The calculated K_{eq_1} and ΔG_1° values are correct for the reaction as defined.   The positive ΔG_1° value says that when all components (including H_2O) are 1 M, the reaction will proceed to the left.   Since the concentration of water is essentially constant in dilute aqueous solutions, we could also define "K_{eq}" as:

$$K_{eq_2} = \frac{[EtOH][HOAc]}{[EtOAc]} = K_{eq_1}[H_2O]$$ (Convention II)

and

$$K_{eq_2} = (0.33)(55.6) = 18.35 \ M$$

$$\Delta G_2^{\circ} = -1364 \log 18.35 = -1724 \ cal/mole$$

The negative ΔG_2° value says that when all components are 1 M *except* water (which is 55.6 M), the reaction will proceed to the right.   Thus, K_{eq_1} and K_{eq_2}

are different numbers, but they say the same thing. For example, we can use K_{eq_1} to calculate the concentrations of EtOH and HOAc present at equilibrium if the equilibrium concentration of EtOAc is known. When K_{eq_1} is used, we must include $[H_2O] = 55.6\ M$. We will get exactly the same answer using K_{eq_2}, but now we do not include the concentration of water in our calculations (because it is built into K_{eq_2}).

Suppose the hydrolysis is catalyzed by an enzyme (called an esterase) at some fixed pH where the HOAc produced ionizes. The reactions occurring are:

$$EtOAc + H_2O \rightleftharpoons EtOH + HOAc \qquad K_{eq_1} = 0.33$$

$$HOAc \rightleftharpoons OAc^- + H^+ \qquad K_a = 1.75 \times 10^{-5}$$

overall: $\quad EtOAc + H_2O \rightleftharpoons EtOH + OAc^- + H^+ \qquad K_{eq_3} = 5.78 \times 10^{-6}$

$\therefore$
$$\boxed{K_{eq_3} = \frac{[EtOH][OAc^-][H^+]}{[EtOAc][H_2O]} = (K_{eq_1})(K_a)} \qquad \text{(Convention III)}$$

$$K_{eq_3} = 5.78 \times 10^{-6}$$

or
$$\boxed{K_{eq_4} = \frac{[EtOH][OAc^-][H^+]}{[EtOAc]} = [H_2O](K_{eq_1})(K_a)} \qquad \text{(Convention IV)}$$

$$K_{eq_4} = 3.21 \times 10^{-4}$$

and

$$\Delta G_3^\circ = -1364 \log 5.78 \times 10^{-6} = +7145\ \text{cal/mole}$$

$$\Delta G_4^\circ = -1364 \log 3.21 \times 10^{-4} = +4765\ \text{cal/mole}$$

K_{eq_4} differs from K_{eq_3} (and ΔG_4° differs from ΔG_3°) because of the way "K_{eq}" is defined. Either set will predict the same correct concentration of one component if the concentrations of the others are known. Note that K_{eq_3} and K_{eq_4} are expressed in terms of the concentration of OAc^-, not HOAc or HOAc plus OAc^- (total analytical concentration of "acetate"). The positive ΔG_4° value says that when all components (including $[H^+]$) are $1\ M$ (but $[H_2O] = 55.6\ M$), the reaction will proceed to the left.

So far, we have seen that the constant $[H_2O]$ can be incorporated into the "K_{eq}" and "ΔG°" values. At a fixed pH, $[H^+]$ is also constant and can be similarly incorporated into the constants yielding a "K_{eq}" and "ΔG°" for one fixed pH. The standard ΔG and K_{eq} at the new pH $\neq 0$ will be designated $\Delta G'$ and K'_{eq}, respectively. For example, at pH 7.0:

$$\boxed{K'_{eq_4} = \frac{[EtOH][OAc^-]}{[EtOAc]} = \frac{K_{eq_4}}{[H^+]}} \qquad \text{(Convention V)}$$

$$K'_{eq_4} = \frac{3.21 \times 10^{-4}}{10^{-7}} = 3.21 \times 10^3$$

$$\Delta G'_4 = -1364 \log 3.21 \times 10^3 = -4783 \text{ cal/mole} \quad \text{(at pH 7.0)}$$

As before, if the equilibrium [EtOH] and [EtOAc] are known at pH 7, the equilibrium [OAc$^-$] can be calculated using K'_{eq_4}. The same [OAc$^-$] results if K_{eq_4} is used provided the calculations include [H$^+$]. The negative $\Delta G'_4$ says that at $1\,M$ EtOH, $1\,M$ OAc$^-$, and $1\,M$ EtOAc (and $55.6\,M$ H$_2$O and $10^{-7}\,M$ H$^+$) the reaction will proceed to the right.

The relationship between $\Delta G'_4$ and ΔG^0_4 is one we have seen before:

$$\Delta G'_4 = -1364 \log \frac{[\text{EtOH}][\text{OAc}^-]}{[\text{EtOAc}]} + 1364 \log [\text{H}^+]$$

$$\boxed{\Delta G'_4 = \Delta G^0_4 - 1364\,\text{pH}} \qquad (25)$$

To reiterate: the fact that some of the ΔG° values calculated above are positive and some are negative is not contradictory. Each standard ΔG value refers to a different state, for example, $1\,M$ H$_2$O (ΔG^0_1), $55.6\,M$ H$_2$O (ΔG^0_2), $1\,M$ H$_2$O and $1\,M$ H$^+$ (ΔG^0_3), $55.6\,M$ H$_2$O and $1\,M$ H$^+$ (ΔG^0_3), $55.6\,M$ H$_2$O and $10^{-7}\,M$ H$^+$ (ΔG^0_4). If we compare the signs and relative magnitudes of the calculated standard ΔG values, we see that the standard ΔG becomes more negative as [H$_2$O] increases and [H$^+$] decreases. This is expected: high [H$_2$O] promotes hydrolysis; low [H$^+$] (high pH) promotes ionization of the HOAc, which pulls the hydrolysis reaction to the right.

At first glance, it would seem that K'_{eq_4} and $\Delta G'_4$ are the values of interest to biochemists. However, these constants are defined in terms of OAc$^-$ concentration. A biochemist studying the esterase reaction at pH 7 would probably find it more convenient to measure the total (analytical) concentration of acetate product (OAc$^-$ *plus* HOAc). Indeed, many compounds of biological interest possess multiple ionizing groups, yielding several ionic species at any given pH. Also, two or more substrates or products may be present in multiple ionic forms. A K_{eq} or $\Delta G'$ value defined in terms of one ionic form of each compound would not be particularly useful. For example, consider the hydrolysis of ATP:

$$\text{ATP} + \text{H}_2\text{O} \rightleftharpoons \text{ADP} + \text{P}_i$$

At pH values around 6 to 7.5, the inorganic phosphate (p$K_a = 6.82$) exists as a mixture of HO-PO$_3$H$^-$ and HO-PO$_3^{2-}$ forms. Similarly, the terminal phosphates of ATP (p$K_a = 6.95$) and ADP (p$K_a = 6.68$) exist as mixtures of the mononegative and dinegative ions. It would be much simpler to define K'_{eq} in terms of total [ATP], [ADP], and [P$_i$]. Let us return to our esterase reaction in order to see how total analytical concentrations are incorporated into the K'_{eq} and $\Delta G'$ expressions. We would like to determine the value for K'_{eq}, where:

$$\boxed{K'_{eq} = \frac{[\text{EtOH}][\text{HOAc} + \text{OAc}^-]}{[\text{EtOAc}]} = \frac{[\text{EtOH}][\text{OAc}]_t}{[\text{EtOAc}]}} \qquad \text{(Convention VI)}$$

where [OAc]$_t$ = total acetate = [HOAc] + [OAc$^-$]

We know that:

$$[OAc]_t = [HOAc] + [OAc^-]$$

$$= [HOAc] + \frac{K_a[HOAc]}{[H^+]}$$

$$= [HOAc]\left(1 + \frac{K_a}{[H^+]}\right)$$

Substituting for $[OAc]_t$ in the expression for K'_{eq}:

$$K'_{eq} = \frac{[EtOH][OAc]_t}{[EtOAc]} = \frac{[EtOH][HOAc]}{[EtOAc]}\left(1 + \frac{K_a}{[H^+]}\right)$$

$$\therefore \quad K'_{eq} = K_{eq_2}\left(1 + \frac{K_a}{[H^+]}\right)$$

At pH 7.0:

$$K'_{eq} = (18.35)\left(1 + \frac{1.75 \times 10^{-5}}{10^{-7}}\right) = (18.35)(176)$$

$$K'_{eq} = 3.23 \times 10^3$$

$$\Delta G' = -1364 \log 3.23 \times 10^3 = -4787 \text{ cal/mole}$$

K'_{eq} and $\Delta G'$ are very close to K'_{eq_4} and $\Delta G'_4$, respectively. This is not unexpected since at pH 7.0, $[OAc^-] \simeq [OAc]_t$. At a lower pH, where all the acetate is not present almost exclusively in one ionic form, the two conventions would yield significantly different values.

Let us backtrack a little and examine the effect of ionization on the standard-state ΔG values. At pH 0 (i.e., $1 M$ H^+), the standard-state ΔG is given by ΔG_2^0, which we shall simply call ΔG°. (From now on the activity of water will be assumed to be unity—that is, all constants have the $55.6 M$ H_2O built in.) At pH 7.0 (i.e., $10^{-7} M$ H^+), the standard-state ΔG is given by $\Delta G'$. Thus:

$$\Delta G^\circ = -1364 \log K_{eq_2} \qquad \Delta G' = -1364 \log K_{eq_2}\left(1 + \frac{K_a}{[H^+]}\right)$$

The difference between the standard ΔG values at the two states is:

$$\Delta G' - \Delta G^\circ = -1364 \log K_{eq_2}\left(1 + \frac{K_a}{[H^+]}\right) + 1364 \log K_{eq_2}$$

$$= -1364 \log K_{eq_2} - 1364 \log\left(1 + \frac{K_a}{[H^+]}\right) + 1364 \log K_{eq_2}$$

$$= -1364 \log\left(1 + \frac{K_a}{[H^+]}\right)$$

or

$$\boxed{\Delta G' = \Delta G^\circ - 1364 \log\left(1 + \frac{K_a}{[H^+]}\right)}$$

(26)

or

$$\Delta G' = \Delta G° + \Delta G_{ion} \qquad (27)$$

where

$$\Delta G_{ion} = -1364 \log \left(1 + \frac{K_a}{[H^+]}\right) \qquad (28)$$

ΔG_{ion} is the additional free energy change per mole resulting from the ionization of a compound *originally present* at $1\,M$ concentration. (After ionization, the *total* concentration of the compound, for example, $HOAc + OAc^-$, is still $1\,M$.)   There seems to be a contradiction here.   Earlier, we saw that the $\Delta G'$ for a reaction that yields H^+ as a product is given by Equation 19 or 25, which is not the same as Equation 26 derived above.   The difference between the two expressions stems from the difference in standard states. Equation 19 or 25 is valid for a standard-state where all reactants (except H_2O and H^+) are $1\,M$.   This includes $[OAc^-] = 1\,M$.   Equation 28 does not assume that $[OAc^-] = 1\,M$ but, instead, that the *total*, $[HOAc] + [OAc^-]$, is $1\,M$.   Since most concentrations that biochemists deal with are *total* (analytical) concentrations, Equation 28 is the one used to predict the effect of a change in pH on the standard-state ΔG when ionizable groups are present.   Again, it should be stressed that if K'_{eq} or $\Delta G'$ is given at a fixed pH, and concentrations include all ionic forms of a given component, then it is not necessary to include H^+ or H_2O in the calculations.   Equations 23 or 24 can then be used directly to calculate ΔG values for nonstandard conditions (total $[S] \neq 1\,M$, total $[P] \neq 1\,M$).

· Problem 3-1

Calculate the standard-state ΔG values at (a) pH 0 and (b) pH 5 for the dissociation of acetic acid: $HOAc \rightleftharpoons OAc^- + H^+$. $K_a = 1.75 \times 10^{-5}$.   (c) Calculate ΔG_{ion} at pH 5.0.

Solution

(a)
$$\Delta G° = -1364 \log K_a$$
$$= -1364 \log (1.75 \times 10^{-5}) = -(1364)(-4.76)$$

$$\boxed{\Delta G° = +6488 \text{ cal/mole}}$$

That is, when $[HOAc] = 1\,M$, $[OAc^-] = 1\,M$, and $[H^+] = 1\,M$, the reaction proceeds from right to left because the $[H^+][OAc^-]/[HOAc]$ ratio is $> K_a$.

(b)
$$\Delta G' = \Delta G° + 1364 \log [H^+]$$
$$= +6488 + 1364 \log (10^{-5}) = +6488 + 1364(-5)$$
$$= +6488 - 6820$$

$$\boxed{\Delta G' = -332 \text{ cal/mole}}$$

That is, when $[HOAc] = 1\ M$, $[OAc^-] = 1\ M$, and $[H^+] = 10^{-5}\ M$, the reaction proceeds from left to right because the $[H^+][OAc^-]/[HOAc]$ ratio is $< K_a$.

(c)
$$\Delta G_{ion} = -1364 \log \left(1 + \frac{K_a}{[H^+]}\right)$$
$$= -1364 \log \left(1 + \frac{1.75 \times 10^{-5}}{10^{-5}}\right) = -1364 \log (1 + 1.75)$$
$$= -1364 \log 2.75$$

$$\boxed{\Delta G_{ion} = -599\ \text{cal/mole}}$$

That is, when 1 mole of HOAc is added to 1 liter of a solution buffered at pH 5, the HOAc will ionize yielding 599 cal.   The total acetate concentration remains 1 M.

· **Problem 3-2**

The hydrolysis of ATP at pH 7 can be written as:

(a) The pK_a of the newly formed ionizable group on ADP is 6.68 at 25°C ($K_a = 2.09 \times 10^{-7}$).   Of the total $\Delta G'$ of -7700 cal/mole, how much can be attributed to the ionization of the ADP?   (b) The terminal phosphate of ATP has pK_a values of 6.95 and 2.3.   The inorganic phosphate derived from the terminal phosphate has pK_a values of 12.5, 6.82, and 2.3.   Does the ionization of the inorganic phosphate contribute to the total $\Delta G'$?

Solution

(a)
$$\Delta G_{ion} = -1364 \log \left(1 + \frac{K_a}{[H^+]}\right) = -1364 \log (1 + 2.09)$$
$$= -1364 \log 3.09 = -1364(0.490)$$

$$\boxed{\Delta G_{ion} = -668\ \text{cal/mole}}$$

(b)   The strongest acid group in the terminal phosphate of ATP has the same pK_a after hydrolysis as before hydrolysis (2.3).   Therefore, the ionization of this group does not contribute to the $\Delta G'$.   The newly formed acid group has a pK_a of 12.5.   At pH 7, it is essentially un-ionized.   Therefore this group contributes nothing to the $\Delta G'$.   The remaining acid group becomes slightly stronger as a result of hydrolysis ($pK_a = 6.95$ or $K_a = 1.12 \times 10^{-7}$ becomes

$pK_a = 6.82$ or $K_a = 1.51 \times 10^{-7}$). The contribution of the further ionization of this group can be calculated from Equation 29:

$$\Delta G_{ion} = -1364 \log \frac{\left(1 + \dfrac{K_{a(P_i)}}{[H^+]}\right)}{\left(1 + \dfrac{K_{a(ATP)}}{[H^+]}\right)} \tag{29}$$

$$= -1364 \log \frac{2.51}{2.12}$$

$$\boxed{\Delta G_{ion} = -100 \text{ cal/mole}}$$

· Problem 3-3

Glucose-6-phosphate was hydrolyzed enzymatically (at pH 7 and 25°C) to glucose and inorganic phosphate. The concentration of glucose-6-phosphate was 0.1 M at the start. At equilibrium, only 0.05% of the original glucose-6-phosphate remained. Calculate (a) K'_{eq} for the hydrolysis of glucose-6-phosphate, (b) $\Delta G'$ for the hydrolysis reaction, (c) K'_{eq} for the reaction by which glucose-6-phosphate is synthesized from inorganic phosphate and glucose, and (d) $\Delta G'$ for the synthesis reaction.

Solution

(a) The equation for the hydrolysis reaction is glucose-6-phosphate + $H_2O \rightleftharpoons$ glucose + P_i. The K'_{eq} for the reaction at unit activity of water is given by:

$$K'_{eq} = \frac{[glucose][P_i]}{[glucose-6-P]}$$

where $[P_i] = [HPO_4^{2-} + H_2PO_4^-]$,

and $[glucose-6-P] = [glucose-6-OPO_3^{2-} + glucose-6-OPO_3H^-]$

At equilibrium:

$$[glucose-6-P] = (0.05\%)(0.10\ M) = (5 \times 10^{-4})(1 \times 10^{-1})$$
$$= 5 \times 10^{-5}\ M$$
$$[glucose] = (99.95\%)(0.10\ M) = (99.95 \times 10^{-2})(1 \times 10^{-1})$$
$$= 99.95 \times 10^{-3}\ M$$
$$[P_i] = [glucose] \quad \therefore \quad [P_i] = 99.95 \times 10^{-3}\ M$$
$$K'_{eq} = \frac{(99.95 \times 10^{-3})(99.95 \times 10^{-3})}{(5 \times 10^{-5})} = \frac{9.99 \times 10^{-3}}{5 \times 10^{-5}}$$

$$\boxed{K'_{eq} = 199.8}$$

(b)
$$\Delta G' = -1364 \log K'_{eq} = -1364 \log 199.8$$
$$= -(1364)(2.301)$$

$$\boxed{\Delta G' = -3138 \text{ cal/mole}}$$

In other words, under standard-state conditions in which the concentrations of glucose-6-phosphate, glucose, and P_i are all maintained at a steady-state level of unit activity, the conversion of 1 mole of glucose-6-phosphate to 1 mole of glucose and 1 mole of P_i liberates 3138 cal.

(c)   The equilibrium constant for a reaction $A \rightarrow B$ is the reciprocal of the equilibrium constant for the reaction $B \rightarrow A$.

1.
$$A \rightarrow B \qquad K'_{eq_1} = \frac{[B]}{[A]}$$

2.
$$B \rightarrow A \qquad K'_{eq_2} = \frac{[A]}{[B]} = \frac{1}{[B]/[A]} = \frac{1}{K'_{eq_1}}$$

Therefore, for the reaction

$$\text{glucose} + P_i \rightleftharpoons \text{glucose-6-P}$$

$$K'_{eq} = \frac{[\text{glucose-6-P}]}{[\text{glucose}][P_i]} = \frac{1}{199.8} = 5 \times 10^{-3}$$

(d)   If the *hydrolysis* of glucose-6-phosphate *yields* 3138 cal/mole, the *synthesis* of glucose-6-phosphate *requires* 3138 cal/mole.

$$\therefore \quad \boxed{\Delta G' = +3138 \text{ cal/mole}}$$

$\Delta G'$ can also be calculated from K'_{eq}.

$$\Delta G' = -1364 \log K'_{eq} = -1364 \log 5 \times 10^{-3}$$
$$\Delta G' = -1364(\log 5 + \log 10^{-3}) = -1364(0.699 - 3)$$

$$\Delta G' = -1364(-2.301) \qquad \boxed{\Delta G' = +3138 \text{ cal/mole}}$$

· Problem 3-4

Calculate the ΔG for the hydrolysis of ATP at pH 7 and 25°C under steady-state conditions (such as might exist in a living cell) in which the concentrations of ATP, ADP, and P_i are maintained at 10^{-3} M, 10^{-4} M, and 10^{-2} M, respectively.

Solution

The equation for the ΔG of the hydrolysis under nonstandard-state conditions is:

$$\Delta G = \Delta G' + 1364 \log \frac{[ADP][P_i]}{[ATP]}$$

$$= -7700 + 1364 \log \frac{(10^{-4})(10^{-2})}{(10^{-3})}$$

$$= -7700 + 1364 \log 10^{-3} = -7700 + 1364(-3)$$

$$= -7700 - 4092 \qquad \boxed{\mathbf{\Delta G = -11,792 \ cal/mole}}$$

ADDITION OF ΔG VALUES FOR COUPLED REACTIONS

Earlier, we saw that if the conversion of A to B releases 15 kcal while the synthesis of X ~ Y utilizes 8 kcal, the coupled, or overall, reaction releases 7 kcal. It is clear that the energy values are additive:

1. $\qquad\qquad$ $A \rightarrow B$ $\qquad$ $\Delta G_1' = -15 \ kcal/mole$

2. $\qquad\qquad$ $X + Y \rightarrow X \sim Y$ $\qquad$ $\Delta G_2' = +8 \ kcal/mole$

---

3. (Sum) $\qquad$ $A + X + Y \rightarrow B + X \sim Y$ $\qquad$ $\Delta G_3' = -7 \ kcal/mole$

The K_{eq}' values are multiplied:

1. $\qquad\qquad$ $A \rightarrow B$ $\qquad$ $K_{eq_1}' = 9.932 \times 10^{10}$

2. $\qquad\qquad$ $X + Y \rightarrow X \sim Y$ $\qquad$ $K_{eq_2}' = 1.364 \times 10^{-6}$

---

3. (Sum) $\qquad$ $A + X + Y \rightarrow B + X \sim Y$ $\qquad$ $K_{eq_3}' = 13.55 \times 10^4$

The additive nature of energy quantities allows us to calculate unknown $\Delta G'$ values if the reaction in question can be expressed as the sum of two or more reactions whose $\Delta G'$ values are known. This is illustrated in Problem 3-5.

· Problem 3-5

The $\Delta G'$ of hydrolysis of ATP at pH 7 and 25°C is $-7700 \ cal/mole$. As shown in Problem 3-2, the $\Delta G'$ of hydrolysis of glucose-6-phosphate at pH 7 and 25°C is $-3138 \ cal/mole$. From this information, calculate the $\Delta G'$ and K_{eq}' for the reaction between glucose and ATP catalyzed by hexokinase.

Solution

Given that:

1. Glucose-6-phosphate + $H_2O \leftrightarrows$ glucose + P_i $\qquad$ $\Delta G' = -3138 \ cal/mole$
$\qquad\qquad\qquad\qquad\qquad\qquad\qquad\qquad\qquad\qquad\qquad K_{eq}' = 199.8$

We can immediately write:

2. glucose + $P_i \rightleftharpoons$ glucose-6-phosphate + H_2O $\qquad$ $\Delta G' = +3138 \ cal/mole$
$\qquad\qquad\qquad\qquad\qquad\qquad\qquad\qquad\qquad\qquad\qquad\qquad K_{eq}' = 5 \times 10^{-3}$

We also know that:

3. $\qquad$ $ATP + H_2O \rightleftharpoons ADP + P_i$ $\qquad$ $\Delta G' = -7700$ cal/mole

$\qquad\qquad\qquad\qquad\qquad\qquad\qquad$ $\therefore$ $K'_{eq} = 4.42 \times 10^5$

The reaction of interest is:

4. $\qquad$ glucose $+ ATP \rightleftharpoons$ glucose-6-phosphate $+ ADP$

Reaction 4 can be expressed as the sum of reactions 2 and 3:

2. glucose $+ P_i \rightleftharpoons$ glucose-6-phosphate $+ H_2O$ $\qquad$ $\Delta G'_2 = +3138$ cal/mole

3. $ATP + H_2O \rightleftharpoons ADP + P_i$ $\qquad\qquad\qquad\qquad$ $\Delta G'_3 = -7700$ cal/mole

---

4. (Sum) $\qquad$ glucose $+ ATP \rightleftharpoons$ glucose-6-phosphate $+ ADP$

$\qquad\qquad\qquad$ $\Delta G'_4 = \Delta G'_2 + \Delta G'_3 = +3138 - 7700$

$$\boxed{\Delta G'_4 = -4562 \text{ cal/mole}}$$

$\qquad\qquad$ $K'_{eq_4} = (K'_{eq_2})(K'_{eq_3}) = (5 \times 10^{-3})(4.42 \times 10^5)$

$$\boxed{K'_{eq_4} = 2.21 \times 10^3}$$

or $\qquad\qquad\qquad$ $\Delta G'_4 = -1364 \log K'_{eq_4} = -4562$

$\qquad\qquad$ $\log K'_{eq_4} = \dfrac{4562}{1364} = 3.3445$ $\qquad$ $\therefore$ $\boxed{K'_{eq_4} = 2.21 \times 10^3}$

· Problem 3-6

Calculate the overall K'_{eq} and $\Delta G'$ at pH 7 and 25°C for the conversion of fumaric acid to citric acid in the presence of the appropriate enzymes, cosubstrates, and cofactors.

Solution

In the previous problem we dealt with a *single reaction* that *could be expressed* as the sum of two or more coupled or consecutive reactions.   In this problem we are dealing with three actual consecutive reactions.   The rules for calculating K'_{eq} and $\Delta G'$ are the same.

The enzyme-catalyzed reactions by which fumaric acid is converted to citric acid are shown below.

1. $\qquad$ fumarate $+ H_2O \xrightleftharpoons{\text{fumarase}}$ malate $\qquad$ $K'_{eq_1} = 4.5$

2. $\qquad$ malate $+ NAD^+ \xrightleftharpoons{\text{malic dehydrogenase}}$ oxalacetate $+ NADH + H^+$

$\qquad\qquad\qquad\qquad\qquad\qquad\qquad\qquad\qquad\qquad\qquad$ $K'_{eq_2} = 1.3 \times 10^{-5}$

3.   oxalacetate + acetyl CoA + H_2O $\xrightleftharpoons{\text{citrate synthetase}}$ citrate + CoASH

$$K'_{eq_3} = 3.2 \times 10^5$$

---

4. (Sum)

   fumarate + $2H_2O$ + acetyl CoA + NAD^+ $\rightleftharpoons$ citrate + NADH + H^+ + CoASH

$$K'_{eq_4} = \frac{[\text{citrate}][\text{NADH}][\text{CoASH}]}{[\text{fumarate}][\text{acetyl CoA}][\text{NAD}^+]}$$

$$K'_{eq_4} = K'_{eq_1} \times K'_{eq_2} \times K'_{eq_3}$$
$$= (4.5)(1.3 \times 10^{-5})(3.2 \times 10^5)$$

$$\boxed{K'_{eq_4} = 18.72}$$

$$\Delta G'_4 = -1364 \log K'_{eq_4} = -1364(1.272)$$

$$\boxed{\Delta G'_4 = -1735 \text{ cal/mole}}$$

We can see that the *overall* conversion of fumarate to citrate is favorable in spite of reaction 2 with its low K'_{eq}. This problem and those preceding it illustrate some general rules and principles summarized below.

General Principles

1.   The overall K'_{eq} for any number of consecutive reactions, 1, 2, 3, 4, ... etc., is $K'_{eq_1} \times K'_{eq_2} \times K'_{eq_3} \times K'_{eq_4}$ ... etc.

2.   The overall $\Delta G'$ for any number of consecutive reactions, 1, 2, 3, 4, ... etc., is $\Delta G'_1 + \Delta G'_2 + \Delta G'_3 + \Delta G'_4$ ... etc.   The $\Delta G'_{\text{overall}}$ can also be calculated from $K'_{eq_{\text{overall}}}$

$$\Delta G'_{\text{overall}} = -2.3 \, RT \log K'_{eq_{\text{overall}}}$$

3.   The K'_{eq} for a single reaction that can be expressed as the sum of two or more consecutive reactions, 1, 2, 3, ... etc., is $K'_{eq_1} \times K'_{eq_2} \times K'_{eq_3}$ ... etc.   Similarly, the $\Delta G'$ for a single reaction that can be expressed as the sum of two or more consecutive reactions is $\Delta G'_1 + \Delta G'_2 + \Delta G'_3$ ... etc.

$$\Delta G'_{\text{overall}} = -2.3 \, RT \log K'_{eq_{\text{overall}}}$$

Some further examples will illustrate these principles.

· Problem 3-7

The cleavage of citrate to acetate and oxalacetate has a $\Delta G'$ of -680 cal/mole.   The K'_{eq} of the citrate synthetase reaction is 3.2×10^5.   From this information, calculate the standard free energy of hydrolysis of acetyl-S-CoA and the K'_{eq} for the hydrolysis.

Solution

The two reactions given are shown below.

1. $\qquad$ citrate $\underset{\text{citrate lyase}}{\rightleftharpoons}$ acetate + oxalacetate $\qquad \Delta G_1' = -680$ cal/mole

2. acetyl-S-CoA + oxalacetate + $H_2O \underset{\text{citrate synthetase}}{\rightleftharpoons}$ citrate + CoASH

$$K_{eq_2}' = 3.2 \times 10^5$$

We can work the problem in terms of $\Delta G'$ or K_{eq}' values. First calculate the missing values.

$$\Delta G_1' = -1364 \log K_{eq_1}' = -680 \text{ cal/mole}$$

$$\log K_{eq_1}' = \frac{-680}{-1364} = 0.4985$$

$$K_{eq_1}' = \text{antilog of } 0.4985$$

$$\boxed{K_{eq_1}' = 3.15}$$

$$\Delta G_2' = -1364 \log K_{eq_2}'$$
$$= -1364 \log 3.2 \times 10^5 = -1364(\log 3.2 + \log 10^5)$$
$$= -1364(0.505 + 5) = -1364(5.505)$$

$$\boxed{\mathbf{\Delta G_2' = -7509 \text{ cal/mole}}}$$

The reaction we are interested in is the hydrolysis of acetyl-S-CoA:

3. $\qquad$ acetyl-S-CoA + $H_2O \rightleftharpoons$ acetate + CoASH

Can this reaction be expressed in terms of the ones given? We can see that reaction 3 is the sum of reactions 1 and 2.

1. citrate $\rightleftharpoons$ acetate + oxalacetate $\qquad \Delta G_1' = -680$ cal/mole $\qquad K_{eq_1}' = 3.15$

2. acetyl-S-CoA + oxalacetate + $H_2O \rightleftharpoons$ citrate + CoASH
$$\Delta G_2' = -7509 \text{ cal/mole} \qquad K_{eq_2}' = 3.2 \times 10^5$$

---

3. (Sum) $\qquad$ acetyl-S-CoA + $H_2O \rightleftharpoons$ acetate + CoASH

$$\Delta G_3' = \Delta G_1' + \Delta G_2'$$
$$= (-680) + (-7509)$$

$$\boxed{\mathbf{\Delta G_3' = -8189 \text{ cal/mole}}}$$

$$\Delta G_3' = -1364 \log K_{eq_3}'$$
$$-8189 = -1364 \log K_{eq_3}'$$

$$\frac{-8189}{-1364} = \log K_{eq_3}' = 6$$

$$\boxed{K_{eq_3}' = 10^6}$$

$$K_{eq_3}' = K_{eq_1}' \times K_{eq_2}'$$
$$= (3.15)(3.2 \times 10^5)$$

$$\boxed{\mathbf{K_{eq_3}' = 10.08 \times 10^5}}$$

$$\Delta G_3' = -1364 \log K_{eq_3}'$$
$$= -1364 \log 10.08 \times 10^5$$

$$\Delta G_3' = -1364(6.003)$$

$$\boxed{\mathbf{\Delta G_3' = -8189 \text{ cal/mole}}}$$

We could also determine the $\Delta G'$ of hydrolysis of acetyl-S-CoA by a different line of reasoning.   If the *cleavage* of citrate to acetate and oxalacetate *liberates* 680 cal/mole, then the *synthesis* of citrate from acetate and oxalacetate *requires* 680 cal/mole.   When the synthesis is carried out from acetyl-S-CoA (an "activated" form of acetate), 7509 cal/mole are released.   The acetyl-S-CoA then must have contained sufficient energy to form the new carbon-carbon bond (680 cal/mole) *plus* have 7509 cal/mole left over; that is, the acetyl-S-CoA is worth 680 *plus* 7509 = 8189 cal/mole in terms of "group-transfer potential" (or "free energy of hydrolysis").   The citrate synthetase reaction may be thought of as the sum of two intimately coupled reactions:

$$\text{acetate} + \text{oxalacetate} + 680\ \text{cal} \rightleftharpoons \text{citrate}$$

$$\text{acetyl-S-CoA} + H_2O \rightleftharpoons \text{acetate} + \text{CoASH} + 8189\ \text{cal}$$

Sum: $\text{acetyl-S-CoA} + H_2O + \text{oxalacetate} \rightleftharpoons \text{citrate} + \text{CoASH} + 7509\ \text{cal}$

The situation is analogous to the hexokinase reaction.   An energy-rich compound provides both the energy to drive an endergonic condensation as well as the particular group that is transferred to an acceptor.

General Principle

Endergonic reactions may be driven toward completion by coupling them to highly exergonic reactions.   The coupling may be intimate so that the overall coupled reaction appears as a single step (e.g., the hexokinase reaction or the citrate synthetase reaction), or the coupling may take place in two or more consecutive steps (e.g., the fumarate → citrate sequence).   In sequential reactions, we can think of a subsequent exergonic reaction as removing the product of a preceding endergonic reaction as it is formed, thereby driving the overall sequence to the right.

· **Problem 3-8**

Estimate the $\Delta G'$ values for the following reactions: (a) ATP + GDP $\rightleftharpoons$ GTP + ADP,   (b) glycerol + ATP $\rightleftharpoons$ α-glycerophosphate + ADP,   and (c) 3-phosphoglycerate + ATP $\rightleftharpoons$ 1,3-diphosphoglycerate + ADP.

Solution

(a)   In this reaction the energy (group-transfer potential) of ATP is utilized to transfer its terminal phosphate to GDP.   The product (GTP) is itself "energy rich"; in fact the group-transfer potential of GTP is as high as that of ATP.   Thus, $\Delta G' = 0$ and $K'_{eq} = 1$.   We can verify these results mathematically by considering the reaction as the sum of two component reactions.

1. $\text{ATP} + H_2O \rightleftharpoons \text{ADP} + P_i + 7700\ \text{cal}$ or $\Delta G'_1 = -7700\ \text{cal/mole}$

2. $\text{GDP} + P_i + 7700\ \text{cal} \rightleftharpoons \text{GTP} + H_2O$ or $\Delta G'_2 = +7700\ \text{cal/mole}$

---

3. (Sum) $\text{ATP} + \text{GDP} \rightleftharpoons \text{GTP} + \text{ADP}$ $\boxed{\Delta G'_3 = 0}$

(b)   In this reaction the energy of ATP is used to transfer its terminal phosphate to glycerol to form an "energy-poor" phosphate ester.   The $\Delta G'$ of hydrolysis of α-glycerophosphate is about -2000 cal/mole while that of ATP (terminal phosphate) is about -7700 cal/mole.   Thus, of the original 7700 cal, only 2000 cal are conserved.   The $\Delta G'$ of the reaction is the difference between -7700 cal and -2000 cal, or -5700 cal/mole.   We can verify this mathematically by considering the reaction as the sum of two component reactions.

1. $ATP + H_2O \rightleftharpoons ADP + P_i + 7700$ cal

$$\text{or} \quad \Delta G_1' = -7700 \text{ cal/mole}$$

2.   glycerol $+ P_i + 2000$ cal $\rightleftharpoons \alpha$-glycerophosphate

$$\text{or} \quad \Delta G_2' = +2000 \text{ cal/mole}$$

3. (Sum) $ATP +$ glycerol $\rightleftharpoons ADP + \alpha$-glycerophosphate $+ 5700$ cal

$$\boxed{\Delta G_3' = -5700 \text{ cal/mole}}$$

(c)   The product of this reaction, 1,3-diphosphoglycerate, is more "energy-rich" than ATP ($\Delta G'$ of hydrolysis $= -12,000$ cal/mole).   Thus, the reaction as written is endergonic.   We can consider the reaction as the sum of two component reactions.

1. $ATP + H_2O \rightleftharpoons ADP + P_i + 7700$ cal

$$\text{or} \quad \Delta G_1' = -7700 \text{ cal/mole}$$

2.   3-PGA $+ P_i + 12,000$ cal $\rightleftharpoons$ 1,3-DiPGA $+ H_2O$

$$\text{or} \quad \Delta G_2' = +12,000 \text{ cal/mole}$$

3. (Sum) $ATP +$ 3-PGA $+ 4300$ cal $\rightleftharpoons ADP +$ 1,3-DiPGA

$$\text{or} \quad \boxed{\Delta G_3' = +4300 \text{ cal/mole}}$$

· Problem 3-9

The ATP/ADP ratio in an actively respiring yeast cell is about 10.   What should the intracellular 3-phosphoglycerate/1,3-diphosphoglycerate ratio be to make the phosphoglycerate kinase reaction thermodynamically favorable in the direction of 1,3-diphosphoglycerate synthesis?

Solution

The reaction catalyzed by phosphoglycerate kinase is:

$$ATP + \text{3-PGA} \underset{\text{phosphoglycerate kinase}}{\rightleftharpoons} ADP + \text{1,3-DiPGA}$$

$$\Delta G' = +4300 \text{ cal/mole} \qquad K_{eq}' = 7.039 \times 10^{-4}$$

The positive $\Delta G'$ value indicates that under *standard-state* conditions the reaction is endergonic in the direction of 1,3-DiPGA synthesis ($\Delta G' = 4300$ cal/mole).   In other words, if the concentrations of the four components of the reaction are maintained at steady-state levels of 1 M, the reaction goes spontaneously in the direction of ATP and 3-PGA formation with the liberation of 4300 cal/mole.   In a living cell, however, the concentrations of ATP, ADP, 3-PGA, and 1,3-DiPGA are not maintained at 1 M.   The reaction can be made to proceed spontaneously in the direction of 1,3-DiPGA and ADP formation if the concentrations of the components are maintained at suitable levels.   Qualitatively, we know that the reaction can be made to go spontaneously from left to right if the concentrations of ATP and 3-PGA are increased sufficiently, or if the concentrations of ADP and 1,3-DiPGA are decreased sufficiently, or a combination of both.   All we need do is calculate the [3-PGA]/[1,3-DiPGA] ratio that would make $\Delta G = 0$ at an [ATP]/[ADP] ratio of 10.   Alternatively, we could calculate the [3-PGA]/[1,3-DiPGA] ratio that would be at equilibrium with the [ATP]/[ADP] ratio of 10.

$$K'_{eq} = \frac{[\text{ADP}][1,3\text{-DiPGA}]}{[\text{ATP}][3\text{-PGA}]} = 7.039 \times 10^{-4}$$

$$\frac{[3\text{-PGA}]}{[1,3\text{-DiPGA}]} = \frac{[\text{ADP}]}{[\text{ATP}]K'_{eq}} = \frac{(1)}{(10)(7.039 \times 10^{-4})} = \boxed{142}$$

Thus, when the ratio of ATP/ADP is 10 and the ratio of 3-PGA/1,3-DiPGA is 142, the phosphoglycerate kinase reaction would be at equilibrium.   Any slight increase in either ratio would force the reaction in the direction of ADP and 1,3-DiPGA formation—it would make ΔG a negative value.   This problem illustrates another general principle.

General Principle

The $\Delta G'$ (or K'_{eq}) values provide a convenient way to classify and tabulate various kinds of reactions but they do not indicate the direction in which a reaction proceeds in a living cell.   The spontaneous direction *in vivo* (the nonstandard-state ΔG value) depends on the intracellular concentrations (activities) of the reaction components.

B. CALCULATIONS OF EQUILIBRIUM CONCENTRATIONS

· Problem 3-10

The K'_{eq} for the fructose-1,6-diphosphate aldolase reaction at 25°C and pH 7, (written in the direction of triose phosphate formation) is about 10^{-4} M. $\Delta G' = +5456$ cal/mole.   Calculate the concentrations of fructose-1,6-diphosphate (FDP), dihydroxyacetone phosphate (DHAP), and glyceraldehyde-3-phosphate (GAP) at equilibrium when the initial FDP concentration is (a) 1 M, (b) 10^{-2} M, (c) 2×10^{-4} M, and (d) 10^{-5} M.

Solution

The reaction catalyzed by FDP-aldolase is shown below.

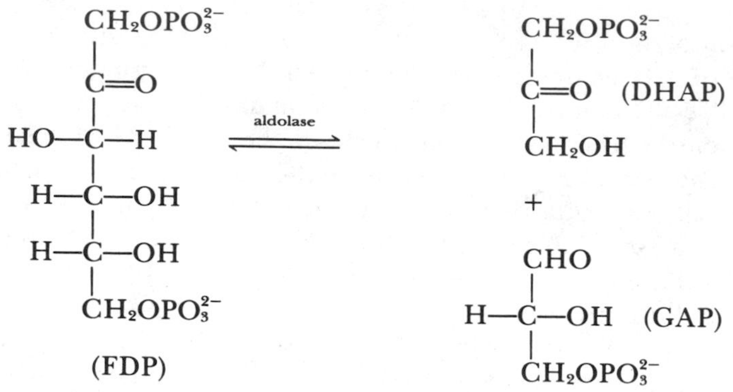

$$
\begin{array}{l}
CH_2OPO_3^{2-} \\
| \\
C{=}O \\
| \\
HO{-}C{-}H \\
| \\
H{-}C{-}OH \\
| \\
H{-}C{-}OH \\
| \\
CH_2OPO_3^{2-} \\
\\
(FDP)
\end{array}
\quad \xrightleftharpoons[]{\text{aldolase}} \quad
\begin{array}{l}
CH_2OPO_3^{2-} \\
| \\
C{=}O \quad (DHAP) \\
| \\
CH_2OH \\
\\
+ \\
\\
CHO \\
| \\
H{-}C{-}OH \quad (GAP) \\
| \\
CH_2OPO_3^{2-}
\end{array}
$$

(a)   Let

$$y = M \text{ FDP that disappears}$$
$$\therefore \quad y = M \text{ DHAP produced}, \; y = M \text{ GAP produced}$$

and

$$(1 - y) = M \text{ FDP remaining}$$

$$FDP \rightleftharpoons DHAP + GAP$$

	FDP	DHAP	GAP
Start:	1	0	0
Change:	$-y$	$+y$	$+y$
Equilibrium:	$1-y$	y	y

$$K'_{eq} = \frac{(y)(y)}{(1-y)} = 10^{-4}$$

First calculate y, assuming y is small compared to $1\,M$.   Therefore, y can be ignored in the denominator.

$$K'_{eq} = \frac{(y)(y)}{(1)} = 10^{-4} \qquad y^2 = 10^{-4} \qquad y = 10^{-2}$$

$\therefore$ At equilibrium:

$$\boxed{\begin{array}{l} \textbf{[DHAP]} = \mathbf{10^{-2}}\,\boldsymbol{M} \\ \textbf{[GAP]} = \mathbf{10^{-2}}\,\boldsymbol{M} \\ \textbf{[FDP]} = \mathbf{99 \times 10^{-2}}\,\boldsymbol{M} \end{array}}$$

Our assumption that y is small compared to $1\,M$ is valid.   We can see that at high initial FDP concentrations the reaction comes to equilibrium when the concentrations of DHAP and GAP are small compared to FDP.

(b)   When the initial FDP concentration is $0.01\,M$:

$$K'_{eq} = \frac{(y)(y)}{(0.01 - y)} = 10^{-4}$$

Again assume that y is small compared to $0.01\ M$.

$$K'_{eq} = \frac{(y)(y)}{(0.01)} = 10^{-4} \qquad y^2 = 10^{-6} \qquad y = 10^{-3}$$

$\therefore$ At equilibrium:

$$\boxed{\begin{aligned} &[\text{DHAP}] = 10^{-3}\ M \\ &[\text{GAP}] = 10^{-3}\ M \\ &[\text{FDP}] = 9 \times 10^{-3}\ M \end{aligned}}$$

Under these conditions y is about 10% of [FDP].   Thus, the elimination of y from the denominator still gives a reasonably correct answer.   (The quadratic solution yields $y = 9.5 \times 10^{-4}\ M$.)

(c)   When the initial FDP concentration is $2 \times 10^{-4}\ M$:

$$K'_{eq} = \frac{(y)(y)}{(2 \times 10^{-4} - y)} = 10^{-4}$$

We have seen from the trend established in parts a and b that y becomes larger as the initial concentration of FDP decreases.   For an accurate solution in part c, we can no longer neglect y in the denominator.

Cross multiplying:

$$y^2 = (10^{-4})(2 \times 10^{-4} - y) = 2 \times 10^{-8} - 10^{-4}y$$
$$y^2 + 10^{-4}y - 2 \times 10^{-8} = 0$$
$$y = \frac{-b \pm \sqrt{b^2 - 4ac}}{2a}$$

where $a = 1$ $b = 10^{-4}$ and $c = -2 \times 10^{-8}$.

$$y = \frac{-10^{-4} \pm \sqrt{(10^{-4})^2 - 4(1)(-2 \times 10^{-8})}}{2}$$

$$y = \frac{-10^{-4} \pm \sqrt{10^{-8} + 8 \times 10^{-8}}}{2} = \frac{-10^{-4} \pm \sqrt{9 \times 10^{-8}}}{2}$$

$$y = \frac{-10^{-4} \pm 3 \times 10^{-4}}{2} = \frac{-4 \times 10^{-4}}{2} \quad \text{and} \quad \frac{+2 \times 10^{-4}}{2}$$

$$= -2 \times 10^{-4} \quad \text{and} \quad 1 \times 10^{-4}$$

The negative value is obviously incorrect.

$$\therefore \quad y = 10^{-4}$$

At equilibrium:

$$\boxed{\begin{aligned} &[\text{DHAP}] = 10^{-4}\ M \\ &[\text{GAP}] = 10^{-4}\ M \\ &[\text{FDP}] = 10^{-4}\ M \end{aligned}}$$

Thus, when the initial concentration of FDP is $2 \times 10^{-4} M$, the reaction comes to equilibrium when the concentrations of all three components are equal at $10^{-4} M$ each.

(d)   When the initial FDP concentration is $10^{-5} M$:

$$K'_{eq} = \frac{(y)(y)}{(10^{-5} - y)} = 10^{-4} \qquad y^2 = 10^{-9} - 10^{-4}y$$

$$y^2 + 10^{-4}y - 10^{-9} = 0 \qquad y = \frac{-b \pm \sqrt{b^2 - 4ac}}{2a}$$

where $a = 1$ $b = 10^{-4}$ and $c = -10^{-9}$.

$$y = \frac{-10^{-4} \pm \sqrt{(10^{-4})^2 - 4(1)(-10^{-9})}}{2} = \frac{-10^{-4} \pm \sqrt{10^{-8} + 4 \times 10^{-9}}}{2}$$

$$y = \frac{-10^{-4} \pm \sqrt{10^{-8} + 0.4 \times 10^{-8}}}{2} = \frac{-10^{-4} \pm \sqrt{1.4 \times 10^{-8}}}{2}$$

$$y = \frac{-10^{-4} \pm 1.183 \times 10^{-4}}{2} = \frac{-2.183 \times 10^{-4}}{2} \quad \text{and} \quad \frac{+0.183 \times 10^{-4}}{2}$$

$y = 0.0915 \times 10^{-4} = 9.15 \times 10^{-6}$, neglecting the negative value.

∴   At equilibrium:

$$\boxed{\begin{aligned} \mathbf{[DHAP]} &= \mathbf{9.15 \times 10^{-6}} \textbf{\textit{M}} \\ \mathbf{[GAP]} &= \mathbf{9.15 \times 10^{-6}} \textbf{\textit{M}} \\ \mathbf{[FDP]} &= \mathbf{8.5 \times 10^{-7}} \textbf{\textit{M}} \end{aligned}}$$

We see from this problem that although K'_{eq} is small (10^{-4}) and $\Delta G'$ is positive ($+5456$ cal/mole), we cannot automatically assume that at equilibrium there will always be more FDP than DHAP or GAP.   The relative proportions of the reaction components depend on the initial concentration of the starting compound(s).   This phenomenon will be observed whenever there is an unequal number of components on both sides of the equation.   (See Chapter 1, "Effect of Concentration on Degree of Dissociation.")

C. OXIDATION-REDUCTION REACTIONS

Many reactions that occur in living cells are oxidation-reduction reactions.   Appendix IX lists several compounds of biological importance and shows their relative tendencies to gain electrons at 25°C and pH 7 under standard conditions.   The numerical values of E'_0 reflect the reduction potentials relative to the $2H^+ + 2e^- \rightarrow H_2$ half-reaction which is taken as -0.414 volt at pH 7. The value for the hydrogen half-reaction at pH 7 was calculated from the arbitrarily assigned value (E_0) of 0.00 volt under true standard-state conditions ($1 M$ H^+ and 1 atm H_2).   For those few half-reactions of biological importance that do not involve H^+ as a reactant, the E_0 and E'_0 values are essentially identical.

Because no substance can gain electrons without another substance losing electrons, a complete oxidation-reduction reaction must be composed of two half-reactions.   When any two of the half-reactions are coupled, the one with the greater tendency to gain electrons (the one with the *more positive* reduction potential) goes as written (as a reduction).   Consequently, the other half-reaction (the one with the lesser tendency to gain electrons as shown by the *less positive* reduction potential) is driven backwards (as an oxidation).   The reduced forms of those substances with highly negative reduction potentials are good *reducing agents* (and are easily oxidized).   The oxidized form of those substances with highly positive reduction potentials are good *oxidizing agents* (and are easily reduced).

	Oxidized Form		Reduced Form		Relative E_0'
	A	$+2H^+ + 2e^- \rightarrow$	AH_2		$+3$
	B	$+2H^+ + 2e^- \rightarrow$	BH_2		$+2$
	C	$+2H^+ + 2e^- \rightarrow$	CH_2		$+1$
Increasing strength as an oxidizing agent	D	$+2H^+ + 2e^- \rightarrow$	DH_2	Increasing strength as a reducing agent	0
	E	$+2H^+ + 2e^- \rightarrow$	EH_2		-1
	F	$+2H^+ + 2e^- \rightarrow$	FH_2		-2
	G	$+2H^+ + 2e^- \rightarrow$	GH_2		-3

For example, EH_2 is a much better reducing agent than DH_2, CH_2, BH_2, or AH_2, but not as good a reducing agent as FH_2 or GH_2.   C is a much better oxidizing agent than D, E, F, or G, but not as good an oxidizing agent as B or A.   In other words, the better oxidizing agent is that substance which has the greater tendency to become reduced; the better reducing agent is that substance which has the greater tendency to become oxidized.

The relative tendency of the *overall* oxidation-reduction reaction to go may be calculated from the difference between the reduction potentials of the component half-reactions:

$$\Delta E_0' = [E_0' \text{ of the half-reaction containing the oxidizing agent}]$$
$$- [E_0' \text{ of the half-reaction containing the reducing agent}] \quad (30)$$

If the oxidizing and reducing agents are identified correctly, Equation 30 always yields a positive $\Delta E_0'$. E_0' values may be thought of as electron pressures and, as such, they are independent of the number of electrons in the half-reaction.   The $\Delta G'$ of the reaction can be calculated from the following relationship:

$$\Delta G' = -n\mathscr{F}\Delta E_0' \quad (31)$$

where n is the number of electrons transferred per mole (i.e., the number of equivalents per mole) and $\mathscr{F}$ is Faraday's constant (23,063 cal $\times$ volt^{-1} $\times$ equivalent^{-1}), a factor converting volts/equivalent to cal/equivalent.   We can see that $\Delta E_0'$ must be positive to yield a "spontaneous" reaction (i.e., a negative $\Delta G'$).

The K'_{eq} of the reaction is related to $\Delta G'$ (and hence to $\Delta E'_0$) in the following ways:

$$\Delta G' = -2.3 \, RT \log K_{eq} \qquad \text{and} \qquad \Delta G' = -n\mathscr{F}\Delta E'_0$$

$$\therefore \quad \boxed{\Delta E'_0 = \frac{2.3 \, RT}{n\mathscr{F}} \log K'_{eq}} \tag{32}$$

or

$$\boxed{\Delta E'_0 = \frac{0.059}{n} \log K'_{eq}} \qquad \text{at } 25°C \tag{33}$$

The reduction potential of a half-reaction in which the oxidized and reduced forms of the substance are present at nonstandard concentrations may be calculated from the following expression, called the Nernst equation.

$$\boxed{E = E'_0 + \frac{2.3 \, RT}{n\mathscr{F}} \log \frac{[\text{oxidized form}]}{[\text{reduced form}]}} \tag{34}$$

or

$$\boxed{E = E'_0 + \frac{0.059}{n} \log \frac{[\text{oxidized form}]}{[\text{reduced form}]}} \qquad \text{at } 25°C \tag{35}$$

For the reaction $A_{ox} + B_{red} \rightarrow A_{red} + B_{ox}$, the nonstandard state ΔE is given by:

$$\boxed{\Delta E = \Delta E'_0 - \frac{2.3 \, RT}{n\mathscr{F}} \log \frac{[A_{red}][B_{ox}]}{[A_{ox}][B_{red}]}} \tag{36}$$

which is analogous to Equation 23. We can see from Equation 34 that E will equal E'_0 even under nonstandard conditions as long as there is only one oxidized form and one reduced form of a compound involved in a half-reaction and their concentrations are equal.

EFFECT OF pH ON E'_0

Equation 35 can be used to predict the E'_0 of a half-reaction at one pH if E'_0 at another pH is known. For example, consider the half-reaction:

$$A + 2H^+ + 2e^- \rightarrow AH_2$$

The nonstandard-state E value at 25°C is related to E_0 by:

$$\boxed{E = E_0 + \frac{0.059}{2} \log \frac{[A][H^+]^2}{[AH_2]}} \tag{37}$$

If [A] and [AH$_2$] are each 1 M (or are equal), Equation 37 gives E_0':

$$E_0' = E_0 + \frac{0.059}{2} \log [H^+]^2 = E_0 + 0.059 \log [H^+]$$

$$= E_0 - 0.059 \log \frac{1}{[H^+]} \qquad \boxed{E_0' = E_0 - 0.059 \text{ pH}} \qquad (38)$$

> **Thus, for each increase of one pH unit, E_0' becomes more negative by 0.059 volt (0.0295 volt for NAD$^+$/NADH and NADP$^+$/NADPH half-reactions.)**

· Problem 3-11

(a) The E_0 of the $2H^+ + 2e^- \rightarrow H_2$ half-reaction is arbitrarily set at zero. Calculate E_0' at pH 8. (b) The E_0' of the $NAD^+ + 2H^+ + 2e^- \rightarrow NADH + H^+$ half-reaction is -0.32 v at pH 7. Calculate E_0' at pH 5.

Solution

(a) $\qquad E_0' = E_0 - 0.059 \text{ pH} = E_0 - (0.059)(8) = 0 - 0.472$

$$\boxed{E_0' = -0.472 \text{ v}}$$

(b) $\qquad E_{0_{pH\,5}}' = E_{0_{pH\,7}}' - (0.0295)(5-7) = (-0.32) - (0.0295)(-2)$
$$= -0.32 + 0.059$$

$$\boxed{E_{0_{pH\,5}}' = -0.261 \text{ v}}$$

· Problem 3-12

Calculate the $\Delta G'$ and K_{eq}' values for the reaction shown below.

1. $\quad FADH_2 + 2 \text{ cytochrome-}c\text{-Fe}^{+3} \rightleftharpoons FAD + 2 \text{ cytochrome-}c\text{-Fe}^{+2} + 2H^+$

Solution

The overall reaction can be expressed as the sum of two half-reactions.

2. $\qquad\qquad\qquad FADH_2 \rightarrow FAD + 2H^+ + 2e^-$

3. $\qquad 2[\text{cytochrome-}c\text{-Fe}^{+3} + 1e^- \rightarrow \text{cytochrome-}c\text{-Fe}^{+2}]$

---

1. (Sum)
$\qquad FADH_2 + 2 \text{ cytochrome-}c\text{-Fe}^{+3} \rightarrow FAD + 2 \text{ cytochrome-}c\text{-Fe}^{+2} + 2H^+$

The two half-reactions written as reductions are shown below.

2a. $\qquad FAD + 2H^+ + 2e^- \rightarrow FADH_2 \qquad\qquad\qquad E_0' = -0.18 \text{ v}$

3. $\qquad 2[\text{cytochrome-}c\text{-Fe}^{+3} + 1e^- \rightarrow \text{cytochrome-}c\text{-Fe}^{+2}] \qquad E_0' = +0.25 \text{ v}$

The E_0' value of the cytochrome-c-Fe^{+3}/cytochrome-c-Fe^{+2} half-reaction is more positive than that of the $FAD/FADH_2$ half-reaction. Thus, half-reaction 3 goes as written, forcing half-reaction 2a to go in reverse—as an oxidation. The overall spontaneous reaction (under standard-state conditions) is in the direction shown in reaction 1. $FADH_2$ (the reducing agent) is oxidized to FAD by cytochrome-c-Fe^{+3} (the oxidizing agent).

$\Delta E_0' = (E_0'$ of the half-reaction containing the oxidizing agent)
$\qquad - (E_0'$ of the half-reaction containing the reducing agent)

$\Delta E_0' = (0.25) - (-0.18) = 0.25 + 0.18$

$$\boxed{\Delta E_0' = +0.43 \text{ v}}$$

Note that the E_0' value for the cytochrome-c-Fe^{+3}/cytochrome-c-Fe^{+2} half-reaction is *not* doubled when calculating $\Delta E_0'$. The fact that 2 moles of cytochrome are required per mole of $FADH_2$ is taken into account in calculating $\Delta G'$ and K_{eq}'.

$$\Delta G' = -n\mathscr{F}\Delta E_0' = -(2)(23,063)(0.43)$$

$$\boxed{\Delta G' = -19,834 \text{ cal/mole } FADH_2 = -9917 \text{ cal/mole cytochrome } c}$$

$$\Delta G' = -1364 \log K_{eq}' \qquad \log K_{eq}' = \frac{-19,834}{-1364} = 14.54$$

$$K_{eq}' = \text{antilog of } 14.54 \qquad \boxed{K_{eq}' = 3.467 \times 10^{14}}$$

where

$$K_{eq}' = \frac{[FAD][\text{cytochrome-}c\text{-}Fe^{+2}]^2}{[FADH_2][\text{cytochrome-}c\text{-}Fe^{+3}]^2}$$

· Problem 3-13

A solution containing $0.2 \, M$ dehydroascorbate and $0.2 \, M$ ascorbate was mixed with an equal volume of a solution containing $0.01 \, M$ acetaldehyde and $0.01 \, M$ ethanol at 25°C and pH 7. (a) Write the equation for a thermodynamically favorable reaction that could occur. (b) Calculate $\Delta E_0'$ and $\Delta G'$.

Solution

(a) The half-reactions involved and their standard reduction potentials are shown below.

1. dehydroascorbate $+ 2H^+ + 2e^- \rightarrow$ ascorbate $\qquad E_0' = +0.06$ v
2. acetaldehyde $+ 2H^+ + 2e^- \rightarrow$ ethanol $\qquad E_0' = -0.163$ v

Under standard-state conditions, dehydroascorbate gains electrons and becomes reduced while ethanol provides the electrons and becomes oxidized.

1.　　　　dehydroascorbate $+ 2H^+ + 2e^- \rightarrow$ ascorbate

2a.　　　　ethanol $\rightarrow 2H^+ + 2e^- +$ acetaldehyde

---

3. (Sum)　dehydroascorbate $+$ ethanol $\rightarrow$ ascorbate $+$ acetaldehyde
　　　　　(oxidizing agent)　　　(reducing agent)

Although the dehydroascorbate and ascorbate concentrations are 0.1 M in the final solution and the ethanol and acetaldehyde concentrations are only 0.005 M, the concentration *ratios* for each oxidized-reduced pair are unity. Consequently, the E values for each half-reaction are identical to their corresponding E_0' values. Any advantage gained by having a high concentration of the oxidized form of a substance on one side of the equation is offset by having an equally high concentration of the reduced form on the other side. The spontaneous reaction is the same as that predicted for standard-state concentrations.

(b)　　$\Delta E_0' = (E_0'$ of half-reaction containing the oxidizing agent)

　　　　　　$- (E_0'$ of half-reaction containing the reducing agent)

　$\Delta E_0' = (+0.060) - (-0.163) = (+0.06) + (0.163)$

$$\boxed{\Delta E_0' = +0.223 \text{ v}}$$

　$\Delta G' = -n\mathscr{F}\Delta E_0' = -(2)(23,063)(0.223)$

$$\boxed{\Delta G' = -10,286 \text{ cal/mole}}$$

· Problem 3-14

Write the spontaneous reaction that occurs and calculate the ΔG of the reaction when the enzyme lactic dehydrogenase is added to a solution containing pyruvate, lactate, NAD^+, and NADH at the following concentration ratios:　(a) lactate/pyruvate $= 1$, NAD^+/NADH $= 1$, (b) lactate/pyruvate $= 159$, NAD^+/NADH $= 159$, and (c) lactate/pyruvate $= 1000$, NAD^+/NADH $= 1000$.

Solution

The half-reactions involved and their standard reduction potentials are shown below.

1.　pyruvate $+ 2H^+ + 2e^- \rightarrow$ lactate　　　　　$E_0' = -0.190$ v

2.　　$NAD^+ + 2H^+ + 2e^- \rightarrow NADH + H^+$　　　$E_0' = -0.320$ v

(a) Under standard-state conditions, or at lactate/pyruvate and NAD^+/NADH ratios of 1, the pyruvate/lactate half-reaction goes as written while the NAD^+/NADH half-reaction goes in reverse (as an oxidation). The overall spontaneous reaction is:

3. (Sum)   pyruvate $+$ NADH $+$ H$^+ \rightarrow$ lactate $+$ NAD$^+$

$$\Delta E_0' = (-0.190) - (-0.320) = +0.130$$

The NADH is the reducing agent; pyruvate is the oxidizing agent.

$$\Delta G' = -n\mathscr{F}\Delta E_0' = -(2)(23{,}063)(0.130)$$

$$\boxed{\Delta G' = -5996 \text{ cal/mole}}$$

(b)   If, instead of standard-state conditions, we set up a reaction mixture containing lactate/pyruvate and NAD$^+$/NADH ratios of 159:1, the half-reactions have new reduction potentials as shown below.

1.
$$E_1 = -0.190 + \frac{0.059}{2} \log \frac{[\text{pyruvate}]}{[\text{lactate}]}$$

$$= -0.190 + 0.0295 \log \frac{1}{159} = -0.190 - 0.0295 \log 159$$

$$= -0.190 - 0.0295(2.201) = -0.190 - 0.065$$

$$\boxed{E_1 = -0.255 \text{ v}}$$

2.
$$E_2 = -0.320 + \frac{0.059}{2} \log \frac{[\text{NAD}^+]}{[\text{NADH}]}$$

$$= -0.320 + 0.0295 \log 159 = -0.320 + 0.0295(2.201)$$

$$= -0.320 + 0.065$$

$$\boxed{E_2 = -0.255 \text{ v}}$$

At these concentration ratios the two half-reactions have the same E values; ΔE is zero, hence $\Delta G = 0$. In other words, at the indicated concentration ratios, the reaction is at equilibrium already so no further net change will occur.

(c)   Now suppose that the lactate/pyruvate and NAD$^+$/NADH ratios are set at 1000:1. Under this condition the reduction potentials of the two reactions are as follows:

1.
$$E_1 = -0.190 + 0.0295 \log \frac{1}{1000}$$

$$= -0.190 - 0.0295 \log 1000$$

$$= -0.190 - 0.0295(3) = -0.190 - 0.089$$

$$\boxed{E_1 = -0.279 \text{ v}}$$

2.
$$E_2 = -0.320 + 0.0295 \log 1000$$
$$= -0.320 + 0.0295(3) = -0.320 + 0.089$$

$$\boxed{E_2 = -0.231 \text{ v}}$$

Now the NAD^+/NADH half-reaction has the more positive reduction potential. Thus, this half-reaction goes as written (as a reduction), driving the pyruvate/lactate half-reaction backwards (as an oxidation). Lactate is now the reducing agent; NAD^+ is the oxidizing agent. The overall spontaneous reaction is:

$$NAD^+ + \text{lactate} \rightarrow NADH + H^+ + \text{pyruvate}$$
$$\Delta E = (-0.231) - (-0.279) = -(0.231) + (0.279)$$
$$\Delta E = +0.048 \text{ v}$$
$$\Delta G = -n\mathscr{F}\Delta E = -(2)(23{,}063)(0.048)$$

$$\boxed{\Delta G = -2214 \text{ cal/mole}}$$

We could also have calculated ΔE from Equation 36. If the reaction is written: pyruvate + NADH + H^+ → lactate + NAD^+ where $\Delta E_0' = 0.130$, then ΔE will come out negative, indicating that the reaction actually goes in reverse.

pH ELECTRODES

Consider the two half-reactions shown below.

		E_0
1.	$X + ne^- \rightarrow X^{-n}$	Zv
2.	$Y + ne^- + nH^+ \rightarrow YH_2$	Mv

Reaction 1 represents any half-reaction whose E_0 value is known (for example, the normal calomel electrode: $Hg_2Cl_2 + 2e^- \rightarrow 2Hg^0 + 2Cl^-$, or the silver/silver chloride electrode: $AgCl + 1e^- \rightarrow Ag^0 + Cl^-$). Reaction 2 represents any other half-reaction that is H^+ ion dependent (e.g., $H^+ + 1e^- \rightarrow \frac{1}{2}H_2$). If the E_0 value of reaction 1 and the concentrations of X and X^{-n} are known, then the E value for half-reaction 1 is known. If the ΔE value of the overall oxidation-reduction reaction can be measured and the E_0 value for reaction 2 and the concentrations of Y and YH_2 are known, then the only remaining unknown value (the H^+ ion concentration) can be calculated. In this manner, oxidation-reduction reactions can be used to measure the pH of a solution.

In practice, commercial pH meters employing a "glass electrode" and a reference electrode are used to make pH measurements. The E value (potential) of the glass electrode does not result from an oxidation-reduction reaction but rather from the transfer of H^+ ions through a thin glass membrane. The electrode assembly is shown in Figure 3-3.

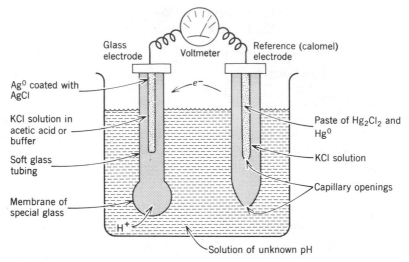

Figure 3-3   Glass electrode for pH measurements.

The glass electrode is composed of a thin-walled bulb of a special H^+ ion-permeable glass blown on the end of a piece of soft glass tubing.   The inner space contains a KCl solution in acetic acid or buffer and some kind of standard electrode such as the $Ag^0/AgCl$ electrode.   The concentration of Ag^+ ions in the solution is fixed by the K_{SP} of AgCl and the concentration of Cl^- ions provided by the KCl.   The H^+ ion concentration is fixed by the buffer or weak acetic acid.   The $Ag^0/AgCl$ electrode has a certain potential relative to the reference electrode.   Superimposed on this potential is an additional potential resulting from the difference in H^+ ion concentration on either side of the glass membrane.   The potential of the glass electrode is given by: $E = E_0 + 0.059 \log [H^+]$.   The E_0 value of the glass electrode depends not only on the potential of the $Ag^0/AgCl$ half-cell, but also depends on the pH of the internal buffer and the particular characteristics of the glass used to construct the membrane.   The ΔE_0 of the overall cell depends on these factors and the E_0 value of the reference cell.   Because the glass electrode has no fixed E_0 value, it does not provide an absolute measure of the H^+ ion concentration (such as the $\frac{1}{2}H_2/H^+$ electrode would).   The glass electrode must be calibrated periodically against buffers of known pH.

We can visualize the glass electrode as operating in the following way: If the electrode is placed into a solution with a pH lower than that of the internal buffer, H^+ ions migrate from the external solution into the thin glass membrane.   This, in turn, causes positive ions to be displaced from the inside surface of the glass membrane to the internal solution.   (These ions may be Na^+ or K^+ or Ca^{++} from the glass.)   The momentary excess of positive charges in the glass electrode compartment causes electrons to flow from the reference electrode compartment ($2Hg^0 + 2Cl^- \rightarrow Hg_2Cl_2 + 2e^-/2AgCl + 2e^- \rightarrow 2Ag^0 + 2Cl^-$).   If the pH of the external solution were higher than that in the glass electrode compartment, then H^+ ions would migrate out of the glass electrode compartment, leaving it with a momentary negative charge.   Electrons would then flow from the glass electrode to the reference electrode ($2Ag^0 + 2Cl^- \rightarrow 2AgCl + 2e^-/Hg_2Cl_2 + 2e^- \rightarrow 2Hg^0 + 2Cl^-$).   The voltmeter measures the potential of the electrons which is proportional to the H^+ ion gradient established across the glass membrane.

D. METABOLISM AND ATP YIELD

· Problem 3-15

The conversion of glucose to lactic acid has an overall $\Delta G'$ of $-52,000$ cal/mole. In an anaerobic cell, this conversion is coupled to the synthesis of 2 moles of ATP per mole of glucose. (a) Calculate the $\Delta G'$ of the overall coupled reaction. (b) Calculate the efficiency of energy conservation in the anaerobic cell. (c) At the same efficiency, how many moles of ATP per mole of glucose could be obtained in an aerobic organism in which glucose is completely oxidized to CO_2 and H_2O ($\Delta G' = -686,000$ cal/mole)? (d) Calculate the $\Delta G'$ for the overall oxidation coupled to ATP synthesis.

Solution

(a)

1. $C_6H_{12}O_6 \rightarrow 2CH_3-CHOH-COOH$ $\Delta G_1' = -52,000$ cal/mole

    glucose   lactic acid

2. $2ADP + 2P_i \rightarrow 2ATP$ $\Delta G_2' = +7700$ cal/mole $\times 2$

                                           $= +15,400$ cal/mole

---

3. (Sum)   glucose $+ 2ADP + 2P_i \rightarrow 2$ lactic acid $+ 2ATP$

$$\Delta G_3' = (-52,000) + (15,400) \qquad \boxed{\Delta G_3' = -36,600 \text{ cal/mole}}$$

(b)   efficiency $= \dfrac{\text{energy conserved}}{\text{energy made available}} \times 100\% = \dfrac{15,400 \text{ cal}}{52,000 \text{ cal}} \times 100\%$

$$\boxed{\textbf{efficiency} = \textbf{29.6\%}}$$

(c)

4. $C_6H_{12}O_6 + 6O_2 \rightarrow 6CO_2 + 6H_2O$ $\Delta G_4' = -686,000$ cal/mole

5. $n\,ADP + n\,P_i \rightarrow n\,ATP$ $\Delta G_5' = n(+7700$ cal/mole$)$

---

6. (Sum) $C_6H_{12}O_6 + 6O_2 + n\,ADP + n\,P_i \rightarrow 6CO_2 + 6H_2O + n\,ATP$

At 29.6% efficiency the total energy that can be conserved is $0.296 \times 686,000$ cal/mole $= 203,000$ cal/mole. If each mole of ATP requires 7700 cal for its synthesis:

$$\frac{203,000}{7700} = 26.4 \text{ moles of ATP}$$

$$= \boxed{\textbf{26 moles of ATP}} \qquad \text{(nearest whole number)}$$

(d) For the overall coupled reaction:

$$\Delta G_6' = \Delta G_4' + \Delta G_5'$$
$$= (-686,000) + 26(+7700) = (-686,000) + (200,200)$$

$$\boxed{\Delta G_6' = -485,000 \text{ cal/mole}}$$

Aerobic cells are actually more efficient than 29.6%. One mole of glucose is capable of yielding 36 moles of ATP when completely oxidized.

36 moles $\times$ 7700 cal/mole = 277,200 cal conserved/mole glucose oxidized

$$\frac{277,200}{686,000} \times 100 = \boxed{\textbf{40.4\% efficiency}}$$

$$\Delta G_6' = (-686,000) + (277,200)$$

$$\boxed{\Delta G_6' = -408,800 \text{ cal/mole}}$$

· Problem 3-16

(a) Calculate the $\Delta G'$ for the complete oxidation of lactic acid to CO_2 and H_2O given the information below. (b) How many moles of ATP could be synthesized in the process at 40% efficiency?

1.   glucose $\rightarrow$ 2 lactic acid $\Delta G_1' = -52,000$ cal/mole

2.   glucose $+ 6O_2 \rightarrow 6CO_2 + 6H_2O$ $\Delta G_2' = -686,000$ cal/mole

Solution

(a) The reaction we are interested in is shown below.

3.   lactic acid $+ 3O_2 \rightarrow 3CO_2 + 3H_2O$

The overall oxidation of glucose to CO_2 and H_2O can be written in two steps.

1.   glucose $\rightarrow$ 2 lactic acid $\Delta G_1' = -52,000$ cal/mole

4.   2 lactic acid $+ 6O_2 \rightarrow 6CO_2 + 6H_2O$ $\Delta G_4' = ?$

---

2. (Sum)   glucose $+ 6O_2 \rightarrow 6CO_2 + 6H_2O$ $\Delta G_2' = -686,000$ cal/mole

The $\Delta G_4'$ can be calculated easily.

$$\Delta G_2' = \Delta G_1' + \Delta G_4'$$
$$-686,000 = -52,000 + \Delta G_4'$$
$$\Delta G_4' = -686,000 + 52,000 = -634,000 \text{ cal}$$

The oxidation of 2 moles of lactic acid yields 634,000 cal. Therefore, the oxidation of 1 mole would yield half as much.

Reaction 3 is half of reaction 4.

$$\therefore \quad \Delta G_3' = \frac{-634,000}{2} = \boxed{-\textbf{317,000 cal/mole of lactic acid}}$$

(b)   At 40% efficiency:

$$0.40 \times 317,000 = 127,000 \text{ cal conserved}$$

Each mole of ATP requires 7700 cal.

$$\frac{127,000}{7700} = \boxed{\sim \textbf{16 moles of ATP}}$$

· Problem 3-17

(a)   *Nitrobacter agilis* plays an important role in the nitrogen cycle in nature by oxidizing soil nitrite to nitrate in the presence of oxygen.   Given the E_0' values shown below, calculate the potential ATP yield per mole of nitrite oxidized assuming an efficiency of 50%.

$$NO_3^- + 2H^+ + 2e^- \rightarrow NO_2^- + H_2O \qquad E_0' = +0.42 \text{ v}$$
$$\tfrac{1}{2}O_2 + 2H^+ + 2e^- \rightarrow H_2O \qquad E_0' = +0.82 \text{ v}$$

(b)   Many facultative microorganisms can use NO_3^- as a terminal electron acceptor (oxidizing agent) under anaerobic conditions.   Compare the ATP yields per mole of NADH oxidized by O_2 and by NO_3^-.   Assume an efficiency of energy conservation of 50%.   The E_0' value of the $NAD^+/NADH$ half-reaction is -0.32 v.   The other relevant E_0' values are given in part a.

Solution

(a)   The overall reaction is:

$$NO_2^- + \tfrac{1}{2}O_2 \rightarrow NO_3^- \qquad \Delta E_0' = 0.40 \text{ v}$$

$$\Delta G' = -(2)(23,063)(0.40) = -18,450 \text{ cal/mole}$$

$$\text{yield} = \frac{(18,450)(0.50)}{(7700)} = 1.19 \qquad \text{or} \qquad \boxed{\sim \textbf{1 ATP/NO}_2^- \textbf{ oxidized}}$$

(b)   When NO_3^- is the terminal electron acceptor, the coupled reaction is:

$$NADH + NO_3^- + H^+ \rightarrow NAD^+ + NO_2^- + H_2O$$

$$\Delta E_0' = (0.42) - (-0.32) = 0.42 + 0.32 = 0.74$$
$$\Delta G' = -(2)(23,063)(0.74) = -34,133 \text{ cal/mole}$$

At an efficiency of energy conservation of 50%:

$$\text{yield} = \frac{(34,133)(0.5)}{(7700)} = 2.2 \qquad \text{or} \qquad \boxed{\sim \textbf{2 ATP/NO}_3^- \textbf{ reduced}}$$

When O_2 is the terminal electron acceptor, the overall reaction is:

$$NADH + \tfrac{1}{2}O_2 + H^+ \rightarrow NAD^+ + H_2O$$

$$\Delta E_0' = (0.82) - (-0.32) = 0.82 + 0.32 = 1.14 \text{ v}$$

$$\Delta G' = -(2)(23,063)(1.14) = -52,584 \text{ cal/mole}$$

$$\text{yield} = \frac{(52,584)(0.5)}{(7700)} = 3.4 \quad \text{or} \quad \boxed{\sim 3 \text{ ATP/O}_2 \text{ reduced}}$$

General Principle

Whether a half-reaction goes as an oxidation or as a reduction depends on the E value of the half-reaction with which it is coupled. Thus, some organisms can oxidize NO_2^- to obtain energy if the oxidizing agent is stronger than NO_3^-. Other organisms reduce NO_3^- by oxidizing a compound that is a stronger reducing agent than NO_2^- (and, in the process, obtain energy). It is all relative:

$$NAD^+/NADH \quad E_0' = -0.32 \text{ v}$$

Part b
NADH is oxidized
NO_3^- is reduced

Part a
NO_2^- is oxidized
O_2 is reduced

$$NO_3^-/NO_2^- \quad E_0' = +0.42 \text{ v}$$

$$\tfrac{1}{2}O_2/H_2O \quad E_0' = +0.82 \text{ v}$$

· **Problem 3-18**

(a) Calculate the number of moles of ATP that is obtained from the complete oxidation of one mole of palmitic acid to $CO_2 + H_2O$ via β-oxidation, the TCA cycle, and the electron transport system. (b) The $\Delta G'$ for the complete oxidation of a long-chain fatty acid is about 9000 cal/g. What fraction of the total ΔG is conserved as ATP in the biological oxidation of palmitic acid?

Solution

(a)   The pathway and ATP yields are shown in Figure 3-4.

(b)   For each mole of palmitic acid oxidized, 130 moles of ATP are made. The MW of palmitic acid is 256.4.

$$\therefore \quad \Delta G' \text{ of oxidation} = (256.4)(9000) = 2.31 \times 10^6 \text{ cal/mole}$$

The energy conserved in 130 moles of ATP is about:

$$(130)(7700) = 1 \times 10^6 \text{ cal}$$

$$\text{efficiency} = \frac{(1.0 \times 10^6)(100)}{(2.31 \times 10^6)} = \boxed{\textbf{43.3\%}}$$

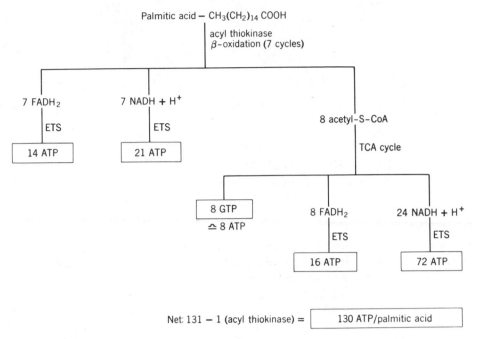

Net: $131 - 1$ (acyl thiokinase) = [130 ATP/palmitic acid]

Figure 3-4 ATP yield from the complete oxidation of palmitic acid.

· **Problem 3-19**

The adenylate pool in a culture of lymphosarcoma cells was found to consist of 10^{-3} M ATP, 3×10^{-4} M ADP, and 10^{-4} M AMP. (a) Calculate the "energy charge" of the cells. (b) Assuming that the adenine nucleotides are at equilibrium for the adenylate kinase reaction, calculate the K'_{eq} of the reaction.

Solution

(a) "Energy charge" is defined as:

$$\text{energy charge} = \frac{[\text{ATP}] + \frac{1}{2}[\text{ADP}]}{[\text{ATP}] + [\text{ADP}] + [\text{AMP}]} \tag{39}$$

That is, "energy charge" is equivalent to the mole fraction of the total adenylate pool represented by ATP *or its equivalent*. One-half the [ADP] is equivalent to [ATP] because of the action of adenylate kinase, which catalyzes the reaction:

$$2\,\text{ADP} \xrightleftharpoons[\text{adenylate kinase}]{} \text{ATP} + \text{AMP}$$

(If $2\,\text{ADP} \rightleftharpoons 1\,\text{ATP}$, then $1\,\text{ADP} \rightleftharpoons \frac{1}{2}\,\text{ATP}$.)

$$\therefore \quad \text{energy charge} = \frac{(10^{-3}) + \frac{1}{2}(3 \times 10^{-4})}{(10^{-3}) + (3 \times 10^{-4}) + (10^{-4})} = \frac{(10^{-3}) + (1.5 \times 10^{-4})}{(10^{-3}) + (3 \times 10^{-4}) + (10^{-4})}$$

$$= \frac{1.15 \times 10^{-3}}{1.4 \times 10^{-3}} \qquad \boxed{\textbf{energy charge} = \textbf{0.82}}$$

(b)
$$K'_{eq} = \frac{[\text{ATP}][\text{AMP}]}{[\text{ADP}]^2} = \frac{(10^{-3})(10^{-4})}{(3 \times 10^{-4})^2} = \frac{10^{-7}}{9 \times 10^{-8}}$$

$$= \frac{10 \times 10^{-8}}{9 \times 10^{-8}} \qquad \boxed{K'_{eq} = 1.11}$$

OXIDATIVE PHOSPHORYLATION AND THE CHEMIOSMOTIC HYPOTHESIS

Several hypotheses have been put forward to explain how electron transport in mitochondria or bacterial membranes can be coupled to ATP formation. The *chemical coupling hypothesis* suggests that oxidative phosphorylation is essentially identical to substrate level phosphorylation. That is, discreet energy-rich intermediates are involved. The *mechanochemical* or *conformational coupling* model suggests that oxidation-reduction reactions within the membrane result in an energized conformation of the membrane or in certain subunits of the membrane. The situation can be considered analogous to the winding of a spring. Somehow, the mechanical energy stored in the energized membrane is dissipated by the formation of ATP. The *chemiosmotic hypothesis* suggests that the electron carriers are asymmetrically positioned within the membrane so that when H^+ ions are produced, they are released on only one side of the membrane (the outer side of the inner membrane). When H^+ ions are utilized, they are obtained from the other side of the membrane (the inner side of the inner membrane) (Fig. 3-5*a*). Thus, the passage of $2e^-$ down the electron transport system results in a ΔpH across the inner membrane. The synthesis of ATP from ADP + P_i involves the removal of water by an ATPase operating in reverse. The driving force is the ΔpH. A simplified model is shown in Figure 3-5*b*. Here, it is assumed that H_2O per se is not removed but, instead, the ions of water (H^+ and OH^-). H^+ is drawn to the high pH (H^+ deficient) side of the inner membrane; OH^- is drawn to the low pH (H^+ excess) side of the inner membrane. Figure 3-5*c* represents a more likely model. It is assumed that the ATPase reaction is obligately coupled to the movement of H^+ ions. Experimental evidence suggests that the reaction catalyzed by the ATPase is:

$$\text{ATP} + \text{H}_2\text{O} + 2\text{H}^+_{\text{inner}} \underset{}{\overset{\text{ATPase}}{\rightleftharpoons}} \text{ADP} + \text{P}_i + 2\text{H}^+_{\text{outer}}$$

Thus, the synthesis of ATP is coupled to the movement of $2H^+$ from the low pH side of the membrane to the high pH side. As shown in Section F, the ΔG for this movement of $2H^+$ is given by:

$$\boxed{\Delta G = 2.3\,RT \log \frac{[\text{H}^+]^2_{\text{high pH side}}}{[\text{H}^+]^2_{\text{low pH side}}}} \qquad \text{or} \qquad \boxed{\Delta G = -(2.3\,RT)(2)\Delta\text{pH}}$$

(40)

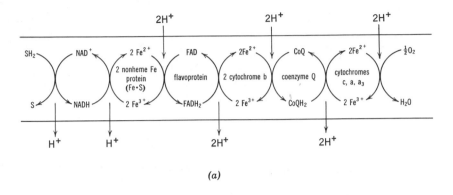

(a)

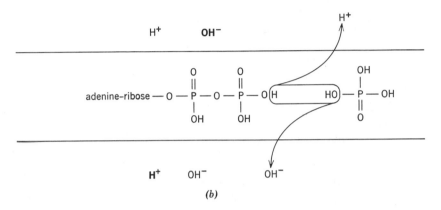

(b)

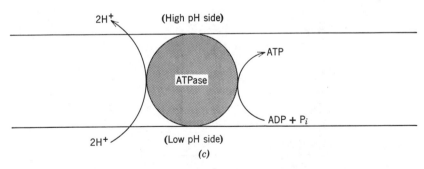

(c)

Figure 3-5 (a) Asymmetric release and uptake of H^+ ions as $2e^-$ move down the electron transport system. This generates a ΔpH across the membrane. (b) Synthesis of ATP from $ADP + P_i$ by a membrane ATPase. The enzyme removes the elements of H_2O. H^+ moves in one direction (into the high pH compartment); OH^- moves in the other direction (into the low pH compartment). (c) A more exact model of ATP synthesis by a membrane ATPase. ATP synthesis is intimately coupled to the translocation of two H^+ ions.

· Problem 3-20

Calculate the ΔpH across the inner mitochondrial membrane that is required at 25°C to drive the synthesis of ATP from $ADP + P_i$ under standard conditions.

Solution

The synthesis of ATP under standard conditions requires 7700 cal/mole. Therefore, the ΔG provided by the ΔpH must be -7700 for the movement of $2H^+$ across the membrane.

$$\Delta G = -(1364)(2)\Delta pH \qquad -7700 = -2728\Delta pH$$

$$\frac{-7700}{-2728} = \Delta pH \qquad \boxed{\Delta pH = 2.82}$$

E. PHOTOSYNTHETIC PHOSPHORYLATION

LIGHT ENERGY

Each photon or quantum of light has an energy of $h\nu$, where:

h = Planck's constant

$\quad = 6.627 \times 10^{-27} \, erg \times sec = 1.58 \times 10^{-34} \, cal \times sec$

and $\quad\quad \nu$ = the frequency of the light in sec^{-1}.

The energy in one einstein (one mole) of photons is given by:

$$\boxed{\mathscr{E} = Nh\nu = Nh\frac{c}{\lambda} \text{ cal/einstein}} \qquad (41)$$

or

$$\boxed{\mathscr{E} = \frac{2.855}{\lambda} \text{ cal/einstein}}$$

where N = Avogadro's number = 6.023×10^{23} photons/einstein
$\quad\quad\quad \lambda$ = the wavelength of the light in centimeters
$\quad\quad\quad c$ = the speed of light in cm/sec if λ is given in centimeters
$\quad\quad\quad\quad = 3 \times 10^{10}$ cm/sec

Thus, the energy of light is inversely proportional to its wavelength.

· Problem 3-21

Calculate the energy per einstein of photons for light of wavelengths (a) 400 nm (violet) and (b) 600 nm (orange).

Solution

(a) $1 \text{ nm} = 10^{-9} \text{ m} = 10^{-7} \text{ cm}$ $\therefore$ $400 \text{ nm} = 4 \times 10^{-5} \text{ cm} = \lambda$

$$\mathscr{E} = \frac{2.855}{4 \times 10^{-5}} = \boxed{\textbf{71,375 cal/einstein}}$$

(b) $$\mathscr{E} = \frac{2.855}{6 \times 10^{-5}} = \boxed{\textbf{47,582 cal/einstein}}$$

· **Problem 3-22**

(a) How many moles of ATP could be synthesized at 100% efficiency by a photosynthetic organism upon absorption of 1 einstein of red light of 700 nm?   (b) How many molecules of ATP could be produced from one photon?   (c) What is the overall efficiency of energy conversion if 1 mole of ATP is formed per 2 equivalents of electrons excited by red light (i.e., per 2 einsteins of photons)?

Solution

(a) $$\mathscr{E} = \frac{2.855}{\lambda} = \frac{2.855}{7 \times 10^{-5}} = 40{,}786 \text{ cal/einstein}$$

$$\frac{40{,}786}{7700} \simeq \boxed{\textbf{5 moles of ATP/einstein}}$$

(b)   The ratio of photons/einstein is the same as the ratio of molecules/mole.

$$\therefore \quad \boxed{\textbf{5 molecules of ATP/photon}}$$

(c)   Assuming that two photons must be absorbed in order to excite two electrons to a sufficient energy level, then:

$$\mathscr{E}_{\text{input}} = (2)(40{,}786) = 81{,}572 \text{ cal}$$
$$\mathscr{E}_{\text{conserved}} = (1)(7700) = 7700 \text{ cal}$$

$$\text{efficiency} = \frac{7700}{81{,}572} \times 100 = \boxed{\textbf{9.4\%}}$$

NONCYCLIC AND CYCLIC PHOTOPHOSPHORYLATION

Green plants possess two distinct pigment systems (PS I and PS II), which provide for a noncyclic flow of electrons (Fig. 3-6).   Light absorption by PS I excites electrons to an energy level capable of reducing ferredoxin via a

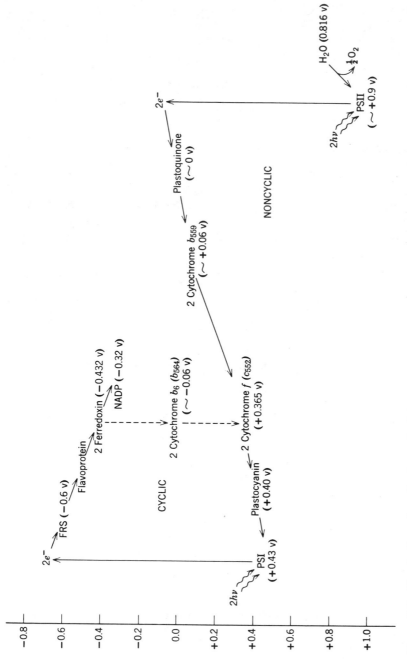

Figure 3-6 Noncyclic and cyclic photophosphorylation in green plants. Pigment system I (PS I) contains chlorophyll *a*, P-700 (a pigment that possesses an absorption maximum at 700 nm), and carotenoids. Pigment system II (PS II) contains chlorophylls *a*, *b*, *c*, and *d*, and phycobilin pigments.

ferredoxin reducing substance (FRS).   Reduced ferredoxin then passes the electrons to NADP providing NADPH for the dark phase (the CO_2 reduction phase) of photosynthesis.   The oxidized PS I is reduced by electrons from PS II.   These electrons are excited to a negative potential somewhere between 0 and -0.2 v.   As the electrons are transferred to oxidized PS I, one or more ATP's are made.   The oxidized PS II is reduced by electrons from H_2O (which results in O_2 evolution). H_2O is a strong enough reducing agent to reduce oxidized PS II, but not oxidized PS I.   It is also believed that electrons from PS I can traverse a cyclic route leading to the synthesis of ATP without the concomitant synthesis of NADPH.   The scheme shown in Figure 3-6 is one of several versions.   The exact sequence of electron transport reactions is unknown.   In fact, the exact standard reduction potentials of some of the carriers are not known for sure.

· **Problem 3-23**

Assuming that the sequence and reduction potentials shown in Figure 3-6 are correct, calculate (a) the ratio of $NADPH/ATP/O_2$ produced in one noncyclic process and (b) the ATP yield of one cyclic process.   (c) Calculate the efficiency of red light ($\lambda = 700$ nm) in the overall synthesis of 1 mole of glucose from $6CO_2 + 6H_2O$.   Assume that each noncyclic process ($4e^-$) yields 1 ATP and 1 NADPH and that each cyclic process ($2e^-$) yields 1 ATP.   (d) What would the efficiency be if each noncyclic process yielded 2 ATP + 1 NADPH?

Solution

(a)   The synthesis of ATP requires 7700 cal/mole under standard conditions at 100% efficiency.   This amount of energy is equivalent to a $\Delta E_0'$ of 0.167 v for a $2e^-$ oxidation-reduction reaction, as shown below:

$$\Delta G' = -n\mathscr{F}\Delta E_0' \quad \text{or} \quad \Delta E_0' = \frac{\Delta G'}{-n\mathscr{F}}$$

$$\Delta E_0' = \frac{-7700}{-(2)(23,063)} = 0.167$$

At a reasonable efficiency of 50%, a $\Delta E_0'$ of 0.334 is required.   Only one oxidation-reduction couple in the noncyclic sequence shown in Figure 3-6 provides a $\Delta E_0'$ of 0.33 or greater—that between cytochrome b_{559} and cytochrome f.   Thus, for each $4e^-$ excited (by 4 quanta), we obtain:

$$\boxed{1 \text{ NADPH: } 1 \text{ ATP: } \tfrac{1}{2} O_2}$$

(b)   The cyclic pathway seems rather incomplete.   It is unlikely that electrons would move directly from ferredoxin to cytochrome b_6 and then directly from cytochrome b_6 to cytochrome f.   The $\Delta E_0'$ values of these reactions are rather large compared to other electron transport reactions.   In fact, the $\Delta G'$ values for these reactions are sufficient to synthesize an ATP at each step for only a $1e^-$ transfer.   Thus, the potential exists for the synthesis of 4 ATP per $2e^-$ in the cyclic process as shown in Figure 3-6.

By a different line of reasoning, we might conclude that no ATP can be formed.   There is no way of generating a ΔpH by passing electrons from ferredoxin to cytochrome f—none of the carriers picks up or releases H^+.   It seems likely then that additional carriers are involved (e.g., flavoproteins, quinones, and such).

(c)   The synthesis of one glucose from $6CO_2 + 6H_2O$ requires $18\,ATP + 12\,NADPH$.   Thus, 12 noncyclic processes ($\approx 12\,ATP + 12\,NADPH$) plus 6 cyclic processes ($\approx 6\,ATP$) are needed for the assumed NADPH and ATP yields.   If each noncyclic process requires the input of 4 quanta and each cyclic process requires the input of 2 quanta, then $48 + 12 = 60$ einsteins of red light will yield the 18 moles of ATP plus 12 moles of NADPH needed.

$$\mathscr{E} = \frac{2.855}{\lambda} = \frac{2.855}{7 \times 10^{-5}} = 40{,}785 \text{ cal/einstein}$$

$$\text{input} = (40{,}785)(60) = 24.47 \times 10^5 \text{ cal}$$

The $\Delta G'$ for the formation of 1 mole of glucose from 6 moles of CO_2 plus 6 moles of H_2O is $+686{,}000$ cal/mole.

$$\therefore \quad \text{efficiency} = \frac{68.6 \times 10^4}{24.47 \times 10^5} \times 100 = \boxed{28\%}$$

(d)   If each noncyclic process yielded $2\,ATP + 1\,NADPH$, then 12 noncyclic processes alone (48 quanta) would yield sufficient NADPH and an excess of ATP.   The efficiency would then be greater:

$$\text{efficiency} = \frac{(68.6 \times 10^4)(100)}{(48)(40{,}785)} = \boxed{35\%}$$

· **Problem 3-24**

Photosynthetic bacteria of the genus *Chlorobium* possess a single pigment system, yet are capable of carrying out noncyclic photophosphorylation in the presence of H_2S (E_0' for the half-reaction $S^0 + 2H^+ + 2e^- \rightarrow H_2S$ is -0.23 v).   In red light ($\lambda = 700$ nm), *Chlorobium* can reduce methyl viologen, an artificial electron acceptor ($E_0' = -0.55$ v).   The organism possesses at least two different cytochromes, one of which is bound to a flavoprotein.   Outline a likely overall photosynthetic phosphorylation process involving natural electron carriers, and estimate the relative yields and ratios of products.   Assume that the efficiency of energy conservation in the reduction of the primary (photo) electron acceptor is 60% and that the efficiency of energy conservation in the formation of ATP when H_2S is oxidized is 50%.

Solution

The overall process (outlined in Figure 3-7) is based on the following considerations.   (a) The electrons from the pigment system must have a reduction potential more negative than -0.55 v in order to reduce methylviologen.   A likely estimate would be -0.60 v, which would be sufficient to reduce a natural electron acceptor, such as ferredoxin, and, in turn, $NADP^+$.   (b) Two einsteins of 700 nm light contains 81,572 cal.   If 60% is

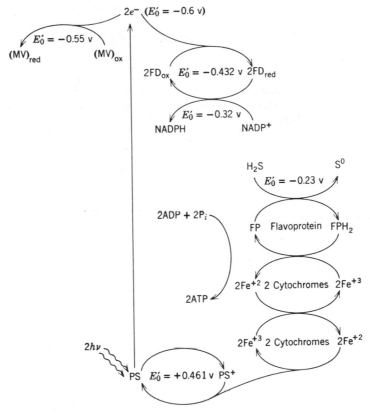

Figure 3-7 Noncyclic photosynthetic phosphorylation in *Chlorobium* in the presence of H_2S.

conserved in the reduction of the primary electron acceptor, then $\Delta G' = -48{,}943$ cal/2 equivalents. This corresponds to a $\Delta E_0'$ of 1.061 v. Therefore, the E_0' of the unexcited pigment system is $+0.461$ v.   (c) H_2S can reduce the oxidized pigment system.   The overall $\Delta E_0'$ is 0.691 v, which corresponds to a $\Delta G'$ of $-31{,}873$ cal/mole.   The amount of energy conserved at 50% efficiency (15,937 cal) is sufficient to make 2 ATP.   The exact sites of coupling cannot be predicted without a knowledge of the E_0' values of the carriers.   The products of the overall process are 1 NADPH, 2 ATP, and 1 free sulfur.   The 2 ATP : 1 NADPH ratio is more than sufficient for CO_2 reduction.   Note that a second pigment system is unnecessary in *Chlorobium* because the reduction potential of H_2S is more negative than that of the pigment system.   Green plants require PS II because H_2O cannot reduce oxidized PS I (but can reduce oxidized PS II).

F. ACTIVE TRANSPORT

The standard free energy change for the movement of an uncharged molecule from one side of a membrane at concentration C_1 to the other at concentration C_2 is given by the usual equation:

$$\Delta G = -2.3 \, RT \, \log K'_{eq} + 2.3 \, RT \, \log \frac{C_2}{C_1} \qquad (42)$$

K'_{eq} for a simple diffusion process is unity.   That is, at diffusion equilibrium, the concentration of the solute is the same on both sides of the membrane.   Thus, the ΔG for the movement of the solute from side 1 to side 2 under nonequilibrium conditions is given by:

$$\Delta G = 2.3 \, RT \, \log \frac{C_2}{C_1} \qquad (43)$$

Equation 43 gives the free energy of dilution or concentration.   That is, the difference in chemical potential of a solute at two different concentrations.   If C_1 is greater than C_2, ΔG is negative.   This says that the molecules of solute will spontaneously move from compartment 1 to compartment 2 (a conclusion we intuitively reach without any equations).   Living cells have the ability to transport and accumulate certain compounds against large concentration gradients.   The ΔG for such transport processes is clearly positive and, consequently, energy must be supplied.   That is, the uphill transport of a molecule must be coupled somehow to an exergonic reaction in order to make the overall ΔG zero or negative.   The mechanisms of energy coupling (as well as the mechanisms of transport itself) are subjects of intensive research.   Most transport processes are mediated by specific membrane carriers that behave in an enzymelike way.   The term "permease" is often used for such membrane transport systems.   Presumably, a protein of the transport system combines with the substrate at the external side of the membrane.   The complex then undergoes a conformational change that results in the release of the substrate internally (Fig. 3-8).   If the transport system simply equilibrates the external and internal substrate, the process is known as *facilitated diffusion*—there is no concentration against a gradient ($\Delta G = 0$), yet, the process is still carrier mediated and enzymelike.   If the transport system promotes an accumulation of the substrate against a gradient, the process is called *active transport* and requires energy input.   The energy could be used to effect translocation of the carrier-substrate complex inward, translocation of the uncharged carrier outward, or to reduce the affinity of the carrier for the substrate at the cytoplasm-membrane boundary, or to increase the affinity of the carrier at the external surface.   In some organisms, energy for active transport may be supplied directly by ATP.   The phosphorylation of the carrier promotes substrate binding, or release, or translocation.   In other organisms (or in other transport systems of the same organism) energy yielding reactions within the membrane create an *electrochemical potential* that then acts as the driving force.   For example, in animal cells, a membrane ATPase promotes the movement of 3 Na^+ ions outward and 2 K^+ ions inward per ATP hydrolyzed.   This results in both a high $[Na^+]_{out}/[Na^+]_{in}$ ratio and a superimposed electrical charge gradient (inside negative with respect to outside).   The ΔG for Na^+ transport inward

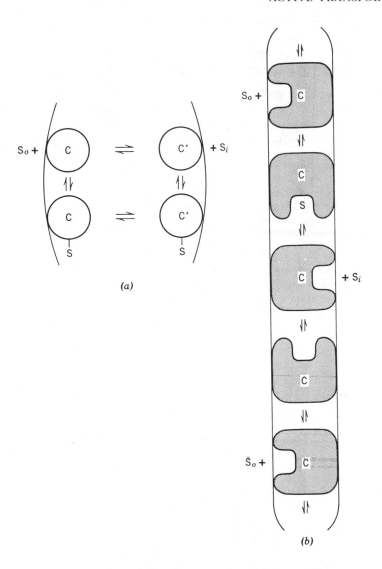

Figure 3-8 Carrier-mediated transport. C is a specific membrane protein with a high affinity for S. The carrier-substrate complex forms at the external surface and then (*a*) diffuses across the membrane or (*b*) undergoes a conformational change that brings the bound substrate to the internal surface, where it is released.

(which equals the energy made available by the concentration and charge gradient) is given by:

$$\Delta G = 2.3\ RT\ \log \frac{C_2}{C_1} + Z\mathscr{F}\Delta\Psi \qquad (44)$$

where $C_2/C_1 =$ the internal/external concentration ratio of the charged molecule (Na^+)

$Z =$ the net charge on the molecule ($+1$ for Na^+)

and $\Delta\Psi =$ the membrane potential in volts (internal relative to external)

The $Z\mathscr{F}\Delta\Psi$ term is analogous to the $n\mathscr{F}\Delta E$ of an oxidation-reduction reaction. (If the ATPase promoted a one-to-one exchange of K^+ for Na^+, then only the concentration gradients would be established and $\Delta\Psi = 0$.) Na^+ ions are carried inward (down the electrochemical gradient) by a membrane protein with a high affinity for Na^+. If this protein also possesses a high affinity for alanine, for example, it simultaneously mediates the transport of alanine. Alanine could be accumulated internally to the extent that the ΔG for further transport (a positive value) equals the ΔG for Na^+ transport (a negative value). In a sense, an endergonic reaction (uphill transport of alanine) has been coupled to an exergonic reaction (the downhill transport of Na^+) so that the sum of the two reactions has a ΔG of 0. In a similar manner, the high $[K^+]_{in}/[K^+]_{out}$ ratio can be used to transport compounds out of the cell. Keep in mind that actively metabolizing cells are in a steady-state. ATP is generated as it is used. Consequently, the transport and accumulation of a solute does not dissipate the membrane potential. For every Na^+ that moves inward (together with its cosubstrate), a Na^+ is moved outward at the expense of ATP.

· Problem 3-25

The concentration of chloride ion in blood serum is about $0.10\ M$. The concentration of chloride ion in urine is about $0.16\ M$. (a) Calculate the energy expended by the kidneys in transporting chloride from plasma to urine. (b) How many moles of Cl^- ions could be transported per mole of ATP hydrolyzed?

Solution

(a)
$$\Delta G = 1362 \log \frac{0.16}{0.10} = 1364(0.204)$$

$$\boxed{\Delta G = 278\ \text{cal/mole}}$$

(b) ATP hydrolysis provides 7700 cal/mole.

$$\frac{7700}{278} = 27.7 \qquad \therefore \qquad \boxed{\sim 28\ Cl^-/\text{ATP}}$$

· Problem 3-26

In *E. coli*, oxidation-reduction reactions (or ATP hydrolysis) within the cell membrane generates a ΔpH of 1 (interior higher by one unit) and a $\Delta\Psi$ of -120 mv (interior negative). Along with H^+ ions, β-galactosides are cotransported in response to the total "proton-motive force." (a) How much energy is made available by the ΔpH and $\Delta\Psi$?   (b) To what concentration gradient could *E. coli* accumulate β-galactosides?

Solution

(a)   A ΔpH of 1 unit is equivalent to an $[H^+]_{in}/[H^+]_{out}$ ratio of $1:10$

$$\Delta G = 1364 \log 0.1 + (1)(23{,}063)(-0.120)$$
$$= -1364 - 2768$$

$$\boxed{\Delta G = -4132 \text{ cal/mole}}$$

Thus, β-galactoside transport is coupled to an exergonic system providing 4132 cal/mole.

(b)   The maximum internal/external β-galactoside ratio is that which is equivalent to a ΔG of $+4132$ cal/mole.   At this ratio, the overall coupled system will be at equilibrium ($\Delta G_{overall} = 0$).

$$\Delta G = 1364 \log \frac{C_2}{C_1} = 4132 \text{ cal/mole}$$

$$\log \frac{C_2}{C_1} = \frac{4132}{1364} = 3.03$$

$$\boxed{\frac{C_2}{C_1} \simeq 1000}$$

G. ENTHALPY AND ENTROPY

ΔH AND ΔS

Suppose a mole of ATP is hydrolyzed at 25°C under standard conditions where $\Delta G' = -7700$ cal/mole.   The hydrolysis is not coupled to any group transfer reaction.   Does this mean then that the entire $\Delta G'$ appears as heat?   The first and second laws of thermodynamics relate the ΔG of a reaction to the heat evolved in the following way:

$$\boxed{\Delta G = \Delta H - T\Delta S} \tag{45}$$

ΔH is called *enthalpy change*, and represents the quantity of heat released (or absorbed) at constant temperature, pressure, and volume. ΔS is the *entropy change* and is a measure of the change in the randomness of the system.   If we measure or calculate ΔH for the hydrolysis of ATP, we obtain a value of about -4000 cal/mole.   The remaining 3700 cal/mole is not released as heat.   This amount of energy is retained by the products in the form of increased random motion. ($T\Delta S = +3700$ cal/mole.) Equation 45 could also be written as

$$\boxed{\Delta H = \Delta G + T\Delta S} \qquad (46)$$

which says that if a quantity of heat is added to a system (ΔH is positive), only a portion of the input can be used to do useful work on the system if ΔS is positive. The $T\Delta S$ term represents that portion of the heat input that becomes unavailable to do useful work.

The sign of ΔH gives no indication of the spontaneous direction of a reaction. We see from Equation 45 that even if ΔH is positive, ΔG can still be negative if ΔS is positive enough. For example, if we mix certain solid salts with water, we observe a marked decrease in the temperature as the salt dissolves. Heat has been absorbed. ΔH is positive (heat must be added to keep the temperature of the system constant). The salt spontaneously dissolves, therefore ΔG is negative. The $T\Delta S$ term is positive—the ions of the salt that were originally in a highly ordered crystalline array are now randomly distributed in solution.

The freezing of a liquid at a temperature below its freezing point (FP) is an example of a process that has a negative ΔH (heat is given off), a negative ΔG (freezing is spontaneous—at a temperature below FP the solid is more stable than the liquid), and a negative $T\Delta S$ (the molecules of the liquid become more highly ordered as a solid). It is obvious that we cannot predict the sign of ΔG from a knowledge of ΔH or vice versa unless we also know $T\Delta S$ (or unless ΔG or ΔH is zero). Similarly, we cannot predict the sign of ΔS from either ΔH or ΔG alone unless one of them is zero. For example, at 0°C, solid ice and liquid water are at equilibrium. Thus $\Delta G = 0$. ΔH for the reaction liquid → solid is negative (-80 cal/g or -1440 cal/mole). Thus, $T\Delta S = -1440$ cal/mole.

Entropy changes are not usually reported in terms of $T\Delta S$, but rather, in terms of ΔS.

$$T\Delta S = \Delta H - \Delta G \qquad \text{or} \qquad \Delta S = \frac{\Delta H - \Delta G}{T}$$

Thus, ΔS has units of $\text{cal} \times \text{mole}^{-1} \times \text{degree}^{-1}$ that are called entropy units (e.u.): 1 e.u. = $1 \text{ cal} \times \text{mole}^{-1} \times \text{degree}^{-1}$. The ΔS for ATP hydrolysis at 25°C can be calculated as

$$\Delta S = \frac{-4000 - (-7700)}{298} = \frac{-4000 + 7700}{298} = \frac{3700}{298}$$

$$\boxed{\Delta S = 12.4 \text{ e.u.}}$$

The ΔS for the freezing of water at 0°C $= -1440/273 = -5.3$ e.u. (The total entropy change—that of the water plus that of the surroundings receiving the heat—will be positive, as required by the second law of thermodynamics.)

In order to appreciate the value of ΔS measurements consider the reversible reaction:

$$\text{enzyme}_{\text{active}} \rightleftharpoons \text{enzyme}_{\text{inactive}}$$

Since ΔS is a measure of the change in orderliness, the ΔS of the above reaction can tell us something about the change in molecular shape that occurs with inactivation. A large positive ΔS would suggest that the inactivation is accompanied by an unfolding of the polypeptide chain into a less highly ordered, more random structure.

Enthalpy and entropy, like free energy, are *functions of state*. Consequently, ΔH and ΔS, like ΔG, depend only on the initial and final states of the system and not the mechanism or path of the reaction.   Thus, it is possible to calculate the ΔH and ΔS of a reaction that is unfeasible to carry out directly if that reaction can be expressed as the sum of two or more reactions where ΔH and ΔS values are known.

· Problem 3-27

The standard heat of combustion of ethanol at 25°C and 1 atm pressure is $-328,000$ cal/mole.   The standard heat of combustion of acetaldehyde is $-279,000$ cal/mole.   The E_0' of the acetaldehyde/ethanol half-reaction is -0.20 v.   The E_0' of the $\frac{1}{2}O_2/H_2O$ half-reaction is $+0.82$ v.   From these values, calculate (a) ΔH, (b) $\Delta G'$, and (c) ΔS for the reaction: ethanol $+\frac{1}{2}O_2 \rightarrow$ acetaldehyde $+ H_2O$.

Solution

(a)   The oxidation of ethanol to acetaldehyde:

1. $\qquad\qquad\qquad C_2H_5OH + \frac{1}{2}O_2 \rightleftharpoons CH_3CHO + H_2O$

can be expressed in terms of the two reactions whose ΔH values we know:

2. $\qquad\quad C_2H_5OH + 3O_2 \rightarrow 2CO_2 + 3H_2O \qquad \Delta H_2 = -328,000$ cal/mole

3. $\qquad\quad CH_3CHO + 2\frac{1}{2}O_2 \rightarrow 2CO_2 + 2H_2O \qquad \Delta H_3 = -279,000$ cal/mole

Writing reaction 3 backwards and then adding reaction 2:

3b. $\qquad 2CO_2 + 2H_2O \rightarrow CH_3CHO + 2\frac{1}{2}O_2 \qquad \Delta H_{3b} = +279,000$ cal/mole

2. $\qquad\quad C_2H_5OH + 3O_2 \rightarrow 2CO_2 + 3H_2O \qquad \Delta H_2 = -328,000$ cal/mole

---

1.   (Sum) $C_2H_5OH + \frac{1}{2}O_2 \rightarrow CH_3CHO + H_2O \qquad \Delta H_1 = \Delta H_{3b} + \Delta H_2$

$$\Delta H_1 = (+279,000) + (-328,000)$$

$$\boxed{\Delta H_1 = -49,000 \text{ cal/mole}}$$

(b) $\qquad\qquad\qquad \Delta G' = -n\mathscr{F}\Delta E_0' = -(2)(23,063)(1.02)$

$$\boxed{\Delta G' = -47,049 \text{ cal/mole}}$$

(c) $\Delta G' = \Delta H - T\Delta S$

$T\Delta S = \Delta H - \Delta G' = (-49,000) - (-47,049)$

$$\boxed{T\Delta S = -1951 \text{ cal/mole}}$$

$$\Delta S = \frac{-1951}{298} \qquad \boxed{\Delta S = -6.5 \text{ e.u}}$$

EFFECT OF TEMPERATURE ON K_{eq}—DETERMINATION OF ΔH

From the two equations for $\Delta G'$:

$$\Delta G' = -2.3\, RT \log K'_{eq} \qquad \text{and} \qquad \Delta G' = \Delta H - T\Delta S$$

we obtain:

$$-2.3\, RT \log K'_{eq} = \Delta H - T\Delta S$$

or

$$\boxed{\log K'_{eq} = -\frac{\Delta H}{2.3\, R}\frac{1}{T} + \frac{\Delta S}{2.3\, R}} \qquad (47)$$

Equation 47 is a linear relationship from which ΔH and ΔS can be obtained. All we need do is determine K'_{eq} at several different temperatures and then plot $\log K'_{eq}$ versus $1/T$ (Fig. 3-9). Alternatively, differentiating

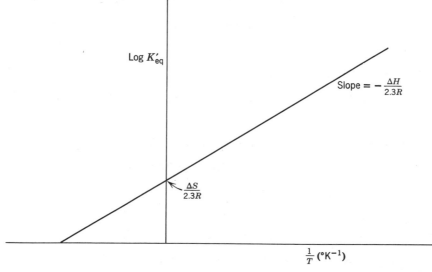

Figure 3-9 Graphical determination of ΔH from K'_{eq} values at two or more different temperatures. (ΔH is negative. If ΔH is positive, the plot has a negative slope.)

Equation 47 we obtain the van't Hoff equation:

$$\frac{d \log K'_{eq}}{d (1/T)} = -\frac{\Delta H}{2.3\,R} \quad \text{or} \quad d \log K'_{eq} = -\frac{\Delta H}{2.3\,R}\, d\,(1/T)$$

Integrating between limits of $\log K'_{eq_1}$ at $1/T_1$ and $\log K'_{eq_2}$ at $1/T_2$, we obtain:

$$\log \frac{K'_{eq_2}}{K'_{eq_1}} = -\frac{\Delta H}{2.3\,R}\left(\frac{1}{T_2} - \frac{1}{T_1}\right)$$

or

$$\log \frac{K'_{eq_2}}{K'_{eq_1}} = \frac{\Delta H}{2.3\,R}\left(\frac{T_2 - T_1}{T_2 T_1}\right) \tag{48}$$

If K'_{eq} is known at one temperature, the K'_{eq} at another temperature can be calculated if ΔH is known. Or, we can calculate ΔH from the K'_{eq} values at two different temperatures. If K'_{eq} increases with increasing temperature, ΔH is positive; if K'_{eq} decreases with increasing temperature, ΔH is negative. The calculations assume that ΔH is constant over the temperature range studied. This assumption is reasonably valid for the small temperature ranges usually employed for enzyme-catalyzed reactions (e.g., 20 to 40°C).

If we substitute $\Delta G'/-2.3\,RT$ for $\log K'_{eq}$, we obtain the integrated Gibbs-Helmholtz equation:

$$\frac{\Delta G'_2}{T_2} = \frac{\Delta G'_1}{T_1} - \Delta H\left(\frac{T_2 - T_1}{T_2 T_1}\right) \tag{49}$$

· Problem 3-28

An amino acid binding protein (presumably involved in membrane transport) was isolated from *E. coli*. Equilibrium dialysis measurements at 25 and 37°C yielded K_S values of 8.8×10^{-6} and 3.0×10^{-5} M, respectively. (K_S is the dissociation constant of the protein-substrate complex.) Calculate (a) $\Delta G'$ for the binding reaction at 25 and 37°C, (b) ΔH for the binding reaction, and (c) ΔS for the binding reaction at 25°C. (d) What does the sign and magnitude of ΔS suggest about the conformational change in the protein that accompanies amino acid binding?

Solution

(a)   The binding reaction is:

$$P + \text{amino acid} \rightleftharpoons P\text{-amino acid}$$

The K_S values are dissociation constants. The corresponding binding constants (called $K'_{eq_{25^\circ}}$ and $K'_{eq_{37^\circ}}$) for the reaction as written are:

$$K'_{eq25°} = \frac{1}{8.8 \times 10^{-6}} = 1.14 \times 10^{5}$$

$$K'_{eq37°} = \frac{1}{3 \times 10^{-5}} = 3.33 \times 10^{4}$$

$$\Delta G'_{25°} = -1364 \log 1.14 \times 10^{5} = -(1364)(5.056)$$

$$\boxed{\Delta G'_{25°} = -6896 \text{ cal/mole}}$$

At 37°C, $2.303\, RT = 1419$.

$$\therefore \quad \Delta G'_{37°} = -1419 \log 3.33 \times 10^{4} = -(1419)(4.522)$$

$$\boxed{\Delta G'_{37°} = -6417 \text{ cal/mole}}$$

(b)

$$\log \frac{K'_{eq37°}}{K'_{eq25°}} = \frac{\Delta H}{2.3\, R}\left(\frac{T_2 - T_1}{T_2 T_1}\right)$$

$$\Delta H = 2.3\, R\left(\log \frac{K'_{eq37°}}{K'_{eq25°}}\right)\left(\frac{T_2 T_1}{T_2 - T_1}\right)$$

$$= (2.3)(1.987)(\log 0.292)\left(\frac{298 \times 310}{12}\right)$$

$$\boxed{\Delta H = -18,809 \text{ cal/mole}}$$

(c)

$$T\,\Delta S = \Delta H - \Delta G' = (-18,809) - (-6896)$$

$$= -18,809 + 6896$$

$$\boxed{T\,\Delta S = -11,913 \text{ cal/mole}} \qquad \text{at } 25°C$$

$$\Delta S = \frac{-11,913}{298}$$

$$\boxed{\Delta S = -40 \text{ e.u.}} \qquad \text{at } 25°C$$

(d) We cannot conclude anything about conformational changes in the protein because the ΔS does not refer to the protein alone, but rather to the total entropy change:

$$\Delta S = (S_{p-aa}) - (S_p + S_{aa})$$
$$= (S_p + S_{aa})_{bound} - (S_p + S_{aa})_{free}$$

A large part of the negative ΔS may result from the restricted random motion of the bound amino acid.

H. ACTIVATION ENERGY

THE COLLISION THEORY AND THE ARRHENIUS EQUATION

The fact that a reaction has a large negative ΔG does not mean that the reaction proceeds at a rapid (or even a measurable) rate. The negative ΔG simply indicates the direction of the reaction, if it goes at all. The velocity of any homogeneous chemical reaction depends on the frequency of collisions between reactant molecules. The collision frequency is influenced by the concentrations of reactant molecules and how fast they move about (i.e., their kinetic energy). The kinetic energy of the molecules depends on the temperature. The collision frequency does not equal the reaction velocity because only a small proportion of the collisions occur with sufficient energy to promote the reaction. This minimum energy required for a fruitful reaction is called the *energy of activation, E_a*.

The relationship between E_a and temperature was formulated empirically by Arrhenius in 1889. The relationship is usually written as the Arrhenius equation:

$$k = Ae^{-E_a/RT} \tag{50}$$

Or, in linear form:

$$\log k = -\frac{E_a}{2.3\,R}\frac{1}{T} + \log A \tag{51}$$

The integrated form of the Arrhenius equation is:

$$\log \frac{k_2}{k_1} = \frac{E_a}{2.3\,R}\left(\frac{T_2 - T_1}{T_2 T_1}\right) \tag{52}$$

where k_2 and k_1 are the specific reaction rate constants at T_2 and T_1, respectively. The equations have the same form as the van't Hoff equations except now we are dealing with rate constants instead of equilibrium constants. At a constant concentration of reactants, the rate constants can be replaced with velocities since v is proportional to k. (See Problem 4-17.)

THE TRANSITION STATE

In order to explain the requirement for a "minimum energy for a fruitful reaction," Eyring, in 1935, proposed that a reactant molecule must overcome an energy barrier and pass through a transition state before proceeding on to the product of the reaction (Fig. 3-10). Reactant molecules that attain only a fraction of the required activation energy simply fall back to the ground

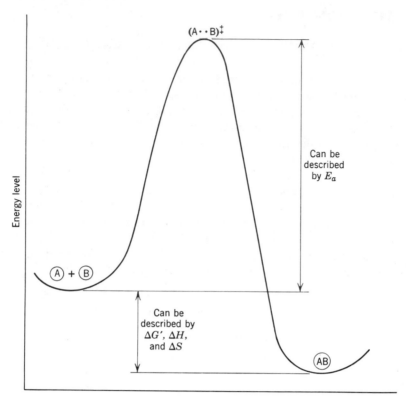

Figure 3-10 "Energy diagram" for the reaction $A + B \rightleftharpoons AB$. As shown, $\Delta G'$, ΔH, and ΔS are negative. E_a is always positive.

state. The transition state is viewed as an unstable, "halfway," transient phase in which bonds and orientations are distorted. Once the reactants overcome the energetic obstacle and attain the transition state, they proceed to form the products of the reaction at a rate independent of temperature and the nature of the reactants, that is, they slide down the other side of the energy barrier to the next ground state.

· Problem 3-29

The velocity of the reaction $A + B \rightarrow P + Q$ at 35°C is twice as great as the velocity at 25°C. Calculate the activation energy.

Solution

$$\log \frac{k_2}{k_1} = \frac{E_a}{2.3\,R} \left(\frac{T_2 - T_1}{T_2 T_1} \right)$$

$$E_a = \frac{2.3\,R T_2 T_1 \log \dfrac{k_2}{k_1}}{T_2 - T_1} = \frac{(2.3)(1.987)(298)(308) \log 2}{10}$$

$$\boxed{E_a = 12{,}627 \text{ cal/mole}}$$

GENERAL REFERENCES

Barrow, G. M., *Physical Chemistry for the Life Sciences*. McGraw-Hill (1974).

Klotz, I. M., *Energy Changes in Biochemical Reactions*. Academic Press (1967).

Lehninger, A. L., *Bioenergetics*. 2nd ed. W. A. Benjamin (1971).

Morris, J. G., *A Biologist's Physical Chemistry*. Addison-Wesley (1968).

Patton, A. R., *Biochemical Energetics and Kinetics*. W. B. Saunders (1965).

Racker, E., *Mechanisms in Bioenergetics*. Academic Press (1965).

PRACTICE PROBLEMS

(Consult Appendices VIII and IX for the necessary $\Delta E_0'$ and $\Delta G'$ values. Assume pH = 7 unless otherwise indicated.) Answers to Practice Problems are given on pages 424–426.

1. The K_{eq}' of the reaction glucose-6-phosphate $\rightleftharpoons$ fructose-6-phosphate is 0.43. Calculate the $\Delta G'$.

2. The K_{eq}' of the reaction G-1-P + ATP $\rightleftharpoons$ adenosine diphosphate glucose (ADPG) + PP_i is 1.0. (a) Calculate ΔG under steady-state conditions where [G-1-P] = 10^{-4} M, [ATP] = 10^{-3} M, [ADPG] = 10^{-5} M, and [PP_i] = 10^{-5} M. (b) In which direction will the reaction proceed spontaneously under these conditions?

3. Which reaction will proceed further to the right at pH 7: the hydrolysis of glucose-6-phosphate or the hydrolysis of glucose-6-sulfate? Why?

4. Fructose-1,6-diphosphate may be converted to glucose-1-phosphate by three consecutive reactions:

(1) fructose-1,6-diphosphate + H_2O

$\overset{\text{frustose}}{\underset{\text{diphosphatase}}{\rightleftharpoons}}$ fructose-6-phosphate + P_i

$$\Delta G_1' = -3800 \text{ cal/mole}$$

(2) fructose-6-phosphate

$\overset{\text{phospho-}}{\underset{\text{hexiosomerase}}{\rightleftharpoons}}$ glucose-6-phosphate

$$K_{eq_2}' = 2.0$$

(3) glucose-6-phosphate

$\overset{\text{phospho-}}{\underset{\text{glucomutase}}{\rightleftharpoons}}$ glucose-1-phosphate

$$K_{eq_3}' = 0.0526$$

From the above information, calculate the K_{eq}' and $\Delta G'$ for the overall reaction.

5. Calculate the $\Delta G'$ of hydrolysis of phosphenolpyruvate (PEP) to P_i and pyruvate given the following information.

(1) PEP + ADP $\overset{\text{pyruvic kinase}}{\rightleftharpoons}$ pyruvate + ATP

$$K_{eq_1}' = 3.2 \times 10^3$$

(2) ATP + H_2O $\overset{\text{ATPase}}{\rightleftharpoons}$ ADP + P_i

$$\Delta G_2' = -7700 \text{ cal/mole}$$

6. The $\Delta G'$ of succinyl-S-CoA hydrolysis at pH 7.0 is about -8000 cal/mole. How much of this negative ΔG results from the ionization of the newly formed carboxyl group? The pK_a of the carboxyl group (pK_{a_2} of succinic acid) is 5.57. (In other words, calculate ΔG_{ion}.)

7. The hydrolysis of asparagine to aspartate + NH_4^+ has a $\Delta G'$ of -3400 cal/mole. The hydrolysis of ATP to AMP + PP_i has a $\Delta G'$ of -8000 cal/mole.

(a) From the above information, calculate the $\Delta G'$ for asparagine biosynthesis via the overall reaction:

aspartate + ATP + NH_4^+
$\rightarrow$ asparagine + PP_i + AMP

(b) The overall biosynthetic reaction occurs in two steps:

(1) aspartate + ATP
$\rightleftharpoons$ β-aspartyladenylate + PP_i

(2) β-aspartyladenylate + NH_4^+
$\rightleftharpoons$ asparagine + AMP

The $\Delta G'$ of β-aspartyladenylate hydrolysis is $-10,000$ cal/mole. Calculate the $\Delta G'$ of each step in the overall synthesis of asparagine.

8. Hexokinase catalyzes the reaction ATP + glucose $\rightleftharpoons$ glucose-6-phosphate + ADP. $\Delta G' = -4562$, $K_{eq}' = 2.21 \times 10^3$. Calculate the

concentration of glucose-6-phosphate required to force the hexokinase reaction backwards (in the direction of glucose and ATP formation) in the presence of $10^{-5}\,M$ glucose, $10^{-3}\,M$ ATP, and $10^{-4}\,M$ ADP.

9. The enzyme ATP sulfurylase catalyzes the reaction:

$$ATP + SO_4^{2-} \rightleftharpoons APS + PP_i$$

APS is adenosine-5′-phosphosulfate (AMP-O-SO$_3^-$). The K'_{eq} of the reaction as written is 10^{-8}. Estimate the $\Delta G'$ of APS hydrolysis to $AMP + SO_4^{2-}$.

10. The K'_{eq} of the adenyl cyclase reaction (ATP $\rightleftharpoons$ cyclic AMP + PP$_i$) was found to be 0.065. If the $\Delta G'$ for ATP hydrolysis to $AMP + PP_i$ is taken to be -8000 cal/mole, calculate the $\Delta G'$ of cyclic AMP hydrolysis (to 5′-AMP).

11. Estimate the $\Delta G'$ values of the following reactions:

(a) Acetyl-S-CoA + butyrate
$\rightleftharpoons$ butyryl-S-CoA + acetate.

(b) Acetyl-S-CoA + ethanol
$\rightleftharpoons$ ethyl acetate + CoASH.

(c) ATP + pyruvate $\rightleftharpoons$ PEP + AMP + P$_i$.

(d) Glycerol + PP$_i$ $\rightleftharpoons$ α-glycerolphosphate $+ P_i$.

12. Calculate the equilibrium concentrations and concentration ratio of glucose-6-phosphate/glucose-1-phosphate in the phosphoglucomutase reaction when the initial concentration of glucose-6-phosphate is (a) $1\,M$, (b) $0.1\,M$, (c) $10^{-2}\,M$, (d) $10^{-3}\,M$, and (e) $10^{-4}\,M$. The K'_{eq} for the reaction written as glucose-1-phosphate $\rightleftharpoons$ glucose-6-phosphate is 19.

13. Calculate the equilibrium concentrations and concentration ratios of all components of the isocitritase reaction (isocitrate $\rightleftharpoons$ glyoxylate + succinate) when the initial concentration of isocitrate is (a) $1\,M$, (b) $0.1\,M$, (c) $0.01\,M$, (d) $10^{-3}\,M$, and (e) $10^{-4}\,M$. The $\Delta G'$ for the isocitritase reaction as written is $+2110$ cal/mole.

14. A solution was made $1\,M$ in the following compounds: acetoacetate, pyruvate, β-hydroxybutyrate, and lactate. (a) Write an equation for a thermodynamically favorable oxidation-reduction reaction that could occur. (b) Identify the compounds that are oxidized and reduced in the reaction. Identify the oxidizing and reducing agents. (c) Calculate $\Delta E'_0, \Delta G'$, and K'_{eq} for the reaction.

15. A solution containing $0.001\,M$ ubiquinone and $0.001\,M$ ubiquinone-H$_2$ was mixed with an equal volume of a solution containing $0.1\,M$ fumarate and $0.1\,M$ succinate. (a) Write the equation for a thermodynamically favorable reaction that could occur. (b) Calculate the ΔE, $\Delta G'$, and K'_{eq} values.

16. Calculate the reduction potential of the flavoprotein/flavoprotein-H$_2$ half-reaction when the oxidized/reduced concentration (activity) ratios are (a) 10^{-3}, (b) 0.2, (c) 1, (d) 3, (e) 25, and (f) 400. $E'_0 = -0.06$ v.

17. (a) Calculate the minimum NADH/NAD$^+$ ratio required to reduce oxalacetate to malate when $[OAA] = 10^{-4}\,M$ and $[malate] = 10^{-4}\,M$. (b) What must the ratio be if $[OAA] = 10^{-6}\,M$ and $[malate] = 10^{-4}\,M$?

18. Calculate the E'_0 values at pH 9.0 for (a) the oxalacetate/malate half-reaction and (b) the Fe^{3+}/Fe^{2+} half-reaction. (The E'_0 values given in Appendix IX are for pH 7.)

19. (a) *Thiobacillus thiooxidans* plays an important role in the sulfur cycle in nature. The organism oxidizes reduced inorganic sulfur compounds, and, in the process, obtains energy for the synthesis of ATP. The oxidation of elemental sulfur proceeds via the reaction:

$$S + 1\tfrac{1}{2}O_2 + H_2O \rightleftharpoons SO_4^{2-} + 2H^+$$
$$\Delta G' = -120,000 \text{ cal/mole}$$

Calculate the potential ATP yield at 40% efficiency of energy conservation under standard conditions. (b) Bacteria of the genus *Nitrosomonas* play a role in the nitrogen cycle in nature by oxidizing NH$_4^+$ to NO$_2^-$. The overall reaction is:

$$NH_4^+ + 1\tfrac{1}{2}O_2 \rightleftharpoons NO_2^- + H_2O + 2H^+$$
$$\Delta G' = -65,400 \text{ cal/mole}$$

Calculate the potential ATP yield at 40% efficiency of energy conservation under standard conditions.

20. Which compound would you expect to yield the greater amount of ATP per mole upon complete oxidation to $CO_2 + H_2O$: a 6-carbon fatty acid (e.g., *n*-hexanoic acid) or a 6-carbon carbohydrate (e.g., fructose)?

21. Given the following information:

$$\text{glucose} \rightleftharpoons 2\text{ ethanol} + 2CO_2$$
$$\Delta G' = -55{,}000 \text{ cal/mole}$$
$$\text{glucose} + 6O_2 \rightleftharpoons 6CO_2 + 6H_2O$$
$$\Delta G' = -686{,}000 \text{ cal/mole}$$

Calculate the number of moles of ATP that could be synthesized from $ADP + P_i$ upon complete oxidation of one mole of ethanol to $2CO_2 + 3H_2O$. Assume an efficiency of energy conservation of 44% under standard conditions.

22. Calculate the ΔpH across the inner mitochondrial membrane that is necessary to drive the ATPase reaction in the direction of ATP synthesis under steady-state conditions at 25°C where $[ATP] = 10^{-6}\,M$, $[ADP] = 10^{-3}\,M$, and $[P_i] = 10^{-2}\,M$.

23. Calculate the energy in one einstein of photons for light of wavelengths (a) 260 nm and (b) 750 nm.

24. In the presence of H_2, *Chromatium okenii*, a purple photosynthetic bacterium, carries out a cyclic photophosphorylation process that yields only ATP. (NADPH is obtained from the hydrogenase-catalyzed reaction: $H_2 + NADP^+ \rightleftharpoons NADPH + H^+$.) How many moles of ATP could theoretically be made when a pair of electrons are transferred from reduced ferredoxin back to the oxidized pigment systems? Assume an efficiency of energy conservation of 40–50%.

25. Blood contains about $0.1\,M$ Cl^-. Brain tissue contains about $0.04\,M$ Cl^-. Calculate (a) the ΔG for the transport of Cl^- from blood into brain cells and (b) the energy expended by brain cells in transporting Cl^- outward against the concentration gradient.

26. The pH of gastric juice is about 1.5. Assuming that the pH inside the cells of the gastric mucosa is 6.8, calculate the amount of energy required to secrete a mole of H^+ ions. Assume $T = 37$°C.

27. (a) Actively respiring *Neurospora crassa* develops a membrane potential of -0.30 v (interior negative). What is the ΔG for Ca^{2+} transport into the mycelium? Assume that under steady-state conditions the $[Ca^{2+}]_{in}/[Ca^{2+}]_{out}$ ratio is maintained at unity by an ATPase that moves Ca^{2+} outward. (b) To what concentration ratio could an amino acid be accumulated if the uptake of the amino acid is coupled to Ca^{2+} influx?

28. The ΔH of combustion of acetaldehyde is $-279{,}000$ cal/mole. The reaction is $CH_3CHO + 2\frac{1}{2}O_2 \rightarrow 2CO_2 + 2H_2O$. The ΔH of combustion of acetate is $-209{,}000$ cal/mole. The reaction is $CH_3COOH + 2O_2 \rightarrow 2CO_2 + 2H_2O$. Using these data, and the standard reduction potentials of the acetaldehyde/acetate and O_2/H_2O half-reactions at 25°C, calculate ΔH, $\Delta G'$, $T\Delta S$, and ΔS for the reaction: $CH_3CHO + \frac{1}{2}O_2 \rightarrow CH_3COOH$.

29. The K'_{eq} of a deamination reaction at 20°C is 185. The K'_{eq} at 37°C is 65. (a) Calculate ΔH and $\Delta G'$, $T\Delta S$ and ΔS at 37°C. (b) What is K'_{eq} at 28°C?

30. The *Neurospora crassa* protein kinase has an activation energy of 10,700 cal/mole. How much faster will the reaction proceed at 37°C compared to 15°C?

4

ENZYMES

A. ENZYMES AS BIOLOGICAL CATALYSTS

ACTIVATION ENERGY

The fact that a reaction has a very large negative free energy change does not mean that it will proceed at a rapid rate. A negative ΔG simply means that the existing [product]/[reactant] ratio is smaller than the equilibrium ratio. For example, the oxidation of glucose has a $\Delta G'$ of $-686,000$ cal/mole. That is, glucose in air is quite unstable in a *thermodynamic* sense. Yet, glucose as solid crystals or in sterile solution does not break down to $CO_2 + H_2O$ at a measurable rate. Glucose is quite stable in a *kinetic* sense.

Before a molecule of reactant, or substrate, S, can become a molecule of product, P, it must possess a certain minimum energy in order to pass into a transition state, $S \cdot \cdot P^{\ddagger}$ (Fig. 4-1). The amount of energy required is called the *activation energy*. As noted in Chapter 3, the transition state represents a halfway point where the bonds of S are distorted sufficiently so that conversion to P becomes possible. The rate of the reaction $S \rightarrow P$ depends on the number of molecules of S that enter the transition state per unit time. There are two ways of increasing the reaction rate. One is to raise the temperature. The other is to lower the activation energy. Living cells exist at relatively low temperatures—between 0 and 100°C. At life temperatures, few, if any, of the reactions of intermediary metabolism would occur at a rate sufficient to permit cell growth and maintenance. Living cells can exist under relatively mild environmental conditions because they possess *enzymes*—biological catalysts that selectively lower the energies of activation of vital chemical reactions. In the presence of an appropriate enzyme, the ambient temperature provides a substantial fraction of the reactant molecules with the required activation energy. An enzyme-catalyzed reaction at 25°C may proceed 10^6 to 10^{15} times faster than the same uncatalyzed reaction.

Enzymes have no effect on the ΔG or K_{eq} of a reaction. They simply speed up the rate at which a reaction approaches equilibrium. For example, in the reaction $S \underset{k_{-1}}{\overset{k_1}{\rightleftharpoons}} P$, k_1 might be 10^{-3} min^{-1} while k_{-1} might be 10^{-5} min^{-1}. At equilibrium, the forward and reverse reactions are equal. Therefore:

$$v_f = k_1[S] = v_r = k_{-1}[P]$$

$$K_{eq} = \frac{[P]}{[S]} = \frac{k_1}{k_{-1}} = \frac{10^{-3}}{10^{-5}} = 100$$

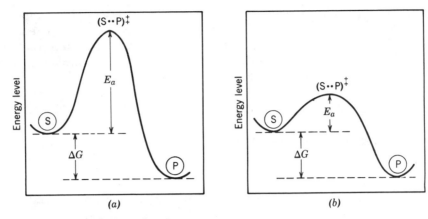

Figure 4-1 ΔG and E_a of (a) nonenzymatic and (b) enzymatic reactions.

In the presence of an appropriate enzyme, both k_1 and k_{-1} are enhanced to the same degree. Thus, if k_1 increases 10^6-fold, k_{-1} must also increase 10^6-fold. ΔG and K_{eq} are unchanged (Fig. 4-1b).

$$K_{eq} = \frac{k_1}{k_{-1}} = \frac{10^3}{10} = 100$$

ENZYMES AS MEDIATORS OF COUPLED REACTIONS

Enzymes do more than accelerate reactions. They also *couple* reactions in a productive manner. For example, consider the two reactions below:

1.   A-B → A + B + energy   (ΔG is negative)

2.   C + D + energy → C-D   (ΔG is positive)

The cleavage of A-B can supply sufficient energy to drive the synthesis of C-D. Yet, if reactions 1 and 2 occurred in different parts of the cell at different times there would be no way of using the energy of reaction 1 to push reaction 2, even if both occurred at a rapid rate. In a living cell, the overall process might be coupled in a productive manner as follows:

3.   A-B + X $\xrightarrow{E_1}$ A + B~X

4.   B~X + Y $\xrightarrow{E_2}$ B + X~Y

5.   C + X~Y $\xrightarrow{E_3}$ C~X + Y

6.   C~X + D $\xrightarrow{E_4}$ C-D + X

---

7. (Sum)   A-B + C + D $\longrightarrow$ C-D + A + B

In reaction 3, A-B is cleaved by enzyme E_1 and a portion of the energy made available is used to condense B with X to yield an activated form of B. The energy conserved in B~X is retained when X is transferred to Y in a reaction catalyzed by E_2.   The resulting $X \sim Y$ is a mobile "energy-rich" compound, such as ATP.   The overall condensation of $C + D$ occurs via two enzyme-catalyzed reactions.   First, C is activated to form $C \sim X$, which then condenses with D to form the final product, C-D.   The overall coupled reaction sequence is catalyzed by four different enzymes.

THE ACTIVE SITE

An enzyme-catalyzed reaction $S \rightarrow P$ can be written

$$S \xrightarrow{E} P$$

but it was recognized quite early that the enzyme, E, and the substrate must combine in some way during the reaction.   The overall reaction sequence can be written as:

$$S + E \longrightarrow ES \longrightarrow EP \longrightarrow E + P$$

Although the enzyme participates in the reaction sequence, it is not used up.   Thus, only a few molecules of E might catalyze the conversion of thousands of molecules of S to P each second.   The existence of an enzyme-substrate complex, ES, was inferred from (a) the high degree of specificity exhibited by enzymes, (b) the shape of the velocity versus substrate concentration curve, and (c) the fact that substrates frequently protect enzymes from inactivation.   The high degree of specificity prompted Emil Fischer in 1894 to suggest the *template* or *lock-and-key* analogy.   This relationship assumes that the enzyme possesses a region, called the active site, which is complementary in size, shape, and chemical nature to the substrate molecule (Fig. 4-2).   The more modern *flexible enzyme* or *induced fit* hypothesis of Koshland suggests that the active site need not be a preexisting rigid geometrical cavity, but rather a specific and precise spatial arrangement of amino acid R-groups that is induced by contact with the substrate (Fig. 4-3).

The active site of an enzyme occupies only a very small portion of the enzyme molecule.   In fact, there may be only a dozen or so amino acid

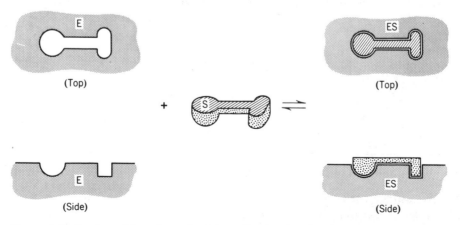

Figure 4-2   Lock-and-key (template) hypothesis of enzyme specificity.

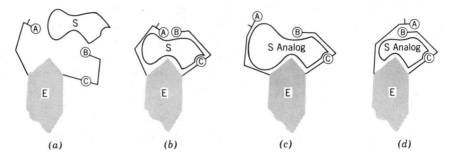

(a)   *(b)*   *(c)*   *(d)*

Figure 4-3   Induced-fit hypothesis of Koshland.   (*a*) The substrate approaches the active site.   (*b*) Substrate binding induces the proper alignment of the catalytic groups, A and B.   (*c*) and (*d*) Substrate analogs (competitive inhibitors) bind to the enzyme (aided by group C) but the catalytic groups are not aligned properly.   [Redrawn from D. E. Koshland, Jr., *Cold Spring Harbor Symposia on Quantitative Biology*, **28**, 473 (1963).]

residues surrounding the absorption pocket and, of these, only two or three may actually participate in substrate binding and/or catalysis.   Why then are enzymes large proteins instead of small tripeptides or dodecapeptides?   The answer is obvious when we consider that the two or three essential R-groups must be perfectly juxtaposed in three dimensional space.   A small linear peptide might contain all the essential binding and catalytic groups, but the

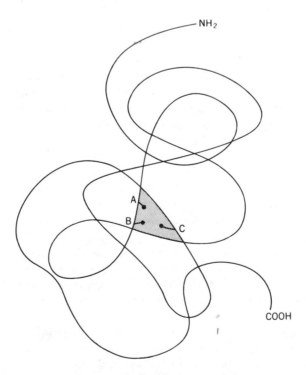

Figure 4-4   The active site (shaded area) occupies only a small region of the enzyme.   A, B, and C are the amino acid R-groups responsible for substrate binding and catalytic activity.

fixed bond distances and angles would not allow the essential R-groups to assume the required spatial relationship. A large protein composed of hundreds of amino acid residues can bend, twist, and fold back on itself and thereby fix the positions of the essential R-groups exactly in space (Fig. 4-4). The three essential groups that compose the catalytic center might be residues number 40, 41, and 92. A great many of the other (noncatalytic) amino acid residues play an equally important role, that of maintaining the enzyme in its tertiary structure (via electrostatic interactions, hydrogen bonds, dipole-dipole interactions, disulfide bonds, and hydrophobic interactions).

FACTORS RESPONSIBLE FOR THE CATALYTIC EFFICIENCY OF ENZYMES

Now that we have examined the concept of the active site, we can return to an important question: how do enzymes lower the activation energy? A number of factors have been suggested that we can examine in a very qualitative way. First of all, it is generally agreed that most enzyme-catalyzed reactions proceed via recognized organic reaction mechanisms (e.g., general acid-base catalysis, nucleophilic and electrophilic displacements) in which the enzyme provides the catalytic groups. Certainly, some of the rate enhancement by an enzyme stems from *proximity and orientation* factors. For two substrate molecules (or a substrate molecule and a catalytic group) to react they must get close enough to each other and the approach must occur at the proper angle. In solution, the random motion of the two molecules would yield a low probability of an effective collision. When the two molecules are adsorbed onto the active site of the enzyme (or when one of the reactants is the substrate and the other reactant is an R-group of the active site), then both the intermolecular distance and orientation may be optimized. The effective "concentration" of a substrate in the volume of the active site is much greater than the concentration in the solution from which the substrate was absorbed. Koshland and co-workers have proposed that the active sites of enzymes are so constructed that they align the orbitals of the substrate and catalytic groups optimally to enter the transition state. This concept is called *orbital steering*. A qualitatively similar concept of *stereopopulation control* has been discussed by Milstien and Cohen. These workers point out that the combined effect of multipoint attachment and the precise fit of the substrate into the active site would tend to restrict the rotational freedom of the substrate and "freeze" it into a unique conformation. Also, we might expect that substrates confined to the active site of an enzyme have a relatively long residence time (compared to the time interval that the same substrates would be within striking distance of each other if they were in random motion in solution). As a consequence of this *substrate anchoring* (as termed by Reuben), the number of substrate molecules attaining the transition state per unit time may be increased tremendously.

The idea that certain bonds of the substrate are distorted upon binding to the enzyme has been suggested by several workers. This so-called *rack mechanism* assumes that the substrate fits loosely into the active site, but the bonds that are formed between the enzyme and the substrate are so strong that a susceptible bond within the substrate is distorted producing the activated transition state (Fig. 4-5). In this mechanism, a portion of the

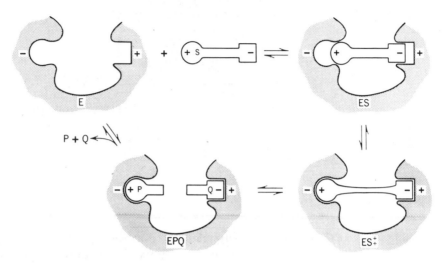

Figure 4-5   Distortion or rack mechanism.

activation energy is provided directly by the binding forces between E and S. Although the mechanism is easily illustrated for a cleavage reaction, similar distortion models can easily be imagined for condensations and group transfer reactions in which the transition states of the substrates are more tightly bound to the enzyme than the unactivated substrates.

· **Problem 4-1**

Under most *in vitro* assay conditions, the enzyme is used in *catalytic* amounts (10^{-12} to 10^{-8} M). Estimate the concentration of an enzyme in a living cell. Assume that (a) fresh tissue is 80% water and all of it is intracellular, (b) the total soluble protein in a cell represents 15% of the wet weight, (c) all the soluble proteins are enzymes, (d) the average molecular weight of a protein is 150,000, and (e) about 1000 different enzymes are present.

Solution

The intracellular concentration of soluble protein (enzymes) is

$$\frac{15 \text{ g protein}}{100 \text{ g wet weight}} = \frac{15 \text{ g protein}}{80 \text{ g water}} = 0.1875 \text{ g protein/g water}$$

or

$$\boxed{187.5 \text{ g/liter}}$$

The total molarity is

$$\frac{187.5 \text{ g/liter}}{150,000 \text{ g/mole}} = \boxed{1.25 \times 10^{-3} \ M}$$

If there are 1000 different enzymes, the average concentration of each is

$$\frac{1.25 \times 10^{-3}}{10^3} = \boxed{\mathbf{1.25 \times 10^{-6}\, M}}$$

Thus, we might expect a range of 10^{-8} to $10^{-4}\, M$ for individual enzymes. (See Problem 4-22 for another way of estimating the intracellular concentration of an enzyme.)

B. ENZYME KINETICS

Enzyme kinetics is that branch of enzymology that deals with the factors affecting the rates of enzyme-catalyzed reactions. The most important factors are those of enzyme concentration, ligand concentrations (substrates, products, inhibitors, and activators), pH, ionic strength, and temperature. When all these factors are analyzed properly, it is possible to learn a great deal about the nature of the enzyme-catalyzed reaction. For example, by varying the substrate and product concentrations, it is possible to deduce the *kinetic mechanism* of the reaction, that is, the order in which substrates add and products leave the enzyme and whether this order is obligate or random. Such studies can establish the kinds of enzyme-substrate and enzyme-product complexes that can form, and in some cases provide evidence for stable, covalently bound intermediates that are undetectable by ordinary chemical analyses. Certain kinetic constants can be determined and from these we can deduce the usual intracellular concentrations of substrates and products and the physiological direction of the reaction. The kinetics of a reaction may indicate the way in which the activity of the enzyme is regulated *in vivo*. A study of the effect of varying pH and temperature on the kinetic constants can provide information concerning the identities of the amino acid R-groups of the active site. A kinetic analysis can lead to a model for an enzyme-catalyzed reaction and, conversely, the principles of enzyme kinetics can be used to write the velocity equation for an attractive model, which can then be tested experimentally.

A SIMPLE UNIREACTANT SYSTEM—RAPID EQUILIBRIUM APPROACH (HENRI, MICHAELIS, AND MENTEN)

The simplest enzyme-catalyzed reaction involves a single substrate going to a single product. The system is called Uni Uni in the commonly used Cleland nomenclature. The reaction sequence is:

$$\mathrm{E} + \mathrm{S} \underset{k_{-1}}{\overset{k_1}{\rightleftharpoons}} \mathrm{ES} \underset{k_{-2}}{\overset{k_2}{\rightleftharpoons}} \mathrm{EP} \underset{k_{-3}}{\overset{k_3}{\rightleftharpoons}} \mathrm{E} + \mathrm{P}$$

ES and EP are called *central complexes*. For simplicity, we will assume that there is only one central complex and that the reverse reaction is insignificant. This latter assumption is valid if we concern ourselves with the *initial velocity* in the forward direction before a significant concentration of P has accumulated.   Thus, the reaction under consideration can be written:

$$E + S \underset{k_{-1}}{\overset{k_1}{\rightleftharpoons}} ES \overset{k_p}{\longrightarrow} E + P$$

The velocity equation can be derived in either of two ways.   The simplest method assumes *rapid equilibrium* conditions.   That is, that E, S, and ES equilibrate very rapidly compared to the rate at which ES breaks down to E + P.   The instantaneous velocity at any time depends on the concentration of ES:

$$v = k_p[ES]$$

k_p is called the catalytic rate constant.   The total enzyme is distributed between E and ES:

$$[E]_t = [E] + [ES]$$

Dividing the velocity-dependence equation by $[E]_t$, where $[E] + [ES]$ is used on the right-hand side, we obtain:

$$\frac{v}{[E]_t} = \frac{k_p[ES]}{[E] + [ES]}$$

Because of the equilibrium assumption, $[ES]$ can be expressed in terms of $[S]$, $[E]$, and K_S, where K_S is the dissociation constant of the ES complex:

$$K_S = \frac{[E][S]}{[ES]} = \frac{k_{-1}}{k_1} \qquad \therefore \quad [ES] = \frac{[S]}{K_S}[E]$$

Substituting for $[ES]$:

$$\frac{v}{[E]_t} = \frac{k_p \dfrac{[S]}{K_S}[E]}{[E] + \dfrac{[S]}{K_S}[E]}$$

Or, cross multiplying k_p and canceling $[E]$:

$$\frac{v}{k_p[E]_t} = \frac{\dfrac{[S]}{K_S}}{1 + \dfrac{[S]}{K_S}}$$

If $v = k_p[ES]$, then $k_p[E]_t = V_{max}$, the maximal velocity that would be observed when all the enzyme is present as ES.

$$\therefore \qquad \boxed{\frac{v}{V_{max}} = \frac{\dfrac{[S]}{K_S}}{1 + \dfrac{[S]}{K_S}}} \qquad (1)$$

All velocity equations for rapid equilibrium systems can be derived in the above manner.   The numerator of the right-hand side of the final equation will contain terms corresponding to the complexes that yield product.   The denominator will contain a term for each enzyme species present.   The term for any given complex contains a numerator and a denominator. The numerator of the term is the product of all ligand concentrations in the complex. The denominator of the term is the product of all dissociation constants between the complex and free E. The "1" in the denominator of the final velocity equation represents free E. In the simple unireactant system, there is only one product-forming complex, ES, and two enzyme species, E and ES. Hence, the numerator of the final velocity equation has only one term, while the denominator has two terms. The velocity equation for the simple unireactant system can be rearranged to yield the more familiar Henri-Michaelis-Menten equation:

$$\frac{v}{V_{max}} = \frac{[S]}{K_S + [S]} \qquad (2)$$

The Henri-Michaelis-Menten equation gives the *instantaneous* or *initial* velocity relative to V_{max} at a given substrate concentration. The equation is valid only if v is measured over a short enough time so that [S] remains essentially constant. This requires that no more than 5% of the substrate be utilized over the assay period.

THE STEADY-STATE APPROACH (BRIGGS AND HALDANE)

If the rate at which ES forms E + P is rapid compared to the rate at which ES dissociates back to E + S, then E, S, and ES will not be at equilibrium. (The equilibrium level of ES cannot accumulate.) If the enzyme is present in "catalytic" amounts (i.e., $[S] \gg [E]_t$), then very shortly after mixing E and S, a *steady-state* will be established in which the concentration of ES remains essentially constant with time (Fig. 4-6). A velocity equation can be derived in a manner very similar to that described earlier. This time, however, the concentration of ES is obtained from steady-state equations instead of equilibrium expressions. The reaction sequence is:

$$E + S \underset{k_{-1}}{\overset{k_1}{\rightleftharpoons}} ES \xrightarrow{k_p} E + P$$

As usual:

$$v = k_p[ES] \qquad \frac{v}{[E]_t} = \frac{k_p[ES]}{[E] + [ES]} \qquad (2a)$$

If the concentration of ES is constant, then the rate at which ES forms equals the rate at which ES decomposes. ES forms by one process:

$$E + S \xrightarrow{k_1} ES$$

ES decomposes by two processes:

$$ES \xrightarrow{k_p} E + P \quad \text{and} \quad ES \xrightarrow{k_{-1}} E + S$$

$\therefore$ rate of ES formation = $k_1[E][S]$

rate of ES decomposition = $k_{-1}[ES] + k_p[ES] = (k_{-1} + k_p)[ES]$

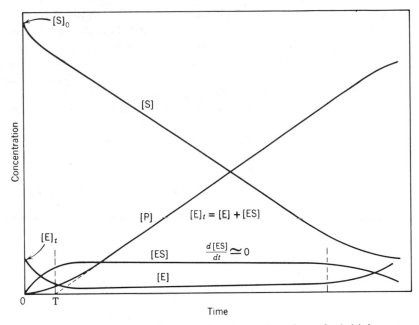

Figure 4-6   Progress curve for a catalyzed reaction where the initial reactant (substrate) concentration, $[S]_0$, is significantly greater than the initial enzyme concentration, $[E]_t$. As the ratio of $[S]_0/[E]_t$ increases, the steady-state region accounts for an increasing fraction of the total reaction time.   T represents the presteady-state interval.

At the steady-state, $d[ES]/dt = 0$, or

$$k_1[E][S] = (k_{-1} + k_p)[ES]$$

Solving for [ES]:

$$[ES] = \frac{k_1[E][S]}{(k_{-1} + k_p)}$$

The group of three rate constants can be defined as a single "Michaelis" constant, K_m:

$$K_m = \frac{k_{-1} + k_p}{k_1} \qquad \therefore \quad [ES] = \frac{[S]}{K_m}[E]$$

which upon substitution into the velocity dependence equation(2a) yields

$$\boxed{\frac{v}{V_{max}} = \frac{\dfrac{[S]}{K_m}}{1 + \dfrac{[S]}{K_m}}} \qquad \text{or} \qquad \boxed{\frac{v}{V_{max}} = \frac{[S]}{K_m + [S]}} \tag{3}$$

Thus, the form of the velocity equation is the same as that derived for rapid equilibrium conditions.   Only the rate constants that compose the final constant, K, differ.   Note that the expression for [ES] can be rearranged to:

$$K_m = \frac{[S][E]}{[ES]} = \frac{k_{-1} + k_p}{k_1}$$

Thus, K_m, the Michaelis constant, is a dynamic or pseudo-equilibrium constant expressing the relationship between the actual steady-state concentrations, rather than the equilibrium concentrations. If k_p is very small compared to k_{-1}, K_m reduces to K_S. A steady-state treatment of the more realistic reaction sequence $E + S \rightleftharpoons ES \rightleftharpoons EP \rightleftharpoons E + P$ yields the same final velocity equation although now K_m is a more complex function, composed of the rate constants of all the steps. Thus, the physical significance of K_m cannot be stated with any certainty in the absence of other data concerning the relative magnitudes of the various rate constants. Nevertheless, K_m represents a valuable constant that relates the velocity of an enzyme-catalyzed reaction to the substrate concentration. Inspection of the Henri-Michaelis-Menten equation shows that K_m is equivalent to the substrate concentration that yields half-maximal velocity:

$$v = \frac{[S]}{K_m + [S]} V_{\max} \qquad \therefore \quad \text{when } [S] = K_m: \quad v = \frac{K_m}{K_m + K_m} V_{\max} = \tfrac{1}{2} V_{\max}$$

WHY DETERMINE K_m?

The numerical value of K_m is of interest for several other reasons: (a) the K_m establishes an approximate value for the intracellular level of the substrate. It is unlikely that this level would be significantly greater or significantly lower than K_m. If $[S]_{intracell} \ll K_m$, v would be very sensitive to changes in [S] but most of the catalytic potential of the enzyme would be wasted since v would be $\ll V_{\max}$. There is also no physiological sense in maintaining $[S] \gg K_m$ since v cannot exceed $V_{\max}$, and the difference between v at $[S] = K_m$ and v at $[S] = 1000 K_m$ is only twofold. Also at $[S] \gg K_m$, v becomes insensitive to small changes in [S]. (b) Since K_m is a constant for a given enzyme, its numerical value provides a means of comparing enzymes from different organisms or from different tissues of the same organism, or from the same tissue at different stages of development. In this way, we might determine whether enzyme A is identical to enzyme B, or whether they are different proteins that catalyze the same reaction. (Note that $V_{\max}$ is not a constant. $V_{\max}$ depends on k_p, which is a constant, and the concentration of enzyme in the assay.) (c) A ligand-induced change in the effective value of K_m is one mode of regulating the activity of an enzyme. If K_m determined *in vitro* seems "unphysiologically" high then we might search for activators that function *in vivo* to lower the effective K_m. By measuring the effects of different compounds on K_m we might identify physiologically important inhibitors as well. (d) If we know K_m, we can adjust the assay conditions so that $[S] \gg K_m$ and thereby determine $V_{\max}$, which is a measure of $[E]_t$. (e) The Michaelis constant indicates the relative "suitability" of alternate substrates of a particular enzyme. That is, the substrate with the lowest K_m value has the highest apparent affinity for the enzyme. (The "best" substrate is that which has the highest $V_{\max}/K_m$ ratio.)

· **Problem 4-2**

Given the reaction $E + S \underset{k_{-1}}{\overset{k_1}{\rightleftharpoons}} ES \xrightarrow{k_p} E + P$ where $k_1 = 1 \times 10^7 \ M^{-1} \times sec^{-1}$, $k_{-1} = 1 \times 10^2 \ sec^{-1}$, and $k_p = 3 \times 10^2 \ sec^{-1}$, calculate (a) K_S and (b) K_m. (c) Can k_p be very much greater than k_1?

Solution

(a)
$$K_S = \frac{k_{-1}}{k_1} = \frac{1 \times 10^2 \ sec^{-1}}{1 \times 10^7 \ M^{-1} \times sec^{-1}} = \boxed{1 \times 10^{-5} \ M}$$

(b)
$$K_m = \frac{k_{-1} + k_p}{k_1} = \frac{(1 \times 10^2 \ sec^{-1}) + (3 \times 10^2 \ sec^{-1})}{1 \times 10^7 \ M^{-1} \times sec^{-1}}$$

$$= \frac{4 \times 10^2 \ sec^{-1}}{1 \times 10^7 \ M^{-1} \times sec^{-1}} = \boxed{4 \times 10^{-5} \ M}$$

(c)   Students often have the mistaken idea that k_p cannot be greater than k_1 because if $k_p \gg k_1$ it would mean that "ES breaks down faster than it forms." First of all, k_p and k_1 have different units and cannot be compared directly; k_1 has units of $M^{-1} \times sec^{-1}$ or $M^{-1} \times min^{-1}$ while k_p has units of sec^{-1} or min^{-1}. Furthermore, k_1 and k_p are *not rates* but, instead, they are *rate constants* (second-order and first-order rate constants, respectively). The *rate* of ES formation is $k_1[E][S]$. The *rate* of ES breakdown to $E + P$ is $k_p[ES]$. Thus it is quite possible for k_p to be numerically much greater than k_1. But for any given substrate and enzyme concentration $k_p[ES]$ cannot exceed $k_1[E][S]$. It is also quite possible for $k_p \gg k_{-1}$ (in which case, K_m reduces to k_p/k_1).

HALDANE RELATIONSHIP BETWEEN KINETIC CONSTANTS AND EQUILIBRIUM CONSTANT

The constants K_m and V_{max} were derived in terms of the various rate constants of the overall reaction. The equilibrium constant for the overall reaction is composed of the same rate constants. Consequently, it should be possible to express K_{eq} in terms of K_m and V_{max}. For example, consider the simple two-step reaction shown below.

$$E + S \underset{k_{-1}}{\overset{k_1}{\rightleftharpoons}} ES \underset{k_{-2}}{\overset{k_2}{\rightleftharpoons}} E + P$$

The reaction may be measured in either direction. We will designate the K_m values for S and P as K_{ms} and K_{mp}, respectively. The maximal initial velocities in the forward and reverse direction will be designated V_{max_f} and V_{max_r}, respectively. As shown earlier:

$$K_{ms} = \frac{k_2 + k_{-1}}{k_1} \qquad and \qquad V_{max_f} = k_2[E]_t$$

By an identical steady-state treatment, we can show that:

$$K_{mp} = \frac{k_2 + k_{-1}}{k_{-2}} \qquad and \qquad V_{max_r} = k_{-1}[E]_t$$

The overall equilibrium constant for the reaction reading left to right is the product of the equilibrium constants for the individual steps which may be expressed in terms of the rate constants:

$$K_{eq} = K_1 K_2 = \frac{k_1 k_2}{k_{-1} k_{-2}}$$

We can express this grouping of rate constants in terms of K_m and V_{max} values as shown below.

$$\frac{V_{max_f}}{K_{m_S}} = \frac{k_1 k_2 [E]_t}{k_2 + k_{-1}} \qquad \frac{V_{max_r}}{K_{m_P}} = \frac{k_{-2} k_{-1} [E]_t}{k_2 + k_{-1}}$$

Now dividing one ratio by the other:

$$\frac{V_{max_f}/K_{m_S}}{V_{max_r}/K_{m_P}} = \frac{(k_1 k_2 [E]_t)(k_2 + k_{-1})}{(k_2 + k_{-1})(k_{-2} k_{-1} [E]_t)}$$

or

$$\boxed{\frac{V_{max_f} K_{m_P}}{V_{max_r} K_{m_S}} = \frac{k_1 k_2}{k_{-1} k_{-2}} = K_{eq}}$$

(4)

The relationship between K_{eq}, K_m, and V_{max} is known as the *Haldane equation.*

REVERSIBLE REACTIONS—EFFECT OF PRODUCT ON FORWARD VELOCITY

Strictly speaking, all enzyme-catalyzed reactions are reversible. The overall reaction can be represented as:

$$E + S \underset{k_{-1}}{\overset{k_1}{\rightleftharpoons}} ES \underset{k_{-2}}{\overset{k_2}{\rightleftharpoons}} EP \underset{k_{-3}}{\overset{k_3}{\rightleftharpoons}} E + P$$

Under the usual assay conditions, velocities are measured very early in the reaction before the product concentration has increased to a significant level. For the reaction sequence shown above, we can calculate the initial velocity for the reaction in either direction from the appropriate Henri-Michaelis-Menten equations.

When $[P] = 0$: $\quad v_f = \dfrac{V_{max_f}[S]}{K_{m_S} + [S]}$ and when $[S] = 0$: $\quad v_r = \dfrac{V_{max_r}[P]}{K_{m_P} + [P]}$

It would be instructive to examine the effect of the product on the initial forward velocity. For example, suppose we have a solution containing a certain concentration of S and a certain concentration of P. In the absence of an appropriate enzyme, the reaction does not occur at a measurable rate. Now we add an enzyme catalyzing the reversible reaction $S \rightleftharpoons P$. In which direction and at what rate will the reaction progress? The direction of the reaction will depend on the ratio of $[P]/[S]$ relative to K_{eq}. An equation for the net velocity can be derived quite easily from rapid equilibrium assumptions (where $K_{m_S} = K_S$, and $K_{m_P} = K_P$).

$$v_{net} = k_2[ES] - k_{-2}[EP]$$

$$\frac{v_{net}}{[E]_t} = \frac{k_2[ES] - k_{-2}[EP]}{[E] + [ES] + [EP]}$$

$$[ES] = \frac{[S]}{K_S}[E] \qquad [EP] = \frac{[P]}{K_P}[E]$$

$$\therefore \quad \frac{v_{net}}{[E]_t} = \frac{k_2 \dfrac{[S]}{K_S}[E] - k_{-2} \dfrac{[P]}{K_P}[E]}{[E] + \dfrac{[S]}{K_S}[E] + \dfrac{[P]}{K_P}[E]}$$

$$v_{net} = \frac{k_2[E]_t \dfrac{[S]}{K_S} - k_{-2}[E]_t \dfrac{[P]}{K_P}}{1 + \dfrac{[S]}{K_S} + \dfrac{[P]}{K_P}}$$

$$v_{net} = \frac{V_{max_f} \dfrac{[S]}{K_S} - V_{max_r} \dfrac{[P]}{K_P}}{1 + \dfrac{[S]}{K_S} + \dfrac{[P]}{K_P}} \qquad (5)$$

or

$$v_{net} = \frac{V_{max_f}\left([S] - \dfrac{[P]}{K_{eq}}\right)}{K_S\left(1 + \dfrac{[P]}{K_P}\right) + [S]} \qquad \text{where} \quad K_{eq} = \frac{V_{max_f} K_P}{V_{max_r} K_S} \qquad (6)$$

A steady-state treatment yields the same final equations with K_{m_S} and K_{m_P} replacing K_S and K_P.   In place of the usual [S] in the numerator of the above equation, we have the difference between [S] and the equilibrium value of [S].   The K_S term in the denominator is modified in a manner consistent for the product acting as a competitive inhibitor with respect to the substrate.   In other words, the initial net velocity depends on the displacement of the system from equilibrium (i.e., the thermodynamic driving force) and the amount of enzyme tied up with product.   A more detailed account of competitive inhibition is given in a later section.

VELOCITY VERSUS SUBSTRATE CONCENTRATION CURVES

The Henri-Michaelis-Menten equation describes the curve obtained when initial velocity is plotted versus substrate concentration.   The curve shown in Figure 4-7 is a right rectangular hyperbola with limits of V_{max} and $-K_m$.   The curvature is fixed regardless of the values of K_m and V_{max}.   Consequently, the ratio of substrate concentrations for any two fractions of V_{max} is constant for all enzymes that obey Henri-Michaelis-Menten kinetics.   For example, the ratio of substrate required for 90% of V_{max} to the substrate required for

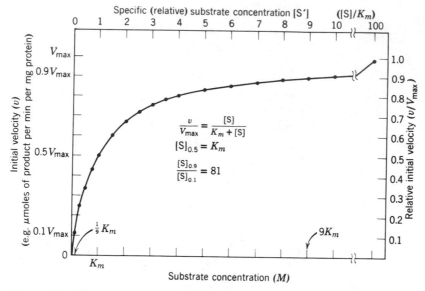

Figure 4-7 The curvature of the v versus [S] plot is constant. $[S]_{0.9}/[S]_{0.1}$ always equals 81 regardless of the absolute values of K_m and V_{max}.

10% of V_{max} is always 81 as shown below:

When $v = 0.9\ V_{max}$: $\quad 0.9 = \dfrac{[S]_{0.9}}{K_m + [S]_{0.9}} \qquad [S]_{0.9} = 9\ K_m$

When $v = 0.1\ V_{max}$: $\quad 0.1 = \dfrac{[S]_{0.1}}{K_m + [S]_{0.1}} \qquad [S]_{0.1} = \dfrac{K_m}{9}$

$$\therefore \qquad \boxed{\dfrac{[S]_{0.9}}{[S]_{0.1}} = 81}$$

· **Problem 4-3**

(a) What fraction of V_{max} is observed at $[S] = 4\ K_m$, $[S] = 5\ K_m$, $[S] = 6\ K_m$, $[S] = 9\ K_m$, and $[S] = 10\ K_m$? (b) Calculate the ratios of $[S]_{0.9}/[S]_{0.5}$ and $[S]_{0.75}/[S]_{0.5}$.

Solution

(a) Without any hesitation or substitution we can state that:

At $[S] = 4\ K_m$, $v = \frac{4}{5}\ V_{max}$; at $[S] = 5\ K_m$, $v = \frac{5}{6}\ V_{max}$; at $[S] = 6\ K_m$, $v = \frac{6}{7}\ V_{max}$; at $[S] = 9\ K_m$, $v = \frac{9}{10}\ V_{max}$; at $[S] = 10\ K_m$, $v = \frac{10}{11}\ V_{max}$.

(b) As shown above, $[S]_{0.9} = 9\ K_m$ and, of course, $[S]_{0.5} = K_m$. Therefore, the $[S]_{0.9}/[S]_{0.5}$ ratio is always 9, regardless of the absolute values of V_{max} and K_m. We observe $0.75\ V_{max}$ at $[S] = 3\ K_m$. Thus, $[S]_{0.9}/[S]_{0.75}$ is always 3.

· **Problem 4-4**

The following data were recorded for the enzyme-catalyzed reaction $S \rightarrow P$.

[S]	v
(M)	(nmoles $\times$ liter^{-1} $\times$ min^{-1})
6.25×10^{-6}	15.0
7.50×10^{-5}	56.25
1.00×10^{-4}	60
1.00×10^{-3}	74.9
1.00×10^{-2}	75

(a) Estimate V_{max} and K_m. (b) What would v be at $[S] = 2.5 \times 10^{-5}\,M$ and at $[S] = 5.0 \times 10^{-5}\,M$? (c) What would v be at $5.0 \times 10^{-5}\,M$ if the enzyme concentration were doubled? (d) The v given in the above table was determined by measuring the concentration of product that had accumulated over a 10-minute period. Verify that v represents a true initial (or "instantaneous") velocity.

Solution

(a)   The best way to obtain V_{max} and K_m is to plot the data by one of the methods described later. However, for the present, we see that v becomes insensitive to changes in [S] above $10^{-3}\,M$.   That is, in the region of $[S] = 10^{-3}$ to $10^{-2}\,M$, v must be very close to V_{max}.

$$V_{max} = 75 \text{ nmoles} \times \text{liter}^{-1} \times \text{min}^{-1}$$

. To solve for K_m, we can pick any v and the corresponding [S]:

$$\frac{v}{V_{max}} = \frac{[S]}{K_m + [S]} \qquad \frac{60}{75} = \frac{10^{-4}}{K_m + 10^{-4}}$$

$$75 \times 10^{-4} = 60\,K_m + 60 \times 10^{-4}$$

$$K_m = \frac{15 \times 10^{-4}}{60} = 0.25 \times 10^{-4}$$

$$K_m = 2.5 \times 10^{-5}\,M$$

Any other set of data should give the same answer if the enzyme obeys the Henri-Michaelis-Menten equation.

(b)   At $[S] = 2.5 \times 10^{-5}\,M = K_m$, $v = 0.5\,V_{max}$, or

$$v = 37.5 \text{ nmoles} \times \text{liter}^{-1} \times \text{min}^{-1}$$

At $[S] = 5.0 \times 10^{-5}\ M$:

$$\frac{v}{75} = \frac{5 \times 10^{-5}}{(2.5 \times 10^{-5}) + (5 \times 10^{-5})} = \frac{5}{7.5}$$

$$v = \frac{(5)(75)}{7.5} \quad \text{or} \quad \boxed{v = 50\ \textbf{nmoles} \times \textbf{liter}^{-1} \times \textbf{min}^{-1}}$$

Note that when $[S] = 2\ K_m$, v is not V_{max}, although when $[S] = K_m$, v is $0.5\ V_{max}$. We are not dealing with a linear relationship but, instead, a hyperbolic relationship.

(c)   The Henri-Michaelis-Menten equation can be written as:

$$v = \frac{[S]}{K_m + [S]}\ k_p [E]_t$$

Thus, v is directly proportional to the enzyme concentration at all substrate concentrations.   Doubling $[E]_t$ at $[S] = 5 \times 10^{-5}\ M$ doubles V_{max} and, hence, v.

$$\therefore \quad \boxed{v = 100\ \textbf{nmoles} \times \textbf{liter}^{-1} \times \textbf{min}^{-1}}$$

(d)   The velocity can be considered as the true initial or instantaneous velocity only if the substrate concentration remains essentially constant over the assay time, that is, only if a small fraction of $[S]$ is utilized.   This is no problem at substrate concentrations that are relatively high compared to K_m, so let us check for substrate depletion at the lowest $[S]$ used.   The apparent v at $6.25 \times 10^{-6}\ M$ $[S]$ is 15 nmoles $\times$ liter^{-1} $\times$ min^{-1} or, in other words, 150 nmoles/liter of P had accumulated (and 150 nmoles of S had disappeared) in 10 minutes.

$$\frac{150 \times 10^{-9}\ \text{moles S utilized per liter}}{6.25 \times 10^{-6}\ \text{moles S originally present per liter}} = \frac{0.150 \times 10^{-6}}{6.25 \times 10^{-6}}$$

$$= \boxed{\textbf{0.024 or 2.4\%}}$$

Only 2.4% of S was utilized.   Anything less than 5% is acceptable.

· **Problem 4-5**

The equilibrium constant for the reaction $S \rightleftharpoons P$ is 5.   Suppose we have a mixture of $[S] = 2 \times 10^{-4}\ M$ and $[P] = 3 \times 10^{-4}\ M$. $K_{m_S} = 3 \times 10^{-5}\ M$, $V_{max_f} = 2\ \mu\text{moles} \times \text{liter}^{-1} \times \text{min}^{-1}$, $V_{max_r} = 4\ \mu\text{moles} \times \text{liter}^{-1} \times \text{min}^{-1}$.   (a) In which direction will the reaction proceed on addition of an appropriate enzyme?   (b) At what initial velocity will the reaction start towards equilibrium?

Solution

(a)   The existing $[P]/[S]$ ratio is 1.5, which is less than K_{eq}.   Consequently, the reaction will proceed from S to P.

(b)   In order to calculate the initial v_{net}, we need to know K_{m_P}.   This can be calculated from the Haldane equation.

$$K_{eq} = \frac{V_{max_f} K_{m_P}}{V_{max_r} K_{m_S}} \qquad 5 = \frac{2 \, K_{m_P}}{(4)(3 \times 10^{-5})}$$

$$\boxed{K_{m_P} = 3.0 \times 10^{-4} \, M}$$

$$v_{net} = \frac{V_{max_f} \dfrac{[S]}{K_{m_S}} - V_{max_r} \dfrac{[P]}{K_{m_P}}}{1 + \dfrac{[S]}{K_{m_S}} + \dfrac{[P]}{K_{m_P}}}$$

$$= \frac{\dfrac{(2)(2 \times 10^{-4})}{(3 \times 10^{-5})} - \dfrac{(4)(3 \times 10^{-4})}{(3 \times 10^{-4})}}{1 + \dfrac{2 \times 10^{-4}}{3 \times 10^{-5}} + \dfrac{3 \times 10^{-4}}{3 \times 10^{-4}}}$$

$$= \frac{13.33 - 4}{1 + 6.67 + 1} = \frac{9.33}{8.67}$$

$$\boxed{v_{net} = 1.08 \; \mu\text{moles} \times \text{liter}^{-1} \times \text{min}^{-1}}$$

C. REACTION ORDER

If we examine the v versus [S] curve, we find three distinct regions where the velocity responds in a characteristic way to increasing [S] (Fig. 4-8a).   At very low substrate concentrations (e.g., $[S] < 0.01 \, K_m$), the v versus [S] curve is essentially linear, that is, the velocity (for all practical purposes) is directly proportional to the substrate concentration (Fig. 4-8b).   This is the region of *first-order kinetics*.   At very high substrate concentrations (e.g., $[S] > 100 \, K_m$), the velocity is essentially independent of the substrate concentration.   This is the region of *zero-order kinetics* (Fig. 4-8c).   At intermediate substrate concentrations, the relationship between v and [S] follows neither first-order nor zero-order kinetics.   The characteristics of the first-order and zero-order regions are described below.

FIRST-ORDER KINETICS

The linear relationship between v and [S] when $[S] \ll K_m$ can be derived from the Henri-Michaelis-Menten equation.

$$v = \frac{V_{max}[S]}{K_m + [S]}$$

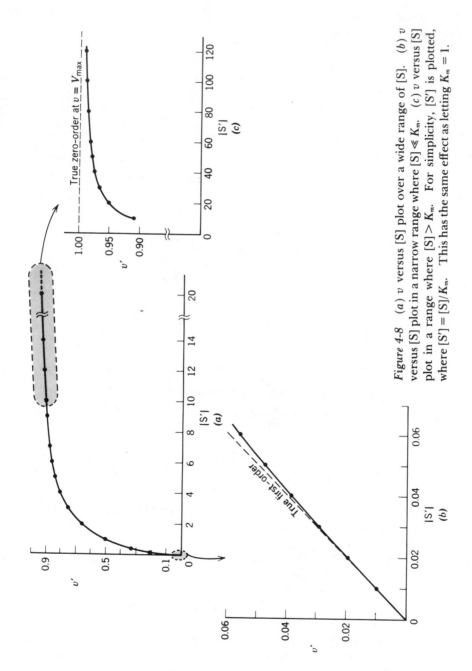

Figure 4-8 (*a*) v versus [S] plot over a wide range of [S]. (*b*) v versus [S] plot in a narrow range where [S] ≪ K_m. (*c*) v versus [S] plot in a range where [S] > K_m. For simplicity, [S'] is plotted, where [S'] = [S]/K_m. This has the same effect as letting $K_m = 1$.

When $[S] \ll K_m$, the $[S]$ in the denominator may be ignored and the equation reduces to:

$$v = \frac{V_{max}}{K_m}[S] \qquad \text{or} \qquad v = k[S] \qquad (7)$$

where k is the first-order rate constant for the overall reaction. The units of k are min^{-1} if v is expressed as $moles \times liter^{-1} \times min^{-1}$ and K_m is expressed as moles/liter:

$$k = \frac{V_{max}}{K_m} = \frac{moles \times liter^{-1} \times min^{-1}}{moles \times liter^{-1}} = \frac{(moles)}{(liter)(min)} \times \frac{liter}{moles} = \frac{1}{min}$$

The equation expresses the fact that when $[S]$ is very small, the absolute velocity decreases from moment to moment as $[S]$ decreases (Fig. 4-9a). However, at any given moment, a constant *fraction* of the substrate present undergoes conversion to product:

$$\frac{-d[S]}{dt} \qquad = \qquad v \qquad = \qquad k \qquad\qquad [S]$$

| The amount of S used up per small increment of time... | that is, the velocity... | is some constant fraction... | of the substrate present at that time. |

or

$$\frac{-d[S]/[S]}{dt} = k$$

Thus, the physical significance of the first-order rate constant is that it approximates the fraction of the substrate present that is converted to product per small increment of time. A k that is greater than $1\ min^{-1}$ means that more than 100% of the substrate present at zero time would be utilized in a minute *if* v remained constant for a minute. It may be more meaningful to express k values in units that yield numerical values that are less than unity.

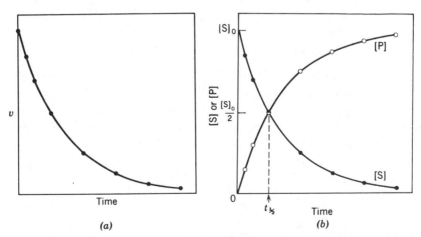

Figure 4-9 First-order region of the velocity curve. (*a*) v decreases continually with time. (*b*) The appearance of P and disappearance of S are not linear with time.

Because v decreases with time in the first-order region, plots of [S] versus time and [P] versus time are curved (Fig. 4-9b). We can determine the amount of substrate utilized or product formed during any given time interval by using the integrated first-order rate equation:

$$v = -\frac{d[S]}{dt} = k[S] \qquad \text{or} \qquad -\frac{d[S]}{[S]} = k\,dt$$

Integrating between any two different substrate concentrations, $[S]_0$ and $[S]$, and the corresponding times, t_0 and t:

$$-\int_{[S]_0}^{[S]} \frac{d[S]}{[S]} = k \int_{t_0}^{t} dt$$

$$\ln \frac{[S]_0}{[S]} = k(t - t_0) \qquad \text{or} \qquad 2.3 \log \frac{[S]_0}{[S]} = k(t - t_0)$$

If $[S]_0$ = the initial substrate concentration, and t_0 = zero time, then the above equation may be written as:

$$2.3 \log \frac{[S]_0}{[S]} = kt \tag{8}$$

where
t = elapsed time
$[S]$ = substrate concentration at time t

In exponential form, the equation may be written:

$$[S] = [S]_0 e^{-kt} \tag{9}$$

Equation 8 may be rearranged to:

$$\log [S] = -\frac{k}{2.3} t + \log [S]_0 \tag{10}$$

Thus, a plot of log [S] versus t is linear with a slope of $-k/2.3$ and an intercept of log $[S]_0$ on the log [S] axis (Fig. 4-10). When $[S] = \frac{1}{2}[S]_0$, t = the "half-life," $t_{1/2}$. In other words, $t_{1/2}$ is the time required to convert half the substrate originally present to product. The $t_{1/2}$ is constant for first-order reactions and is related to k as shown below.

$$2.3 \log \frac{1}{0.5} = kt_{1/2} \qquad \frac{2.3 \log 2}{k} = t_{1/2}$$

$$\frac{0.693}{k} = t_{1/2} \tag{11}$$

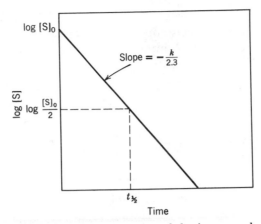

Figure 4-10 Semilog plot of the integrated
first-order velocity equation.

ZERO-ORDER KINETICS

When $[S] \gg K_m$, the K_m in the denominator of the Henri-Michaelis-Menten
equation may be ignored and the equation simplifies as shown below.

$$v = \frac{V_{max}[S]}{K_m + [S]} \xrightarrow{\ [S] \gg K_m\ } \frac{V_{max}[S]}{[S]}$$

or

$$v = V_{max}$$

For all practical purposes, the velocity is constant and independent of [S] (Fig.
4-11).   Plots of [S] versus time and [P] versus time are linear.

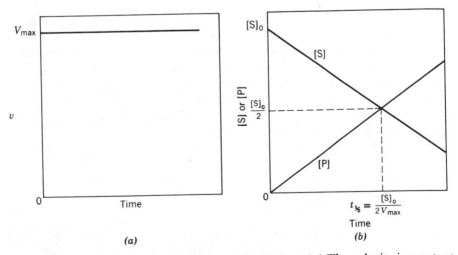

Figure 4-11 Zero-order region of the velocity curve. (a) The velocity is constant
over time. (b) P appears and S disappears linearly with time.

· Problem 4-6

Estimate k, the first-order rate constant, for an enzyme preparation with a V_{max} of $4.6\ \mu$moles $\times$ liter$^{-1}\times$ min^{-1} under the given experimental conditions. $K_m = 2 \times 10^{-6}\ M$.

Solution

$$k = \frac{V_{max}}{K_m} = \frac{4.6 \times 10^{-6}\ M \times \text{min}^{-1}}{2 \times 10^{-6}\ M} = \boxed{2.3\ \text{min}^{-1}}$$

or

$$\frac{2.3\ \text{min}^{-1}}{60\ \text{sec} \times \text{min}^{-1}} = \boxed{0.0383\ \text{sec}^{-1}}$$

Thus, a k of 2.3 min^{-1} means that approximately 3.83% of [S] is utilized each second.

· Problem 4-7

An enzyme was assayed at an initial substrate concentration of $10^{-5}\ M$. The K_m for the substrate is $2 \times 10^{-3}\ M$. At the end of 1 min, 2% of the substrate had been converted to product. (a) What percent of the substrate will be converted to product at the end of 3 min? What will be the product and substrate concentrations after 3 min? (b) If the initial concentration of substrate were $10^{-6}\ M$, what percent of the substrate will be converted to product after 3 min? (c) What is the maximum attainable velocity (V_{max}) with the enzyme concentration used? (d) At about what substrate concentration will V_{max} be observed? (e) At this S concentration, what percent of the substrate will be converted to product in 3 min?

Solution

(a) At an initial S concentration of $10^{-5}\ M$ (which is less than 0.01 K_m), the reaction will be *first-order*—the velocity will be directly proportional to the S concentration. Because [S] keeps decreasing with time, v will also decrease with time. For first-order reactions, a constant *proportion* of the substrate is converted to product per unit time (*not* a constant *amount*). We can solve the problem in two ways. One way is an approximate method that assumes the velocity is constant over a small increment of time. The exact method takes note of the fact that v is constantly changing.

Approximate Method

At the end of 1 min, 2% of the substrate is used, leaving 98%. *During the second minute, 2% of the remaining 98% will be used up; 2% of 98% is 1.96% more.

∴ By the *end* of 2 min, 2% + 1.96% = 3.96% of the substrate will be gone, leaving 100% − 3.96% = 96.04%. During the third minute, 2% of the remaining 96.04% will be used up; 2% of 96.04% is 1.92% more.

∴ By the *end* of the third minute, 3.96% + 1.92% = 5.88% of the substrate will be gone, leaving 100% − 5.88% = 94.12%.

∴ [product] = 5.88% of 10^{-5} M
= 0.0588×10^{-5} M

$$[\text{product}] = 5.88 \times 10^{-7}\ M$$

[substrate] = 94.12% of 10^{-5} M
= 0.9412×10^{-5} M

$$[\text{substrate}] = 9.412 \times 10^{-6}\ M$$

General Principle

The amount of substrate used in a given time interval (as calculated by the approximate method) will always be larger than the true value. This results from our assumption that the velocity is constant over a short time interval when in fact it is constantly decreasing.

The true rate constant, k, in terms of min^{-1} will always be larger than the fraction of the substrate used per minute for the same reason.

Exact Method

First calculate k knowing that at the end of 1 min 98% of the original substrate remains.

$$2.3 \log \frac{[S]_0}{[S]} = kt$$

Let:

$$[S]_0 = 100 \quad \text{and} \quad [S] = 98$$

$$2.3 \log \tfrac{100}{98} = (k)(1)$$

$$2.3(\log 1.02) = k$$

$$2.3(0.009) = k$$

$$k = 0.0207\ \text{min}^{-1}$$

Next calculate [S] at $t = 3$ min.

$$2.3 \log \frac{100}{[S]} = (0.0207)(3) = 0.0621$$

$$2.3(\log 100 - \log [S]) = 0.0621$$

$$4.6 - 2.3 \log [S] = 0.0621$$

$$-2.3 \log [S] = 0.0621 - 4.60$$
$$= -4.54$$

$$\log [S] = \frac{-4.54}{-2.3} = 1.975$$

[S] = antilog of 1.975

[S] = 94.4

or   [S] = 94.4% of original conc.

[S] = 0.944×10^{-5} M

$$[S] = 9.44 \times 10^{-6}\ M$$

∴ [product] = 100% − 94.4%
= 5.6%
= 5.6% of 10^{-5} M
= 0.056×10^{-5} M

$$[\text{product}] = 5.60 \times 10^{-7}\ M$$

When using the integrated form of the first-order rate equation, $[S]_0$ and [S] can be expressed in terms of percents (100 and 94.4 in the above example), or as decimals (1.0 and 0.944), or as actual concentrations (10^{-5} M and 0.944×10^{-5} M).

(b) If the initial substrate concentration were 10^{-6} M, the reaction would still be first-order. The proportion of the substrate converted to product would still be 5.6% by 3 min. The *amount* of product formed would, of course, be less than in part a.

(c) V_{max} can be estimated since we know K_m and an initial velocity (v) at a given substrate [S] concentration. At [S] = 10^{-5} M,

$$v = 2\% \text{ of } 10^{-5} \ M/\text{min}$$

$$2\% \text{ of } 10^{-5} \ M/\text{min} = 2\% \times 10^{-5} \text{ moles} \times \text{liter}^{-1} \times \text{min}^{-1}$$
$$= (0.02)(10^{-5} \text{ moles} \times \text{liter}^{-1} \times \text{min}^{-1})$$
$$v = 2 \times 10^{-7} \text{ moles} \times \text{liter}^{-1} \times \text{min}^{-1}$$

$$\frac{v}{V_{max}} = \frac{[S]}{K_m + [S]}$$

$$\frac{2 \times 10^{-7}}{V_{max}} = \frac{10^{-5}}{(2 \times 10^{-3}) + (10^{-5})} = \frac{10^{-5}}{201 \times 10^{-5}}$$

$$V_{max} = \frac{(201 \times 10^{-5})(2 \times 10^{-7})}{(10^{-5})} = 402 \times 10^{-7} = 40.2 \times 10^{-6}$$

$$\boxed{V_{max} = 40.2 \ \mu\text{moles} \times \text{liter}^{-1} \times \text{min}^{-1}}$$

(d) V_{max} will be observed at about 100 K_m.

$$100(2 \times 10^{-3} \ M) = 2 \times 10^{-1} \ M = 0.2 \ M$$

$$\boxed{[S] \simeq 0.2 \ M}$$

(e) At 0.2 M, the reaction will be essentially zero-order.

$$[\text{product}] = V_{max} \times t$$
$$= 4.02 \times 10^{-5} \text{ moles} \times \text{liter}^{-1} \times \text{min}^{-1} \times 3 \text{ min}$$
$$= 12.06 \times 10^{-5} \ M \text{ at 3 min}$$

$$\frac{12.06 \times 10^{-5} \ M}{0.2 \ M} \times 100 = \% \text{ conversion of substrate to product}$$

$$\frac{12.06 \times 10^{-3}}{2 \times 10^{-1}} \% = 6.03 \times 10^{-2}\% = \boxed{0.0603\%}$$

Now that we know V_{max} and K_m, we can obtain another estimate of the first-order rate constant, k.

$$k = \frac{V_{max}}{K_m} = \frac{4.02 \times 10^{-5} \text{ moles} \times \text{liter}^{-1} \times \text{min}^{-1}}{2 \times 10^{-3} \text{ moles/liter}}$$

$$\boxed{k = 2.01 \times 10^{-2} \ \text{min}^{-1}}$$

· Problem 4-8

An enzyme was assayed at an initial substrate concentration of $2 \times 10^{-5}\ M$. In 6 min, half of the substrate had been used. The K_m for the substrate is $5 \times 10^{-3}\ M$. Calculate (a) k, (b) V_{max}, and (c) the concentration of product produced by 15 min.

Solution

The [S] is $<0.01\ K_m$. $\therefore$ The reaction is first-order.

(a)
$$\frac{0.693}{k} = t_{\frac{1}{2}} \qquad \frac{0.693}{k} = 6\ \text{min} \qquad k = \frac{0.693}{6}$$

$$\boxed{k = 0.115\ \text{min}^{-1}}$$

(b)
$$k = \frac{V_{max}}{K_m} \qquad V_{max} = (k)(K_m)$$

$$V_{max} = (0.115\ \text{min}^{-1})(5 \times 10^{-3}\ M) = 0.575 \times 10^{-3}\ M\ \text{min}^{-1}$$
$$= 0.575 \times 10^{-3}\ \text{mole} \times \text{liter}^{-1} \times \text{min}^{-1}$$

$$\boxed{V_{max} = 575\ \mu\text{moles} \times \text{liter}^{-1} \times \text{min}^{-1}}$$

(c)
$$2.3 \log \frac{[S]_0}{[S]} = kt \qquad 2.3 \log \frac{2 \times 10^{-5}}{[S]} = (0.115)(15)$$

$$2.3 \log (2 \times 10^{-5}) - 2.3 \log [S] = 1.725$$

$$2.3 \log [S] = 2.3 \log (2 \times 10^{-5}) - 1.725$$

$$\log [S] = \frac{-10.81 - 1.725}{2.3} = \frac{-12.533}{2.3}$$

$$\log [S] = -5.45 \qquad \therefore \quad [S] = 3.55 \times 10^{-6}\ M$$

$$[P] = [S]_0 - [S] = (2 \times 10^{-5}) - (0.355 \times 10^{-5})$$

$$\boxed{[P] = 1.645 \times 10^{-5}\ M}$$

D. METHODS OF PLOTTING ENZYME KINETICS DATA

Because the v versus [S] curve is a hyperbola, it is rather difficult to determine V_{max} and, hence, the [S] that yields $\frac{1}{2} V_{max}$ (i.e., K_m). To facilitate the determination of the kinetic constants, the data are usually plotted in one of the linear forms described below.

LINEWEAVER-BURK RECIPROCAL PLOT: 1/v VERSUS 1/[S]

This plot is based on the rearrangement of the Henri-Michaelis-Menten equation into a linear $(y = mx + b)$ form:

$$\frac{v}{V_{max}} = \frac{[S]}{K_m + [S]}$$ Inverting: $\frac{V_{max}}{v} = \frac{K_m + [S]}{[S]}$

Cross multiplying: $\frac{1}{v} = \frac{K_m + [S]}{V_{max}[S]}$ Separating terms: $\frac{1}{v} = \frac{K_m}{V_{max}[S]} + \frac{[S]}{V_{max}[S]}$

or

$$\boxed{\frac{1}{v} = \frac{K_m}{V_{max}} \frac{1}{[S]} + \frac{1}{V_{max}}}$$ (13)

Thus, if we plot $1/v$ versus $1/[S]$, the slope $= K_m/V_{max}$ and the intercept on the y axis $= 1/V_{max}$ (Fig. 4-12). We can also see that when $1/v = 0$, $(K_m/V_{max}) \times (1/[S]) = -1/V_{max}$ and, therefore, $1/[S] = -1/K_m$. Thus, the intercept on the $1/[S]$ axis is $-1/K_m$. As we shall see later, any factor that multiplies the K_m term of the original Henri-Michaelis-Menten equation will turn out to be a factor of the slope (i.e., of K_m/V_{max}) in the reciprocal equation. Any factor that multiplies the [S] term of the original equation will turn out to be a factor of the $1/v$-axis intercept (i.e., of $1/V_{max}$) in the reciprocal equation.

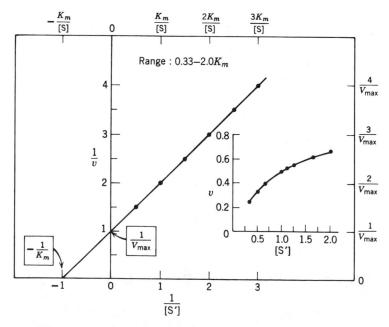

Figure 4-12 Double reciprocal ($1/v$ versus $1/[S]$) Lineweaver-Burk plot. The [S] range chosen is optimal for the determination of K_m and V_{max}.

SUBSTRATE CONCENTRATION RANGE

The concentrations of substrate chosen to generate the reciprocal plot should be in the neighborhood of the K_m value.   If the concentrations chosen are very high relative to the K_m value, the curve will be essentially horizontal.   This will allow V_{max} to be determined, but the slope of the line will be near zero.   Consequently, it will be difficult to determine K_m accurately.   If the substrate concentrations chosen are very low relative to the K_m value, the curve will intercept both axes too close to the origin to allow either V_{max} or K_m to be determined accurately.   (At very low substrate concentrations, the reaction is essentially first-order.   There is no hint of saturation. V_{max} and K_m appear to be infinite.)

LABELING THE AXES OF RECIPROCAL PLOTS

The beginning student sometimes is uncertain about the units used in labeling axes or columns of data.   The uncertainty arises because there are two ways of interpreting units containing factors.   For example, we might find a column headed "substrate concentration" with units of "mM $\times 10^2$."   Below the heading we might find the figure "0.1."   Some people interpret the heading as "the units of the data" and, hence, the "0.1" really represents 0.1×10^2 mM or 10 mM.   Others interpret the heading as "the numbers shown below are mM concentrations that have already been increased 100-fold."   The figure "0.1" then represents 0.1×10^{-2} mM or 0.001 mM.   Most biochemists use the latter convention.   To avoid confusion it is desirable, wherever possible, to reduce the data to units that do not include factors.   In the above example, the column may have been headed "substrate concentration, μM."   Then, 0.001 mM could have been entered as 1.0.

The Lineweaver-Burk reciprocal plot is the most widely used primary diagnostic plot.   However, the use of the plot has been criticized on two grounds: first, equal increments of [S] that yield equally spaced points on the basic v versus [S] plot do not yield equally spaced points on the reciprocal plot.   For example, relative values of [S] equal to 1, 2, 3, . . . 10, and so on, will yield reciprocal values that tend to cluster close to the $1/v$ axis.   Thus, there will be relatively few points at the high end of the 1/[S] scale and it is these points that are most heavily weighted in the subjective visual fitting of the line.   The second, and more important, criticism is that small errors in the determination of v are magnified when reciprocals are taken.   Errors in the determination of v are most likely to be significant at low substrate concentrations (and low values v).   One or two "bad" points at high $1/v$—1/[S] values can introduce a marked error to the slope of the plot.   The first criticism is dealt with simply by including assay points that yield equal reciprocal increments.   This means that relative substrate concentrations of 1.00, 1.11, 1.25, 1.43, 1.67, 2.0, 2.5, 3.33, 5, and 10 must be used.

The Lineweaver-Burk reciprocal plot is not the only linear transformation of the basic velocity (or ligand binding) equation.   Indeed, under some circumstances one of the other linear plots described below may be more suitable or may yield more reliable estimates of the kinetic constants.   For example, the Hanes-Woolf plot of [S]/v versus [S] may be more convenient for data obtained at equally spaced increments of [S]. The Woolf-Augustinsson-Hofstee plot of v versus $v/[S]$ and the Eadie-Scatchard plot of $v/[S]$ versus v do not involve reciprocals of v and, consequently, may be more reliable when the error in v is significant. These latter two plots have the further advantage of calling attention to points that deviate significantly from the theoretical relationship because both plotted variables are influenced in the same direction by an error in v.

HANES-WOOLF PLOT: [S]/v VERSUS [S]

The Lineweaver-Burk equation may be rearranged to yield the linear equation for the Hanes-Woolf plot:

$$\frac{1}{v} = \frac{K_m}{V_{max}} \frac{1}{[S]} + \frac{1}{V_{max}}$$

Multiplying both sides of the equation by [S]:

$$\frac{[S]}{v} = \frac{[S]K_m}{V_{max}} \frac{1}{[S]} + \frac{[S]}{V_{max}}$$

or

$$\boxed{\frac{[S]}{v} = \frac{1}{V_{max}}[S] + \frac{K_m}{V_{max}}}$$

(14)

Thus, a plot of [S]/v versus [S] is linear with a slope of $1/V_{max}$. The intercept on the [S]/v axis gives K_m/V_{max}. When [S]/$v = 0$, the intercept on the [S] axis gives $-K_m$. As usual, care should be exercised in choosing the range of substrate concentration. If the substrate concentration range is very low compared to K_m, the plot will be nearly horizontal. If the substrate range is very high compared to K_m, the plot will intersect the axes very close to the origin. The Hanes-Woolf plot is shown in Figures 4-15 and 4-33.

THE WOOLF-AUGUSTINSSON-HOFSTEE PLOT: v VERSUS $v/[S]$

Another linear form is obtained by rearranging the basic velocity equation as shown below.

$$v = \frac{V_{max}[S]}{K_m + [S]}$$

Dividing numerator and denominator by [S]:

$$v = \frac{V_{max}}{\dfrac{K_m}{[S]} + 1} \qquad V_{max} = \frac{vK_m}{[S]} + v$$

$$v = V_{max} - \frac{vK_m}{[S]} \qquad \text{or} \qquad \boxed{v = -K_m\frac{v}{[S]} + V_{max}}$$

(15)

The plot of v versus $v/[S]$ is linear with a slope of $-K_m$.   The intercept on the v axis gives V_{max}.   When $v = 0$, the intercept on the $v/[S]$ axis gives V_{max}/K_m.   If the substrate concentration range is very low compared to K_m, the plot will have an extremely steep slope, approaching a vertical line that intersects the horizontal axis at V_{max}/K_m (i.e., the first-order rate constant for the reaction).   If the substrate concentration range is very high compared to K_m, the plot will be nearly horizontal at a height of V_{max} above the $v/[S]$ axis. The Woolf-Augustinsson-Hofstee plot is shown in Figures 4-14 and 4-34.

THE EADIE-SCATCHARD PLOT: $v/[S]$ VERSUS v

If the Henri-Michaelis-Menten equation is rearranged as described above for the v versus $v/[S]$ plot, we obtain:

$$V_{max} = \frac{vK_m}{[S]} + v$$

Dividing both sides of the equation by K_m:

$$\frac{V_{max}}{K_m} = \frac{v}{[S]} + \frac{v}{K_m} \qquad \text{or} \qquad \boxed{\frac{v}{[S]} = -\frac{1}{K_m}v + \frac{V_{max}}{K_m}} \tag{16}$$

Thus, a plot of $v/[S]$ versus v is linear with a slope of $-1/K_m$ and an intercept of V_{max}/K_m on the $v/[S]$ axis.   When $v/[S] = 0$, the intercept on the v axis gives V_{max}. The Eadie-Scatchard plot is shown in Figure 4-35.

· Problem 4-9

The following data were obtained for an enzyme that catalyzes the reaction $S \rightarrow P$.   The substrate concentrations have been spaced to allow use of any of the linear plots.   Plot the data and determine K_m and V_{max}.

[S]	v
(M)	$(\text{nmoles} \times \text{liter}^{-1} \times \text{min}^{-1})$
8.33×10^{-6}	13.8
1.00×10^{-5}	16.0
1.25×10^{-5}	19.0
1.67×10^{-5}	23.6
2.00×10^{-5}	26.7
2.50×10^{-5}	30.8
3.33×10^{-5}	36.3
4.00×10^{-5}	40.0
5.00×10^{-5}	44.4
6.00×10^{-5}	48.0
8.00×10^{-5}	53.4
1.00×10^{-4}	57.1
2.00×10^{-4}	66.7

Solution

The data are arranged in Table 4-1 in a manner suitable for analysis by the four different linear plots. In order to avoid problems with units when dividing v by [S] or [S] by v, all concentrations are expressed in molarity, and all velocities are expressed in terms of moles per liter per minute. In order to simplify the numbers to be plotted, the decimal point has been moved appropriately. Thus, $[S] = 2.50 \times 10^{-5} M$ is indicated simply as 2.5. The

Table 4-1

[S]	v	1/[S]	1/v	v/[S]	[S]/v
$(M) \times 10^5$	$(M \times min^{-1}) \times 10^9$	$(M^{-1}) \times 10^{-4}$	$(M \times min^{-1})^{-1} \times 10^{-7}$	$(min)^{-1} \times 10^5$	$(min) \times 10^{-3}$
0.833	13.8	12.00	7.24	1.66	0.602
1.00	16.0	10.00	6.25	1.60	0.625
1.25	19.0	8.00	5.26	1.52	0.658
1.67	23.6	6.00	4.24	1.41	0.709
2.00	26.7	5.00	3.74	1.34	0.746
2.50	30.8	4.00	3.25	1.23	0.812
3.00	34.3	3.33	2.91	1.14	0.875
3.33	36.3	3.00	2.75	1.09	0.917
4.00	40.0	2.50	2.50	1.00	1.00
5.00	44.4	2.00	2.25	0.89	1.13
6.00	48.0	1.67	2.08	0.80	1.25
8.00	53.3	1.25	1.88	0.67	1.50
10.00	57.1	1.00	1.75	0.57	1.75
20.00	66.7	0.50	1.50	0.334	2.99

column heading "$(M) \times 10^5$" means that the numbers shown represent *the concentrations multiplied by 10^5*. The v observed at $[S] = 2.50 \times 10^{-5} M$ was 30.8 nmoles $\times$ liter^{-1} $\times$ min^{-1} or 30.8×10^{-9} moles $\times$ liter^{-1} $\times$ min^{-1}. This is indicated simply as 30.8 under the heading $(M \times min^{-1}) \times 10^9$, that is, the number shown is 10^9 times the observed v. The corresponding $1/[S] = 1/(2.5 \times 10^{-5}) = 0.400 \times 10^5 = 4.00 \times 10^4$. This value is entered simply as 4.00 and the column is headed $(M^{-1}) \times 10^{-4}$, that is, the value shown is 10^{-4} times the true reciprocal concentration. $1/v = 1/(30.8 \times 10^{-9}$ moles $\times$ liter^{-1} $\times$ min$^{-1}) = 0.0325 \times 10^9 = 3.25 \times 10^7$. Only 3.25 is entered. The column heading shows that this represents the true reciprocal velocity multiplied by 10^{-7}. With the units chosen, the v/[S] and [S]/v entries come out relatively simple numbers. For example at $[S] = 2.5 \times 10^{-5} M$ and $v = 30.8 \times 10^{-9} M \times$ min^{-1}, $[S]/v = (2.5 \times 10^{-5})/(30.8 \times 10^{-9}) = 812$ min, which is entered as 0.812 in the column headed $(min) \times 10^{-3}$. The $1/v$ versus $1/[S]$, [S]/v versus [S], and v versus v/[S] plots are shown in Figures 4-13 to 4-15. The v/[S] versus v plot is identical to the v versus v/[S] plot with the axes reversed.

· **Problem 4-10**

Serum lipase activity can be used as an indicator of acute disease of the pancreas. However, the interpretation of the data is often uncertain because more than one lipase active on a given triglyceride may be present. Which

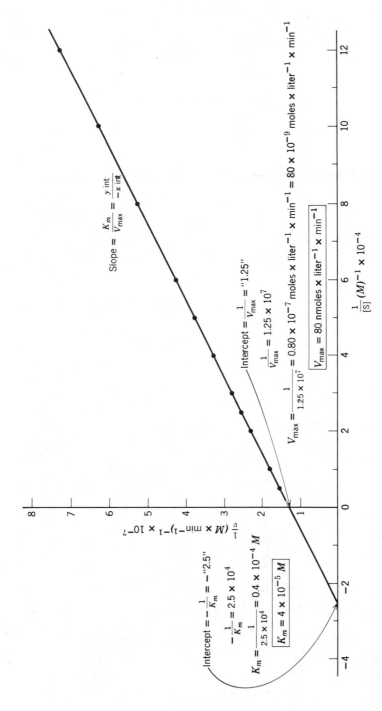

Figure 4-13 Plot of $1/v$ versus $1/[S]$ of data given in Problem 4-9.

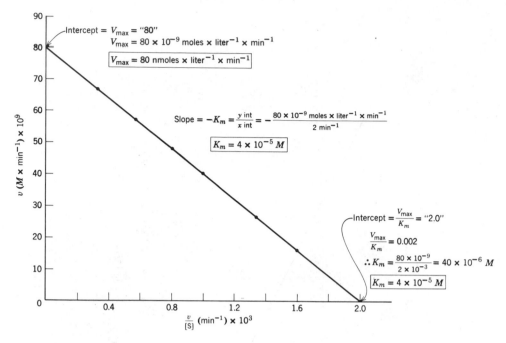

Figure 4-14 Plot of v versus $v/[S]$ of data given in Problem 4-9.

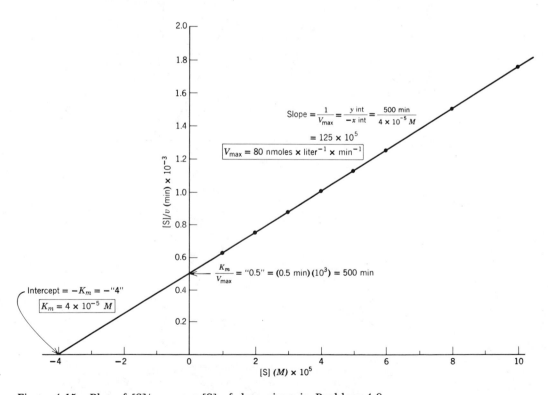

Figure 4-15 Plot of $[S]/v$ versus $[S]$ of data given in Problem 4-9.

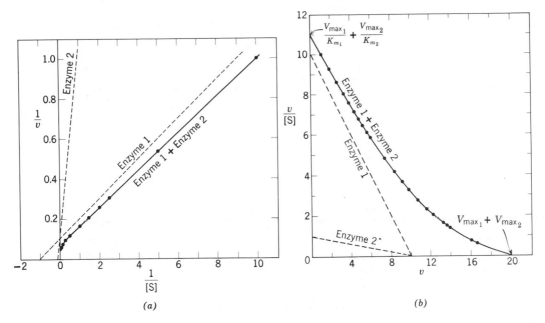

Figure 4-16 (*a*) Plot of $1/v_{obs}$ versus $1/[S]$. Two enzymes are present: $K_{m_1} = 1$, $V_{max_1} = 10$; $K_{m_2} = 10$, $V_{max_2} = 10$. The [S] range plotted is 0.1 to 25. (*b*) Plot of $v_{obs}/[S]$ versus v_{obs} for the same two enzymes described in (*a*).

linear plot would best disclose the presence of multiple enzymes? To decide, assume that the serum sample contains two enzymes with the following kinetic constants: $K_{m_1} = 1$, $V_{max_1} = 10$; $K_{m_2} = 10$, $V_{max_2} = 10$.

Solution

The observed v would be the sum of the activities of the two enzymes.

$$v_{obs} = v_1 + v_2 = \frac{[S]V_{max_1}}{K_{m_1} + [S]} + \frac{[S]V_{max_2}}{K_{m_2} + [S]}$$

Calculate v_{obs} over a wide range of substrate concentrations, for example, $[S] = 0.1$ to $[S] = 25$. Plot $1/v_{obs}$ versus $1/[S]$, $v_{obs}/[S]$ versus v_{obs}, and so on. All the plots are linear if only one enzyme is present. If more than one enzyme is present, the plots will deviate from linearity. Figures 4-16*a* and *b* show two of the plots. The $v_{obs}/[S]$ versus v_{obs} obviously provides the better indication that the data do not conform to a single Henri-Michaelis-Menten equation. (The plot is curved over a wider range of points.) The v_{obs} versus $v_{obs}/[S]$ and $[S]/v_{obs}$ versus [S] plots are also better than the $1/v_{obs}$ versus $1/[S]$ plot for detecting multiple enzymes that catalyze the same reaction.

SCATCHARD PLOT FOR EQUILIBRIUM BINDING DATA: $[S]_b/[S]_f$ VERSUS $[S]_b$ OR $[S]_b/[S]_f[E]_t$ VERSUS $[S]_b/[E]_t$

Many proteins bind small molecules but do not catalyze a reaction of the ligand. Some examples include the binding of oxygen by myoglobin or hemoglobin and the binding of hormones and drugs by specific receptor

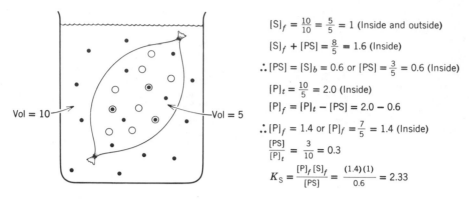

$[S]_f = \frac{10}{10} = \frac{5}{5} = 1$ (Inside and outside)

$[S]_f + [PS] = \frac{8}{5} = 1.6$ (Inside)

$\therefore [PS] = [S]_b = 0.6$ or $[PS] = \frac{3}{5} = 0.6$ (Inside)

$[P]_t = \frac{10}{5} = 2.0$ (Inside)

$[P]_f = [P]_t - [PS] = 2.0 - 0.6$

$\therefore [P]_f = 1.4$ or $[P]_f = \frac{7}{5} = 1.4$ (Inside)

$\frac{[PS]}{[P]_t} = \frac{3}{10} = 0.3$

$K_S = \frac{[P]_f[S]_f}{[PS]} = \frac{(1.4)(1)}{0.6} = 2.33$

Figure 4-17   A simple representation of an equilibrium dialysis experiment where the volumes inside and outside the dialysis bag are unequal. The large circles inside the dialysis bag represent a nondiffusible protein, P, which may be an enzyme or a noncatalytic binding protein. The small dots represent a diffusible ligand, S, which may be a substrate, inhibitor, or activator, and so on. At equilibrium, the concentration of free ligand, $[S]_f$, is the same on both sides of the membrane. The excess ligand inside the dialysis bag represents bound ligand, $[S]_b$ or $[PS]$. Note that if no protein were present inside the dialysis bag, the equilibrium $[S]_f$ would be 1.2 (6/5 = 1.2 inside the bag, 12/10 = 1.2 outside the bag, or 18/15 = 1.2 overall). To minimize the amount of protein and ligand required, equilibrium dialysis is usually carried out with specially-made plastic chambers of equal volumes (e.g., 0.1 to 1.0 ml), separated by a semipermeable membrane (see Fig. 6-2).

proteins. The interaction of the protein and the ligand can be studied by equilibrium dialysis (Fig. 4-17) and other suitable techniques. The binding of substrates, inhibitors, and activators to enzymes can also be studied by equilibrium dialysis if no catalytic reaction occurs. (This is feasible with a substrate if the reaction requires two substrates and only one is present.) Equilibrium binding data are usually analyzed by a Scatchard plot. The equation for the Scatchard plot can be derived directly from the equilibrium expression for K_S or obtained by modifying the Eadie-Scatchard equation (Equation 16). The modification involves substituting $[ES]$ or $[S]_b$ for v, and $n[E]_t$ for V_{max} since:

$$v \propto [S]_b \qquad \text{and} \qquad V_{max} \propto n[E]_t$$

where $[S]_b = [ES]$ = the concentration of bound ligand
= the concentration of occupied sites
$[E]_t$ = the total enzyme concentration
n = the number of identical and independent ligand binding sites per molecule of enzyme
$n[E]_t$ = the total concentration of ligand binding sites

The equation becomes:

$$\frac{[S]_b}{[S]_f} = -\frac{1}{K_S}[S]_b + \frac{n[E]_t}{K_S} \qquad (17)$$

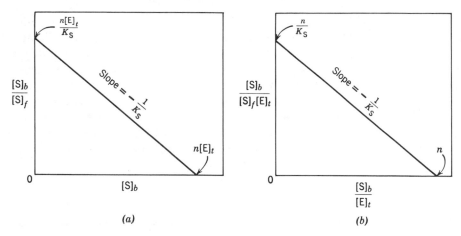

Figure 4-18 Scatchard plots for equilibrium ligand binding.

where K_S is the intrinsic substrate dissociation constant of a site. The $[S]_f$ term in the above equation stands for the concentration of *free* substrate. (Usually, it is indicated simply as [S].) $[S]_b/[S]_f$ represents the ratio of *bound* to *free* substrate. In most *in vitro* initial velocity studies, the concentration of enzyme is many orders of magnitude lower than the concentration of substrate. Consequently, the formation of ES does not significantly decrease the concentration of S and it is safe to assume that the concentration of *free* substrate, $[S]_f$, is the same as the total added substrate concentration, $[S]_t$. In equilibrium binding studies, relatively high enzyme concentrations must be used. Consequently, a relatively large proportion of the added substrate is bound and $[S]_f$ will be significantly less than $[S]_t$.

A plot of the ratio of bound to free ligand versus the concentration of bound ligand is linear with a slope of $-1/K_S$ (Fig. 4-18a). The horizontal axis intercept gives $n[E]_t$ (i.e., the total concentration of ligand binding sites). If the molar concentration of enzyme is known, the data can be plotted as:

$$\frac{[S]_b}{[S]_f[E]_t} = -\frac{1}{K_S}\frac{[S]_b}{[E]_t} + \frac{n}{K_S} \qquad (18)$$

A plot of $[S]_b/[S]_f[E]_t$ (i.e., moles of ligand bound per mole of enzyme divided by the concentration of *free* substrate) versus $[S]_b/[E]_t$ (i.e., moles of ligand bound per mole of enzyme) is linear with a slope of $-1/K_S$ (Fig. 4-18b). The intercept on the vertical axis gives n/K_S. The intercept on the horizontal axis gives n, the number of ligand binding sites per molecule of enzyme. If the enzyme possesses multiple independent binding sites with different affinities for the ligand, the plot will be curved.

· **Problem 4-11**

An amino acid binding protein, P (presumably involved in membrane transport and chemotaxis), was released from *E. coli* by osmotic shock. The protein was purified to homogeneity and found to have a molecular weight of

35,000.   A solution of the protein (0.5 mg/ml) was placed in one compartment of a dialysis chamber.   An equal volume of buffer containing 4×10^{-5} M L-leucine-C^{14} was placed in the other compartment.   The compartments are separated by a semipermeable membrane through which the labeled amino acid can move freely.   The protein, however, is restricted to one compartment.   After equilibration, the compartment containing the protein had 2.3×10^{-5} M total (bound + free) L-leucine-C^{14}.   The compartment without protein contained 1.7×10^{-5} M L-leucine-C^{14}.   Calculate (a) the concentration of bound L-leucine-C^{14}, [PS] or $[S]_b$, (b) the concentration of free protein, $[P]_f$, and (c) the dissociation constant for the protein-leucine complex assuming one binding site per molecule of protein.

Solution

First, note that if no protein were present on one side of the dialysis membrane, the concentration of L-leucine-C^{14} would be 2×10^{-5} M in both compartments after equilibration.

(a) $$[S]_b + [S]_f = 2.3 \times 10^{-5}$$
$$[S]_f = 1.7 \times 10^{-5} \ M$$

$$\therefore \quad \boxed{[S]_b = [PS] = 0.6 \times 10^{-5} \ M}$$

Note that $[S]_b$ represents the concentration of bound substrate *in one compartment* of the dialysis chamber.

(b)   The total protein concentration, $[P]_t$, is:

$$[P]_t = \frac{0.5 \text{ g/liter}}{3.5 \times 10^4 \text{ g/mole}} = 1.43 \times 10^{-5} \ M$$

$[P]_t$ also equals the total concentration of binding sites when there is only one site per molecule of protein.

$$[P]_f = [P]_t - [PS] = (1.43 \times 10^{-5}) - (0.6 \times 10^{-5})$$

$$\boxed{[P]_f = 0.83 \times 10^{-5} \ M}$$

(c) $$K_S = \frac{[P]_f[S]_f}{[PS]} = \frac{(0.83 \times 10^{-5})(1.7 \times 10^{-5})}{(0.6 \times 10^{-5})}$$

$$\boxed{K_S = 2.35 \times 10^{-5} \ M}$$

Figure 6-2 shows the equilibrium distribution of radioactive S and protein in a dialysis chamber composed of two equal-volume compartments separated by a semipermeable membrane.   Problem 6-18 illustrates the use of radioactive substrates in equilibrium binding studies.

INTEGRATED FORM OF THE HENRI-MICHAELIS-MENTEN EQUATION

Under some conditions it might be difficult to measure initial velocities although it is still possible to determine the substrate or product concentration during the course of a reaction. If K_{eq} is very high and the product has a very low affinity for the enzyme, then the decreasing velocity with time results only from the decreasing saturation of the enzyme. K_m and V_{max} may be determined by using the integrated form of the velocity equation.

$$v = -\frac{d[S]}{dt} = \frac{V_{max}[S]}{K_m + [S]}$$

Rearranging

$$V_{max}dt = -\frac{K_m + [S]}{[S]} d[S]$$

Integrating between any two times (e.g., zero time, t_0, and any other time, t) and the corresponding two substrate concentrations ($[S]_0$ and $[S]$):

$$V_{max}\int_{t_0}^{t} dt = -\int_{[S]_0}^{[S]} \frac{K_m + [S]}{[S]} d[S]$$

Separating the terms in the right-hand expression:

$$V_{max}\int_{t_0}^{t} dt = -K_m\int_{[S]_0}^{[S]} \frac{d[S]}{[S]} - \int_{[S]_0}^{[S]} d[S]$$

or

$$V_{max}t = -K_m \ln \frac{[S]}{[S]_0} - ([S] - [S]_0)$$

$$\boxed{V_{max}t = 2.3 K_m \log \frac{[S]_0}{[S]} + ([S]_0 - [S])} \tag{19}$$

where $[S]$ = the concentration of substrate at any time t
                  $= [S]_0 - [P]$
$([S]_0 - [S])$ = concentration of substrate utilized by time t
                  $= [P]$, the concentration of product produced by time t

Note that the right-hand part of Equation 19 is composed of a first-order term and a zero-order term. Equation 19 may be divided throughout by t and then rearranged to:

$$\boxed{\frac{2.3}{t} \log \frac{[S]_0}{[S]} = -\frac{1}{K_m} \frac{([S]_0 - [S])}{t} + \frac{V_{max}}{K_m}} \tag{20}$$

This is the equation for a straight line. Thus, K_m and V_{max} may be determined by measuring the concentration of substrate utilized (or product produced) several times during the reaction and then plotting the appropriate values as shown in Figure 4-19.

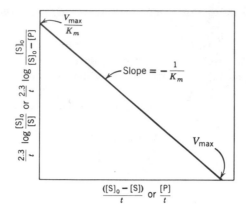

Figure 4-19 Plot of the integrated Henri-Michaelis-Menten equation.

E. ENZYME INHIBITION

Any substance that reduces the velocity of an enzyme-catalyzed reaction can be considered to be an "inhibitor." The inhibition of enzyme activity is one of the major regulatory devices of living cells, and one of the most important diagnostic procedures of the enzymologist. Inhibition studies often tell us something about the specificity of an enzyme, the physical and chemical architecture of the active site, and the kinetic mechanism of the reaction. In our everyday life, enzyme inhibitors can be found masquerading as drugs, antibiotics, preservatives, poisons, and toxins. In this section we examine a few simple types of enzyme inhibitors.

COMPETITIVE INHIBITION

A competitive inhibitor is a substance that combines with free enzyme in a manner that prevents substrate binding. That is, the inhibitor and the substrate are *mutually exclusive*, often because of true competition for the same site. A competitive inhibitor might be a nonmetabolizable analog or derivative of the true substrate, or an alternate substrate of the enzyme, or a product of the reaction.

Malonic acid is a classical example of a true competitive inhibitor. Malonic acid inhibits succinic dehydrogenase, which catalyzes the oxidation of succinic acid to fumaric acid, as shown below.

$$
\begin{array}{ccc}
\text{CH}_2\text{—COOH} & & \text{H—C—COOH} \\
| & +\text{FAD} \rightleftharpoons & \| \qquad\qquad +\text{FADH}_2 \\
\text{CH}_2\text{—COOH} & & \text{HOOC—CH} \\
\text{\small succinic acid} & & \text{\small fumaric acid}
\end{array}
$$

Malonic acid resembles succinic acid sufficiently to combine with the enzyme at the active site.

$$
\begin{array}{c}
\text{COOH} \\
| \\
\text{CH}_2 \\
| \\
\text{COOH} \\
\text{\small malonic acid}
\end{array}
$$

However, because malonic acid has only one methylene group, it is obvious that no oxidation-reduction can take place. Only association of the enzyme and inhibitor, and dissociation of the EI complex, can occur. Another classical example of a competitive inhibitor is the sulfa drug sulfanilamide, which interferes with the biosynthesis of folic acid from the precursor p-aminobenzoic acid (PABA).

PABA

sulfanilamide

Model 1 (Fig. 4-20) illustrates classical competitive inhibition in which an inhibitor competes with a substrate for a single binding site. Models 2–4 represent other ways in which an inhibitor and substrate would be mutually exclusive: steric hindrance (Model 2); steric hindrance or competition for a common binding group (Model 3); and overlapping binding sites (Model

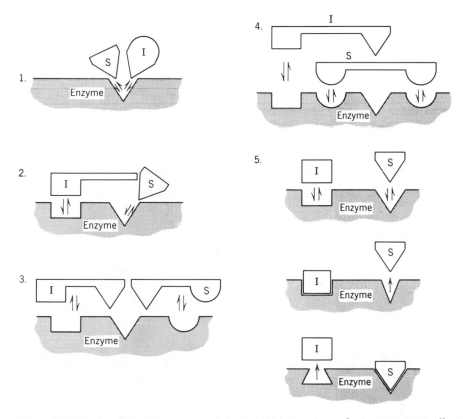

Figure 4-20 Models of competitive inhibition: S and I are mutually exclusive. (1) Classical model. S and I compete for the same binding site. I must resemble S structurally. (2) I and S are mutually exclusive because of steric hindrance. (3) I and S share a common binding group on the enzyme. (4) The binding sites for I and S are distinct, but overlapping. (5) The binding of I to a distinct inhibitor site causes a conformational change in the enzyme that distorts or masks the substrate binding site (and vice versa).

4). There are many examples of "competitive" inhibition by compounds that bear no structural relationship to the substrate. The inhibitor is generally an end product or near end product of a metabolic pathway; the enzyme is one that catalyzes an early reaction (or a branch-point reaction) in the pathway. The phenomenon is called feedback inhibition. The inhibitor (effector, modulator, or regulator) combines with the enzyme at a position other than the active (substrate) site. The combination of the inhibitor with the enzyme causes a change in the conformation (tertiary or quaternary structure) of the enzyme that distorts the substrate site and thereby prevents the substrate from binding (Model 5).

The inhibition of the hexokinase-catalyzed reaction between glucose and ATP by fructose or mannose is an example of competitive inhibition by alternate substrates. Glucose, fructose, and mannose are all substrates of hexokinase and can be converted to product (hexose-6-phosphate). All three hexoses combine with the enzyme at the same active site. Consequently, the utilization of any one of the hexoses is inhibited in the presence of either of the other two. The reaction scheme describing "dead-end" competitive inhibition is:

$$E + S \underset{K_s}{\rightleftharpoons} ES \xrightarrow{k_p} E + P$$

$$E + S \underset{K_s}{\rightleftharpoons} ES \xrightarrow{k_p} E + P$$
$$+$$
$$I$$
$$K_i \Updownarrow \qquad\qquad K_i = \frac{[E][I]}{[EI]}$$
$$EI$$

The initial velocity of the reaction is proportional to the steady-state concentration of the enzyme-substrate complex, ES. All the reactions are reversible. Consequently, we can predict that at any fixed subsaturating concentration of inhibitor (a) v_i (the velocity in the presence of a competitive inhibitor) can be made to equal v (the velocity in the absence of the inhibitor), but that a higher substrate concentration will be required (in order to obtain the same ES concentration), and (b) in the presence of an infinitely high (saturating) substrate concentration all the enzyme can be driven to the ES form. Consequently, the maximal initial velocity in the presence of the competitive inhibitor equals V_{max} (the maximal initial velocity in the absence of inhibitor). The apparent K_m (measured as [S] required for $\frac{1}{2} V_{max}$) will increase in the presence of a competitive inhibitor because at any inhibitor concentration a portion of the enzyme exists in the EI form, which has no affinity for S. The velocity equation can be derived in the usual manner from rapid equilibrium conditions. This time we recognize that the enzyme is distributed among three species:

$$v = k_p [ES] \qquad \frac{v}{[E]_t} = \frac{k_p [ES]}{[E] + [ES] + [EI]}$$

$$[ES] = \frac{[S]}{K_s} [E] \qquad \text{and} \qquad [EI] = \frac{[I]}{K_i} [E]$$

$$\therefore \quad \frac{v}{k_p [E]_t} = \frac{\dfrac{[S]}{K_s} [E]}{[E] + \dfrac{[S]}{K_s} [E] + \dfrac{[I]}{K_i} [E]} \qquad \text{or} \qquad \boxed{\frac{v}{V_{max}} = \frac{\dfrac{[S]}{K_s}}{1 + \dfrac{[S]}{K_s} + \dfrac{[I]}{K_i}}} \qquad (21)$$

If we compare the above equation to the usual velocity equation we see that the denominator has gained an additional $[I]/K_i$ term representing the EI complex. The numerator still has one term indicating that there is still only one product-forming complex (ES). To obtain a more familiar form, the numerator and denominator of the right-hand part of the above equation can be multiplied by K_s and factored:

$$\frac{v}{V_{\max}} = \frac{[S]}{K_s\left(1 + \dfrac{[I]}{K_i}\right) + [S]} \tag{22}$$

We obtain the same final velocity equation for steady-state conditions, except K_m replaces K_s. This is not surprising since the steady-state assumption does not change the form of the velocity equation for the uninhibited reaction while the reaction between E and I to yield EI must be at equilibrium. (There is nowhere for EI to go but back to E + I.) The velocity equation differs from the usual Michaelis-Menten equation in that the K_m term is multiplied by the factor $[1 + ([I]/K_i)]$. The above derivation confirms our original prediction that $V_{\max}$ is unaffected by a competitive inhibitor, but that the *apparent K_m* value is increased. The increase in the K_m value does not mean that the EI complex has a lower affinity for the substrate. EI has no affinity at all for the substrate, while the affinity of E (the only form that can bind substrate) is unchanged. The apparent increase in K_m results from a distribution of available enzyme between the "full affinity" and "no affinity" forms. The factor $[1 + ([I]/K_i)]$ may be considered as an [I]-dependent statistical factor describing the distribution of enzyme between the E and EI forms. Figure 4-21 shows the effect of a competitive inhibitor on the v versus [S] plot.

The velocity equation for competitive inhibition in reciprocal form is:

$$\frac{1}{v} = \frac{K_m}{V_{\max}}\left(1 + \frac{[I]}{K_i}\right)\frac{1}{[S]} + \frac{1}{V_{\max}} \tag{23}$$

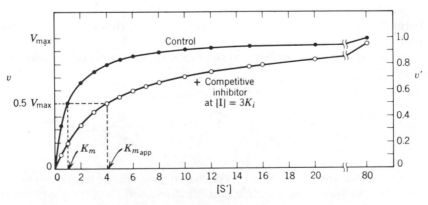

Figure 4-21 v versus [S] plot in the presence and in the absence of a fixed concentration of a competitive inhibitor.

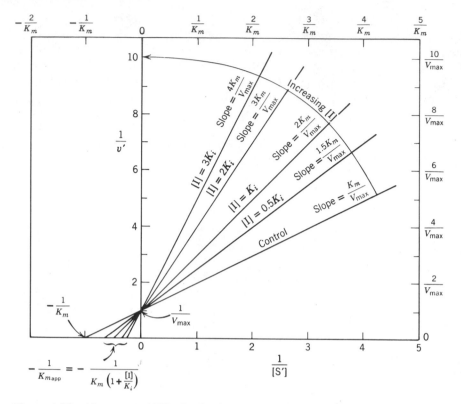

Figure 4-22 $1/v$ versus $1/[S]$ plot in the presence of different fixed concentrations of a competitive inhibitor.

Thus, the slope of the plot increases by the factor $[1+([I]/K_i)]$ (which multiplied K_m in the original equation), but the $1/v$-axis intercept remains $1/V_{max}$. K_i can be calculated from either the slope of the plot, or the $1/[S]$-axis intercept. When $1/v = 0$, the $1/[S]$-axis intercept gives $-1/K_{m_{app}}$, where $K_{m_{app}} = K_m[1+([I]/K_i)]$. For each inhibitor concentration, a new reciprocal plot can be drawn. As [I] increases, the "plus inhibitor" curves increase in slope (Fig. 4-22) pivoting counterclockwise about the point of intersection with the control curve (at $1/V_{max}$ on the $1/v$ axis). Because the initial velocity can be driven to zero by a saturating inhibitor concentration, the limiting plot will be a vertical line on the $1/v$ axis. As [I] increases, the intercept on the $1/[S]$ axis moves closer to the origin, that is, $K_{m_{app}}$ continually increases.

The slope of the reciprocal plot in the presence of a competitive inhibitor is given by:

$$slope_{1/S} = \frac{K_m}{V_{max}}\left(1 + \frac{[I]}{K_i}\right) \quad \text{or} \quad \boxed{slope_{1/S} = \frac{K_m}{V_{max}K_i}[I] + \frac{K_m}{V_{max}}} \quad (24)$$

Thus, a replot of the slope of each reciprocal plot versus the corresponding inhibitor concentration at which it was obtained will be a straight line (Fig. 4-23*a*). For convenience the slope of a reciprocal plot can be read off directly

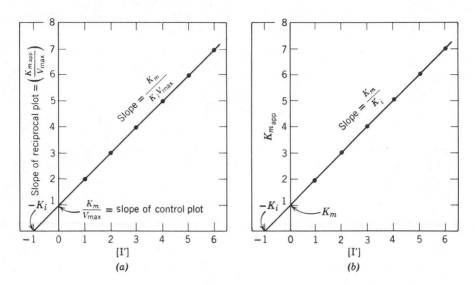

Figure 4-23 Replots of data taken from the reciprocal plot. (*a*) *Slope*_{1/S} versus
[I]. (*b*) $K_{m_{app}}$ versus [I].

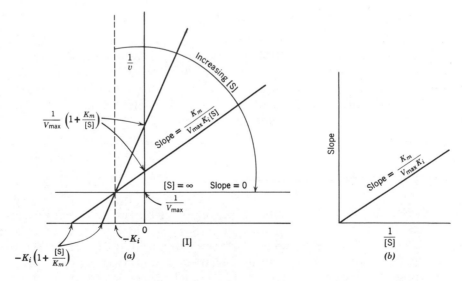

Figure 4-24 (*a*) Dixon plot for a competitive inhibitor: $1/v$ versus [I] in the
presence of different fixed concentrations of substrate. If V_{max} is known, a
horizontal line at a height of $1/V_{max}$ can be drawn directly. (*b*) Replot of the
slopes of the Dixon plots. A linear mixed-type inhibitor (Fig. 4-32) yields the
same type of Dixon plot. However, the slope replot does not go through the
origin.

as the ratio (absolute values) of the vertical-axis intercept to horizontal-axis intercept. $K_{m_{app}}$ is also a linear function of [I] (Fig. 4-23b):

$$K_{m_{app}} = \frac{K_m}{K_i}[I] + K_m \qquad (25)$$

The Dixon plot of $1/v$ versus [I] provides another way of identifying the type of inhibition and of determining K_i. The equation for the plot is obtained by multiplying out the equation for the reciprocal plot and then regrouping terms:

$$\frac{1}{v} = \frac{K_m}{V_{max}K_i[S]}[I] + \frac{1}{V_{max}}\left(1 + \frac{K_m}{[S]}\right) \qquad (26)$$

The plot is shown in Figure 4-24. The alternate linear plots are shown in Figures 4-33 to 4-35.

General Principles

A competitive inhibitor acts only to increase the apparent K_m for the substrate. As [I] increases, $K_{m_{app}}$ increases. The V_{max} remains unchanged, but in the presence of a competitive inhibitor a much greater substrate concentration is required to attain any fraction of V_{max}. The *degree of inhibition* caused by a competitive inhibitor depends on [S], [I], K_m, and K_i. An increase in [S] at constant [I] decreases the degree of inhibition. An increase in [I] at constant [S] increases the degree of inhibition. The lower the value of K_i, the greater is the degree of inhibition at any given [S] and [I]. K_i is equivalent to the concentration of I that doubles the slope of the $1/v$ versus $1/[S]$ plot. (K_i is *not* equivalent to the [I] that yields 50% inhibition.)

NONCOMPETITIVE INHIBITION

A classical noncompetitive inhibitor has no effect on substrate binding and vice versa. S and I bind reversibly, randomly, and independently at different sites. That is, I binds to E and to ES; S binds to E and to EI. However, the resulting ESI complex is catalytically inactive. I might prevent the proper positioning of the catalytic center. The equilibria are:

$$E + S \underset{}{\overset{K_S}{\rightleftharpoons}} ES \overset{k_p}{\longrightarrow} E + P$$

$$K_S = \frac{[E][S]}{[ES]} = \frac{[EI][S]}{[ESI]}$$

$$K_i = \frac{[E][I]}{[EI]} = \frac{[ES][I]}{[ESI]}$$

$$EI + S \underset{}{\overset{K_S}{\rightleftharpoons}} ESI$$

We can see from the equilibria that, at any inhibitor concentration, an infinitely high substrate concentration cannot drive all the enzyme to the productive ES form.   At any [I] a portion of the enzyme will remain as the nonproductive ESI complex.   Consequently, we can predict that the V_{max} in the presence of a noncompetitive inhibitor (V_{max_i}) will be less than the V_{max} observed in the absence of inhibitor.   The K_m value (measured as the [S] required for 0.5 V_{max_i}) will be unchanged by a noncompetitive inhibitor because, at any inhibitor concentration, the enzyme forms that can combine with S (E and EI) have equal affinities for S.   The net effect of a noncompetitive inhibitor is to make it appear as if less total enzyme is present.   The velocity equation is derived in the usual way.

$$v = k_p [ES] \qquad \frac{v}{[E]_t} = \frac{k_p [ES]}{[E] + [ES] + [EI] + [ESI]}$$

$$[ES] = \frac{[S]}{K_S}[E] \qquad [EI] = \frac{[I]}{K_i}[E]$$

$$[ESI] = \frac{[S]}{K_S}[EI] = \frac{[S][I]}{K_S K_i}[E]$$

$$\therefore \; \frac{v}{k_p [E]_t} = \frac{\dfrac{[S]}{K_S}[E]}{[E] + \dfrac{[S]}{K_S}[E] + \dfrac{[I]}{K_i}[E] + \dfrac{[S][I]}{K_S K_i}[E]}$$

$$\frac{v}{V_{max}} = \frac{\dfrac{[S]}{K_S}}{1 + \dfrac{[S]}{K_S} + \dfrac{[I]}{K_i} + \dfrac{[S][I]}{K_S K_i}} \tag{27}$$

or

$$\frac{v}{V_{max}} = \frac{[S]}{K_S\left(1 + \dfrac{[I]}{K_i}\right) + [S]\left(1 + \dfrac{[I]}{K_i}\right)} \tag{28}$$

We can better appreciate the effect of a noncompetitive inhibitor by dividing both sides of the velocity equation by the parenthetical factor:

$$\frac{\dfrac{v}{V_{max}}}{\left(1 + \dfrac{[I]}{K_i}\right)} = \frac{[S]}{K_S + [S]} \tag{29}$$

or

$$\frac{v}{V_{max_i}} = \frac{[S]}{K_S + [S]} \qquad \text{where} \quad V_{max_i} = \frac{V_{max}}{\left(1 + \dfrac{[I]}{K_i}\right)}$$

$$= \text{the apparent } V_{max} \text{ at the given [I]}$$

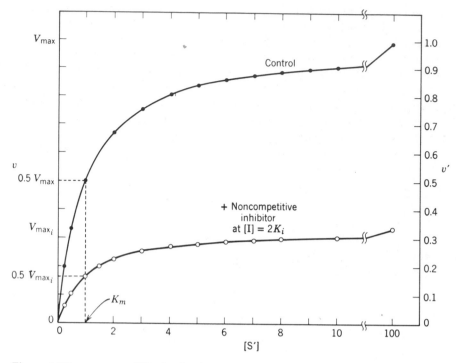

Figure 4-25 v versus [S] plot in the presence of a noncompetitive inhibitor.

As predicted, we see that the only effect of a noncompetitive inhibitor is to decrease V_{max}. The K_s value remains unchanged (Fig. 4-25). (The student should not be confused by the fact that the K_s term in the original equation was multiplied by $[1 + ([I]/K_i)]$. Before deciding whether or not K_s is affected, we must first modify the equation by removing any factor of the variable [S].) The decrease in V_{max} does not mean that the inhibitor has decreased the rate constant for the breakdown of ES to E + P. This constant, k_p, is unchanged. It is the equilibrium level of ES that is decreased. At any [S] and [I], the enzyme-substrate complex is present as a mixture of productive ES and nonproductive ESI forms (and $k_p = 0$ for the ESI complex). The factor $[1 + ([I]/K_i)]$ may be considered to be an [I]-dependent statistical factor describing the distribution of the enzyme-substrate complexes between the ES and ESI forms. *Classical noncompetitive inhibition is obtained only under rapid equilibrium conditions. Thus, $K_m = K_s$.* A steady-state treatment does not yield an equation of the Henri-Michaelis-Menten form, but rather a complex expression containing squared terms. The reciprocal plots would theoretically be nonlinear.

The reciprocal equation is:

$$\frac{1}{v} = \frac{K_m}{V_{max}}\left(1 + \frac{[I]}{K_i}\right)\frac{1}{[S]} + \frac{1}{V_{max}}\left(1 + \frac{[I]}{K_i}\right) \tag{30}$$

The equation indicates that both the slope and the $1/v$-axis intercept of the reciprocal plot are increased by the factor $(1 + [I]/K_i)$ compared to the

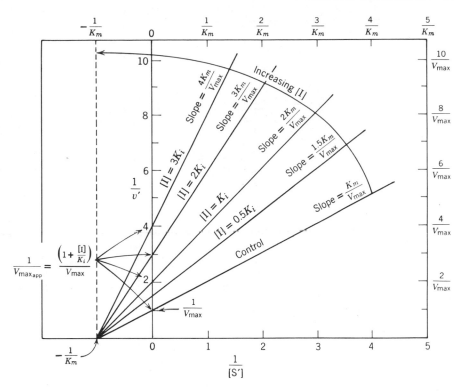

Figure 4-26 $1/v$ versus $1/[S]$ plot in the presence of different fixed concentrations of a noncompetitive inhibitor.

"control" plot. If the slope *and* the $1/v$-axis intercept increase by the same factor, then the $1/[S]$-axis intercept will remain the same (equal to $-1/K_m$). K_i can be calculated from the slope or the $1/v$-axis intercept. For each inhibitor concentration, a new reciprocal plot can be drawn. As [I] increases, the "plus inhibitor" curves increase in slope and $1/v$-axis intercept (Fig. 4-26), pivoting counterclockwise about the point of intersection with the control curve (at $-1/K_m$ on the $1/[S]$ axis). As [I] increases, $1/V_{max_i}$ continually increases, that is, V_{max_i} continually decreases. Because the initial velocity can be driven to zero at a saturating inhibitor concentration, the limiting slope will be a vertical line through $-1/K_m$ and parallel to the $1/v$ axis.

   The slope of the reciprocal plot in the presence of a pure noncompetitive inhibitor is a linear function of [I] as shown earlier for pure competitive inhibition. The $1/v$-axis intercept ($1/V_{max_i}$) is also a linear function of [I] as shown below.

$$\frac{1}{V_{max_i}} = \frac{1}{V_{max}}\left(1 + \frac{[I]}{K_i}\right) \quad \text{or} \quad \boxed{\frac{1}{V_{max_i}} = \frac{1}{V_{max}K_i}[I] + \frac{1}{V_{max}}} \quad (31)$$

Thus, a replot of $1/V_{max_i}$ for each reciprocal plot versus the corresponding inhibitor concentration at which it was obtained will be a straight line with slope $1/V_{max}K_i$ and intercepts of $1/V_{max}$ (at [I] = 0 on the $1/V_{max_i}$ axis) and $-K_i$ (at $1/V_{max_i} = 0$ on the [I] axis).

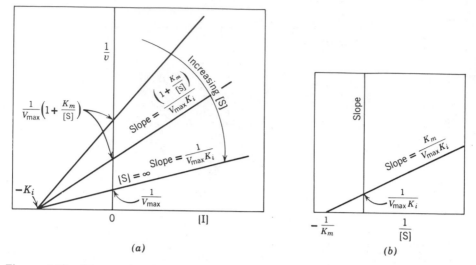

Figure 4-27 Dixon plot for a noncompetitive inhibitor: $1/v$ versus $[I]$ in the presence of different fixed concentrations of substrate.

The reciprocal equation for noncompetitive inhibition can be rearranged to the equation for the Dixon plot.

$$\frac{1}{v} = \frac{\left(1 + \dfrac{K_m}{[S]}\right)}{V_{\max} K_i}[I] + \frac{1}{V_{\max}}\left(1 + \frac{K_m}{[S]}\right) \tag{32}$$

The plot is shown in Figure 4-27. Noncompetitive inhibition may be distinguished from competitive inhibition by the intersection of plots at all values of $[S]$ on the $[I]$ axis at $[I] = -K_i$. The alternate linear plots are shown in Figures 4-33 to 4-35.

IRREVERSIBLE INHIBITION

A substance that combines irreversibly with an enzyme may resemble a noncompetitive inhibitor because $V_{\max}$ is decreased but K_m remains unchanged. The reactions are

$$\begin{array}{c} E + S \xrightleftharpoons{K_S} ES \xrightarrow{k_p} E + P \\ + \\ I \\ \downarrow \\ EI \end{array}$$

Diisopropylfluorophosphate, which irreversibly binds to active serine residues on some hydrolytic enzymes, is an example of this type of inhibitor. $V_{\max}$ decreases because some enzyme is completely removed from the system. (Remember, $V_{\max} = k_p[E]_t$.) An irreversible inhibitor can be dis-

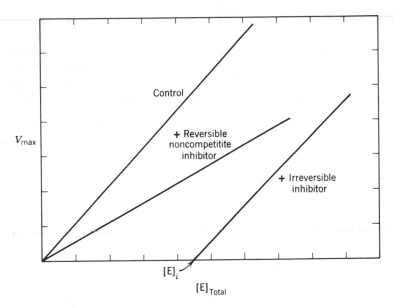

Figure 4-28 A plot of V_{max} versus amount of enzyme added will distinguish between a reversible and an irreversible noncompetitive inhibitor. $[E]_i$ represents the amount of enzyme titrated by the irreversible inhibitor.

tinguished from a classical noncompetitive inhibitor by plotting V_{max} versus the total amount of enzyme added to the assay mixture in the presence of I (Fig. 4-28).

UNCOMPETITIVE INHIBITION

A classical uncompetitive inhibitor is a compound that binds reversibly to the enzyme-substrate complex yielding an inactive ESI complex. The inhibitor does not bind to the free enzyme. Pure uncompetitive inhibition may be rare in unireactant systems. Nevertheless, it is worth considering because it is a simple example of the sequential addition of two enzyme ligands in an obligate order. Uncompetitive inhibition is common in multireactant systems for reasons similar to those described here. That is, I will be uncompetitive with respect to a given substrate if I binds to the enzyme only after the substrate binds (although I rarely binds to a central complex where all the substrate binding sites are filled). Classical uncompetitive inhibition is described by the following equilibria:

$$E + S \underset{K_S}{\rightleftharpoons} ES \xrightarrow{k_p} E + P$$
$$+$$
$$I$$
$$K_i \updownarrow$$
$$ESI$$

The equilibria show that at any [I] an infinitely high substrate concentration will not drive all of the enzyme to the ES form; some nonproductive ESI complex will always be present. Consequently, we can predict that V_{max} in the presence of an uncompetitive inhibitor (V_{max_i}) will be lower than V_{max} in the absence of inhibitor. Unlike noncompetitive inhibition, however, the apparent K_m decreases. The decrease results from the reaction $ES + I \rightleftharpoons ESI$ that uses up some ES causing the substrate binding reaction $E + S \rightleftharpoons ES$ to proceed further to the right. The velocity equation is derived below:

$$v = k_p[ES] \qquad \frac{v}{[E]_t} = \frac{k_p[ES]}{[E] + [ES] + [ESI]}$$

$$[ES] = \frac{[S]}{K_S}[E]$$

and

$$[ESI] = \frac{[I]}{K_i}[ES] = \frac{[S][I]}{K_S K_i}[E]$$

$$\therefore \quad \frac{v}{k_p[E]_t} = \frac{\dfrac{[S]}{K_S}[E]}{[E] + \dfrac{[S]}{K_S}[E] + \dfrac{[S][I]}{K_S K_i}[E]}$$

$$\frac{v}{V_{max}} = \frac{\dfrac{[S]}{K_S}}{1 + \dfrac{[S]}{K_S} + \dfrac{[S][I]}{K_S K_i}} \tag{33}$$

or

$$\frac{v}{V_{max}} = \frac{[S]}{K_S + [S]\left(1 + \dfrac{[I]}{K_i}\right)} \tag{34}$$

A steady-state treatment yields the same equation with K_m replacing K_S. If we remove the parenthetical factor from the variable [S], the equation becomes:

$$\frac{\dfrac{v}{V_{max}}}{\left(1 + \dfrac{[I]}{K_i}\right)} = \frac{[S]}{\dfrac{K_m}{\left(1 + \dfrac{[I]}{K_i}\right)} + [S]} \tag{35}$$

or

$$\frac{v}{V_{max_i}} = \frac{[S]}{K_{m_{app}} + [S]} \tag{36}$$

where $V_{max_i} = \dfrac{V_{max}}{\left(1 + \dfrac{[I]}{K_i}\right)}$ = the apparent V_{max} at the given [I]

and $K_{m_{app}} = \dfrac{K_m}{\left(1 + \dfrac{[I]}{K_i}\right)}$ = the apparent K_m at the same given [I]

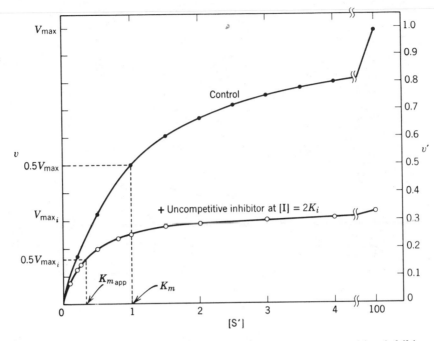

Figure 4-29 v versus [S] plot in the presence of an uncompetitive inhibitor.

In other words, an uncompetitive inhibitor decreases V_{max} and K_m to the same extent. The v versus [S] plot is shown in Figure 4-29. The reciprocal form of the velocity equation for uncompetitive inhibition is:

$$\frac{1}{v} = \frac{K_m}{V_{max}}\frac{1}{[S]} + \frac{1}{V_{max}}\left(1 + \frac{[I]}{K_i}\right)$$ (37)

The slope of the plot is still K_m/V_{max}, but the $1/v$-axis intercept is increased by the factor $[1 + ([I]/K_i)]$ that multiplied the [S] term in the original equation. Consequently, the "plus inhibitor" and control curves will be parallel. As [I] increases, the $1/v$-axis intercepts increase, yielding a series of parallel curves (Fig. 4-30). A saturating inhibitor concentration will drive the velocity to zero. Consequently, the displacement of the "plus inhibitor" curves from the control curve increases without limit. A replot of $1/V_{max_i}$ versus [I] will be linear with intercepts of $1/V_{max}$ and $-K_i$ as shown for noncompetitive inhibition. $K_{m_{app}}$ varies inversely with [I]:

$$\frac{1}{K_{m_{app}}} = \frac{1}{K_i K_m}[I] + \frac{1}{K_m}$$ (38)

Thus, a replot of $1/K_{m_{app}}$ versus [I] will be a straight line with a slope of $1/K_i K_m$ and an intercept on the $1/K_{m_{app}}$ axis of $1/K_m$. When $1/K_{m_{app}} = 0$, the intercept

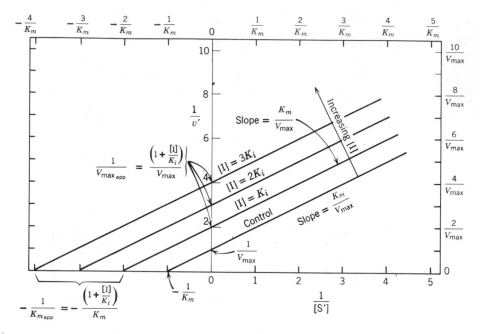

Figure 4-30 $1/v$ versus $1/[S]$ plot in the presence of different fixed concentrations of an uncompetitive inhibitor.

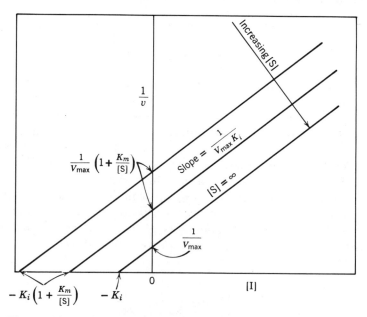

Figure 4-31   Dixon plot for an uncompetitive inhibitor: $1/v$ versus $[I]$ at different fixed concentrations of S.

on the [I] axis gives $-K_i$. The equation for the Dixon plot is:

$$\frac{1}{v} = \frac{1}{V_{\max}K_i}[I] + \frac{1}{V_{\max}}\left(1 + \frac{K_m}{[S]}\right)$$ (39)

The slope expression does not contain an [S] term. Consequently, the family of plots are parallel (Fig. 4-31). The alternate linear plots are shown in Figures 4-33 to 4-35.

LINEAR MIXED-TYPE INHIBITION

The equilibria shown below represent the simplest scheme for mixed-type inhibition (actually a form of noncompetitive inhibition):

$$E + S \xrightarrow{K_s} ES \xrightarrow{k_p} E + P \qquad K_s = \frac{[E][S]}{[ES]} \qquad \alpha K_s = \frac{[EI][S]}{[ESI]}$$

$$EI + S \xrightarrow{\alpha K_s} ESI \qquad K_i = \frac{[E][I]}{[EI]} \qquad \alpha K_i = \frac{[ES][I]}{[ESI]}$$

The presence of I on the enzyme changes the dissociation constant for S from K_s to αK_s. Note that the dissociation constant of I from ESI must also change by the factor α for the four enzyme species to be at equilibrium, that is, the overall K_{eq} of the reaction between E and ESI must be the same regardless of the path. Thus the path $E \rightarrow ES \rightarrow ESI$ has an overall K_{eq} of $1/K_s\alpha K_i$. The path $E \rightarrow EI \rightarrow ESI$ has an overall K_{eq} of $1/K_i\alpha K_s$. (See Fig. 1-11 for a similar situation.) ESI is catalytically inactive. The velocity equation for rapid equilibrium conditions is obtained in the usual manner:

$$v = k_p[ES] \qquad \frac{v}{[E]_t} = \frac{k_p[ES]}{[E] + [ES] + [EI] + [ESI]}$$

From past experience, we can write the terms for each complex without deriving them.

$$\frac{v}{[E]_t} = \frac{k_p\dfrac{[S]}{K_s}}{1 + \dfrac{[S]}{K_s} + \dfrac{[I]}{K_i} + \dfrac{[S][I]}{\alpha K_s K_i}}$$

Letting $k_p[E]_t = V_{\max}$:

$$\frac{v}{V_{\max}} = \frac{\dfrac{[S]}{K_s}}{1 + \dfrac{[S]}{K_s} + \dfrac{[I]}{K_i} + \dfrac{[S][I]}{\alpha K_s K_i}}$$ (40)

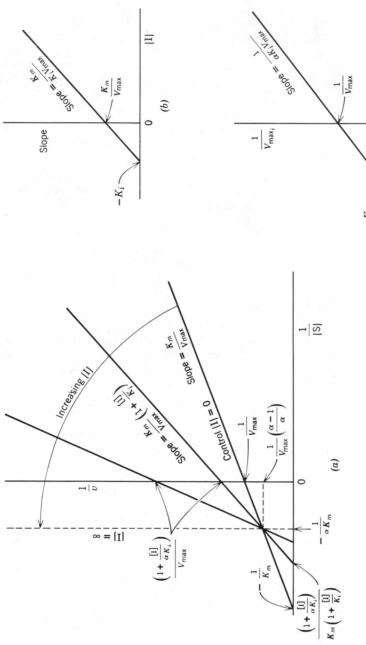

Figure 4-32 $1/v$ versus $1/[S]$ plot in the presence of different fixed concentrations of a mixed-type inhibitor ($1 < \alpha < \infty$). ESI is catalytically inactive. $K_m = K_S$ for rapid equilibrium conditions.

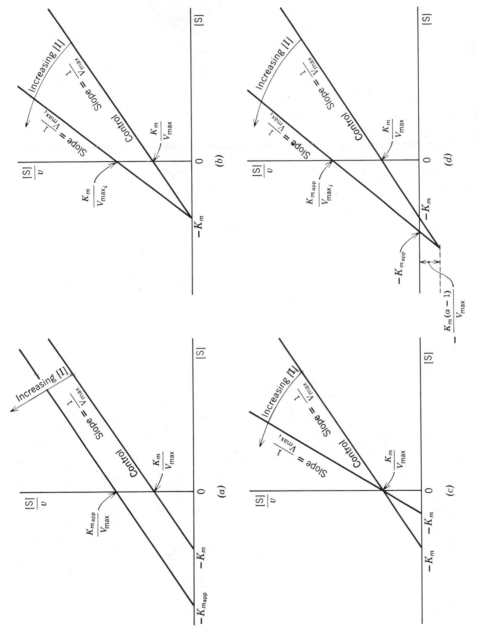

Figure 4-33 Hanes-Woolf Plot: $[S]/v$ versus $[S]$. (*a*) Competitive inhibition. (*b*) Noncompetitive inhibition. (*c*) Uncompetitive inhibition. (*d*) Mixed-type inhibition.

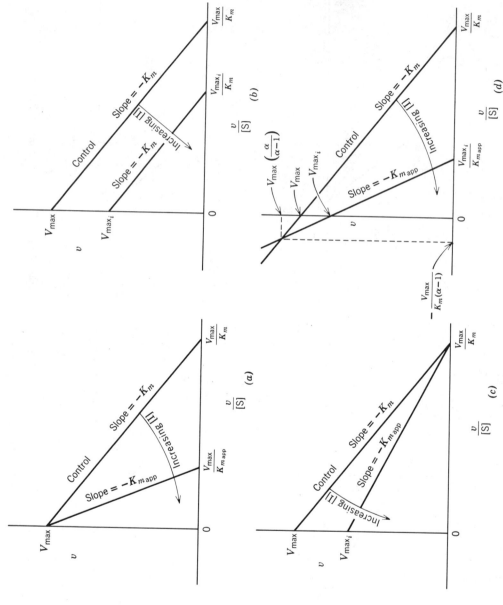

Figure 4-34 Woolf-Augustinsson-Hofstee Plot: v versus v/S. (*a*) Competitive inhibition. (*b*) Noncompetitive inhibition. (*c*) Uncompetitive inhibition. (*d*) Mixed-type inhibition.

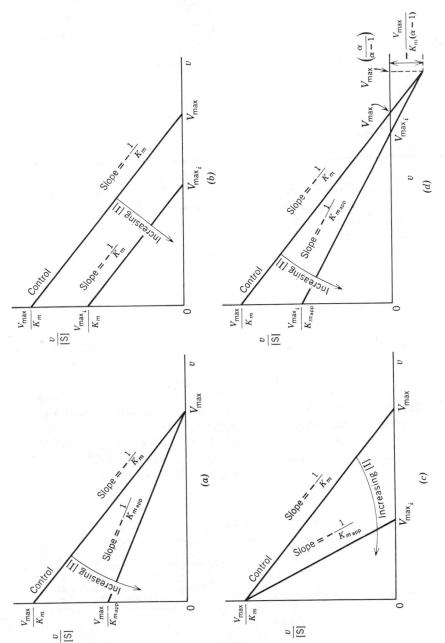

Figure 4-35 Eadie-Scatchard plot: $v/[S]$ versus v. (*a*) Competitive inhibition. (*b*) Noncompetitive inhibition. (*c*) Uncompetitive inhibition. (*d*) Mixed-type inhibition.

or

$$\frac{v}{V_{max}} = \frac{[S]}{K_s\left(1 + \frac{[I]}{K_i}\right) + [S]\left(1 + \frac{[I]}{\alpha K_i}\right)} \qquad (41)$$

The family of reciprocal plots intersect to the left of the $1/v$ axis. The intersection point is above the $1/[S]$-axis if $\alpha > 1$ (Fig. 4-32a). The slope replot gives K_i (Fig. 4-32b); the $1/v$-axis intercept replot gives αK_i (Fig. 4-32c). The alternate liner plots are shown in Figures 4-33 to 4-35.

General Rules for Unireactant Inhibition Systems Where ESI is not Catalytically Active

1. If I and S are mutually exclusive, I will be a competitive inhibitor.
2. If I and S are not mutually exclusive and both ligands bind independently of each other, I will be a noncompetitive inhibitor. If I affects the affinity of E for S, I will be a mixed-type inhibitor with linear slope and intercept replots.
3. If I binds only after S, I will be an uncompetitive inhibitor.

FEEDBACK INHIBITION*

Many biosynthetic pathways are regulated by feedback inhibition; that is, the end product(s) or near end product(s) control the metabolic flux by inhibiting one or more early reactions of the pathway. Often, maximum feedback inhibition is attained only by the combined action of multiple end products. This prevents one end product of a branched biosynthetic pathway from shutting down the pathway completely and thereby starving the organism for the other end products. In the pathway shown below, end product X might inhibit enzyme E_5; end product I might inhibit enzyme E_9. Both I and X together might inhibit enzyme E_1 in a cooperative, concerted, cumulative, or additive way as described in Table 4-2.

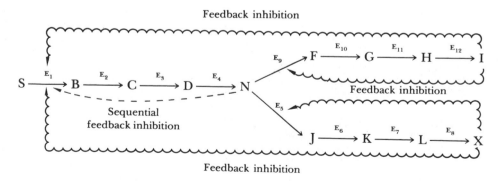

* For a more complete discussion of the various types of inhibition and feedback systems, including partial and mixed-type systems where the ESI complex is catalytically active, the student is referred to the author's *Enzyme Kinetics: Behavior and Analysis of Rapid Equilibrium and Steady-State Enzyme Systems*, Wiley-Interscience (1975).

Table 4-2   Patterns of Feedback Inhibition

Cooperative (synergistic) inhibition *Example:* Phosphoribosylamine synthetase	Each end product, I and X, inhibits E_1. At saturating levels of either one, the velocity can be driven to zero. Mixtures of I and X at low concentrations are more inhibitory than the same total specific concentration of I or X alone. That is, if 35% inhibition is observed at $[I] = 6\,K_i$ and $[X] = 0$ and if 35% inhibition is observed at $[X] = 6\,K_X$ and $[I] = 0$, then at $[I] = 3\,K_i$ plus $[X] = 3\,K_X$, the inhibition will be greater than 35%. Cooperative inhibition implies that I and X are not mutually exclusive. Both end products can combine with E_1 simultaneously to form dead-end EIX and/or EIXS complexes.
Concerted (multivalent) inhibition *Example:* β-aspartyl-kinase of some microorganisms	Either end product alone has no effect at all on E_1, but when both are present, enzyme activity is markedly reduced. Concerted inhibition may be an extreme case of cooperative inhibition where the binding of one end product greatly increases the affinity of the enzyme for the other end product. That is, at low levels of I alone only a negligible amount of EI forms; at low levels of X alone only a negligible amount of EX forms. At low levels of I plus X a large proportion of the enzyme becomes tied up as the dead-end EIX and/or EIXS complexes. An alternate model for concerted inhibition is as follows: EI and EX bind S just as well as E; EIS and EXS function just as well as ES, but EIX (formed only when I and X are both present) does not bind S, or EIXS is catalytically inactive.
Cumulative (partial) inhibition *Example:* Glutamine synthetase	Each end product is a partial inhibitor. That is, a saturating level of I alone or X alone cannot drive the velocity to zero. This implies that EI and EX can bind S but not as well as E (yielding partial competitive inhibition), or that EIS and EXS are catalytically active, but not as active as ES (yielding partial noncompetitive inhibition), or both (yielding partial mixed-type inhibition). If I and X are partial noncompetitive inhibitors, then true cumulative inhibition is observed. That is, if I alone at a given concentration yields 40% inhibition (leaving 60% of the original activity) and X alone at a given concentration yields 20% inhibition (leaving 80% of the original activity), then the same concentrations of I + X will leave 48% (i.e., 80% of 60% of the original activity and yield 52% inhibition (i.e., 40% + [20% of 60%] *or* 20% + [40% of 80%] = 52%).
Additive inhibition *Example:* Multiple β-aspartylkinases in some microorganisms	True additive inhibition by I and X implies the presence of two distinct enzymes (or catalytic sites) each sensitive to only one of the feedback inhibitors. Thus, a saturating level of I or X alone will yield partial inhibition (as only the sensitive

267

Table 4-2   (Cont.)

	enzyme or site is affected). Mixtures of I plus X will appear to act synergistically. That is, the inhibition observed at [I] = 3 K_i plus [X] = 3 K_x will be greater than the inhibition observed at [I] = 6 K_i and [X] = 0, or vice versa. However, if X at 6 K_x inhibits 35%, while I at 6 K_i inhibits 35%, then X at 6 K_x *plus* I at 6 K_i will yield exactly 70% inhibition. None of the other types of feedback inhibition described above yields additive inhibitions.
Sequential inhibition *Example:* DHAP synthetase (of aromatic amino acid biosynthetic pathway)	End product I inhibits only E_9; end product X inhibits only E_5. When I and X are *both* at high levels, both pathways of N utilization are inhibited and the concentration of N builds up. N inhibits E_1.

· Problem 4-12

An enzyme has a K_m of $4.7 \times 10^{-5} M$. If the V_{max} of the preparation is 22 μmoles $\times$ liter^{-1} $\times$ min^{-1}, what velocity would be observed in the presence of $2 \times 10^{-4} M$ substrate and $5 \times 10^{-4} M$ of (a) a competitive inhibitor, (b) a noncompetitive inhibitor, (c) an uncompetitive inhibitor. K_i in all three cases is $3 \times 10^{-4} M$.   (d) What is the degree of inhibition in all three cases?

Solution

(a)
$$v_i = \frac{[S]V_{max}}{K_m\left(1 + \frac{[I]}{K_i}\right) + [S]} = \frac{(2 \times 10^{-4})(22)}{(4.7 \times 10^{-5})\left(1 + \frac{5 \times 10^{-4}}{3 \times 10^{-4}}\right) + (2 \times 10^{-4})}$$

$$= \frac{44 \times 10^{-4}}{(4.7 \times 10^{-5})(2.67) + (2 \times 10^{-4})} = \frac{44 \times 10^{-4}}{3.25 \times 10^{-4}}$$

$$\boxed{v_i = 13.54 \ \mu\text{moles} \times \text{liter}^{-1} \times \text{min}^{-1}}$$

(b)
$$v_i = \frac{[S]V_{max}}{([S] + K_m)\left(1 + \frac{[I]}{K_i}\right)} = \frac{(2 \times 10^{-4})(22)}{[(2 \times 10^{-4}) + (4.7 \times 10^{-5})]\left(1 + \frac{5 \times 10^{-4}}{3 \times 10^{-4}}\right)}$$

$$= \frac{44 \times 10^{-4}}{(2.47 \times 10^{-4})(2.67)} = \frac{44 \times 10^{-4}}{6.59 \times 10^{-4}}$$

$$\boxed{v_i = 6.68 \ \mu\text{moles} \times \text{liter}^{-1} \times \text{min}^{-1}}$$

(c) $\quad v_i = \dfrac{[S]V_{max}}{K_m + [S]\left(1 + \dfrac{[I]}{K_i}\right)} = \dfrac{(2 \times 10^{-4})(22)}{(4.7 \times 10^{-5}) + (2 \times 10^{-4})\left(1 + \dfrac{5 \times 10^{-4}}{3 \times 10^{-4}}\right)}$

$\qquad = \dfrac{44 \times 10^{-4}}{(4.7 \times 10^{-5}) + (2 \times 10^{-4})(2.67)} = \dfrac{44 \times 10^{-4}}{5.81 \times 10^{-4}}$

$$\boxed{v_i = 7.57 \; \mu\text{moles} \times \text{liter}^{-1} \times \text{min}^{-1}}$$

(d)   In the absence of inhibitors the velocity, v_0, is given by:

$$v_0 = \dfrac{[S]V_{max}}{K_m + [S]} = \dfrac{(2 \times 10^{-4})(22)}{(4.7 \times 10^{-5}) + (2 \times 10^{-4})} = \dfrac{44 \times 10^{-4}}{2.47 \times 10^{-4}}$$

$$\boxed{v_0 = 17.8 \; \mu\text{moles} \times \text{liter}^{-1} \times \text{min}^{-1}}$$

We can express the inhibition in a number of ways:

$$a = \dfrac{v_i}{v_0} = \text{relative activity as a fraction}$$

$$a_\% = \dfrac{100 \, v_i}{v_0} = \text{relative activity in terms of percent}$$

$$i = \left(1 - \dfrac{v_i}{v_0}\right) = 1 - a = \text{fractional inhibition}$$

$$i_\% = 100(1 - a) = \text{degree of inhibition as a percent (“percent inhibition”)}$$

Thus for the three inhibitors described above:

Competitive inhibitor

$$a = \dfrac{v_i}{v_0} = \dfrac{13.54}{17.8} = 0.76 \qquad \text{or} \qquad 76\% \text{ of the original activity}$$

$$i = 1 - 0.76 = 0.24 \qquad \text{or} \qquad \boxed{\textbf{24\% inhibition}}$$

Noncompetitive inhibitor

$$a = \dfrac{6.68}{17.8} = 0.375 \qquad \text{or} \qquad 37.5\% \text{ of the original activity}$$

$$i = 1 - 0.375 = 0.625 \qquad \text{or} \qquad \boxed{\textbf{62.5\% inhibition}}$$

Uncompetitive inhibitor

$$a = \frac{7.57}{17.8} = 0.425 \qquad \text{or} \qquad 42.5\% \text{ of the original activity}$$

$$i = 1 - 0.425 = 0.575 \qquad \text{or} \qquad \boxed{\textbf{57.5\% inhibition}}$$

· Problem 4-13

What is the relative activity and the degree of inhibition caused by a competitive inhibitor when $[S] = K_m$ and $[I] = K_i$?

Solution

In general, the relative activity, $a = v_i/v_0$, in the presence of a competitive inhibitor is given by:

$$a = \frac{v_i}{v_0} = \frac{\dfrac{[S]V_{max}}{K_m\left(1 + \dfrac{[I]}{K_i}\right) + [S]}}{\dfrac{[S]V_{max}}{K_m + [S]}} \qquad \text{or} \qquad \boxed{a = \frac{K_m + [S]}{K_m\left(1 + \dfrac{[I]}{K_i}\right) + [S]}} \qquad (42)$$

The inhibition, i, is $1 - a$:

$$i = 1 - a = 1 - \frac{K_m + [S]}{K_m\left(1 + \dfrac{[I]}{K_i}\right) + [S]}$$

$$= \frac{K_m + \dfrac{K_m[I]}{K_i} + [S] - K_m - [S]}{K_m + \dfrac{K_m[I]}{K_i} + [S]} = \frac{K_m[I]}{K_iK_m + K_m[I] + K_i[S]}$$

$$\boxed{i = \frac{[I]}{K_i\left(1 + \dfrac{[S]}{K_m}\right) + [I]}} \qquad \text{and} \qquad \boxed{i_\% = \frac{100[I]}{K_i\left(1 + \dfrac{[S]}{K_m}\right) + [I]}} \qquad (43)$$

Thus, when $[S] = K_m$ and $[I] = K_i$:

$$a = \frac{K_m + K_m}{K_m(1 + 1) + K_m} = \frac{2\,K_m}{3\,K_m} = \boxed{\textbf{0.667}}$$

$$i = 1 - a = 1 - 0.667 = \boxed{\textbf{0.333}}$$

$$i_\% = 100\,i = \boxed{\textbf{33.3\%}}$$

or

$$i_\% = \frac{100\,K_i}{K_i(1+1)+K_i} = \frac{100\,K_i}{3\,K_i} = \boxed{\textbf{33.3\%}}$$

Note that the value of V_{max} is not needed to calculate either a or i.
  Similar derivations yield:

Noncompetitive inhibition

$$\boxed{a = \frac{K_i}{K_i + [\text{I}]}} \quad \text{and} \quad \boxed{i = \frac{[\text{I}]}{K_i + [\text{I}]}} \tag{44}$$

Uncompetitive inhibition

$$\boxed{a = \frac{K_m + [\text{S}]}{K_m + [\text{S}]\left(1 + \dfrac{[\text{I}]}{K_i}\right)}} \quad \text{and} \quad \boxed{i = \frac{[\text{I}]}{K_i\left(1 + \dfrac{K_m}{[\text{S}]}\right) + [\text{I}]}} \tag{45}$$

Equations 43, 44, and 45 tell us that at a fixed inhibitor concentration, increasing the substrate concentration (a) decreases the degree of competitive inhibition, (b) has no effect on the degree of noncompetitive inhibition, and (c) increases the degree of uncompetitive inhibition. The effect of increasing [S] on the degree of inhibition caused by a mixed-type inhibitor depends on the interaction factor, α. In the usual case of $\alpha > 1$, the degree of inhibition decreases as [S] increases at a fixed [I].

· Problem 4-14

A marine microorganism contains an enzyme that hydrolyzes glucose-6-sulfate (S). The assay is based on the rate of glucose formation. The enzyme in a cell-free extract has kinetic constants of $K_m = 6.7 \times 10^{-4}\,M$ and $V_{max} = 300$ nmoles $\times$ liter$^{-1} \times$ min^{-1}. Galactose-6-sulfate is a competitive inhibitor (I). At $10^{-5}\,M$ galactose-6-sulfate and $2 \times 10^{-5}\,M$ glucose-6-sulfate, v was 1.5 nmole $\times$ liter$^{-1} \times$ min^{-1}. Calculate K_i for galactose-6-sulfate.

Solution

$$\frac{v}{V_{max}} = \frac{[\text{S}]}{K_{m_{app}} + [\text{S}]} \qquad K_{m_{app}} = K_m\left(1 + \frac{[\text{I}]}{K_i}\right)$$

$$\frac{1.5}{300} = \frac{(2 \times 10^{-5})}{K_{m_{app}} + (2 \times 10^{-5})}$$

$$1.5\,K_{m_{app}} + 3 \times 10^{-5} = 600 \times 10^{-5}$$

$$1.5\,K_{m_{app}} = (600 \times 10^{-5}) - (3 \times 10^{-5})$$

$$K_{m_{app}} = \frac{597 \times 10^{-5}}{1.5} = 398 \times 10^{-5}$$

$$\boxed{K_{m_{app}} = 39.8 \times 10^{-4}\,M}$$

$$K_{m_{app}} = K_m\left(1 + \frac{[I]}{K_i}\right) = 39.8 \times 10^{-4}$$

$$6.7 \times 10^{-4}\left(1 + \frac{10^{-5}}{K_i}\right) = 39.8 \times 10^{-4}$$

$$6.7 \times 10^{-4} + \frac{6.7 \times 10^{-9}}{K_i} = 39.8 \times 10^{-4}$$

$$\frac{6.7 \times 10^{-9}}{K_i} = (39.8 \times 10^{-4}) - (6.7 \times 10^{-4}) = 33.1 \times 10^{-4}$$

$$33.1 \times 10^{-4}\, K_i = 6.7 \times 10^{-9}$$

$$K_i = \frac{6.7 \times 10^{-9}}{33.1 \times 10^{-4}} = 0.202 \times 10^{-5}$$

$$\boxed{K_i = 2.02 \times 10^{-6}\ M}$$

In practice, the data would be analyzed by one of the linear plots described earlier. The K_i would be obtained from an appropriate replot.

· Problem 4-15

Ethylene glycol is oxidized to toxic oxalic acid in the liver. The first enzyme responsible for the multistep reaction sequence is alcohol dehydrogenase (ADH). Ethylene glycol poisoning can be counteracted by providing large doses of ethanol, the true substrate of ADH. The ethanol displaces ethylene glycol from the enzyme. In this case, S acts as an inhibitor, when I is the substrate. Suppose that the K_m for ethanol is 10^{-5} M, and the K_m for ethylene glycol is 10^{-4} M at the fixed NAD^+ concentration in the liver. Let us assume that the concentrations of ethanol and ethylene glycol in the liver cells are the same as those in the blood stream (which can be measured). Analysis gave [ethylene glycol] $= 5 \times 10^{-5}$ M in the blood. (a) To what concentration must the blood ethanol be raised in order to inhibit the activity of ADH with ethylene glycol by 95% (thereby providing sufficient time for the unchanged ethylene glycol to be excreted in the urine)? (b) Proinebrium is a new drug that acts as a noncompetitive inhibitor of ADH ($K_i = 2 \times 10^{-6}$ M). What concentration of Proinebrium is required to achieve 95% inhibition of ADH?

Solution

(a) If we assume that K_i for ethanol is equal to its K_m, then to obtain 95% inhibition:

$$i_\% = \frac{100\,[\text{EtOH}]}{K_{i_{EtOH}}\left(1 + \frac{[\text{EG}]}{K_{m_{EG}}}\right) + [\text{EtOH}]} = \frac{100\,[\text{EtOH}]}{10^{-5}\left(1 + \frac{5 \times 10^{-5}}{10^{-4}}\right) + [\text{EtOH}]} = 95\%$$

$$95 \times 10^{-5}(1 + 0.5) + 95\,[\text{EtOH}] = 100\,[\text{EtOH}]$$

$$142.5 \times 10^{-5} = 5\,[\text{EtOH}] \qquad [\text{EtOH}] = 28.5 \times 10^{-5}\ M$$

$$\boxed{\text{EtOH} = 2.85 \times 10^{-4}\ M}$$

(b)   Although v_0 and v_i vary with [S], the degree of inhibition is independent of [S] for a noncompetitive inhibitor (Equation 44).   Thus, to achieve 95% inhibition:

$$0.95 = \frac{1.00[I]}{K_i + [I]} \qquad 0.95\,K_i + 0.95[I] = 1.00[I]$$

$$0.05[I] = 0.95\,K_i \qquad [I] = 19\,K_i = (19)(2 \times 10^{-6})$$

$$\boxed{[I] = 3.8 \times 10^{-5}\ M}$$

If Proinebrium is toxic, then we can reduce its blood concentration 19-fold (to $[I] = 2 \times 10^{-6}\ M$) and still achieve 50% inhibition.

F. EFFECT OF pH ON ENZYME STABILITY AND ACTIVITY

It is not surprising that pH will influence the velocity of an enzyme-catalyzed reaction.   The active sites on enzymes are frequently composed of ionizable groups that must be in the proper ionic form in order to maintain the conformation of the active site, bind the substrates, or catalyze the reaction.   Furthermore, one or more of the substrates themselves may contain ionizable groups and only one ionic form of that substrate may bind to the enzyme or undergo catalysis.   The pK values of the prototropic groups of the active site can often be determined by measuring the pH-dependence of $slope_{1/S}$ and $V_{max_{app}}$.   Once the pK values are known, we can make an educated guess as to the identities of the groups involved.   The effects of pH on the stability of an enzyme must be taken into account in any study of the effect of pH on substrate binding and catalysis.   Figure 4-36 shows an experimental v versus pH curve for an enzyme (curve A).   The pH "optimum" is at 6.8.   Curve A gives no indication why the velocity declines above and below pH 6.8.   The decline could result from the formation of an improper ionic form of the substrate or enzyme (or both), or from inactivation of the enzyme, or from a combination of these effects.   Curve B shows the effect of pH on enzyme stability.   We see that preincubation of the enzyme at pH 5 or pH 8 has no effect on the activity measured at pH 6.8.   Thus, the decline in activity between pH 6.8 and 8 and between pH 6.8 and 5 must result from the formation of an improper ionic form of the enzyme and/or substrate.   When the enzyme is preincubated at pH >8 or pH <5, full activity is not regained at pH 6.8.   Thus, part of the decline in activity above pH 8 and below pH 5 results from irreversible enzyme inactivation.   A pH stability study, such as that shown in curve B, is an essential part of any enzyme characterization.   Unfortunately, it is frequently omitted and only the curve A data presented.   The stability curve B can be obtained by preincubating the enzyme at the indicated pH for a time at least as long as the usual assay time.   Enzyme activity is then measured at the optimum pH.   The pH stability of an enzyme depends on many factors including (a) temperature, (b) ionic strength, (c) chemical nature of the buffer, (d) concentration of various preservatives (e.g., glycerol, sulfhydryl compounds), (e) concentration of

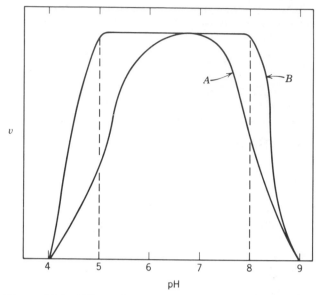

Figure 4-36 Effect of pH on activity and stability of an enzyme. Curve *A*: *v* versus pH plot. Curve *B*: *v* at pH 6.8 after preincubating the enzyme at the indicated pH values. The decline in activity between pH 6.8 and 5.0 and between 6.8 and 8.0 can be ascribed to the effect of pH on ionizable groups of the active site or substrate. The decline in activity above pH 8.0 and below 5.0 can be ascribed to irreversible denaturation of the enzyme.

contaminating metal ions, (f) concentration of substrates or cofactors of the enzyme, and (g) enzyme concentration. In many cases a substrate may induce a conformational change in the enzyme to a form that is more resistant or less resistant to pH or temperature denaturation. The concentration of the enzyme itself may be a factor. At low concentrations, the enzyme may dissociate into smaller oligomers or monomers, which may be less stable than the original oligomer. It should also be kept in mind that an enzyme may be more stable over a long period of time at a pH significantly different from the "optimum" used in the assay. In the following discussion it is assumed that preliminary studies have established that the enzyme is stable over the pH range studied.

V VERSUS pH—*A SIMPLE MONOPROTIC MODEL*

Consider a system in which the substrate is a weak acid, HA, but only the ionized form, A^-, binds to the enzyme. The true substrate then is A^-. At a fixed concentration of total weak acid, the proportion that is in the proper ionic form can be calculated from the Henderson-Hasselbalch equation. Thus, at $pH = pK_a$, half of the total concentration is present as A^-. When the pH is one unit above the pK_a, 10/11 of the total is present as A^- (i.e., the ratio of $[A^-]/[HA] = 10$). When $pH = pK_a + 2$, 100/101 of the total is present as A^-. When the pH is one unit less than pK_a, 1/11 of the total is present as A^-, and so on.   Thus, we might expect the velocity to increase as the pH increases, as shown in curve A of Figure 4-37 where the pK_a of the substrate is assumed to be 4.0.   However, suppose that the active site of the enzyme contains a basic group that must be protonated in order to bind the negatively charged substrate.   The proportion of the total enzyme concentration in the proper ionic form decreases as the pH increases, as shown in curve B of Figure 4-37 where the pK_a of the active site is assumed to be 7.0.   The theoretical maximum activity would occur at a pH where all the enzyme exists as EH^+ and all the substrate exists as A^-.   However, the two pK_a values are only three units

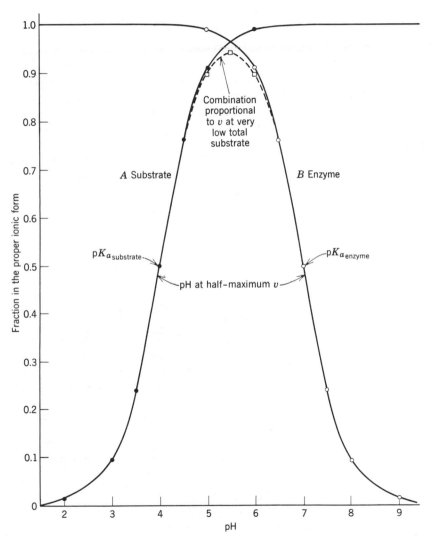

Figure 4-37   Curve A: Fraction of the total substrate that is in the proper ionic form (A^-) as a function of pH.   Curve B: fraction of the total enzyme that is in the proper ionic form (EH^+) as a function of pH. pK_a of the substrate = 4.0; pK_a of the enzyme = 7.0.   The broken line represents the product of the two fractions, which is proportional to v at $[A^-] \ll K_{m_A}$.   Because the two pK_a values are widely separated (by 3 pH units), the pH at half-maximal v will be close to the pK_a values.

apart and, consequently, there is no pH where this is possible. A simple calculation will show that the maximum combination of EH^+ and A^- occurs at pH 5.5 where $31.6/32.6 = 97\%$ of the total substrate and total enzyme are each in the proper ionic form. Thus, for the two pK_a values chosen, we can obtain only 97% of 97% = 94% of the theoretical maximum combination of EH^+ and A^-. If the enzyme and substrate are stable over the pH range plotted and the A^- concentration is very low compared to K_m even at pH 5.5, the v versus pH curve will resemble the curve formed by the overlap of the two curves shown in Figure 4-37 (except the maximum will occur at 0.94). If either the enzyme or substrate is unstable at extreme pH's, the curve will be narrower than that shown. If the A^- concentration is high compared to K_m, then the curve will not decrease as rapidly below pH 5.5. For example, if $K_m = 10^{-5} M$ and $[S]_t = 10^{-3} M$, then even at pH 3.0 the A^- concentration will still be almost 10 times K_m. At first glance, it may seem an easy matter to determine the two pK values as the pH's at half-maximum velocity. However, this is true only if the two pK's are widely separated so that the maximum velocity occurs close to the theoretical maximum. This is almost true for the values chosen to generate Figure 4-38. If the pK values are quite close, the velocity will still be maximal at the average of the two pK's. However, the pH's at the half-maximum velocities will not correspond to the pK values.

A complete analysis of pH effects can be quite complicated, especially when we consider that many biological compounds possess multiple ionizable groups and that the active site of an enzyme may also possess two or more ionizable groups that must be in the proper ionic form before substrate binding or catalysis occurs.

· **Problem 4-16**

The active site of an enzyme contains a single ionizable group that must be in the negative form before the substrate can bind and catalysis occur. The pK of this group is 5.0. The substrate is a positively charged compound and remains completely ionized over the pH range studied. (a) Write the reactions showing the effect of pH on the distribution of enzyme species. (b) Derive a velocity equation for this system. (c) What would a plot of v versus pH look like?

Solution

(a) If we indicate the active form of the enzyme as E^- and the substrate as S^+, the reactions are:

$$E^- + S^+ \rightleftharpoons E^-S^+ \xrightarrow{\ k_p\ } E^- + P^+$$
$$+$$
$$H^+$$
$$\kappa_e \big\Updownarrow$$
$$EH$$

Thus, for this simple monoprotic system, H^+ acts as a competitive inhibitor. As $[H^+]$ increases (i.e., pH decreases), the enzyme is driven to the inactive, dead-end EH form.

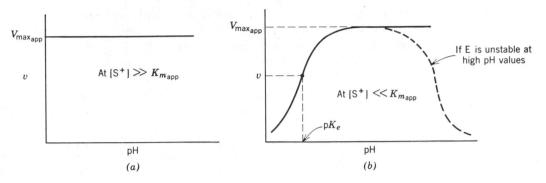

Figure 4-38 (*a*) Plot of v versus pH at $[S^+] \gg K_{m_{app}}$. (*b*) Plot of v versus pH at $[S^+] \ll K_{m_{app}}$.

(b)   The velocity equation is the usual equation for competitive inhibition with $[H^+]/K_e$ replacing $[I]/K_i$:

$$\frac{v}{V_{max}} = \frac{[S^+]}{K_m\left(1 + \dfrac{[H^+]}{K_e}\right) + [S^+]}$$ and $$\frac{1}{v} = \frac{K_m}{V_{max}}\left(1 + \frac{[H^+]}{K_e}\right)\frac{1}{[S]} + \frac{1}{V_{max}}$$

(46)

Reciprocal plots of $1/v$ versus $1/[S^+]$ at different fixed H^+ concentrations will yield different values of $K_{m_{app}}$, from which K_e (hence, pK_e) can be obtained in the usual way.

(c)   If $[S^+]$ is saturating (very high compared to $K_{m_{app}}$ at all pH values) then $v = V_{max}$ and the plot of V_{max} versus pH will be a straight line (no obvious optimum) (Fig. 4-38*a*).   If $[S^+]$ is low compared to $K_{m_{app}}$, v will increase with increasing pH (decreasing $[H^+]$) and then approach a straight line at pH values high compared to pK_e (as almost all the enzyme is converted to the active E^- form) (Fig. 4-38*b*).   If the enzyme is unstable at high pH values, v may decrease, giving an apparent "optimum" pH region.

G. EFFECT OF TEMPERATURE ON ENZYME STABILITY AND ACTIVITY

Most chemical reactions proceed at a faster velocity as the temperature is raised.   An increase in T imparts more kinetic energy to the reactant molecules resulting in more productive collisions per unit time.   Enzyme catalyzed reactions behave similarly, up to a point.   Enzymes are complex protein molecules.   Their catalytic activity results from a precise, highly ordered tertiary structure that juxtaposes specific amino acid R-groups in such a way as to form the stereospecific substrate binding sites and the catalytic center.   The tertiary structure of an enzyme is maintained primarily by a large number of weak noncovalent bonds.   In practical terms, an enzyme molecule is a very delicate and fragile structure.   If the molecule absorbs too much energy, the tertiary structure will disrupt and the enzyme will be denatured, that is, lose catalytic activity. Thus, as the temperature increases, the expected increase in v resulting from increased E + S collisions is offset by the increasing rate of denaturation. Consequently, a plot of v versus T usually shows a peak, sometimes referred to as the "optimum temperature." The "optimum temperature" depends on the assay time chosen. The true "optimum" temperature for an assay is the maximum temperature at which the enzyme exhibits a constant activity over a time period at least as long as the assay time. This can easily be established by preincubating the enzyme at different temperatures for one or two times the desired assay time and then measuring the activity at a temperature low enough to cause no denaturation. The temperature stability of an enzyme depends on a number of factors including the pH and ionic strength of the medium and the presence or absence of ligands. Substrates frequently protect against temperature denaturation. Low molecular weight enzymes composed of single polypeptide chains and possessing disulfide bonds are usually more heat stable than high molecular weight, oligomeric enzymes. In general, an enzyme will be more heat stable in crude cell-free preparations containing a high concentration of other proteins (provided no proteases are present).

THE ARRHENIUS EQUATION—ENERGY OF ACTIVATION

The relationship between the rate constant of a reaction, k, and the activation energy, E_a, is given by the Arrhenius equation (Chapter 3):

$$k = Ae^{-E_a/RT} \qquad \text{or} \qquad \log k = -\frac{E_a}{2.3RT}\frac{1}{T} + \log A \qquad (47)$$

A is a constant for the particular reaction. A plot of $\log k$ versus $1/T$ is linear (Fig. 4-39). The integrated form of the Arrhenius equation is:

$$\log \frac{k_2}{k_1} = \frac{E_a}{2.3R}\left(\frac{T_2 - T_1}{T_2 T_1}\right) \qquad \text{or} \qquad E_a = \frac{2.3R\, T_2 T_1}{(T_2 - T_1)}\log \frac{k_2}{k_1} \qquad (48)$$

where k_2 and k_1 are the specific reaction rate constants at two different temperatures, T_2 and T_1, respectively.

In a simple rapid equilibrium system, $V_{max}/[E]_t = k_p$, a first order rate constant. Thus, a plot of $\log V_{max}/[E]_t$ versus $1/T$ yields E_a for the catalytic step. In practice, just $\log V_{max}$ can be plotted since the V_{max} of a given preparation is proportional to k_p. Because the K_m varies with T, it cannot be assumed that a given concentration of substrate will be saturating at all temperatures. Ideally, V_{max} should be determined from a reciprocal plot at each temperature. For most enzyme catalyzed reactions, V_{max} depends on several rate constants, each of which may be affected differently by changing temperature. As a result, the E_a calculated from the Arrhenius plot will be

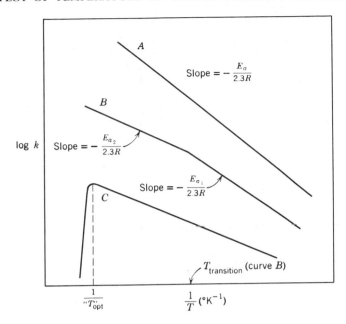

Figure 4-39 The activation energy, E_a, for a reaction can be determined by measuring the reaction rate constant at different temperatures and plotting $\log k$ versus $1/T$. For enzyme-catalyzed reactions, $\log V_{max}/[E]_t$ or just $\log V_{max}$ can be plotted. Curve A: The usual plot. Curve B: Sometimes the plot will show a definite change in slope if at some temperature a different step becomes rate-limiting. Curve C: A sudden drop in the plot indicates enzyme inactivation.

an apparent or "average" value. The Arrhenius plot itself may be nonlinear if different steps become rate limiting at different temperatures. In some cases, the plot may show a sharp change in slope at some temperature ("transition temperature") where the dependency of V_{max} changes from one rate-limiting step to another (Fig. 4-39, Curve B). A sudden drop in the Arrhenius plot at low $1/T$ (high T) indicates protein denaturation (Fig. 4-39, Curve C).

The effect of temperature on the rate of reactions is frequently expressed in terms of a temperature coefficient, Q_{10}, which is the factor by which the rate increases by raising the temperature 10°C.

$$E_a = \frac{2.3R\,T_2T_1\,\log Q_{10}}{10} \tag{49}$$

A Q_{10} of 2 is equivalent to an E_a of about 12,600 cal/mole in the region of 25 to 35°C.

· Problem 4-17

The effect of temperature on the hydrolysis of lactose by a β-galactosidase is shown below. Calculate the activation energy, E_a, and also Q_{10}.

$T\ (°C)$	$V_{\max}\ (\mu moles \times min^{-1} \times mg\,protein^{-1})$
20	4.50
30	8.65
35	11.80
40	15.96
45	21.36

Solution

The best way to obtain E_a is to plot $\log V_{\max}$ versus $1/T$ where T is given in °K. The data are arranged below.

$T\ (°C)$	$T\ (°K)$	$(1/T) \times 10^3$	$V_{\max}$	$\log V_{\max}$
20	293	3.413	4.50	0.653
30	303	3.300	8.65	0.937
35	308	3.247	11.80	1.071
40	313	3.195	15.96	1.203
45	318	3.145	21.36	1.330

The plot has a slope of:

$$\frac{1.330 - 0.653}{(3.145 - 3.413) \times 10^{-3}} = \frac{0.677}{-0.268 \times 10^{-3}} = -2.53 \times 10^3$$

This slope equals $-E_a/2.3R$.

$$\therefore \quad E_a = -2.3(1.98)(-2.53 \times 10^3)$$

$$\boxed{E_a \simeq 11{,}500 \text{ cal/mole}}$$

We can calculate E_a directly from any two points:

$$\log \frac{V_{\max_2}}{V_{\max_1}} = \frac{E_a}{2.3R}\left(\frac{T_2 - T_1}{T_2 T_1}\right)$$

$$E_a = \frac{2.3R \log \dfrac{V_{\max_2}}{V_{\max_1}} T_1 T_2}{(T_2 - T_1)}$$

Choosing 30 and 40°C:

$$E_a = \frac{(2.3)(1.98) \log 1.845(293)(303)}{10}$$

$$\boxed{E_a \simeq 11{,}500 \text{ cal/mole}}$$

The Q_{10} between 30 and 40°C is:

$$Q_{10} = \frac{15.96}{8.65} = \boxed{1.85}$$

Slightly different Q_{10} values are obtained for different 10° temperature differences.

H. ENZYME ASSAYS

INITIAL VELOCITY AS A FUNCTION OF $[E]_t$

Under the usual *in vitro* assay conditions, the enzyme is present in limiting or "catalytic" amounts in the neighborhood of 10^{-12} to 10^{-7} M while [S] is generally 10^{-6} to 10^{-2} M. At any substrate concentration, the initial velocity is given by:

$$v = \frac{[S]V_{max}}{K_m + [S]} = \frac{[S]k_p[E]_t}{K_m + S} = \frac{k_p}{\left(1 + \dfrac{K_m}{[S]}\right)}[E]_t$$

Thus, v is always directly proportional to $[E]_t$ and this fact can be used to quantitate the amount of enzyme present. It should be stressed that the relationship between v and $[E]_t$ is linear only if true initial velocities are measured. Since v varies with [S], the assay period must be short enough to insure that only a small fraction of the substrate is utilized (5% or less). Figure 4-40a shows the appearance of product at different concentra-

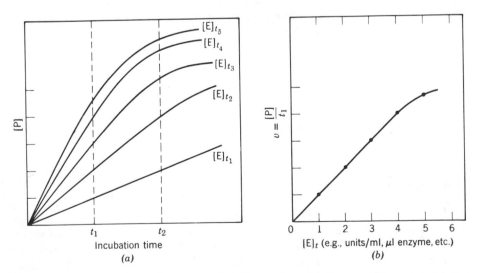

Figure 4-40   Enzyme assays.   (a) Product formation with time at different concentrations of enzyme.   (b) Initial velocity (calculated as [P]/t_1) as a function of enzyme concentration.

tions of enzyme and a fixed [S].   The rate of product formation, $d[P]/dt$, is constant for $[E]_{t_1}$ to $[E]_4$ up until time $= t_1$ (Fig. 4-40b).   If a longer assay time is chosen (e.g., t_2), the response would not be linear over the entire range of $[E]_t$.   Similarly if an enzyme concentration greater than $[E]_4$ is used, the response would not be linear for an assay time t_1.   Thus, the first thing to do in any assay or kinetic study is to establish the limits of linearity, that is, establish the maximum concentration of product that can accumulate before the [P] versus time and v versus $[E]_t$ responses become nonlinear.

ENZYME UNITS AND SPECIFIC ACTIVITIES— QUANTITATING $[E]_t$

In most preparations the actual molar concentration of enzyme is unknown.   Consequently, the amount of enzyme present can be expressed only in terms of its activity.   In order to standardize the reporting of enzyme activities, the Commission on Enzymes of the International Union of Biochemistry has defined a standard unit:

> **One International Unit (1 U) of enzyme is that amount that catalyzes the formation of 1 μ mole of product per minute under defined conditions.**

The concentration of enzyme in an impure preparation is expressed in terms of units/ml.   The specific activity of the preparation is reported as units/mg protein.   As the enzyme is purified, the specific activity will increase to a limit (that of the pure enzyme).   Since v varies with [S], pH, ionic strength, temperature, and the like, a given preparation can have an infinite number of specific activities.   Consequently, specific activities are usually reported for optimal assay conditions at a fixed temperature (usually 25, 30, or 37°C), with all substrates present at saturating concentrations.

A new unit of enzyme activity, the *katal*, has been proposed.   One katal is that amount of enzyme that catalyzes the conversion of one mole of substrate per second.   Thus, one International Unit $= 1/60 \mu$katal $= 16.67$ nkatal. One katal $= 6 \times 10^7$ Units.   Specific activity can be expressed as katals/kg protein or μkatals/mg protein.   The molar activity of an enzyme is the katals/mole protein with units of sec^{-1}.

TURNOVER NUMBER

The term "turnover number" can be used in two ways.   One way, which has been redefined as "molecular activity" or "molar activity," is *the number of moles of substrate transformed per minute per mole of enzyme (units per micromole of enzyme) under optimum conditions*.   Since many enzymes are oligomers containing n subunits, another possible "turnover number" is *the number of moles of substrate transformed per minute per mole of active subunit or catalytic center (under optimum conditions)*.   This latter definition of catalytic power is called "catalytic center activity." Both values are sometimes given simply as a number with units of min^{-1}.

$$k_p = \frac{V_{max}}{[E]_t} = \frac{\mu \text{ moles (of } S \rightarrow P) \, min^{-1} \times ml^{-1}}{\mu \text{ moles (of E) } ml^{-1}} = min^{-1}$$

The k_p ranges from about 50 to about $10^5 \, min^{-1}$. Carbonic anhydrase has one of the highest turnover numbers known ($36 \times 10^6 \, min^{-1}$). The reciprocal of k_p gives the time required for a single catalytic cycle. Thus, for carbonic anhydrase, $1/k_p = 1/(36 \times 10^6) = 0.028 \times 10^{-6} \, min = 1.7 \, \mu sec$.

QUANTITATION OF $[E]_t$ USING THE INTEGRATED VELOCITY EQUATION

One form of the integrated velocity equation is:

$$V_{max} t = 2.3 \, K_m \, \log \frac{[S]_0}{[S]_0 - [P]} + [P] \tag{50}$$

Thus, for any fixed $[S]_0$ the time required for the formation of a given $[P]$ is inversely proportional to V_{max}, that is, $V_{max} t$ is constant for a given $[S]_0$ and $[P]$. Keep in mind that V_{max} is a measure of $[E]_t$. Thus, the rule is usually stated as:

$$[E]_t \times \text{time} = \text{constant} \tag{51}$$

If n units of enzyme produce 1 mM product in 5 min, then $2n$ units will yield 1 mM product in 2.5 min while $0.5n$ units will require 10 min. This relationship holds for all regions of $[S]_0$. Although $2n$ units yields 1 mM product in 2.5 min, it is not necessarily true that $2n$ units will yield 2 mM product in 5 min (unless $[S]_0 \gg K_m$ so that $v = V_{max}$ and the reaction is zero-order). The relationship $V_{max} t = $ constant can be used to determine V_{max} if K_m is known. We simply measure the time required for the appearance of a certain $[P]$ starting with a fixed $[S]_0$. However, the procedure is practical only where the appearance of P (or disappearance of S) can be monitored continuously (such as in spectrophotometric assays involving NADH production or utilization).

· Problem 4-18

A crude cell-free extract contained 20 mg of protein per milliliter. Ten microliters of this extract in a standard total reaction volume of 0.5 ml catalyzed the formation of 30 nmoles of product in 1 min under optimum assay conditions (optimum pH and ionic strength, saturating concentrations of all substrates, coenzymes, activators, and the like). (a) Express v in terms of nmoles/assay, nmoles $\times ml^{-1} \times min^{-1}$, nmoles $\times liter^{-1} \times min^{-1}$, μ moles $\times liter^{-1} \times min^{-1}$, $M \times min^{-1}$. (b) What would v be if the same 10 μl of extract were assayed in a total volume of 1.0 ml? (c) What is the concentration of the enzyme in the assay mixture and in the extract (in terms of units/ml)? (d) What is the specific activity of the preparation?

Solution

(a) $v = \boxed{\textbf{30 nmoles/assay}} = \dfrac{30 \text{ nmoles/min}}{0.5 \text{ ml}}$

$$
\begin{aligned}
&= \boxed{\begin{aligned}
&\textbf{60 nmoles} \times \textbf{ml}^{-1} \times \textbf{min}^{-1} &&= &&\textbf{60} \times \textbf{10}^{9} \textbf{ nmoles} \times \textbf{liter}^{-1} \times \textbf{min}^{-1} \\
&\textbf{60 } \boldsymbol{\mu}\textbf{moles} \times \textbf{liter}^{-1} \times \textbf{min}^{-1} &&= &&\textbf{6} \times \textbf{10}^{-5} \textbf{ } M \times \textbf{min}^{-1}
\end{aligned}}
\end{aligned}
$$

(b) The same total amount of product would be formed. This would halve the velocity in concentration terms:

$$
v = \boxed{\textbf{30 nmoles} \times \textbf{ml}^{-1} \times \textbf{min}^{-1} \quad = \quad \textbf{3} \times \textbf{10}^{-5} \textbf{ } M \times \textbf{min}^{-1}}
$$

(c) $60 \text{ nmoles} \times \text{ml}^{-1} \times \text{min}^{-1} = 0.060 \text{ } \mu \text{moles} \times \text{ml}^{-1} \times \text{min}^{-1}$

$$
\boxed{\textbf{[E]}_t = \textbf{0.06 units/ml assay mixture}}
$$

The actual assay volume of 0.5 ml contained 0.03 units. The 0.03 units came from $10 \text{ } \mu \text{l} = 0.01 \text{ ml}$ of extract.

$$
\therefore \quad [E]_t = \dfrac{0.03 \text{ units}}{0.01 \text{ ml}} = \boxed{\textbf{3 units/ml extract}}
$$

(d) $\text{S.A.} = \dfrac{3 \text{ units/ml}}{20 \text{ mg protein/ml}} = \boxed{\textbf{0.15 units/mg protein}}$

· **Problem 4-19**

An enzyme preparation has a specific activity of 42 units/mg protein and contains 12 mg of protein per milliliter. Calculate the initial velocity of the reaction in a standard 1 ml reaction mixture containing: (a) 20 μl and (b) 5 μl of the preparation. (c) Should the preparation be diluted before an assay?

Solution

First calculate the enzyme concentration in the preparation in terms of units/ml.

$$[\text{enzyme}] = 42 \text{ units/mg protein} \times 12 \text{ mg protein/ml}$$

$$\boxed{\textbf{[enzyme]} = \textbf{504 units/ml}}$$

Now the velocities in terms of μmoles/min can be calculated, taking into consideration that the velocities will be directly proportional to the enzyme concentration in the reaction mixture.

(a)   Twenty microliters of preparation contain

$$504 \text{ units/ml} \times 0.02 \text{ ml} = 10.08 \text{ units enzyme}$$

$\therefore$

$$v = 10.08 \, \mu \text{ moles} \times \text{min}^{-1} \times \text{ml}^{-1}$$

(b)   Five microliters of preparation contain

$$504 \text{ units/ml} \times 0.005 \text{ ml} = 2.52 \text{ units enzyme}$$

$\therefore$

$$v = 2.52 \, \mu \text{ moles} \times \text{min}^{-1} \times \text{ml}^{-1}$$

(c)   In all likelihood, the preparation will have to be diluted so that the substrate is not depleted during a reasonable assay time.   For example, in 10 min 5 μl of preparation in a 1.0 ml assay volume would use up

$$(2.52 \times 10^{-6} \text{ moles} \times \text{min}^{-1} \times \text{ml}^{-1})(10 \text{ min}) = 2.52 \times 10^{-5} \text{ moles/ml}$$

$$= 2.52 \times 10^{-2} \text{ moles/liter} = \boxed{2.52 \times 10^{-2} \, M}$$

[S] would have to be greater than 0.5 M in order to keep the fraction of S utilized below 5%.

· Problem 4-20

Fifteen microliters of an enzyme preparation catalyzed the production of 0.52 μmole of product in 1 min under standard optimum assay conditions.   (a) How much product will be produced in 1 min by 150 μl of the preparation under the same reaction conditions?   (b) How long will it take 150 μl of the preparation to produce 0.52 μmole of product under the same assay conditions?

Solution

(a)   The initial velocity of the reaction was 0.52 μmole/min.   Our first tendency is to say that 10 times as much enzyme will produce 10 times as much product in the same period of time.   However, this will only be true if the velocity remains constant for the entire minute, that is, if the reaction remains zero-order over the minute interval.   In many instances, however, the decrease in the substrate concentration will cause the velocity of the reaction to drop back out of the zero-order region.   The amount of product formed by 150 μl of enzyme in 1 min will be $(10)(0.52) = 5.2 \, \mu$moles only if the substrate concentration at the end of the 1 min is still at least 100 K_m and the product of the reaction is not an inhibitor of the enzyme.

(b)   If the enzyme concentration is increased 10-fold, then only one-tenth as much time is needed to produce a given amount of product.   Therefore the 0.52 μmole of product will be produced by 150 μl of enzyme preparation in one-tenth as much time as it took 15 μl of enzyme preparation.

$$\therefore \quad \boxed{t = 0.1 \text{ min} = 6 \text{ sec}}$$

or $[E]_t \times \text{time} = k$ $(15\ \mu\text{l})(1\ \text{min}) = k$ $k = 15\ \mu\text{l-min}$

Then $(150\ \mu\text{l})(t) = 15\ \mu\text{l-min}$

$$t = \frac{15\ \mu\text{l-min}}{150\ \mu\text{l}}$$

$$\boxed{t = 0.1 \text{ min} = 6 \text{ sec}}$$

· Problem 4-21

One microgram of a pure enzyme (MW = 92,000) catalyzed a reaction at a rate of 0.50 μmoles/min under optimum conditions.   Calculate (a) the specific activity of the enzyme in terms of units/mg protein and units/mole, and (b) the turnover number.   (c) How long is one catalytic cycle?

Solution

(a)   S.A. $= \dfrac{V_{max}}{\text{mg}} = \dfrac{0.5\ \mu\text{moles/min}}{10^{-3}\ \text{mg}} = \boxed{\textbf{500 units/mg protein}}$

S.A. $= (5 \times 10^5 \text{ units/g protein})(9.2 \times 10^4 \text{ g/mole}) =$

$$\boxed{\textbf{4.6} \times \textbf{10}^{\textbf{10}} \textbf{ units/mole enzyme}}$$

(b)   turnover number $= (4.6 \times 10^{10}\ \mu\text{moles} \times \text{min}^{-1} \times \text{mole enzyme}^{-1}) \times$
$$(10^{-6}\ \text{moles} \times \mu\text{mole}^{-1})$$

$$\boxed{\textbf{turnover number} = \textbf{4.6} \times \textbf{10}^{\textbf{4}} \textbf{ min}^{-1}}$$

or $1\ \mu\text{g} = \dfrac{10^{-6}\ \text{g}}{9.2 \times 10^4\ \text{mole/g}} = 1.09 \times 10^{-11}\ \text{moles enzyme}$

turnover number $= k_p = \dfrac{V_{max}}{[E]_t} = \dfrac{0.5 \times 10^{-6}\ \text{moles S} \rightarrow \text{P/min}}{1.09 \times 10^{-11}\ \text{moles enzyme}} =$

$$\boxed{\textbf{4.6} \times \textbf{10}^{\textbf{4}} \textbf{ min}^{-1}}$$

(c)   The time required for one catalytic cycle is the reciprocal of the turnover number.   Therefore, one molecule of enzyme will convert one molecule of S to P in:

$$\text{time} = \frac{1}{4.6 \times 10^4 \text{ min}^{-1}} = \boxed{2.17 \times 10^{-5} \text{ min}}$$

· **Problem 4-22**

One gram fresh weight of muscle contains 40 units of an enzyme with a turnover number of 6×10^4 min^{-1}.   Estimate the intracellular concentration of the enzyme.

Solution

Assume that 1 g fresh weight of muscle contains about 0.80 ml of intracellular water:

$$\frac{40 \text{ units}}{0.80 \text{ ml}} = 50 \text{ units/ml} = 50 \times 10^3 \text{ units/liter}$$

$$= 50 \times 10^{-3} \text{ moles} \times \text{liter}^{-1} \times \text{min}^{-1}$$

$$V_{\max} = k_p [E]_t \qquad \text{or} \qquad [E]_t = \frac{V_{\max}}{k_p}$$

$$\frac{50 \times 10^{-3} \text{ moles} \times \text{liter}^{-1} \times \text{min}^{-1}}{6 \times 10^4 \text{ moles} \times \text{mole}^{-1} \times \text{min}^{-1}} = \boxed{8.33 \times 10^{-7} \ M}$$

Or, if we knew that the specific activity of the pure enzyme is 500 units/mg and its molecular weight is 120,000, we could proceed as shown below:

$$\frac{50 \times 10^3 \text{ units/liter}}{(5.0 \times 10^5 \text{ units/g enzyme})(1.2 \times 10^5 \text{ g enzyme/mole})} = \boxed{8.33 \times 10^{-7} \ M}$$

ENZYME PURIFICATION

Enzymes are purified by employing successive chemical or physical fractionation procedures.   The object of each step is to retain as much of the desired enzyme as possible while getting rid of as much of the other proteins, nucleic acids, and the like, as possible.   The efficiency of each step is given by the "yield" or "recovery" (the percent of the total enzyme activity originally present that is retained) and the "purification" or "purification factor" (the factor by which the specific activity of the preparation has increased).   The object is to optimize both factors.   Sometimes a good yield is sacrificed for the sake of an excellent purification step; sometimes a good purification step is not used because the yield is too low.   If the crude cell-free extract contains inhibitors, yields greater than 100% may be observed in the early stages of purification.   A hypothetical purification scheme is shown in Table 4-3.   The crude cell-free extract may be prepared by a number of means depending on the nature of the starting tissue or cells and the size of the

Table 4-3 Hypothetical Enzyme Purification Scheme

Step	Volume of Fraction (ml)	Protein		Enzyme				
		Conc. (mg/ml)	Total Amount (mg)	Conc. (units/ml)	Specific Activity (units/mg protein)	Total Amount (units)	Yield (percent)	Purification Factor (Fold)
Crude cell-free extract	1000	12	12,000	5	0.416	5000	"100"	"1.00"
Heat step: 50°C for 5 min, then remove denatured protein	1000	8	8,000	4.8	0.60	4800	96	1.44
Ammonium sulfate precipitation: 30–50% saturation fraction	250	3	750	11.0	3.67	2730	55	8.83
Ion-exchange chromatography: *DEAE-Sephadex:* elution via pH gradient. Fractions 50–60, 5 ml each, pooled, dialyzed, and concentrated	25	9	225	88	9.8	2200	44	23.6
Ion exchange chromatography: *DEAE-Sephadex:* elution via KCl gradient. Fractions 21–31, 2 ml each, pooled and concentrated	5	7	35	364	52	1820	36.4	125
Gel filtration: BioGel P-100. Fractions 30–40, 1 ml each, pooled	10	0.92	9.2	170	185	1700	34	444
Hydroxyl apatite chromatography: elution via phosphate buffer gradient. Fractions 15–18, 1 ml each, pooled	4	0.75	3	375	500	1500	30	1200

preparation. Some common cell-breakage methods include autolysis, freeze-thaw, sonic oscillation, mechanical grinding (with or without an abrasive), ballistic homogenization, or disruption in any one of a number of pressure cells (X-press, French press). The resulting homogenate is usually centrifuged to remove unbroken cells and large debris. There are no general rules concerning the order of the purification steps although heat treatment (where possible) and ammonium sulfate precipitations (Appendices II and III) are usually done early in the purification sequence. Gel filtration can follow ammonium sulfate precipitation and, thereby, serve to desalt the preparation as well as fractionate the proteins according to size. If ion-exchange chromatography is to follow the ammonium sulfate step, then it is a good idea to dialyze the preparation first, or pass the preparation through a rapid gel filtration column (e.g., Sephadex G-25). The removal of the ammonium sulfate will facilitate the binding of the proteins to the ion-exchange column. Other steps not shown in the purification table that may be highly effective for certain enzymes include differential centrifugation (for mitochondria, chloroplasts, nuclei, microsomes, ribosomes), pH precipitation, organic solvent precipitation (e.g., ethanol, acetone), protamine sulfate or streptomycin sulfate precipitation (to precipitate nucleic acids and acidic proteins), affinity chromatography, and preparative gel electrophoresis. The purity of the final preparation should be checked by several methods before concluding that the preparation is homogenous. Suitable methods include analytical disc gel electrophoresis at several pH values and gel concentrations, and ultracentrifugation. A homogeneous enzyme preparation should elute from an ion-exchange or gel filtration column as a single symmetrical activity and protein peak with a constant specific activity throughout. A homogeneous preparation is by no means necessary for kinetic analyses, but the purer the enzyme, the less the complications from the competing reactions that may use up the substrate or the product.

· Problem 4-23

A crude cell-free extract of skeletal muscle contained 32 mg protein/ml. Ten microliters of the extract catalyzed a reaction at a rate of 0.14 μmole/min under standard optimum assay conditions. Fifty milliliters of the extract were fractionated by ammonium sulfate precipitation. The fraction precipitating between 20% and 40% saturation was redissolved in 10 ml. This solution was found to contain 50 mg protein/ml. Ten microliters of this purified fraction catalyzed the reaction at a rate of 0.65 μmole/min. Calculate (a) the percent recovery of the enzyme in the purified fraction, and (b) the degree of purification obtained by the fractionation (the purification factor).

Solution

The crude cell-free extract contained:

$$\frac{0.14 \ \mu\text{mole/min}}{0.01 \ \text{ml}} = 14 \ \mu\text{moles} \times \text{ml}^{-1} \times \text{min}^{-1} = \boxed{\textbf{14 units/ml}}$$

$$14 \text{ units/ml} \times 50 \text{ ml total volume} = \boxed{\textbf{700 total units}}$$

and $$32 \text{ mg protein/ml} \times 50 \text{ ml total volume} = \boxed{\textbf{1600 mg total protein}}$$

The specific activity of the crude cell-free extract was:

$$\frac{14 \text{ units/ml}}{32 \text{ mg protein/ml}} = \boxed{\textbf{0.4375 unit/mg protein}}$$

The purified fraction contained:

$$\frac{0.65 \ \mu\text{mole/min}}{0.01 \text{ ml}} = 65 \ \mu\text{moles} \times \text{ml}^{-1} \times \text{min}^{-1} = \boxed{\textbf{65 units/ml}}$$

$$65 \text{ units/ml} \times 10 \text{ ml} = \boxed{\textbf{650 total units}}$$

and $$50 \text{ mg protein/ml} \times 10 \text{ ml} = \boxed{\textbf{500 mg total protein}}$$

The specific activity of the purified fraction was:

$$\frac{65 \text{ units/ml}}{50 \text{ mg protein/ml}} = \boxed{\textbf{1.30 units/mg protein}}$$

(a) $$\text{recovery} = \frac{\text{total units in purified fraction}}{\text{total units in crude extract}} \times 100\%$$

$$\text{recovery} = \frac{650}{700} \times 100\% = \boxed{\textbf{93.8\%}}$$

(b) $$\text{purification} = \frac{\text{specific activity of purified fraction}}{\text{specific activity of crude extract}}$$

$$\text{purification} = \frac{1.30}{0.4375} \qquad \boxed{\textbf{purification = 2.97-fold}}$$

ASSAYS WITH AUXILIARY ENZYMES

Frequently, the product of a reaction cannot be detected and quantitated directly but it is possible to add an auxiliary enzyme that converts the product quantitatively to another substance that can be measured. The overall reaction sequence is:

$$A \xrightarrow{\ E_1\ } S \xrightarrow{\ E_2\ } P$$

where E_1 is the enzyme being assayed and E_2 is the auxiliary enzyme. P is usually a compound that can be observed spectrophotometrically (e.g., NADH). The conditions for the conversion of S to P by E_2 may not be compatible with those for the conversion of A to S. (One of the cosubstrates for the E_2 reaction may be an inhibitor of E_1, the pH optimum for the E_2 reaction may be quite different from that of the E_1 reaction, and so on.) In this case, the assay is run in two stages. First A is incubated with E_1 (plus any cosubstrates) for a time sufficient to accumulate a detectable concentration of S. The reaction is then stopped (by boiling or changing the pH, for example). Then E_2 and all necessary cosubstrates are added and the reaction is allowed to proceed until all the S accumulated in stage 1 is converted to P, which is then measured. The question is: How long should the second-stage incubation time be in order to convert the maximum level of S to P for a fixed amount of E_2, and saturating cosubstrates of E_2? The incubation time for any level of S can be calculated from the integrated velocity equation:

$$t = \frac{2.3\,K_{m_S}}{V_{max_{E2}}} \log \frac{[S]_0}{[S]_0 - [P]} + \frac{[P]}{V_{max_{E2}}} \tag{52}$$

Since part of the reaction will be first-order, a 100% conversion of S to P will take an infinite time. We can settle for 98% conversion, which would not introduce any significant error.

If the auxiliary enzyme is very expensive, then we may wish to calculate the minimum amount needed to "complete" the reaction in a reasonable time by solving for $V_{max_{E2}}$.

If the K_{m_S} is 100 times or more greater than the maximum concentration of S that is allowed to accumulate, then the reaction $S \rightarrow P$ will always be first-order with respect to S. In this case, the concentration of S can be determined by measuring the *initial velocity* of the second-stage reaction:

$$v = k\,[S] = \frac{V_{max_{E2}}}{K_{m_S}}\,[S] \qquad \therefore \quad [S] = \frac{v K_{m_S}}{V_{max_{E2}}} \tag{53}$$

COUPLED ASSAYS

If none of the conditions required for the $S \xrightarrow{\ E_2\ } P$ reaction is detrimental to the reaction $A \xrightarrow{\ E_1\ } S$ (and vice versa), then both stages of the assay can be carried out simultaneously. The amount of E_1 present can be determined by measuring the velocity of P formation. The following conditions are necessary for a valid coupled assay: (a) the primary reaction must be zero-order with respect to [A] over the assay time and irreversible, and (b) the second-stage reaction must be first-order with respect to [S] and irreversible.   Condition a is easily met if only a small fraction of $[A]_0$ is utilized during the assay period or if $[A]_0 \gg K_{m_A}$ for E_1 and all cosubstrates are saturating.   Irreversibility is assumed by the removal of S in the second-stage reaction.   Condition b is met if $[S]_{ss}$, the steady-state concentration of S, is $\ll K_{m_S}$ for E_2. $[S]_{ss}$ can be maintained $\ll K_{m_S}$ by using a sufficient excess of E_2.   Irreversibility can be assumed if the equilibrium of the E_2 reaction lies far to the right, or if one of the coproducts of the reaction is continuously removed, or if the reaction proceeds only to a small extent.   Under these conditions, A will yield a certain $[S]_{ss}$ *after a short lag* and thereafter the rate of P formation will be constant and proportional to $[E_1]$.   If $[E_1]$ is doubled, $[S]_{ss}$ will double, and the rate of P formation will double.   Doubling $[E_2]$ halves $[S]_{ss}$ but since the velocity of the reaction $S \rightarrow P$ is given by $v_2 = k_2[S]_{ss}[E_2]$, v_2 is unchanged.   Thus once a sufficient excess of E_2 is present, the rate of P formation will be independent of $[E_2]$.

· Problem 4-24

A cardiologist is studying the effect of alcohol on triglyceride accumulation in rat heart.   The triglycerides are saponified and the glycerol released is determined by the coupled enzyme system shown below.

$$\text{glycerol} + \text{MgATP} \xrightarrow{\alpha\text{-glycerolkinase}} \alpha\text{-glycerol phosphate} + \text{MgADP}$$

$$+$$
$$\text{PEP}$$
$$\Big\downarrow \text{pyruvate kinase}$$

$$\text{NAD}^+ + \text{lactate} \xleftarrow{\text{lactate dehydrogenase}} \text{NADH} + \text{pyruvate}$$
$$+$$
$$\text{MgATP}$$

The concentration of glycerol is calculated from the decrease in the concentration of NADH (which can be measured spectrophotometrically, as described in Chapter 5). α-Glycerolkinase is rather expensive.   What is the minimum amount (units) of this enzyme needed in a 1.0 ml reaction mixture in order to "complete" the reaction in 15 min if the maximum concentration of glycerol that could be present is 0.3 μmoles/ml?   All other enzymes and substrates are present in excess.   The K_m of α-glycerolkinase for glycerol is 1.5×10^{-4} M.

Solution

The integrated velocity equation can be rearranged to:

$$V_{max} = \frac{2.3\, K_m}{t} \log \frac{[S]_0}{[S]_0 - [P]} + \frac{[P]}{t}$$

If we assume "completion" of the reaction when 98% of $[S]_0$ (glycerol) has been converted to P (α-glycerolphosphate), then

$$[S]_0 = 3 \times 10^{-4}\,M \qquad [P] = 2.94 \times 10^{-4}\,M \qquad [S]_0 - [P] = 6 \times 10^{-6}\,M$$

$$V_{max} = \frac{(2.3)(1.5 \times 10^{-4})M}{15\ \text{min}} \log \frac{(3 \times 10^{-4})M}{(6 \times 10^{-6})M} + \frac{2.94 \times 10^{-4}\,M}{15\ \text{min}}$$

$$= (2.3 \times 10^{-5})(1.7) + 1.96 \times 10^{-5}$$

$$= 5.87 \times 10^{-5}\,M/\text{min} = 0.0587\ \text{mM/min} = 0.0587\ \mu\text{moles} \times \text{ml}^{-1} \times \text{min}^{-1}$$

or

$$\boxed{\textbf{0.0587 units/ml}}$$

Since $[E]_t \times t = \text{constant}$, she can use half as much enzyme and read the decrease in NADH after 30 min.

We would not have obtained an accurate answer assuming only first-order kinetics or only zero-order kinetics because [S] starts at above K_m but far below $100\ K_m$.

I. MULTISUBSTRATE ENZYMES AND KINETIC MECHANISMS

Most enzymes catalyze reactions between two or more substrates to yield two or more products. Yet, most introductory biochemistry texts restrict their discussion to unireactant systems. As a result, it is not always appreciated that the K_m for a particular substrate at one fixed set of cosubstrate concentrations may not be the "real" K_m, but, instead, an apparent value that changes as the cosubstrate concentrations vary. Similarly, the observed V_{max} of a preparation at a saturating concentration of one substrate may not be the same "V_{max}" observed when another substrate is saturating. The true K_m for a particular substrate is that observed when all other substrates are saturating. The true V_{max} is observed when all substrates are present at saturating concentrations. Inhibition constants are also affected by the fixed concentrations of substrates. Thus, an observed "K_i" may not represent the true inhibitor dissociation constant.

In this section the subject of multireactant enzymes will be introduced by examining some common bireactant systems. We will indicate the ligands as A and B, where B is a substrate and A can be a cosubstrate or coenzyme or an essential activator.

Consider the reaction catalyzed by hexokinase:

$$\text{glucose} + \text{MgATP} \rightleftharpoons \text{glucose-6-phosphate} + \text{MgADP}$$

Theoretically, the reaction might proceed by a number of kinetic mechanisms. These are described below.

RAPID EQUILIBRIUM RANDOM BI BI

The two substrates (shown as A and B) might add randomly to the enzyme (Fig. 4-41) exactly as S and I do in a classical noncompetitive or mixed-type

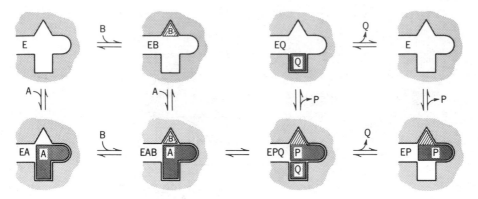

Figure 4-41   A schematic representation of a random kinetic mechanism for a group transfer reaction.   The binding sites for both substrates are available on free E.

inhibition system.   The products (shown as P and Q) might also leave randomly.

$$
\begin{array}{ccccccc}
\mathrm{E} + \mathrm{A} & \xrightleftharpoons{K_A = K_{ia}} & \mathrm{EA} & \qquad & \mathrm{EP} & \xrightleftharpoons{K_P = K_{ip}} & \mathrm{P} + \mathrm{E} \\
+ & & + & & + & & + \\
\mathrm{B} & & \mathrm{B} & & \mathrm{Q} & & \mathrm{Q} \\
{\scriptstyle K_B = K_{it}}\updownarrow & & \updownarrow{\scriptstyle\alpha K_B = K_{m_B}} & \xrightleftharpoons{\beta K_Q = K_{m_Q}} & \updownarrow & & \updownarrow{\scriptstyle K_Q = K_{iq}} \\
\mathrm{EB} + \mathrm{A} & \xrightleftharpoons[\alpha K_A = K_{m_A}]{} & \mathrm{EAB} & \underset{k_{-p}}{\overset{k_p}{\rightleftharpoons}} & \mathrm{EPQ} & \xrightleftharpoons{\beta K_P = K_{m_p}} & \mathrm{P} + \mathrm{EQ}
\end{array}
$$

A binds to free E with a dissociation constant K_A (also called K_{ia} in the Cleland nomenclature).   B binds to free E with a dissociation constant K_B (or K_{ib}).   The binding of one substrate may alter the affinity of the enzyme for the other.   Thus, A binds to EB with a dissociation constant αK_A.   Since the overall equilibrium constant between EAB and E must be the same regardless of the path taken, B binds to EA with a dissociation constant αK_B. αK_A is the same as K_{m_A} (the K_m for A at saturating [B]). αK_B is the same as K_{m_B} (the K_m for B at saturating [A]).   If the rate-limiting step is the slow conversion of EAB to EPQ, we can derive the velocity equation for the forward reaction in the absence of P and Q in the usual manner.   In fact, the only difference between the rapid equilibrium random bireactant system and noncompetitive or linear mixed-type inhibition is that now the ternary complex (EAB) is catalytically active, while ESI was not.

$$
v = k_p [\mathrm{EAB}] \qquad \text{and} \qquad \frac{v}{[\mathrm{E}]_t} = \frac{k_p [\mathrm{EAB}]}{[\mathrm{E}] + [\mathrm{EA}] + [\mathrm{EB}] + [\mathrm{EAB}]}
$$

Expressing the concentrations of each enzyme species in terms of free E we obtain:

$$
\boxed{\frac{v}{V_{\max}} = \frac{\dfrac{[\mathrm{A}][\mathrm{B}]}{\alpha K_A K_B}}{1 + \dfrac{[\mathrm{A}]}{K_A} + \dfrac{[\mathrm{B}]}{K_B} + \dfrac{[\mathrm{A}][\mathrm{B}]}{\alpha K_A K_B}}}
\tag{54}
$$

or

$$\frac{v}{V_{\max}} = \frac{[A][B]}{\alpha K_A K_B + \alpha K_B [A] + \alpha K_A [B] + [A][B]}$$ (55)

where $V_{\max} = k_p [E]_t$.

The equation can be rearranged to show either A or B as the varied substrate at a fixed level of the other substrate.   For example, when [A] is varied:

$$\frac{v}{V_{\max}} = \frac{[A]}{\alpha K_A \left(1 + \dfrac{K_B}{[B]}\right) + [A] \left(1 + \dfrac{\alpha K_B}{[B]}\right)}$$ (56)

At a fixed [B], the velocity equation can be written as:

$$\frac{v}{V_{\max_{app}}} = \frac{[A]}{K_{m_{A_{app}}} + [A]}$$ (57)

where $V_{\max_{app}}$ and $K_{m_{A_{app}}}$ are *apparent* constants at the fixed [B].   We see from Equation 56 that when [B] is saturating, the constants are αK_A (i.e., K_{m_A}) and the true $V_{\max}$.   (When [B] = ∞, the $K_B/[B]$ and $\alpha K_B/[B]$ terms go to zero.)   The reciprocal plots for varied [A] at different fixed concentrations of B are shown in Figure 4-42.   The plots for varied [B] are symmetrical to those shown for varied [A].

ORDERED BI BI

It may be impossible for B to bind until after A binds and promotes a conformational change in the enzyme that exposes the B binding site (Fig. 4-43).   The reaction sequence is:

$$E + A \underset{k_{-1}}{\overset{k_1}{\rightleftharpoons}} EA \qquad EQ \underset{k_{-5}}{\overset{k_5}{\rightleftharpoons}} Q + E$$

$$+ \qquad\qquad +$$
$$B \qquad\qquad P$$
$$k_{-2} \Big\Updownarrow k_2 \qquad k_{-4} \Big\Updownarrow k_4$$

$$EAB \underset{k_{-3}}{\overset{k_3}{\rightleftharpoons}} EPQ$$

Or, in the abbreviated scheme of Cleland:

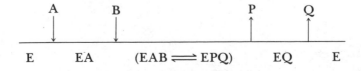

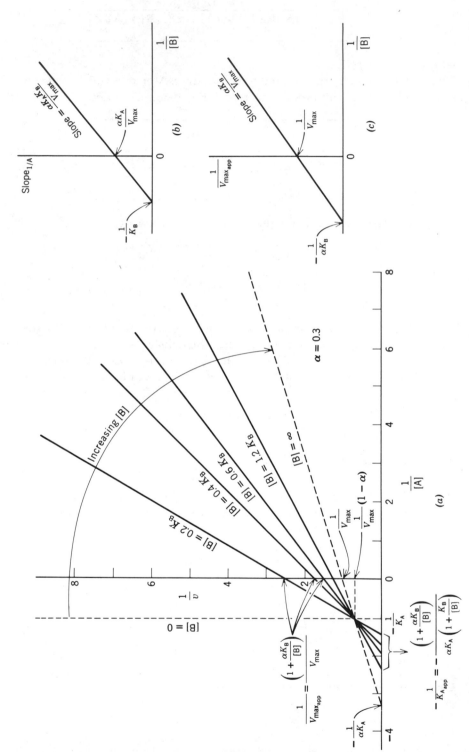

Figure 4-42 (*a*) Plot of $1/v$ versus $1/[A]$ at different fixed concentrations of B for a rapid equilibrium random system where α, the interaction factor, is 0.3. (*b*) $Slope_{1/A}$ versus $1/[B]$ replot to obtain K_B. (*c*) $1/V_{max_{app}}$ versus $1/[B]$ replot to obtain αK_B and V_{max}.

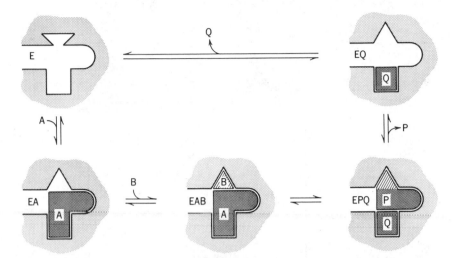

Figure 4-43 A schematic representation of an ordered kinetic mechanism for a group transfer reaction.   The binding of A causes a conformational change in the enzyme that exposes the binding site for B.   Thus, the B-P site is accessible only when the A-Q site is occupied.

If the conversion of EAB to EPQ is rate-limiting so that E, A, EA, B, and EAB are at equilibrium, the equation is:

$$\frac{v}{V_{max}} = \frac{\dfrac{[A][B]}{K_A K_B}}{1 + \dfrac{[A]}{K_A} + \dfrac{[A][B]}{K_A K_B}} \tag{58}$$

or

$$\frac{v}{V_{max}} = \frac{[A][B]}{K_A K_B + K_B[A] + [A][B]} \tag{59}$$

If the conversion of EAB to EPQ is as rapid as the dissociation reactions, then steady-state assumptions must be used to derive the velocity equation.   In multireactant systems, the rapid equilibrium and steady-state approaches do not yield the same final equation.   For the ordered Bi Bi system, a steady-state derivation yields:

$$\frac{v}{V_{max}} = \frac{[A][B]}{K_{ia}K_{m_B} + K_{m_B}[A] + K_{m_A}[B] + [A][B]} \tag{60}$$

where K_{ia} is the dissociation constant for A and K_{m_A} is the concentration of A that yields half-maximal velocity at saturating [B].   The reciprocal plots are shown in Figures 4-44 and 4-45.

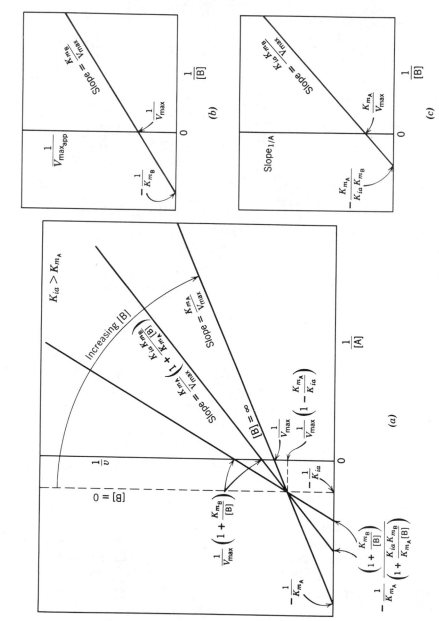

Figure 4-44 **Ordered Bi Bi system.** (*a*) $1/v$ versus $1/[A]$ at different fixed [B] concentrations. (*b*) $1/v$-axis intercept replot. (*c*) *Slope*$_{1/A}$ replot. Although the nomenclature is different, Figure 4-44 is identical to Figure 4-42. Thus, initial velocity measurements alone will not discriminate between random and ordered mechanisms.

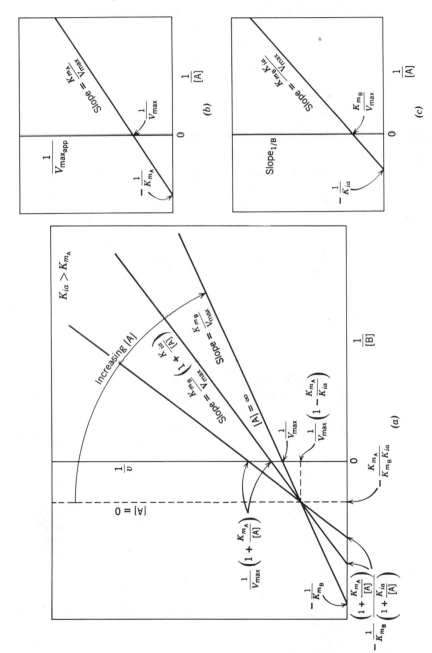

Figure 4-45 Ordered Bi Bi system. (*a*) $1/v$ versus $1/[B]$ at different fixed A concentrations. (*b*) $1/v$-axis intercept replot. (*c*) *Slope*$_{1/B}$ replot.

PING PONG BI BI

A third possible mechanism involves a transfer of phosphate from MgATP to the enzyme, followed by release of MgADP before the glucose binds and picks up the phosphate (Fig. 4-46).   This type of mechanism is called *Ping Pong* because the enzyme oscillates between two stable forms, E and F.   The reaction sequence is:

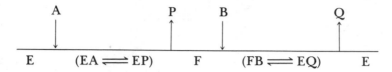

where   A = MgATP,   P = MgADP,   B = glucose,   and   Q = glucose-6-phosphate.   EA represents enzyme-MgATP.   FP represents enzyme-phosphate-MgADP.   Enzyme species F represents the stable (possible covalent) enzyme-phosphate.   FB represents enzyme-phosphate-glucose, and EQ represents enzyme-glucose-6-phosphate.   A steady-state treatment yields:

$$\frac{v}{V_{\text{max}}} = \frac{[A][B]}{K_{m_B}[A] + K_{m_A}[B] + [A][B]}$$

(61)

When rearranged to show A as the varied substrate at different fixed concentrations of B, the equation becomes:

$$\frac{v}{V_{\text{max}}} = \frac{[A]}{K_{m_A} + [A]\left(1 + \dfrac{K_{m_B}}{[B]}\right)}$$

(62)

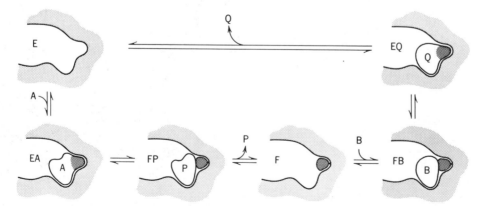

Figure 4-46   A schematic representation of a Ping Pong kinetic mechanism for a group transfer reaction.   The donor A reacts with E to form a substituted enzyme; P leaves before the acceptor, B, adds.

Since the K_{m_A} term in the denominator is not multiplied by any factor, the family of reciprocal plots obtained at different fixed B concentrations will be parallel. A symmetrical equation results when [B] is varied at different fixed A concentrations. The reciprocal plots and replots for varied [A] are shown in Figure 4-47.

Hexokinase does not yield parallel reciprocal plots, so the Ping Pong mechanism can be discarded. However, initial velocity studies alone will not discriminate between the rapid equilibrium random and steady-state ordered mechanisms. Both yield the same velocity equation and families of intersecting reciprocal plots. Other diagnostic procedures must be used (e.g., product inhibition, dead-end inhibition, equilibrium substrate binding, and isotope exchange studies). These procedures are described in detail in the author's *Enzyme Kinetics: Behavior and Analysis of Rapid Equilibrium and Steady-State Enzyme Systems*, Wiley-Interscience (1975).

· Problem 4-25

If the laws of the thermodynamics cannot be violated, then the overall $\Delta G'$ or K'_{eq} from E to EAB must be the same regardless of the path taken. Explain then how it is possible to have an ordered bireactant sequence $E \to EA \to EAB$. That is, shouldn't the sequence $E \to EB \to EAB$ be equally likely?

Solution

The reaction sequence between E and EAB is shown below with some arbitrary values inserted for the rate constants.

$$E + A \underset{k_{-1}=10^{-1}}{\overset{k_1=10^5}{\rightleftharpoons}} EA$$

$$EB + A \underset{k_4=10^2}{\overset{k_{-4}=10^{-7}}{\rightleftharpoons}} EAB \underset{k_{-p}}{\overset{k_p}{\rightleftharpoons}} EPQ$$

with vertical equilibria: $k_3 = 10^2$, $k_{-3} = 1$ (left), $k_{-2} = 10^{-1}$, $k_2 = 10^4$ (right).

$$K_A = \frac{k_{-1}}{k_1} = 10^{-6} \qquad \alpha K_B = \frac{k_{-2}}{k_2} = 10^{-5}$$

$$K_B = \frac{k_{-3}}{k_3} = 10^{-2} \qquad \alpha K_A = \frac{k_{-4}}{k_4} = 10^{-9}$$

The overall K_{eq} between free E and EAB is 10^{11} by either route. Consequently, no rules of thermodynamics have been violated. However, the product of the forward rate constants for one route to EAB is much greater than the product of rate constants for the other route:

$$k_1 k_2 \gg k_3 k_4$$

even though

$$\frac{k_1 k_2}{k_{-1} k_{-2}} = \frac{k_3 k_4}{k_{-3} k_{-4}}$$

Thus, the overall reaction sequence has a kinetically preferred path and, for all practical purposes, the sequence appears obligately ordered: $E \to EA \to$

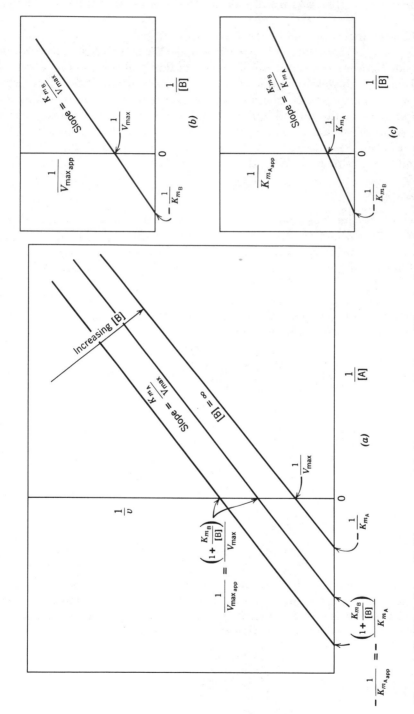

Figure 4-47 Ping Pong Bi Bi system. (*a*) $1/v$ versus $1/[A]$ at different fixed B concentrations. (*b*) $1/v$-axis intercept replot. (*c*) $1/K_{m_{A(app)}}$ replot. The plots and replots for B as the varied substrate are symmetrical to those shown for A.

EAB.   Furthermore, binding studies with [A] alone in the range $10^{-7} M$ to $10^{-5} M$ will yield substantial EA.   Binding studies with [B] alone in the range 10^{-6} to $10^{-4} M$ will yield little or no detectable EB.   Yet, in this [B] range in the presence of A, v is an appreciable fraction of V_{max}.

J. MULTISITE AND ALLOSTERIC ENZYMES

Many enzymes are oligomers composed of distinct subunits or mono-mers.   Often, the subunits are identical, each bearing a catalytic site.   If the sites are identical and completely independent of each other, then the presence of substrate at one site will have no effect on the binding properties of the vacant sites nor on the catalytic activities of other occupied sites.   If the enzyme is a tetramer, then at any fixed substrate concentration $[E]_t$ will be distributed among five different species: E, ES_1, ES_2, ES_3, and ES_4.   Yet, as we shall soon see, the S binding or velocity curve will be the usual hyperbola.   In other words, n molecules of a one-site enzyme behave identically to one molecule of an n-site enzyme.   Although there are no obvious interactions between the sites, the isolated monomers are often completely inactive.   Association to a tetramer may cause small changes in the tertiary structure of each monomer resulting in the formation of the substrate binding site or the proper juxtaposition of the substrate binding site and the catalytic groups.   Oligomerization may also contribute to the stability of enzymes *in vivo*.

If the presence of substrate on one site does influence the binding of substrate to vacant sites, or the rate of product formation at other occupied sites, then we have a situation where the substrate itself acts as a modifier or effector yielding substrate activation (including sigmoidal v versus [S] responses) or substrate inhibition.

NONCOOPERATIVE SITES

Let us first examine a dimer (two-site) model in which both sites are identical and independent.   The substrate binding sequence is shown below.

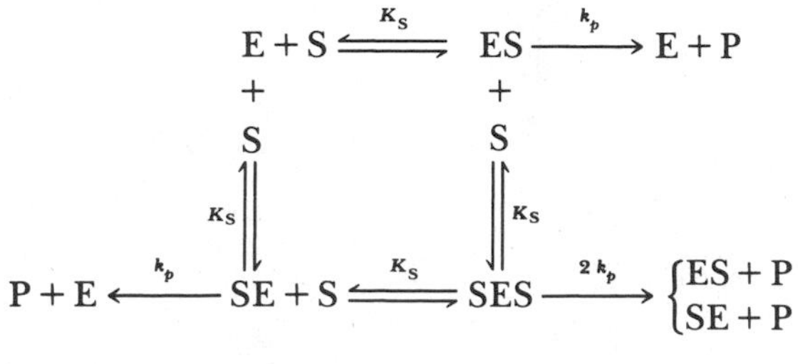

$$E + S \underset{}{\overset{K_S}{\rightleftharpoons}} ES \overset{k_p}{\longrightarrow} E + P$$

The velocity is given by:

$$v = k_p[ES] + k_p[SE] + 2 k_p[SES]$$

SES is twice as active as ES or SE because both sites are filled.   The velocity equation is obtained in the usual way.

$$\frac{v}{[E]_t} = \frac{k_p[ES] + k_p[SE] + 2\,k_p[SES]}{[E] + [ES] + [SE] + [SES]}$$

$$[ES] = \frac{[S]}{K_S}[E] \qquad [SE] = \frac{[S]}{K_S}[E] \qquad [SES] = \frac{[S]}{K_S}[ES] = \frac{[S]^2}{K_S^2}[E]$$

$$\therefore \quad \frac{v}{[E]_t} = \frac{k_p\dfrac{[S]}{K_S} + k_p\dfrac{[S]}{K_S} + 2\,k_p\dfrac{[S]^2}{K_S^2}}{1 + \dfrac{[S]}{K_S} + \dfrac{[S]}{K_S} + \dfrac{[S]^2}{K_S^2}} = \frac{2\,k_p\dfrac{[S]}{K_S} + 2\,k_p\dfrac{[S]^2}{K_S^2}}{1 + \dfrac{2[S]}{K_S} + \dfrac{[S]^2}{K_S^2}}$$

V_{max} will be observed when both sites are filled.   Therefore we can designate $2\,k_p[E]_t$ as V_{max}:

$$\frac{v}{V_{max}} = \frac{\dfrac{[S]}{K_S} + \dfrac{[S]^2}{K_S^2}}{1 + \dfrac{2[S]}{K_S} + \dfrac{[S]^2}{K_S^2}} \tag{63}$$

The numerator contains two terms because there are two kinds of product-forming complexes.   The denominator reflects the fact that there are three enzyme species present (free E, singly occupied enzyme, and doubly occupied enzyme).   The coefficient 2 shows that there are, in effect, two singly occupied species (substrate at one site and substrate at the other site).   The velocity equation for a tetramer is obtained in the same way, as shown in

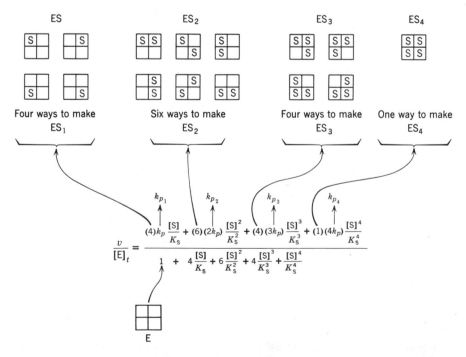

Figure 4-48   Distribution of species for an enzyme with four identical subunits.

Figure 4-48.   If $4\,k_p\,[E]_t$ is taken as V_{max}, the equation becomes:

$$\frac{v}{V_{max}} = \frac{\dfrac{[S]}{K_S} + \dfrac{3[S]^2}{K_S^2} + \dfrac{3[S]^3}{K_S^3} + \dfrac{[S]^4}{K_S^4}}{1 + \dfrac{4[S]}{K_S} + \dfrac{6[S]^2}{K_S^2} + \dfrac{4[S]^3}{K_S^3} + \dfrac{[S]^4}{K_S^4}} \tag{64}$$

In general, the velocity equation for an enzyme with n identical sites is:

$$\frac{v}{V_{max}} = \frac{\dfrac{[S]}{K_S}\left(1 + \dfrac{[S]}{K_S}\right)^{n-1}}{\left(1 + \dfrac{[S]}{K_S}\right)^{n}} \tag{65}$$

which reduces to:

$$\frac{v}{V_{max}} = \frac{\dfrac{[S]}{K_S}}{1 + \dfrac{[S]}{K_S}} = \frac{[S]}{K_S + [S]}$$

Thus, the v versus $[S]$ plots are hyperbolic regardless of the value of n.   It is impossible to tell from the kinetics of the reaction whether we are dealing with 1 pmole of an enzyme with n identical sites or n pmoles of an enzyme with one site.

ALLOSTERIC ENZYMES—COOPERATIVE BINDING

So far, we have considered enzymes that possess multiple, but independent, substrate binding sites, that is, the binding of one molecule of substrate has no effect on the intrinsic dissociation constants of the vacant sites.   Such enzymes yield normal hyperbolic velocity curves.   However, if the binding of one substrate molecule induces structural or electronic changes that result in altered affinities for the vacant sites, the velocity curve will no longer follow Henri-Michaelis-Menten kinetics and the enzyme will be classified as an allosteric enzyme.   In all likelihood, the multiple substrate binding sites of allosteric enzymes reside on different protein subunits.   Generally, allosteric enzymes yield sigmoidal velocity curves.   The binding of one substrate molecule facilitates the binding of the next substrate molecule by increasing the affinities of the vacant binding sites.   The phenomenon has been called "cooperative binding," or "positive cooperativity" with respect to substrate binding, or a "positive homotropic response."   Interactions between different ligands (e.g., substrate and activator, substrate and inhibitor, inhibitor and activator) are called "heterotropic responses" and may be either positive or negative.

The potential advantage of a sigmoidal response to varying substrate is illustrated in Figure 4-49.   For comparison, a normal hyperbolic velocity

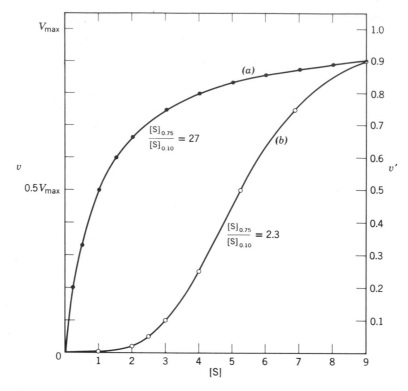

Figure 4-49 Comparison of velocity curves for two different enzymes that coincidentally have the same v at $[S] = 9$. (*a*) Hyperbolic response. (*b*) Sigmoidal response.

curve with the same $[S]_{0.9}$ is shown. Between $[S] = 0$ and $[S] = 3$, the hyperbolic response curve decelerates, but still rises to $0.75\ V_{max}$. The sigmoidal curve accelerates exponentially, but only attains $0.10\ V_{max}$ between the same limits of $[S]$. However, the sigmoidal curve increases from $0.10\ V_{max}$ to $0.75\ V_{max}$ with only an additional 2.3-fold increase in $[S]$. In order to cover the same specific velocity range, the hyperbolic curve requires a 27-fold increase in $[S]$. Thus, the sigmoidal response acts, in a sense, as an "off-on switch." Also, at intermediate specific velocities, the sigmoidal response provides a much more sensitive control of the reaction rate by variations in the substrate concentration.

The term "allosteric" was originally applied by Monod, Changeux, and Jacob to enzymes that display altered kinetic properties (usually a change in $[S]_{0.5}$) in the presence of ligands ("effectors" or "modifiers") that have no structural resemblance to the substrate. The allosteric response is usually quite understandable in terms of metabolic control and cellular economy (e.g., the feedback inhibition of the first reaction of a sequence by the ultimate product). Most allosteric enzymes display sigmoidal ligand saturation curves. Consequently, allosterism has become synonymous with sigmoidal responses. (However, not all sigmoidal binding or velocity curves result from allosteric interactions.)

Two major models for allosteric enzymes have been proposed. These are the "sequential interaction" model and the "concerted-symmetry" model. As the name suggests, the "sequential" model assumes sequential or progressive changes in the affinities of vacant sites as sites are occupied. The "concerted-symmetry" model assumes that the enzyme preexists as an equilibrium mixture of a high affinity oligomer and a low affinity oligomer. Ligands, including the substrate, act by displacing the equilibrium in favor of one state or the other. During the transition, the conformation of all subunits changes at the same time.

THE SIMPLE SEQUENTIAL INTERACTION MODEL

The simplest model of an allosteric enzyme is an extension of the "flexible enzyme" or "induced fit" model of Koshland. This model assumes that significant changes in the conformation of an enzyme can result from the binding of a substrate or other ligand. These conformational changes can result in altered affinities of vacant sites (including the uncovering of a previously buried site). In effect, each substrate molecule that binds makes it easier for the next substrate molecule to bind. The resulting velocity curve has a marked acceleration phase followed by the normal sloping off as the enzyme approaches saturation. Let us examine an allosteric enzyme with two cooperative sites (Fig. 4-50a). The reactions are:

$$
\begin{array}{ccc}
\text{E} + \text{S} \xrightleftharpoons{K_S} & \text{ES} \xrightarrow{k_p} & \text{E} + \text{P} \\
+ & + & \\
\text{S} & \text{S} & \\
\Big\updownarrow K_S & \Big\updownarrow aK_S & \\
\text{P} + \text{E} \xleftarrow{k_p} \text{SE} \xrightleftharpoons{aK_S} & \text{SES} \xrightarrow{2k_p} & \begin{cases} \text{SE} + \text{P} \\ \text{ES} + \text{P} \end{cases}
\end{array}
\tag{66}
$$

When one site is occupied, the dissociation constant of the vacant site changes to aK_S, where $a < 1$. The velocity equation is:

$$
\frac{v}{V_{\max}} = \frac{\dfrac{[S]}{K_S} + \dfrac{[S]^2}{aK_S^2}}{1 + \dfrac{2[S]}{K_S} + \dfrac{[S]^2}{aK_S^2}}
\tag{67}
$$

The velocity equation does not reduce to the Henri-Michaelis-Menten equation.

Next, consider an allosteric tetramer (Fig. 4-50b). The first molecule of S binds to any of the four vacant sites with a dissociation constant K_S. As a result, the dissociation constants of all three vacant sites change to aK_S. When the second molecule of S binds, the dissociation constants of the remaining two sites are changed by a factor b to abK_S. When the third molecule of S binds, the dissociation constant of the last unfilled site is changed by a factor c to $abcK_S$. Note that effects are *progressive* and *cumulative*.

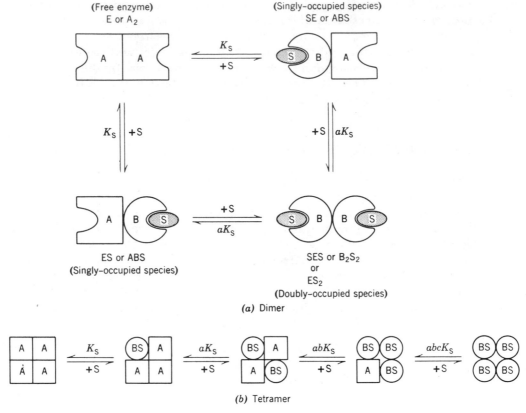

Figure 4-50   The sequential interaction model of allosteric enzymes.   As each site is occupied, the subunit carrying the site undergoes a change from the A conformation to the B conformation.   As a result, new interactions between subunits are established and the affinities of the vacant sites change. K_S represents a dissociation constant.   Thus, if the affinities of vacant sites increase, a, b, and c (the interaction factors) are <1 and we observe positive cooperativity (a sigmoidal velocity curve).   The sequential interaction model also provides for negative cooperativity (a, b, and c are >1).   (*a*) Dimer model.   The two ways of arranging S to form a singly-occupied species is shown.   (*b*) Tetramer model.   For simplicity, only one arrangement of each occupied species is shown.

The velocity is given by:

$$v = k_p[ES_1] + 2\,k_p[ES_2] + 3\,k_p[ES_3] + 4\,k_p[ES_4]$$

Dividing both sides of the equation by $[E]_t$ and introducing the statistical factors as shown in Figure 4-48, we obtain:

$$\frac{v}{V_{\max}} = \frac{\dfrac{[S]}{K_S} + \dfrac{3[S]^2}{aK_S^2} + \dfrac{3[S]^3}{a^2bK_S^3} + \dfrac{[S]^4}{a^3b^2cK_S^4}}{1 + \dfrac{4[S]}{K_S} + \dfrac{6[S]^2}{aK_S^2} + \dfrac{4[S]^3}{a^2bK_S^3} + \dfrac{[S]^4}{a^3b^2cK_S^4}} \tag{68}$$

where $V_{\max} = 4\,k_p[E]_t.$

A SIMPLIFIED VELOCITY EQUATION FOR ALLOSTERIC ENZYMES—THE HILL EQUATION

Consider an enzyme with n equivalent substrate binding sites. If the cooperativity in substrate binding is very marked (i.e., the factors a, b, c, etc., are very small numbers), then the concentrations of all enzyme-substrate complexes containing less than n molecules of substrate will be negligible at any [S] that is appreciable compared to K_S. Under this condition, the velocity equation will be dominated by the $[S]^n$ term.

For example, the equation for the four-site enzyme reduces to:

$$\frac{v}{V_{max}} = \frac{\dfrac{[S]^4}{a^3 b^2 c K_S^4}}{1 + \dfrac{[S]^4}{a^3 b^2 c K_S^4}} = \frac{\dfrac{[S]^4}{K'}}{1 + \dfrac{[S]^4}{K'}}$$

or
$$\frac{v}{V_{max}} = \frac{[S]^4}{K' + [S]^4} \qquad \text{where} \quad K' = a^3 b^2 c K_S^4$$

In general:

$$\boxed{\frac{v}{V_{max}} = \frac{[S]^n}{K' + [S]^n}} \tag{69}$$

The above equation is known as the Hill equation.

n = the number of substrate binding sites per molecule of enzyme

K' = a constant comprising the interaction factors a, b, c, etc., and the intrinsic dissociation constant, K_S
$$= K_S^n (a^{n-1} b^{n-2} c^{n-3} \ldots \ldots z^1)$$

The constant K' in the above equation no longer equals the substrate concentration that yields half-maximal velocity (except when $n = 1$, when the equation reduces to the Henri-Michaelis-Menten equation).

When $v = 0.5\ V_{max}$: $0.5\ K' + 0.5\ [S]_{0.5}^n = [S]_{0.5}^n$

$K' = [S]_{0.5}^n$

$$\boxed{[S]_{0.5} = \sqrt[n]{K'}} \qquad \text{or} \qquad \boxed{n \log [S]_{0.5} = \log K'} \tag{70}$$

If the cooperativity is not very high, the velocity equation will not reduce to the Hill equation. Nevertheless, velocity curves can be expressed in terms of the Hill equation, although n will no longer equal the number of sites. In this case, the "n" should be designated n_{app} or n_H. For example, if the cooperativity is such that the major species present between 10 and 90% of V_{max} are ES_3 and ES_4, then the velocity data can be made to fit the Hill equation if n is taken as some nonintegral value between 3 and 4 (e.g., $n = 3.6$). To put it another way, if experimental velocity data are analyzed in terms of the Hill equation, the calculated value of n will almost always be less than the actual number of sites. The next highest integer above this *apparent*

n value represents the minimum number of actual sites.   Therefore, if the experimental data yield an n_{app} value of 1.8 based on the Hill equation, we are in effect saying that the enzyme behaves as if it possesses exactly 1.8 substrate binding sites with very strong cooperativity.   We know that there are at least two sites with relatively strong cooperativity but there could just as well be four sites with poor cooperativity, or many sites that act in highly cooperative pairs.

SIGMOIDICITY OF THE VELOCITY CURVE

The shape of the v versus [S] curve can be expressed in terms of the ratio of substrate concentration required for any two fractions of V_{max}, for example, 0.9 V_{max} and 0.1 V_{max}.   This ratio, called the cooperativity index, depends on the value of n as shown below.

$$\frac{v}{V_{max}} = \frac{[S]^n}{K' + [S]^n}$$

When $v = 0.9\ V_{max}$: $0.9 = \dfrac{[S]_{0.9}^n}{K' + [S]_{0.9}^n}$ $[S]_{0.9} = \sqrt[n]{9\ K'}$

When $v = 0.1\ V_{max}$: $0.1 = \dfrac{[S]_{0.1}^n}{K' + [S]_{0.1}^n}$ $[S]_{0.1} = \sqrt[n]{\dfrac{K'}{9}}$

$$\therefore\quad \frac{[S]_{0.9}}{[S]_{0.1}} = \frac{\sqrt[n]{9\ K'}}{\sqrt[n]{K'/9}} = \sqrt[n]{\frac{9\ K'}{K'/9}}$$

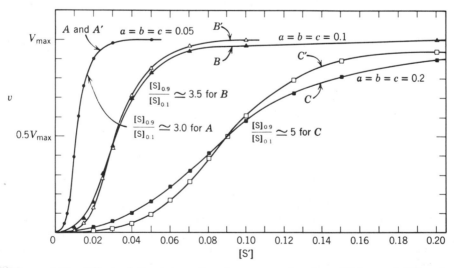

Figure 4-51   Effect of the interaction factors on the sigmoidicity and $[S]_{0.5}$ of a four-site enzyme.   Curve A: $a = b = c = 0.05$.   Curve B: $a = b = c = 0.1$. Curve C: $a = b = c = 0.2$.   Curves A, B, and C were calculated using the full velocity equation.   Curves A', B', and C' were calculated from the corresponding Hill equation (i.e., only the terms corresponding to E and ES$_4$ are taken into account).

$$\frac{[S]_{0.9}}{[S]_{0.1}} = \sqrt[n]{81} \qquad \text{or} \qquad n = \frac{\log 81}{\log \dfrac{[S]_{0.9}}{[S]_{0.1}}} \tag{71}$$

Figure 4-51 shows the effect of different interaction factors on the velocity curve of an allosteric tetramer. As the interaction factors decrease (i.e., as the cooperativity increases), the curves become more sigmoidal and $[S]_{0.5}$ decreases.

THE HILL PLOT—LOGARITHMIC FORM OF THE HILL EQUATION

The Hill equation can be converted to a useful linear form as shown below:

$$\frac{v}{V_{max}} = \frac{[S]^n}{K' + [S]^n} \qquad V_{max}[S]^n = vK' + v[S]^n$$

$$[S]^n(V_{max} - v) = vK' \qquad \frac{[S]^n(V_{max} - v)}{v} = K'$$

$$n \log [S] + \log \frac{V_{max} - v}{v} = \log K'$$

$$\log \frac{V_{max} - v}{v} = \log K' - n \log [S]$$

or

$$\log \frac{v}{V_{max} - v} = n \log [S] - \log K' \tag{72}$$

Thus, a plot of $\log v/(V_{max} - v)$ versus $\log [S]$ is a straight line with a slope of n (Fig. 4-52). When $\log v/(V_{max} - v) = 0$, $v/(V_{max} - v) = 1$ and the corresponding position on the $\log [S]$ axis gives $\log [S]_{0.5}$. K' may be calculated from the relationship $K' = [S]_{0.5}^n$. Theoretically, the Hill plot is linear over the entire range of substrate concentration (by virtue of the derivation that assumes no intermediates between E and ES_n). With experimental data, the Hill plot usually deviates from linearity at low specific velocities, where complexes containing less than n molecules of substrate contribute significantly to the initial velocity. The limiting slope at very low substrate concentrations (which may never be observed experimentally) is 1.0. On the other hand, if the enzyme contains noncatalytic regulatory sites that must be occupied before the substrate can bind to the catalytic site, the slope of the Hill plot will increase as the substrate concentration decreases. At very low [S], the slope will approach the number of sites that must be occupied before any reaction occurs.

Hill plots can be constructed by plotting $\log v/(V_{max} - v)$ versus $\log [S]$ on a linear scale. The slope, n, can then be read directly from the plot. However, it is usually more convenient to plot $v/(V_{max} - v)$ versus [S] directly on a log-log scale. If the decades of the log-log scale are the same size on both axes the slope can be determined by measuring suitable vertical and horizontal distances with a ruler.

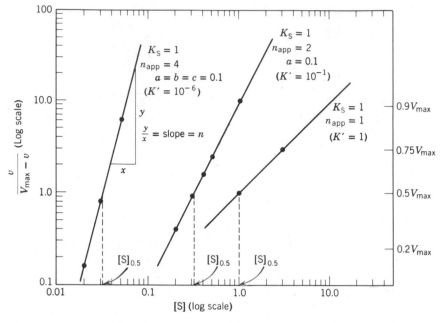

Figure 4-52   Hill plots for enzymes with different *n* values and the same intrinsic K_S.

THE CONCERTED TRANSITION OR SYMMETRY MODEL

In 1965, Monod, Wyman, and Changeux proposed a unique model for allosteric proteins.   The features of this model are: (a) allosteric proteins are polymeric ("oligomers") containing identical minimal units ("protomers") arranged in a symmetrical fashion.   (b) Each identical protomer possesses one, and only one, binding site for any given ligand (substrate, inhibitor, activator).   (c) The oligomer can exist in two different conformations that are in equilibrium.   The different conformations can arise from a rearrangement of the quaternary structure or from a change in the tertiary structure of the protomers (or both).   The transition between one conformation and another is an all-or-nothing event, that is, the symmetry of the oligomer is conserved in the transition.   Thus, there are no hybrid states where some protomers have rearranged in space or changed in conformation while others have not.   (d) The affinity of a binding site for a given ligand depends on the conformation of the protomer (hence, on the conformation of the oligomer).   Some ligands bind preferentially to one oligomer conformation, while other ligands bind preferentially to the other oligomer conformation.   The binding of a ligand to one particular conformation will cause the equilibrium to shift in favor of the conformation with the bound ligand.   Because each oligomer possesses more than one ligand binding site (one per protomer) and the transition from the lower affinity to the higher affinity conformation occurs simultaneously for all protomers, the number of higher affinity binding sites made available by the transition exceeds the one used up.   As a result, the ligand binding curve or velocity curve is sigmoidal.

   Figure 4-53 illustrates the concerted-symmetry model for a tetramer.   The "T" ("taut" or "tight") state represents the conformation with the lower

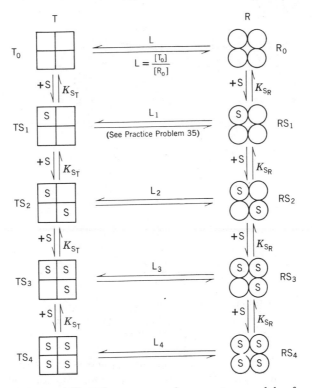

Figure 4-53 The concerted-symmetry model of Monod, Wyman, and Changeux. T represents a low affinity form of an oligomeric enzyme which is in equilibrium with R, a high affinity form of the enzyme. This model allows only positive cooperativity.

affinity for the ligand, S. The "R" ("relaxed") state represents the conformation with the higher affinity for S. The equilibrium constant for the transition $R_0 \rightleftharpoons T_0$ is designated L.

$$L = \frac{[T_0]}{[R_0]} \tag{73}$$

The intrinsic dissociation constant for the S binding site on a protomer in the T state is designated K_{S_T}. The intrinsic dissociation constant for the S binding site on a protomer in the R state is designated K_{S_R}. The ratio K_{S_R}/K_{S_T} is designated c and called the "nonexclusive binding coefficient."

The cooperativity of substrate binding depends on L and c. The velocity curves become more sigmoidal as L increases (i.e., as the $R_0 \rightleftharpoons T_0$ equilibrium favors T_0) and as c decreases (i.e., as the affinity of the T state decreases relative to the affinity of the R state for S). Allosteric inhibitors are assumed to bind preferentially to the T state thereby displacing the $T_0 \rightleftharpoons R_0$ equilibrium in favor of T_0 (Fig. 4-54). In effect, the allosteric constant, L, increases and the velocity curves become more sigmoidal with n_{app} (from Hill plots) approaching the actual number of sites. An activator is assumed to bind preferentially to the R state and thus mimics the substrate by shifting the $T_0 \rightleftharpoons R_0$ equilibrium to the right. As a result, the velocity curves become less

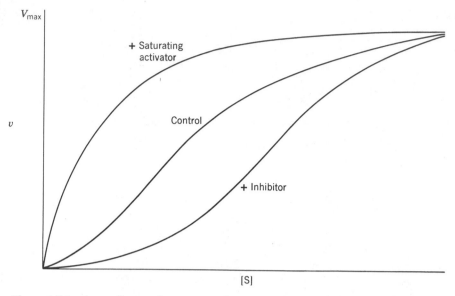

Figure 4-54 According to the concerted-symmetry model, an allosteric inhibitor binds preferentially to the T form.   This causes the velocity curve to become more sigmoidal with a higher $[S]_{0.5}$.   An allosteric activator mimics the substrate by binding preferentially to the R form.   As a result, the velocity curve becomes less sigmoidal (hyperbolic at saturating activator) and $[S]_{0.5}$ decreases.   These observations can also be explained in terms of the sequential interaction model.

sigmoidal.   At an infinitely high activator concentration, all the enzyme will be driven to the R state and the v versus [S] curves become hyperbolic.

Figure 4-55 illustrates the simplest version of the concerted symmetry model for a dimer.   It is assumed that the "T" state has absolutely no affinity for the substrate, S (i.e., $c = 0$).   That is, S binds exclusively to the "R" state

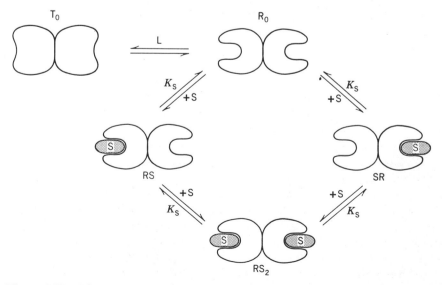

Figure 4-55 The concerted-symmetry model for an allosteric dimer where S binds exclusively to the R form ($c = 0$).

with a dissociation constant that can be indicated simply as K_S.   The velocity is given by:

$$v = k_p [RS] + k_p [SR] + 2 k_p [RS_2]$$

Dividing by $[E]_t$:

$$\frac{v}{[E]_t} = \frac{k_p [RS] + k_p [SR] + 2 k_p [RS_2]}{[T_0] + [R_0] + [RS] + [SR] + [RS_2]}$$

Substituting for [RS], etc., in terms of $[R_0]$:

$$\frac{v}{[E]_t} = \frac{k_p \dfrac{[S]}{K_S} [R_0] + k_p \dfrac{[S]}{K_S} [R_0] + 2 k_p \dfrac{[S]^2}{K_S^2} [R_0]}{L[R_0] + [R_0] + \dfrac{[S]}{K_S} [R_0] + \dfrac{[S]}{K_S} [R_0] + \dfrac{[S]^2}{K_S^2} [R_0]}$$

$$= \frac{2 k_p \dfrac{[S]}{K_S} + 2 k_p \dfrac{[S]^2}{K_S^2}}{L + 1 + 2 \dfrac{[S]}{K_S} + \dfrac{[S]^2}{K_S^2}}$$

or

$$\frac{v}{V_{max}} = \frac{\dfrac{[S]}{K_S}\left(1 + \dfrac{[S]}{K_S}\right)}{L + \left(1 + \dfrac{[S]}{K_S}\right)^2} \tag{74}$$

where $V_{max} = 2 k_p [E]_t$.

In general:

$$\frac{v}{V_{max}} = \frac{\dfrac{[S]}{K_S}\left(1 + \dfrac{[S]}{K_S}\right)^{n-1}}{L + \left(1 + \dfrac{[S]}{K_S}\right)^{n}} \tag{75}$$

Equation 75 (as all others derived from rapid equilibrium assumptions) is really an equilibrium binding equation that gives the ratio of occupied to total sites.   We obtain a velocity equation by assuming that the velocity is proportional to the concentration of occupied sites.   In other words, a velocity equation is obtained when we equate Y_S to v/V_{max}:

$$Y_S = \frac{[\text{occupied sites}]}{[\text{total sites}]} = \frac{[\text{bound S}]}{n [E]_t} = \frac{v}{V_{max}} \tag{76}$$

· Problem 4-26

The $[S]_{0.9}/[S]_{0.1}$ ratio for an enzyme that obeys sigmoidal kinetics is 6.5.   What is the n_{app} value?

Solution

$$\frac{[S]_{0.9}}{[S]_{0.1}} = \sqrt[n]{81} \qquad n = \frac{\log 81}{\log \dfrac{[S]_{0.9}}{[S]_{0.1}}}$$

$$n = \frac{\log 81}{\log 6.5} = \frac{1.91}{0.813} = 2.35$$

$$\boxed{n_{app} = 2.35}$$

· Problem 4-27

Calculate the $[S]_{0.9}/[S]_{0.1}$ ratio for an enzyme that obeys sigmoidal kinetics with an n_{app} value of 2.6.

Solution

$$\text{When } n_{app} = 2.6; \quad \frac{[S]_{0.9}}{[S]_{0.1}} = \sqrt[2.6]{81} = 81^{1/2.6}$$

$$\log \frac{[S]_{0.9}}{[S]_{0.1}} = \frac{1}{2.6} \log 81$$
$$= (0.385)(1.91) = 0.735$$

$$\boxed{\frac{[S]_{0.9}}{[S]_{0.1}} = 5.43}$$

· Problem 4-28

An allosteric dimer has an interaction factor, a, of 0.2 when analyzed according to the sequential interaction model (i.e., the binding of the first molecule of S increases the binding constant of the vacant site by a factor of 5—the dissociation constant of the vacant site decreases to 0.2 of the original value). (a) What is the relative distribution of enzyme species at $[S] = 0.3\ K_S$? (b) What is the specific velocity at $[S] = 0.3\ K_S$? (c) Will the calculated value of n_{app} equal 2?

Solution

(a) The equilibria between enzyme species are shown in scheme 66. The relative distribution is given below where each term represents the concentration of a species relative to free E.

$$\frac{[E]}{[E]_t} = \frac{1}{1 + \dfrac{2[S]}{K_S} + \dfrac{[S]^2}{aK_S^2}} = \frac{1}{1 + 0.6 + \dfrac{0.09}{0.2}}$$

$$= \frac{1}{2.05} = \boxed{0.488} = \boxed{48.8\%}$$

$$\frac{[ES]}{[E]_t} = \frac{\dfrac{[S]}{K_S}}{1 + \dfrac{2[S]}{K_S} + \dfrac{[S]^2}{aK_S^2}} = \frac{0.3}{1 + 0.6 + 0.45} = \frac{0.3}{2.05} = \boxed{0.146} = \boxed{14.6\%}$$

$$\frac{[SE]}{[E]_t} = \frac{[ES]}{[E]_t} = \frac{0.3}{2.05} = \boxed{0.146} = \boxed{14.6\%}$$

$$\frac{[SES]}{[E]_t} = \frac{\dfrac{[S]^2}{aK_S^2}}{1 + \dfrac{2[S]}{K_S} + \dfrac{[S]^2}{aK_S^2}} = \frac{0.45}{2.05}$$

$$= \boxed{0.220} = \boxed{22.0\%}$$

(b)
$$\frac{v}{V_{max}} = \frac{\dfrac{[S]}{K_S} + \dfrac{[S]^2}{aK_S^2}}{1 + \dfrac{2[S]}{K_S} + \dfrac{[S]^2}{aK_S^2}} = \frac{0.3 + 0.45}{2.05}$$

$$= \frac{0.75}{2.05} = 0.366 \qquad \boxed{v = 0.366\ V_{max}}$$

(c) A velocity curve plotted according to the equation given in part b is sigmoidal with $[S]_{0.1} \simeq 0.09\ K_S$ and $[S]_{0.9} \simeq 2.5\ K_S$. The calculated n_{app} from this ratio (or from the slope of the Hill plot between the points corresponding to 10% and 90% of V_{max}, or the slope in the region of $0.5\ V_{max}$) is about 1.3.   The n_{app} value is less than the true n because throughout most of the velocity curve, the ES and SE complexes contribute to a substantial portion of the observed velocity.   (For example, at $v = 0.366\ V_{max}$, ES + SE account for 29.2% of the total enzyme while SES accounts for only 22%.)   If a were much smaller (e.g., 0.02), then most of the enzyme will be present as either E or SES and n_{app} would approach 2.

K. ENZYME TURNOVER

Feedback inhibition, activation, and allosteric phenomena are extremely rapid modes of regulating enzyme *activity* in all types of cells. Repression and induction (or depression) of enzyme synthesis represent slower, long-term regulatory devices whereby the *amount* of a particular enzyme in a cell is optimized. In microbial cells, an enzyme that is no longer needed is repressed and diluted out as new cells grow. In the relatively slow-growing cells of higher organisms, particularly animal cells, direct enzyme degradation often replaces dilution. In fact, the levels of many enzymes in animal cells are controlled by the balance between enzyme synthesis and degradation.   The constant synthesis and degradation is called *turnover*.   The synthesis of an enzyme is a zero-order process.   The degradation of an enzyme usually follows first-order kinetics.   That is, the rate of degradation is proportional to the concentration of enzyme present.   The following problem illustrates the relationship between the rates of synthesis and degradation and the steady-state level of an enzyme.

· Problem 4-29

A particular enzyme of liver is synthesized at a constant rate of 12.5 units per gram tissue per minute.   The steady-state level of the enzyme is 250 units per gram tissue.   (a) Calculate the first-order rate constant for the degradation phase of the turnover.   (b) Administration of a hormone caused the rate of enzyme synthesis to increase sixfold without affecting the first-order rate constant of degradation.   After a lag, the enzyme attained a new steady-state level.   What is the new rate of degradation?   What is the new steady-state level of the enzyme?

Solution

(a)   In order to maintain $[E]_t$ at 250 units per gram tissue, the rate of degradation must equal the constant rate of synthesis.

$$v_{syn} = v_{degrad} = k[E]_t$$

$$12.5 = 12.5 = k[E]_t$$

$$k = \frac{12.5}{[E]_t} = \frac{12.5}{250}$$

$$\boxed{k = 0.05 \text{ min}^{-1}}$$

That is, 5% of the steady-state enzyme level is turned over each minute.

(b)   If v_{syn} increases sixfold to $6 \times 12.5 = 75$ units $\times$ g tissue$^{-1} \times$ min^{-1}, then v_{degrad} must also be 75 units $\times$ g tissue$^{-1} \times$ min^{-1} to maintain a new steady-state level.   If k is unchanged:

$$75 = 0.05[E]_t \qquad \text{or} \qquad [E]_t = \frac{75}{0.05}$$

$$\boxed{[E]_t = 1500 \text{ units/g tissue}}$$

Thus, a sixfold increase in the rate of enzyme synthesis results in a sixfold higher steady-state enzyme level if k remains constant.

GENERAL REFERENCES

Enzyme Kinetics

Segel, I. H., *Enzyme Kinetics: Behavior and Analysis of Rapid Equilibrium and Steady-State Enzyme Systems.* Wiley-Interscience (1975). This book starts at the same elementary level as *Biochemical Calculations* and progresses to the modern subjects of steady-state kinetics of multireactant enzymes, allosteric enzymes, isotope exchange, and membrane transport.

General Enzymology

Whitaker, J. R., *Principles of Enzymology for the Food Sciences.* Marcel Dekker (1972).

Enzyme Mechanisms

Bernhard, S., *The Structure and Function of Enzymes.* Benjamin (1968).

Gray, C. J., *Enzyme-Catalyzed Reactions.* Van Nostrand-Reinhold (1971).

Jencks, W. P., *Catalysis in Chemistry and Enzymology.* McGraw-Hill (1969).

Westley, J., *Enzyme Catalysis.* Harper & Row (1969).

Zeffren, E. and P. L. Hall, *The Study of Enzyme Mechanisms.* Wiley-Interscience (1973).

PRACTICE PROBLEMS

Answers to Practice Problems are given on pages 426–429.

1. The concentration-velocity data shown below were obtained for an enzyme catalyzing a reaction $S \rightarrow P$. (a) Calculate K_m and V_{max}. (b) Verify that the enzyme obeys hyperbolic saturation kinetics. (c) Calculate the first-order rate constant for the enzyme concentration employed.

[S]	v
M	$nmoles \times liter^{-1} \times min^{-1}$
2.50×10^{-6}	24
3.33×10^{-6}	30
4.0×10^{-6}	34
5×10^{-6}	40
1×10^{-5}	60
2×10^{-5}	80
4×10^{-5}	96
1×10^{-4}	109
2×10^{-3}	119
1×10^{-2}	120

2. An enzyme with a K_m of $2.4 \times 10^{-4} M$ was assayed at the following substrate concentrations: (a) $2 \times 10^{-7} M$, (b) $6.3 \times 10^{-5} M$, (c) $10^{-4} M$, (d) $2 \times 10^{-3} M$, and (e) $0.05 M$. The velocity observed at $0.05 M$ was 128 nmoles × liter^{-1} × min^{-1}. Calculate the initial velocities at the other substrate concentrations.

3. If the enzyme concentration in practice problem 2 was increased fivefold, what would the initial velocities be at each of the given substrate concentrations?

4. The equilibrium constant for the reaction $S \overset{E}{\rightleftharpoons} P$ is 2×10^3. Enzyme E catalyzes the reaction ($K_{m_S} = 2.5 \times 10^{-5} M$, $V_{max_f} = 4.2 \mu$moles × liter^{-1} × min^{-1}). (a) What is the first-order rate constant for the forward reaction? (b) What is the first-order rate constant for the reverse reaction? (c) What is the ratio of V_{max_r}/K_{m_P}?

5. An enzyme catalyzes the reaction $S \rightleftharpoons P$. ($V_{max_f} = 22$ μmoles × liter^{-1} × min^{-1}, $V_{max_r} = 14$ μmoles × liter^{-1} × min^{-1}). In which direction and how fast will the reaction proceed if $[S] = 2 K_{m_S}$ and $[P] = 7 K_{m_P}$?

6. An enzyme with a K_m of $1.2 \times 10^{-4} M$ was assayed at an initial substrate concentration of $0.02 M$. By 30 sec, 2.7 μmoles/liter of product had been produced. How much product will be present at (a) 1 min, (b) 95 sec, (c) 3 min, and (d) 5.3 min? (e) What percent of the original substrate will be utilized by the times indicated?

7. An enzyme with a K_m of $2.6 \times 10^{-3} M$ was assayed at an initial substrate concentra-

tion of $0.3\ M$. The observed velocity was 5.9×10^{-5} moles $\times$ liter$^{-1} \times$ min^{-1}. If the initial substrate concentration were $2 \times 10^{-5}\ M$, what would the product concentration be after (a) 5 min and (b) 10 min?

8. An enzyme with a K_m of $3 \times 10^{-4}\ M$ was assayed at an initial substrate concentration of $10^{-6}\ M$. By 1 min, 5.0% of the substrate had been utilized. (a) What percent of the substrate will be utilized by 5 min? (b) If the initial substrate concentration were $8 \times 10^{-7}\ M$, what percent of the substrate will be utilized by 5 min? (c) Calculate V_{max}. (d) At $8 \times 10^{-7}\ M$, how long will it take for 50% of the substrate to be utilized? (e) At $10^{-6}\ M$, how long will it take for 75% of the substrate to be utilized?

against an equal volume of a solution of ^{3}H-morphine. At equilibrium, the chamber containing the glycoprotein contained $1.43 \times 10^{-6}\ M$ total (bound + free) ^{3}H-morphine. The chamber without the glycoprotein contained $0.78 \times 10^{-6}\ M$ ^{3}H-morphine. Calculate (a) the concentration of bound ^{3}H-morphine, (b) the concentration of free protein, and (c) the dissociation constant for the glycoprotein-morphine complex. Assume one binding site per protein molecule.

12. Embryonic liver tissue contains an enzyme that catalyzes the reaction $S \rightarrow P$. Adult liver also displays $S \rightarrow P$ activity. Some kinetic data are shown below. What conclusions can you draw concerning the identity of the two enzymes?

(Data for Practice Problem 12)

[S]	Observed Initial Velocity (μmoles $\times$ mg Protein$^{-1} \times$ Min^{-1})	
(M)	Extract of Adult Liver (E_1)	Extract of Embryonic Liver (E_2)
1.67×10^{-5}	1.05	5.00
2.5×10^{-5}	1.54	6.66
3.33×10^{-5}	1.98	8.00
5.0×10^{-5}	2.86	10.00
7.0×10^{-5}	3.78	11.67
1.0×10^{-4}	5.00	13.33
1.5×10^{-4}	6.67	15.0
1.67×10^{-4}	7.15	15.4
2.0×10^{-4}	8.00	16.0
3.0×10^{-4}	10.00	17.1

9. Calculate (a) $[S]_{0.95}/[S]_{0.05}$, (b) $[S]_{0.80}/[S]_{0.20}$, (c) $[S]_{0.75}/[S]_{0.25}$ and (d) $[S]_{0.75}/[S]_{0.5}$ for an enzyme that obeys hyperbolic saturation kinetics.

10. The $1/v$ axis of a reciprocal plot is labeled v^{-1}: (nmoles $\times$ liter$^{-1} \times$ min^{-1})$^{-1} \times 10^2$. The $1/[S]$ axis is labeled $[S]^{-1}$: (M)$^{-1} \times 10^{-4}$. The plot intersects the two axes at "2" and "-4," respectively. What are V_{max} and K_m?

11. A morphine-binding substance was isolated from brain tissue. The material was purified to homogeneity and identified as a glycoprotein of MW 260,000. A solution of the glycoprotein (0.30 mg/ml) was dialyzed

13. During severe liver damage, an enzyme (E_1 of practice problem 12) is released into the bloodstream. After severe exercise, a muscle enzyme, E_3, that catalyzes the same

(Data for Practice Problem 13)

[S]	v
(M)	μmoles $\times$ ml serum$^{-1} \times$ min^{-1}
5×10^{-5}	43
7×10^{-5}	57
1×10^{-4}	75
1.5×10^{-4}	100
2×10^{-4}	120
3×10^{-4}	150
6×10^{-4}	200

reaction is released into the blood stream. E_1 and E_3 can be differentiated easily because they have different K_m values. (The K_m of the muscle enzyme is $2 \times 10^{-5} M$.)   An assay of a blood sample of a patient gave the results shown on p. 320.   Is the patient suffering from liver disease, or has he simply been exercising strenuously?   (The patient arrived at the hos-

ence of a competitive inhibitor, $[S]_i$, to the substrate concentration required in the absence of inhibitor, $[S]_0$, in order to observe a given velocity?

19.   The following velocity data were obtained.   Determine the nature of each inhibitor and calculate K_i.

Initial Velocity (nmoles/min) (Data for Practice Problem 19)

$[S](mM)$	(Control)	+I at 6 μM	+X at 30 μM	+Y at 4 mM	+Z at 0.2 mM
0.200	16.67	6.25	5.56	10.00	8.89
0.250	20.00	7.69	6.67	11.11	10.81
0.333	24.98	10.00	8.33	12.50	13.78
0.500	33.33	14.29	11.11	14.29	19.05
1.00	50.00	25.00	16.67	16.67	30.77
2.00	66.67	40.00	22.22	18.18	44.44
2.50	71.40	45.45	23.81	18.52	48.78
3.33	76.92	52.63	25.64	18.87	54.06
4.00	80.00	57.14	26.67	19.00	57.14
5.00	83.33	62.50	27.77	19.23	60.60

pital unconscious, so you can't ask him any questions.)

14.   Calculate v_i and the degree of inhibition caused by a competitive inhibitor under the following conditions: (a) $[S] = 2 \times 10^{-3} M$ and $[I] = 2 \times 10^{-3} M$, (b) $[S] = 4 \times 10^{-4} M$ and $[I] = 2 \times 10^{-3} M$, and (c) $[S] = 7.5 \times 10^{-3} M$ and $[I] = 10^{-5} M$.   Assume that $K_m = 2 \times 10^{-3} M$, $K_i = 1.5 \times 10^{-4} M$,   and $V_{max} = 270$ nmoles $\times$ liter$^{-1} \times$ min^{-1}.

15.   (a) What concentration of competitive inhibitor is required to yield 75% inhibition at a substrate concentration of $1.5 \times 10^{-3} M$ if $K_m = 2.9 \times 10^{-4} M$ and $K_i = 2 \times 10^{-5} M$?   (b) To what concentration must the substrate be increased to reestablish the velocity at the original uninhibited value?

16.   Calculate K_i for a noncompetitive inhibitor if $2 \times 10^{-4} M$ [I] yields 75% inhibition of an enzyme-catalyzed reaction.

17.   Calculate (a) the velocity and (b) the degree of inhibition of an enzyme-catalyzed reaction in the presence of $6 \times 10^{-4} M$ substrate $(K_m = 10^{-3} M)$ and $2.5 \times 10^{-4} M$ noncompetitive inhibitor $(K_i = 3 \times 10^{-5} M)$.   The $V_{max} = 515$ nmoles $\times$ liter$^{-1} \times$ min^{-1}.

18.   What is the relationship between the substrate concentration required in the pres-

20.   The product of a Uni Uni reaction, P, acts as a competitive inhibitor with respect to S (both P and S compete for free E).   If K_{eq} is very large ($V_{max, r}$ is very small), the equation for the forward velocity is:

$$\frac{v}{V_{max}} = \frac{[S]}{K_{m_S}\left(1 + \dfrac{[P]}{K_{m_P}}\right) + [S]}$$

Consider a system where the total pool of $[S] + [P]$ is constant and equal to $10^{-3} M$, $K_S = 10^{-4} M$, $K_P = 10^{-5} M$, $V_{max} = 100$ nmoles $\times$ liter$^{-1} \times$ min^{-1}.   What will the v versus $[S]$ curve look like?   Keep in mind that $[S] + [P] = 10^{-3} M$.   Thus, at $[S] = 10^{-4} M$, $[P] = 9 \times 10^{-4} M$; at $[S] = 2 \times 10^{-4} M$, $[P] = 8 \times 10^{-4} M$, and so on.

21.   The substrate of an enzyme is the A^- ion of a weak acid ($pK_a = 4.5$).   The active site of the enzyme contains a histidine residue ($pK_e = 6.5$) that must be protonated for activity to occur.   What is the pH optimum of the reaction?

22.   The active site of an esterase contains an acidic and a basic amino acid residue.   Substrate binding occurs only when the site exists as ^+HN-E-COO$^-$.   Thus, the productive species is ^+HN-ES-COO$^-$, while the ^+HN-ES-COOH and N-ES-COO$^-$ species

do not exist.   The pK's of the two residues are 4.0 (pK_{r_1}) and 7.0 (pK_{r_2}).   (a) What is the pH optimum?   (b) Write a velocity equation expressing the effect of [H^+] on v.

23. V_{max} at 21 and 37°C was 140 nmoles × liter^{-1} × min^{-1} and 400 nmoles × liter^{-1} × min^{-1}, respectively.   Calculate   (a)   the   activation energy and (b) the Q_{10} value between 25 and 35°C.

24.   A cell-free extract of *Escherichia coli* contains   24 mg   protein   per   milliliter. Twenty microliters of this extract in a standard incubation volume of 0.1 ml catalyzed the incorporation of glucose-^{14}C from glucose-1-phosphate-^{14}C into glycogen at a rate of 1.6 nmole/min.   Calculate the velocity of the reaction in terms of (a) μmoles/min, (b) μmoles × liter^{-1} × min^{-1},   (c) μmoles × mg protein^{-1} × min^{-1}.   Also calculate the phosphorylase activity of the extract in terms of (d) units/ml and (e) units/mg protein.

25.   Fifty milliliters of the cell-free extract described above was fractionated by ammonium sulfate precipitation.   The fraction precipitating between 30 and 50% saturation was redissolved in a total volume of 10 ml and dialyzed.   The solution after dialysis occupied   12 ml   and   contained   30 mg protein/ml.   Twenty   microliters   of   the purified fraction catalyzed the phosphorylase reaction at a rate of 5.9 nmoles/min under the standard assay conditions.   Calculate (a) the recovery of the enzyme and (b) the degree of purification obtained in the ammonium sulfate step.

26.   A pure enzyme has a specific activity of 120 units/mg protein.   (a) Calculate the turnover number if MW = 360,000.   (b) Calculate the time required for one catalytic cycle.

27.   (a) How many units of hexokinase must be added to 1 ml of reaction volume to use   up   95%   of $8 \times 10^{-3} M$ glucose in 25 min? $K_m = 4.7 \times 10^{-4} M$.   (b)   How   long would it take 1.0 unit/ml to use up the same amount of glucose?

28.   Calculate the velocity of an enzyme-catalyzed   reaction $A + B \rightarrow P + Q$ at   [A] = $2 \times 10^{-5} M$ and   [B] = $6.7 \times 10^{-5} M$.   Assume that A and B add randomly to the enzyme and (a) $\alpha = 1$ (the binding of one substrate has no effect on the binding of the other), (b) $\alpha = 0.1$ (the binding of one substrate decreases the

dissociation constant for the other by a factor of 10).   Preliminary experiments established that $K_A = 2.2 \times 10^{-4} M$, $K_B = 1.9 \times 10^{-5} M$, and $V_{max} = 72.7$ nmoles × liter^{-1} × min^{-1}.

29.   Rearrange the velocity equation for a steady-state ordered bireactant system to show v/V_{max} as (a) [A] is varied at different fixed concentrations of B and (b) [B] is varied at different fixed concentrations of A.

30.   How do the reciprocal plots for a rapid equilibrium ordered bireactant system differ from those of a steady-state ordered bireactant system?

31.   What relative values of a, b, and c (interaction factors) would yield a velocity curve for an allosteric tetramer that exhibited positive-negative-positive cooperativity?

32.   The following data were obtained for an   enzyme-catalyzed   reaction.   Determine whether the enzyme obeys hyperbolic or sigmoidal kinetics and calculate or estimate the appropriate kinetic constants (K_m and V_{max}, or K', [S]$_{0.5}$, n_{app}, and V_{max}).

Initial Substrate Concentration	Initial Velocity
($M \times 10^4$)	($\mu moles \times liter^{-1} \times min^{-1}$)
6.25	1.54
12.5	5.88
25.0	20.0
50.0	50.0
100.0	80.0
200.0	94.12
400.0	98.46
800.0	99.61

33.   (a) What is the n_{app} value of an enzyme if [S]$_{0.9}$/[S]$_{0.1}$ is 9?   (b) Calculate the [S]$_{0.9}$/[S]$_{0.1}$ ratio of an allosteric enzyme where $n_{app} = 4$.

34.   What is the value of L for an allosteric dimer ($n = 2$) if $v/V_{max} = 0.35$ when [S]/$K_S = 5$?   Assume $c = 0$.

35.   The concerted-symmetry model does not forbid T $\rightleftharpoons$ R transitions among partially filled   complexes.   If   L$_1$ is   defined   as [TS]/[RS], and L$_2$ is defined as [TS$_2$]/[RS$_2$], what are the values of L$_1$ and L$_2$ in terms of L and c?   Consult Fig. 4-53.

36. The level of a particular enzyme is 1500 units/g tissue in the liver of rats on a natural diet. Studies indicate that the first-order rate constant for the degradation of the enzyme is 0.03 min^{-1}. (a) Calculate the zero-order rate of enzyme synthesis. (b) When rats are raised on a completely synthetic diet, the steady-state level of the above enzyme decreases to 848 units/g tissue. If the rate of enzyme synthesis is unaffected by diet, what must the new first-order rate constant for enzyme degradation be? (c) If the first-order rate constant of enzyme degradation is unaffected by diet (and remains 0.03 min^{-1}) what must the new zero-order rate of enzyme synthesis be in order to maintain $[E]_t$ at the new level of 848 units/g tissue?

Also see the practice problems on enzyme assays in the chapters on "Spectrophotometry and Other Optical Methods" and "Isotopes in Biochemistry."

5

SPECTROPHOTOMETRY AND OTHER OPTICAL METHODS

A. SPECTROPHOTOMETRY

ABSORPTION OF ELECTROMAGNETIC ENERGY

Spectrophotometry (the measurement of light absorption or transmission), is one of the most valuable analytical techniques available to biochemists. Unknown compounds may be identified by their characteristic absorption spectra in the ultraviolet, visible, or infrared. Concentrations of known compounds in solutions may be determined by measuring the light absorption at one or more wavelengths. Enzyme-catalyzed reactions frequently can be followed by measuring spectrophotometrically the appearance of a product or disappearance of a substrate.

The physical phenomena underlying light absorption in the various regions of the electromagnetic spectrum are shown in Table 5-1.

Table 5-1

Region	X rays	Ultraviolet 100–	Visible 400–	Infrared 800 nm–	Microwave 100 μm–
Wavelength	0.1–100 nm	400 nm	800 nm	100 μm	30 cm
Effect on molecule	Subvalence electrons excited to higher energy levels	Valence electrons excited to higher energy levels		Molecular vibration	Molecular rotation

In some books, a nm (nanometer = 10^{-9} m) is called a mμ (millimicron); a μm (micrometer = 10^{-6} m) is called a micron. An Angstrom, Å, is 10^{-10} m = 10^{-8} cm. Sometimes, infrared radiation is described in terms of *wave number*, cm^{-1}, which is the reciprocal of the wavelength in centimeters. Radiation is also described by its frequency. The frequency and wavelength are related to speed of light, which is a constant:

$$\lambda \nu = c \qquad (1)$$

where λ = wavelength (e.g., in cm)
       ν = frequency (e.g., in sec^{-1})
and c = the speed of light (3×10^{10} cm/sec)
The units of ν are sec^{-1}, which means vibrations per second.

· **Problem 5-1**

An optical filter passes only far red light with an average wavelength of 6500 Å.   Calculate (a) the wavelength in nanometers and centimeters, (b) the wave number in centimeters^{-1}, and (c) the frequency.

Solution

(a) $6500 \text{ Å} = 6500 \times 10^{-10} \text{ m} = 650 \times 10^{-9} \text{ m}$

$$\boxed{\lambda = 650 \text{ nm}} \quad \text{or} \quad \boxed{\lambda = 6.50 \times 10^{-5} \text{ cm}}$$

(b)   wave number = $1/\lambda = 1/(6.50 \times 10^{-5} \text{ cm})$

$$\boxed{\text{wave number} = 15{,}384 \text{ cm}^{-1}}$$

(c) $\lambda\nu = c \qquad \nu = \dfrac{c}{\lambda}$

$$\nu = \frac{3 \times 10^{10}}{6.50 \times 10^{-5}} \qquad \boxed{\nu = 4.61 \times 10^{14} \text{ sec}^{-1}}$$

SPECTROPHOTOMETERS

A spectrophotometer is an instrument used to measure the amount of light of a given wavelength that is transmitted by a sample.   The essential compo-

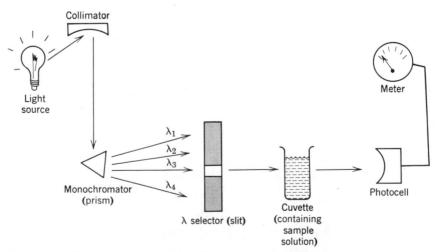

Figure 5-1   Essential components of a spectrophotometer.

nents of a spectrophotometer (Fig. 5-1) include: (a) a light source, (b) a collimator or focusing device that transmits an intense straight beam of light, (c) a monochromator (prism or grating) to divide the light beam into its component wavelengths, (d) a device for selecting the desired wavelength, (e) a compartment in which the sample (in a test tube or cuvette) is placed, (f) a photoelectric detector, and (g) an electrical meter to record the output of the detector.

LAMBERT–BEER LAW

The fraction of the incident light that is absorbed by a solution depends on the thickness of the sample, the concentration of the absorbing compound in the solution, and the chemical nature of the absorbing compound. Light absorption follows an exponential rather than a linear law. The relationship between concentration, length of the light path, and the light absorbed by a particular substance are expressed mathematically as shown below.

$$-\frac{dI}{I} \propto c\,dl \qquad \text{and} \qquad -\frac{dI}{I} \propto l\,dc$$

or

$$\frac{dI}{I} = -kc\,dl \qquad \text{and} \qquad \frac{dI}{I} = -kl\,dc$$

where $-dI$ = the small decrease in light transmission caused by increasing the thickness a small increment, dl, at constant concentration, or the small decrease in light transmission caused by increasing the concentration a small increment, dc, at constant thickness

$\frac{dI}{I}$ = the fraction of the incident light absorbed

k = proportionality constant, specific for the particular compound under consideration

The above differential relationships may be integrated between any two thicknesses (e.g., 0 and l) or any two concentrations (e.g., 0 and c).

$$\int_{I_0}^{I} \frac{dI}{I} = -kc \int_0^l dl \qquad\qquad \int_{I_0}^{I} \frac{dI}{I} = -kl \int_0^c dc$$

$$\ln \frac{I}{I_0} = -kcl \qquad\qquad\qquad \ln \frac{I}{I_0} = -klc$$

$$\ln \frac{I_0}{I} = kcl \qquad\qquad\qquad\quad \ln \frac{I_0}{I} = klc$$

or

$$2.3 \log \frac{I_0}{I} = kcl \qquad\qquad\qquad 2.3 \log \frac{I_0}{I} = klc$$

$$\log \frac{I_0}{I} = \frac{k}{2.3} cl \qquad\qquad\qquad \log \frac{I_0}{I} = \frac{k}{2.3} lc$$

$$\boxed{\log \frac{I_0}{I} = acl} \qquad\qquad \boxed{\log \frac{I_0}{I} = alc} \qquad\qquad (2)$$

where a = the "absorbancy index," or "extinction coefficient," or "absorption coefficient" for the particular absorbing compound.

If the concentration is expressed in molarity, a becomes the "molar absorption coefficient" or "molar extinction coefficient," a_m or E. If the concentration is given in g/liter, a becomes the "specific absorption coefficient," a_s; $a_m = a_s \times$ MW. If the concentration is expressed in % w/v, then the absorption coefficient is given the symbol $a_{1\%}$ or $E_{1\%}$. In most biochemical calculations, molarities and molar absorption coefficients are employed. Furthermore, the sample thickness is almost always 1 cm. Then a_m has units of $M^{-1} \times cm^{-1}$. Absorption coefficients vary with varying wavelengths. Thus, the symbol a_{m340} refers to the molar absorption coefficient at 340 nm. The log I_0/I term is called "absorbance," A, or "optical density," O.D.

$$A = a_m c l \qquad (3)$$

The absorbance of a $1\,M$ solution of a given substance at a given wavelength in a 1 cm cuvette would be equal numerically to a_m. Note that absorbance is a linear function of concentration.

The exponential nature of the Lambert–Beer law is illustrated below. Consider a beam of light passing through a 1 cm cuvette containing 1 mg/liter of a light-absorbing compound. Suppose 80% of the incident light is transmitted (20% of the incident light is absorbed).

$$I_0 = 1.00 \longrightarrow I_1 = 0.8$$
$$c = 1 \text{ mg/liter}$$
$$\leftarrow l = 1 \text{ cm} \rightarrow$$

Now let us place a second identical cuvette in the light path directly behind the first cuvette. What is the intensity of the light transmitted through both cuvettes? The Lambert–Beer law is not a linear relationship. Thus I_2 is not 0.6. Each centimeter of path length does not absorb a constant amount of light. Instead, each centimeter absorbs 20% of the incident light. However, the incident light on the second cuvette is I_1 or 0.8. Thus, the second cuvette absorbs 20% of 0.8, which is 0.16. It transmits 80% of 0.8, which is 64% of the original incident light.

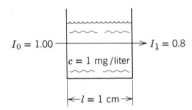

$$I_0 = 1.00 \longrightarrow I_1 = 0.80 \longrightarrow I_2 = 0.64$$
$$c = 1 \text{ mg/liter} \qquad c = 1 \text{ mg/liter}$$
$$\leftarrow l = 1 \text{ cm} \rightarrow \qquad \leftarrow l = 1 \text{ cm} \rightarrow$$

Similarly, if we place three 1 cm cuvettes in series, each one absorbs 20% of the incident light and transmits 80%.

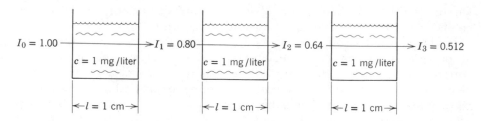

We would obtain exactly the same results if we used single cuvettes with 2 cm and 3 cm thicknesses or increased the concentration of the absorbing compound, maintaining l constant.

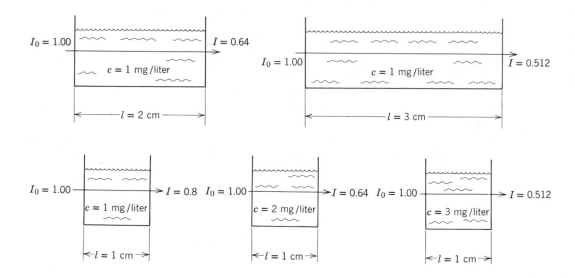

General Principles

1. If a solution of concentration c has a transmission of I (as a decimal fraction), then a solution of concentration $2c$ has a transmission of I^2, a solution of concentration $3c$ has a transmission of I^3, and a solution of concentration nc has a transmission of I^n.

2. Similarly, if the light path thickness is increased n-fold, the new transmission is I^n.

3. The absorbance of a solution is a linear function of concentration. Doubling c results in a doubling of the absorbance, tripling c results in a tripling of the absorbance, and so on.

General Procedure

All light absorption measurements should be made relative to a blank solution that contains all the components of the assay, except the compound being measured.

SOLUTIONS CONTAINING ONLY ONE ABSORBING COMPOUND

· Problem 5-2

A solution containing 2 g/liter of a light-absorbing substance in a 1 cm cuvette transmits 75% of the incident light of a certain wavelength. Calculate the transmission of a solution containing (a) 4 g/liter, (b) 1 g/liter, (c) 6 g/liter, and (d) 5.4 g/liter. (e) If the molecular weight of the compound is 250, calculate a_m.

Solution

(a) $$\log \frac{I_0}{I} = acl$$

First solve the equation for a, which will be a specific absorption coefficient, a_s, because the concentration is given in g/liter.

$$\log \frac{1.00}{0.75} = (a_s)(2)(1) \qquad \log 1.333 = 2a_s$$

$$0.125 = 2a_s \qquad a_s = \frac{0.125}{2}$$

$$\boxed{a_s = 0.0625}$$

Now that we have the specific absorption coefficient, the I values for solutions of any concentration may be calculated.

$$\log \frac{I_0}{I} = a_s cl$$

Calling I_0 100%

$$\log \frac{100}{I} = (0.0625)(4)(1)$$

$$\log 100 - \log I = 0.25$$

$$\log 100 - 0.25 = \log I$$

$$2 - 0.25 = \log I$$

$$1.75 = \log I$$

$$\boxed{I = 56.2\% \text{ or } 0.562}$$

Calling I_0 1.00

$$\log \frac{1.00}{I} = (0.0625)(4)(1)$$

$$\log 1.00 - \log I = 0.25$$

$$\log I = -0.25$$

$$= -1 + 0.75$$

$$I = 5.62 \times 10^{-1}$$

$$\boxed{I = 0.562}$$

We could also solve the problem by recalling that if a solution of concentration c has a transmission of I, then a solution of concentration nc has a transmission of I^n. When $c = 2$, $I = 0.75$. ∴ When $c = 4$ (i.e., $2c_{orig}$), $I = 0.75^2 = 0.562$.

(b)

By Formula

$$\log \frac{I_0}{I} = a_s c l$$

$$\log \frac{100}{I} = (0.0625)(1)(1)$$

$$\log 100 - \log I = 0.0625$$

$$2 - \log I = 0.0625$$

$$2 - 0.0625 = \log I$$

$$1.9375 = \log I$$

$$\boxed{I = 86.6\% \text{ or } 0.866}$$

By Inspection

When $c = 2$, $I = 0.75$. $\therefore$ When $c = 1$ (i.e., $\frac{1}{2}c_{\text{orig}}$), $I = 0.75^{1/2}$.

$$I = \sqrt{0.75}$$

$$\boxed{I = 0.866}$$

(c)

By Formula

$$\log \frac{I_0}{I} = a_s c l$$

$$\log \frac{100}{I} = (0.0625)(6)(1)$$

$$\log 100 - \log I = 0.375$$

$$2 - 0.375 = \log I$$

$$1.625 = \log I$$

$$\boxed{I = 42.2\% \text{ or } 0.422}$$

By Inspection

When $c = 2$, $I = 0.75$. $\therefore$ When $c = 6$ (i.e., $3c_{\text{orig}}$), $I = 0.75^3$.

$$\boxed{I = 0.422}$$

(d)   Because 5.4 g/liter is not an even multiple of 2 g/liter it is far easier to solve this problem by formula than by inspection.

$$\log \frac{100}{I} = a_s c l \qquad \log \frac{100}{I} = (0.0625)(5.4)(1)$$

$$\log 100 - \log 1 = 0.3375 \qquad \text{or} \qquad 2 - 0.3375 = \log I$$

$$1.6625 = \log I \qquad I = \text{antilog of } 1.6625 \qquad \boxed{I = 46.0\% \text{ or } 0.46}$$

(e) $\qquad a_m = a_s \times \text{MW} = (0.0625)(250) \qquad \boxed{a_m = 15.63}$

· Problem 5-3

A solution containing 10^{-5} M ATP has a transmission 0.702 (70.2%) at 260 nm in a 1 cm cuvette.   Calculate the (a) transmission of the solution in a 3 cm cuvette, (b) absorbance of the solution in the 1 cm and 3 cm cuvettes, and (c) absorbance and transmission of a 5×10^{-5} M ATP solution in a 1 cm cuvette.

Solution

(a)   We can calculate the transmission by the formula after first calculating a_m, or by inspection.

By Formula	By Inspection

By Formula

$$\log \frac{I_0}{I} = a_m cl$$

$$\log \frac{100}{70.2} = (a_m)(10^{-5})(1)$$

$$\log 1.425 = 10^{-5}\, a_m$$

$$0.154 = 10^{-5} a_m$$

$$a_m = \frac{0.154}{10^{-5}}$$

$$= 0.154 \times 10^5$$

$$\boxed{a_m = 1.54 \times 10^4}$$

$$\log \frac{I_0}{I} = a_m cl$$

$$\log \frac{100}{I} = (1.54 \times 10^4)$$

$$\times (10^{-5})(3)$$

$$\log 100 - \log I = 4.62 \times 10^{-1}$$

$$2 - 0.462 = \log I$$

$$1.538 = \log I$$

$$\boxed{I = 34.5\% \text{ or } 0.345}$$

By Inspection

When $l = 1$ cm, $I = 0.702$.   ∴   When $l = 3$ cm (i.e., $3\, l_{orig}$), $I = 0.702^3$.

$$\boxed{I = 0.345}$$

(b)

1 cm Cuvette

$$A = \log \frac{I_0}{I}$$

$$A = \log \frac{100}{70.2} = \log 1.425$$

$$\boxed{A = 0.154}$$

3 cm Cuvette

$$A = \log \frac{I_0}{I}$$

$$A = \log \frac{100}{34.5} = \log 2.9$$

$$\boxed{A = 0.462}$$

or

$$A = a_m c l$$
$$A = (1.54 \times 10^4)(10^{-5})(1)$$

$$\boxed{A = 0.154}$$

or

$$A = a_m c l$$
$$A = (1.54 \times 10^4)(10^{-5})(3)$$

$$\boxed{A = 0.462}$$

or

$$A_{3\,cm} = 3 \times A_{1\,cm}$$
$$A_{3\,cm} = (3)(0.154)$$

$$\boxed{A = 0.462}$$

(c)

$$A_{5\times10^{-5}\,M} = 5 \times A_{1\times10^{-5}\,M} \qquad \text{or} \qquad A = a_m c l$$
$$A = (5)(0.154) \qquad\qquad\qquad A = (1.54 \times 10^{-4})(5 \times 10^{-5})(1)$$

$$\boxed{A = 0.77} \qquad\qquad\qquad \boxed{A = 0.77}$$

$$\log \frac{I_0}{I} = A \qquad \log \frac{100}{I} = 0.77$$

$$\log 100 - \log I = 0.77 \qquad \text{or} \qquad 2 - 0.77 = \log I$$

$$1.23 = \log I \qquad I = \text{antilog of } 1.23$$

$$\boxed{I = 17.0\% \text{ or } 0.17}$$

· Problem 5-4

The specific absorption coefficient ($a_{1\%}^{1\,cm}$) of a glycogen-iodine complex at 450 nm is 0.20. Calculate the concentration of glycogen in a solution of the iodine complex, which has an absorbance of 0.38 in a 3 cm cuvette.

Solution

$$A = a_{1\%}^{1\,cm} \, c_\% \, l_{cm} \qquad 0.38 = (0.20)(c_\%)(3)$$

$$c_\% = \frac{0.38}{(0.2)(3)} = \frac{0.38}{0.6} \qquad \boxed{c = 0.633\%}$$

· Problem 5-5

A suspension of bacteria containing 400 mg dry weight per liter has an absorbance of 1.00 in a 1 cm cuvette at 450 nm. What is the cell density in a suspension that has a transmission of 30% in a 3 cm cuvette?

Solution

First calculate the absorbance of the suspension in a 1 cm cuvette.

$$A_{3\,cm} = \log \frac{I_0}{I} = \log \frac{100}{30} = \log 3.333 = 0.523$$

$$A_{1\,cm} = \frac{A_{3\,cm}}{3} = \frac{0.523}{3} = 0.1743$$

Because we know that an A of 1.00 is equivalent to 400 mg/liter of bacterial cells, the density equivalent to an A of 0.174 can be determined by simple proportions.

$$\frac{1.00 \ A \ \text{unit}}{400 \ \text{mg/liter}} = \frac{0.1743 \ A \ \text{unit}}{X \ \text{mg/liter}}$$

$$\boxed{X = 69.7 \ \text{mg/liter}}$$

An alternative way of solving the problem is to define a specific absorption coefficient for the bacteria.

If an A of $1.00 \doteq 400$ mg/liter bacteria, calculate the A of 1 g/liter bacteria.

$$\frac{1.00 \ A \ \text{unit}}{0.40 \ \text{g/liter bacteria}} = \frac{a_s}{1.0 \ \text{g/liter bacteria}}$$

$$\boxed{a_s = 2.5}$$

Now use the usual formula.

$$A = (a_s)(c_{\text{g/liter}})(l_{\text{cm}}) \qquad 0.523 = (2.5)(c_{\text{g/liter}})(3)$$

$$c_{\text{g/liter}} = \frac{0.523}{7.5} = 0.0697 \ \text{g/liter}$$

$$\boxed{c = 69.7 \ \text{mg/liter}}$$

PROTEIN DETERMINATIONS

Several spectrophotometric methods are available for the determination of protein in solution. The biuret method is based on the reaction of Cu^{2+} with peptides in alkaline solution to yield a purple complex that has an absorption maximum at 540 nm. The biuret method is used for solutions containing 0.5 to 10 mg protein/ml. Interfering substances such as thiols, NH_4^+, and the like, can be removed by precipitating the protein with an equal volume of cold 10% trichloroacetic acid, discarding the supernate, then redissolving the protein in a known volume of 1 N NaOH.

The color produced in the Lowry method results from the biuret reaction plus the reduction of the phosphomolybdate-phosphotungstate reagent (Folin-Ciocalteu phenol reagent) by tyrosine residues. The Lowry method is suitable for solutions containing 20 to 400 μg protein/ml.

Most proteins have a distinct absorption maximum at 280 nm, due primarily to the presence of tyrosine, tryptophan, and phenylalanine. The absorbance at 280 nm can be used to measure protein at levels of 0.1 to 0.5 mg/ml in the absence of interfering substances. Partially purified preparations may contain nucleic acids that have an absorption maximum at 260 nm. An equation for calculating the protein concentration in the presence of nucleic acids is:

$$[\text{protein}]_{\text{mg/ml}} = 1.55\, A_{280}^{1\,\text{cm}} - 0.76\, A_{260}^{1\,\text{cm}} \tag{4}$$

The equation was derived for enolase ($A_{280}/A_{260} = 1.75$) in the presence of yeast nucleic acid ($A_{280}/A_{260} = 0.49$) and thus may not be precise for other proteins and other nucleic acids. The $a_{280}^{0.1\%}$ of different proteins ranges from 0.5 to 2.5 (depending on the aromatic amino acid content).

All proteins absorb strongly below 230 nm. For example, the $a^{0.1\%}$ values of bovine serum albumin are 5.0 and 11.7 at 225 and 215 nm, respectively (compared to 0.58 at 280 nm). The absorption below 230 nm is due to the peptide bond. Consequently, the $a^{0.1\%}$ values are essentially the same for all proteins. Protein concentrations in the region 10 to 100 μg/ml can be determined from the *difference* in absorbances at 215 and 225 nm. A standard curve is prepared plotting ΔA versus [protein]. As a first approximation, the protein concentration is given by:

$$[\text{protein}]_{\mu\,\text{g/ml}} = 144(A_{215\,\text{nm}}^{1\,\text{cm}} - A_{225\,\text{nm}}^{1\,\text{cm}}) \tag{5}$$

The absorption *difference* is used to minimize errors resulting from nonprotein compounds in the solution. High concentrations of certain buffer components interfere. Some inorganic compounds may also interfere (e.g., 0.1 N NaOH cannot be used to dissolve the protein, but 5 mM NaOH causes no problems).

· Problem 5-6

A protein solution (0.3 ml) was diluted with 0.9 ml of water. To 0.5 ml of this diluted solution, 4.5 ml of biuret reagent were added and the color was allowed to develop. The absorbance of the mixture at 540 nm was 0.18 in a 1 cm diameter test tube. A standard solution (0.5 ml, containing 4 mg of protein/ml) plus 4.5 ml of biuret reagent gave an absorbance of 0.12 in the same-size test tube. Calculate the protein concentration in the undiluted unknown solution.

Solution

From the absorbance of the standard reaction mixture, we could calculate a specific absorption coefficient.

$$A = a_{1\,\text{mg/ml}} \times c_{\text{mg/ml}} \times l_{\text{cm}}$$

However, because the light path length is the same for the standard and the unknown, we can neglect the l term in the equation. (In a sense, we are incorporating l into the specific absorption coefficient.) Similarly, because the total volume of both reaction mixtures is 5 ml, we can replace the concentration term with the weight of the protein.

$$A = a_{1\,\text{mg}} \times \text{wt}_{\text{mg}} \qquad 0.12 = a_{1\,\text{mg}} \times 2 \text{ mg}$$

$$a_{1\,\text{mg}} = \frac{0.12}{2} \qquad \boxed{a_{1\,\text{mg}} = 0.06}$$

That is, 1 mg of protein (in a standard sample size of 0.5 ml) plus 4.5 ml of biuret reagent will yield an absorbance of 0.06 at 540 nm in the particular test tube used.

The weight of protein in 0.5 ml of the diluted unknown can now be calculated.

$$A = a_{1\,\text{mg}} \times \text{wt}_{\text{mg}} \qquad \text{wt}_{\text{mg}} = \frac{A}{a_{1\,\text{mg}}} = \frac{0.18}{0.06}$$

$$\boxed{\text{wt}_{\text{mg}} = 3 \text{ mg}}$$

Because the absorbance of the reaction mixture is directly proportional to the amount of protein present, we can also solve for the unknown weight by setting up a simple proportion.

$$\frac{A_s}{\text{wt}_s} = \frac{A_u}{\text{wt}_u} \qquad \frac{0.12}{2 \text{ mg}} = \frac{0.18}{\text{wt}_u}$$

$$\text{wt}_u = \frac{0.18}{0.12} \times 2 = 1.5 \times 2$$

$$\boxed{\text{wt}_u = 3 \text{ mg}}$$

where $A_s = A$ of standard solution
$A_u = A$ of unknown solution
wt_s = weight of protein in standard sample
wt_u = weight of protein in unknown sample

The 3 mg of protein were present in 0.5 ml of diluted unknown. The concentration of protein in the diluted unknown was:

$$\frac{3 \text{ mg}}{0.5 \text{ ml}} = \boxed{6 \text{ mg/ml}}$$

The sample that was analyzed was a fourfold dilution of the original unknown solution. The concentration of protein in the original solution can be calculated from the dilution factor.

$$c_{orig} = c_{final} \times \text{dilution factor} = 6 \text{ mg/ml} \times 4$$

$$\boxed{c_{orig} = 24 \text{ mg/ml}}$$

Note that the addition of 0.9 ml of water to 0.3 ml of original solution increased the total volume to 1.2 ml—the dilution was fourfold, not threefold.

· Problem 5-7

Estimate the protein concentrations of the (a) undiluted and (b) diluted solution shown below. The measurements were made using a 1 cm cuvette.

Solution	$A_{280 nm}$	$A_{260 nm}$	$A_{225 nm}$	$A_{215 nm}$
Undiluted	0.35	0.20	—	—
Diluted 1:10	—	—	0.20	0.47

Solution

(a)
$$[\text{protein}]_{mg/ml} = 1.55 \, A_{280} - 0.76 \, A_{260}$$
$$= (1.55)(0.35) - (0.76)(0.20)$$
$$= 0.5425 - 0.1520$$

$$\boxed{[\text{protein}] = 0.391 \text{ mg/ml}}$$

(b)
$$[\text{protein}]_{\mu g/ml} = 144(A_{215} - A_{225})$$
$$= 144(0.47 - 0.20)$$
$$= 144(0.27)$$

$$\boxed{[\text{protein}] = 38.9 \, \mu\text{g/ml}}$$

The diluted solution contained $(38.9)(10) = 389 \, \mu\text{g/ml} = 0.389$ mg/ml. Thus, the two methods yield essentially the same result.

· Problem 5-8

A pure molybdenum-containing enzyme has an $a_{280\,nm}^{0.1\%}$ of 1.5 in a 1 cm cuvette. A concentrated solution of the protein was found to contain 10.56 μg of Mo/ml. A 1:50 dilution of the same solution has an $A_{280\,nm}$ of 0.375. Calculate the minimum molecular weight of the enzyme. The atomic weight of Mo is 95.94.

Solution

The concentrated solution contains:

$$\frac{10.56 \times 10^{-6} \text{ g Mo/ml}}{95.94 \text{ g/g-atom Mo}} = 1.1 \times 10^{-7} \text{ g-atom Mo/ml}$$

An $a^{0.1\%}$ is the absorbance of a solution containing 1 mg protein/ml. The absorbance of the concentrated solution is:

$$(50)(0.375) = 18.75$$

$$A = (a^{0.1\%})(c_{\text{mg/ml}})(l) \quad \text{or} \quad c_{\text{mg/ml}} = \frac{A}{(a^{0.1\%})(l)}$$

$$\therefore \quad c = \frac{18.75}{1.5} = 12.5 \text{ mg/ml} = 12.5 \times 10^{-3} \text{ g/ml}$$

The minimum molecular weight is that amount of enzyme containing one gram-atom of Mo.

$$\frac{12.5 \times 10^{-3} \text{ g protein}}{10.56 \times 10^{-6} \text{ g-atom Mo}} = \frac{MW_g}{95.94 \text{ g Mo}}$$

or

$$\frac{12.5 \times 10^{-3} \text{ g protein}}{1.1 \times 10^{-7} \text{ g-atom Mo}} = \frac{MW_g}{1 \text{ g-atom}}$$

$$\boxed{MW \simeq 113,600}$$

SOLUTIONS CONTAINING TWO ABSORBING COMPOUNDS

· Problem 5-9

A solution containing NAD^+ and NADH had an optical density in a 1 cm cuvette of 0.311 at 340 nm and 1.2 at 260 nm. Calculate the concentrations of the oxidized and reduced forms of the coenzyme in the solution. Both NAD^+ and NADH absorb at 260 nm, but only NADH absorbs at 340 nm. The extinction coefficients are given below.

Compound	$a_m (M^{-1} \times cm^{-1})$ 260 nm	340 nm
NAD^+	18,000	~0
NADH	15,000	6220

Solution

The concentration of each form may be calculated as follows. First calculate the concentration of NADH from its absorbance at 340 nm where the NAD^+ does not absorb.

$$A = (a_m)(c)(l)$$
$$0.311 = (6.22 \times 10^3)(c)(1)$$
$$c = \frac{3.11 \times 10^{-1}}{6.22 \times 10^3} = 0.5 \times 10^{-4}$$

$$\boxed{c_{\text{NADH}} = 5 \times 10^{-5} \; M}$$

Next calculate the absorbance at 260 nm resulting from the NADH.

$$A = (a_m)(c)(l)$$
$$A = (15.0 \times 10^3)(5 \times 10^{-5})(1) = 75 \times 10^{-2}$$

$$\boxed{A_{260[\text{NADH}]} = 0.75}$$

The remainder of the absorbance at 260 nm must result from the NAD^+.

$$\begin{array}{r} 1.20 \text{ total } A \text{ at } 260 \text{ nm} \\ - 0.75 \text{ } A \text{ of NADH at } 260 \text{ nm} \\ \hline 0.45 \text{ } A \text{ of NAD}^+ \text{ at } 260 \text{ nm} \end{array}$$

Finally, from the absorbance of the NAD^+ at 260 nm, calculate the concentration of NAD^+.

$$A = (a_m)(c)(l) \qquad 0.45 = (18.0 \times 10^3)(c)(1)$$
$$c = \frac{4.5 \times 10^{-1}}{18.0 \times 10^3} = 0.250 \times 10^{-4}$$

$$\boxed{c_{\text{NAD}^+} = 2.50 \times 10^{-5} \; M}$$

· Problem 5-10

Ten grams of butter were saponified; the nonsaponifiable fraction was extracted into 25 ml of chloroform. The absorbance of the chloroform solution in a 1 cm cuvette was 0.53 at 328 nm and 0.48 at 458 nm. Calculate the carotene and vitamin A content of the butter. The extinction coefficients for carotene and vitamin A at the above two wavelengths are given below.

Compound	$a_{1\%}^{1 \text{ cm}}$ in $CHCl_3$	
	328 nm	458 nm
Carotene	340	2200
Vitamin A	1550	~ 0

Solution

The concentration of carotene may be obtained from the absorbance at 458 nm where the vitamin A has no absorption.

$$A = (a_{1\%})(c_{g/100\,ml})(l) \qquad 0.48 = (2200)(c)(1)$$

$$c = \frac{0.48}{2200} = \frac{4.8 \times 10^{-1}}{2.2 \times 10^3}$$

$$\boxed{c_{\text{carotene}} = 2.18 \times 10^{-4} \text{ g/100 ml}}$$

The total carotene content of the chloroform extract (25 ml) is

$$\frac{2.18 \times 10^{-4} \text{ g/100 ml}}{4} = 0.545 \times 10^{-4} \text{ g} = 0.0545 \text{ mg}$$

The carotene content of the butter is

$$\frac{0.0545 \text{ mg}}{10 \text{ g}} = 5.45 \times 10^{-3} \text{ mg carotene/g butter}$$

or $\qquad$ $$\boxed{\text{5.45 } \mu\text{g carotene/g butter}}$$

The vitamin A content of the butter may be calculated from the absorbance of the chloroform solution at 328 nm after correction of the absorbance for the carotene present.

$$A_{328_{\text{carotene}}} = (a_{1\%})(c_{g/100\,ml})(l)$$
$$= (340)(2.18 \times 10^{-4})(1) = 741 \times 10^{-4}$$
$$A_{328_{\text{carotene}}} = 0.0741$$

$$\begin{array}{r} 0.530 \text{ total } A \text{ at 328 nm} \\ -0.074 \text{ } A \text{ of carotene at 328 nm} \\ \hline 0.456 \text{ } A \text{ of vitamin A at 328 nm} \end{array}$$

$$A_{\text{vit.A}} = (a_{1\%})(c_{g/100\,ml})(l) \qquad 0.456 = (1550)(c)(1)$$

$$c = \frac{0.456}{1550} = \frac{45.6 \times 10^{-2}}{15.5 \times 10^2}$$

$$\boxed{c_{\text{vit.A}} = 2.94 \times 10^{-4} \text{ g/100 ml}}$$

The vitamin A content of the chloroform extract is:

$$\frac{2.94 \times 10^{-4} \text{ g/100 ml}}{4} = 0.735 \times 10^{-4} \text{ g} = 0.0735 \text{ mg}$$

The vitamin A content of the butter is:

$$\frac{73.5 \text{ } \mu\text{g}}{10 \text{ g}} = \boxed{\text{7.35 } \mu\text{g vitamin A/g butter}}$$

· Problem 5-11

A solution containing two substances, A and B, has an absorbance in a 1 cm cuvette of 0.36 at 350 nm and 0.225 at 400 nm. The molar absorption coefficients of A and B at the two wavelengths are given below. Calculate the concentrations of A and B in the solution.

Compound	$a_m\,(M^{-1} \times cm^{-1})$ 350 nm	400 nm
A	15,000	3000
B	7,000	6500

Solution

Because both compounds absorb at both wavelengths, we can set up two simultaneous equations.

$$A_{350\,nm} = (a_{m\,350[A]} \times c_A) + (a_{m\,350[B]} \times c_B)$$
$$0.36 = (15 \times 10^3 c_A) + (7 \times 10^3 c_B)$$
$$A_{400\,nm} = (a_{m\,400[A]} \times c_A) + (a_{m\,400[B]} \times c_B)$$
$$0.225 = (3 \times 10^3 c_A) + (6.5 \times 10^3 c_B)$$

Next, solve for c_A in terms of c_B or vice versa.

$$0.36 = 15 \times 10^3 c_A + 7 \times 10^3 c_B$$
$$0.36 - 7 \times 10^3 c_B = 15 \times 10^3 c_A$$
$$c_A = \frac{0.36 - 7 \times 10^3 c_B}{15 \times 10^3}$$

Next, substitute the above value for c_A into the second A expression.

$$0.225 = (3 \times 10^3 c_A) + (6.5 \times 10^3 c_B)$$
$$0.225 = (3 \times 10^3)\frac{0.36 - 7 \times 10^3 c_B}{15 \times 10^3} + 6.5 \times 10^3 c_B$$
$$0.225 = \frac{1.08 \times 10^3 - 21 \times 10^6 c_B}{15 \times 10^3} + \frac{6.5 \times 10^3 c_B}{1}$$

Multiply the numerator and denominator of the second right-hand term by 15×10^3, and collect terms:

$$0.225 = \frac{1.08 \times 10^3 - 21 \times 10^6 c_B}{15 \times 10^3} + \frac{(15 \times 10^3)(6.5 \times 10^3 c_B)}{15 \times 10^3}$$
$$0.225 = \frac{1.08 \times 10^3 - 21 \times 10^6 c_B + 97.5 \times 10^6 c_B}{15 \times 10^3}$$
$$3.38 \times 10^3 = 1.08 \times 10^3 - 21 \times 10^6 c_B + 97.5 \times 10^6 c_B$$
$$2.30 \times 10^3 = 76.5 \times 10^6 c_B$$

$$c_B = \frac{2.3 \times 10^3}{76.5 \times 10^6} = \frac{23 \times 10^2}{7.65 \times 10^7} \qquad \boxed{c_B = 3 \times 10^{-5}\ M}$$

Now that c_B is known, c_A may be calculated by substituting the value for c_B into the expression for $A_{350\,nm}$ or $A_{400\,nm}$.

$$0.36 = (15 \times 10^3 c_A) + (7 \times 10^3 c_B) \qquad 0.36 = 15 \times 10^3 c_A + (7 \times 10^3)(3 \times 10^{-5})$$

$$0.36 = 15 \times 10^3 c_A + 0.21 \qquad 0.15 = 15 \times 10^3 c_A$$

$$c_A = \frac{0.15}{15 \times 10^3} = \frac{15 \times 10^{-2}}{15 \times 10^3} \qquad \boxed{c_A = 1 \times 10^{-5}\ M}$$

COUPLED ASSAYS

Many compounds of biological importance do not have a distinct absorption maximum. Nevertheless, their concentrations can be determined if they stoichiometrically promote the formation of another compound that does have a characteristic absorption peak.

· Problem 5-12

To 2.0 ml of a glucose solution, 1.0 ml of solution containing excess ATP, $NADP^+$, $MgCl_2$, hexokinase, and glucose-6-phosphate dehydrogenase was added. The absorbance of the final solution (in a 1 cm cuvette) increased to 0.91 at 340 nm. Calculate the concentration of glucose in the original solution.

Solution

The reactions that take place are shown below.

$$\text{glucose} + \text{ATP} \xrightarrow[\text{Mg}^{2+}]{\text{hexokinase}} \text{glucose-6-phosphate}$$

$$\text{glucose-6-phosphate} + \text{NADP}^+ \xrightarrow{\text{glucose-6-phosphate dehydrogenase}}$$

$$\text{6-phosphogluconic acid-}\delta\text{-lactone} + \text{NADPH} + \text{H}^+$$

Although glucose has no absorption at 340 nm, NADPH does. Because the K_{eq} values of the hexokinase reaction and the glucose-6-phosphate dehydrogenase reaction lie very far to the right, and excess ATP and $NADP^+$ are present, 1 mole of NADPH will be produced for every mole of glucose originally present. From the absorbance at 340 nm, we can calculate the concentration of NADPH present. After correction for dilution, we can calculate the concentration of glucose in the original solution.

$$A = a_m c l$$

$$0.91 = (6.22 \times 10^3)(c)(1)$$

$$c_{NADPH} = \frac{0.91}{6.22 \times 10^3} = \frac{9.1 \times 10^{-1}}{6.22 \times 10^3}$$

$$c_{NADPH} = 1.463 \times 10^{-4}\ M\ \text{NADPH}$$

$$c_{glucose\ orig} = (1.463 \times 10^{-4}) \times \text{dilution factor}$$

$$= (1.463 \times 10^{-4})\tfrac{3}{2}$$

$$\boxed{c_{glucose\ orig} = 2.2 \times 10^{-4}\ M}$$

· **Problem 5-13**

Describe an assay based on light absorption at 340 nm by which the concentrations of glucose, glucose-6-phosphate, glucose-1-phosphate, and fructose-6-phosphate in a mixture may be determined.

Solution

All four compounds can give rise to a stoichiometric amount of NADPH by the reactions shown below.

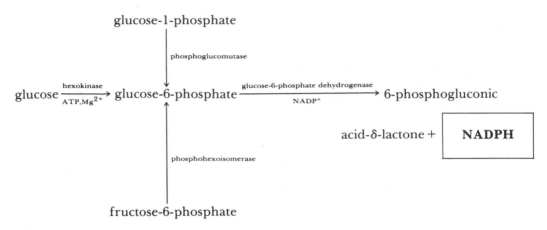

(a)   First add glucose-6-phosphate dehydrogenase and a large excess of NADP$^+$ and MgCl$_2$. The glucose-6-phosphate present is converted to 6-phosphogluconic acid-δ-lactone and a stoichiometric amount of NADPH appears. Measure the absorbance at 340 nm.

(b)   When no further increase in A occurs, add phosphoglucomutase. This enzyme catalyzes the conversion of the glucose-1-phosphate to glucose-6-phosphate that produces more NADPH in an amount equivalent to the amount of glucose-1-phosphate originally present. Measure the increase in $A_{340\,nm}$.

(c)   When no further increase in absorbance occurs, add phosphohexoisomerase. This enzyme catalyzes the conversion of the fructose-6-phosphate to glucose-6-phosphate that yields another increment of NADPH. Measure the increase in $A_{340\,nm}$.

(d)   Finally, add hexokinase and ATP. The glucose present yields a stoichiometric amount of NADPH as described previously. Again measure the increase in $A_{340\,nm}$.

   Although the glucose-6-phosphate dehydrogenase, phosphoglucomutase, and phosphohexoisomerase reactions do not have large K_{eq} values, the overall conversion to 6-phosphogluconic acid-δ-lactone and NADPH can be forced far to the right by using a large excess of NADP$^+$. Furthermore, if any of the enzymes are contaminated with 6-phosphogluconolactonase, the overall reaction sequence becomes irreversible. The $A_{340\,nm}$ time course of the assay is shown in Figure 5-2.

   In calculating the ΔNADPH concentrations at any point from the $\Delta A_{340\,nm}$ values corrections must be made for the dilution of any preexisting NADPH. Also, in calculating the concentrations of glucose, glucose-6-

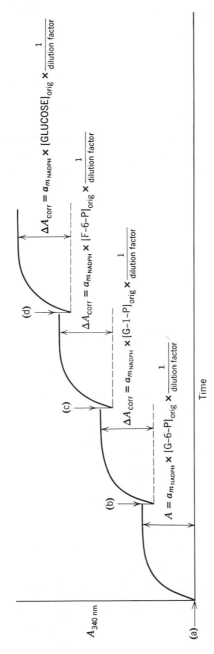

Figure 5-2 Absorbance changes during the NADPH-linked assay of glucose and derivatives.

343

phosphate, and the like, in the original solution from the ΔNADPH values, the total dilution of the assay mixture must be taken into account.

If the enzymes are delivered in very small volumes compared to the total assay volume, the dilution will be negligible. For example, the assay volume might be 1.0 or 3.0 ml (usual cuvette sizes) while the assay enzymes might be added in 10 μl volumes.

· Problem 5-14

To 1.0 ml of a solution containing a mixture of glucose-6-phosphate and glucose-1-phosphate, 1.0 ml of a solution containing excess $NADP^+$, $MgCl_2$, and glucose-6-phosphate dehydrogenase was added. The absorbance in a 1 cm cuvette increased to 0.57 at 340 nm. When no further increase in absorbance was observed, an additional 1.0 ml of a phosphoglucomutase solution was added. The absorbance then decreased to 0.50. Calculate the (a) glucose-6-phosphate and (b) glucose-1-phosphate concentrations in the original solution.

Solution

(a)
$$A = a_m \times c_{\text{NADPH}} \times l \qquad 0.57 = (6.22 \times 10^3)(c_{\text{NADPH}})(1)$$

$$c_{\text{NADPH}} = \frac{0.57}{6.22 \times 10^3} = \frac{57 \times 10^{-2}}{6.22 \times 10^3} = 9.16 \times 10^{-5} \, M$$

$$c_{\text{G-6-P}_{\text{orig}}} = (9.16 \times 10^{-5}) \times \text{dilution factor} = (9.16 \times 10^{-5})(2)$$

$$\boxed{c_{\text{G-6-P}_{\text{orig}}} = 1.83 \times 10^{-4} \, M}$$

(b)
$$\Delta A_{\text{corrected}} = A_{\text{final}} - A_{\text{orig, corrected}}$$

$$A_{\text{orig, corrected}} = (0.57)\tfrac{2}{3} = 0.38$$

$$\Delta A_{\text{corrected}} = 0.50 - 0.38 = 0.12$$

$$\Delta A = a_m \times \Delta c_{\text{NADPH}} \times l$$

$$0.12 = (6.22 \times 10^3) \times (\Delta c_{\text{NADPH}}) \times (1)$$

$$\Delta c_{\text{NADPH}} = \frac{0.12}{6.22 \times 10^3} = \frac{12 \times 10^{-2}}{6.22 \times 10^3}$$

$$\Delta c_{\text{NADPH}} = 1.93 \times 10^{-5} \, M$$

$$c_{\text{G-1-P}_{\text{orig}}} = 1.93 \times 10^{-5} \times \text{dilution factor} = (1.93 \times 10^{-5})(3)$$

$$\boxed{c_{\text{G-1-P}_{\text{orig}}} = 5.79 \times 10^{-5} \, M}$$

ENZYME ASSAYS

Enzymes can be assayed spectrophotometrically by following the rate at which a product appears or a substrate disappears. If neither substrate nor product has a distinct absorption peak, then it is often possible to couple the reaction of interest to another that does yield a light-absorbing product. In such coupled assays, it is imperative that all auxiliary enzymes be present in excess so that the overall rate depends solely on the activity of the enzyme being assayed. If the optimum conditions of the enzyme being assayed are inconsistent with the optimum conditions for production of the final light-absorbing product, then the overall assay can be run in two stages. The problems below illustrate the continuous and two-stage assay methods.

· **Problem 5-15**

Glutamic-oxalacetate transaminase (GOT) is released into the blood stream as a result of myocardial infarction. The enzyme is assayed in serum by following the decrease in the absorbance of NADH in the malic dehydrogenase (MDH)-coupled reaction sequence shown below.

$$\text{aspartate} + \alpha\text{-ketoglutarate} \xrightleftharpoons{\text{GOT}} \text{glutamate} + \text{oxalacetate}$$

$$+$$
$$\text{NADH}$$
$$\text{MDH} \downarrow\uparrow$$
$$\text{malate}$$
$$+$$
$$\text{NAD}^+$$

A reaction mixture contained excess aspartate (i.e., 100 times its K_m value), 0.1 ml of serum, 0.3 μmole of NADH, and an excess of malic dehydrogenase in a total volume of 0.9 ml. The reaction was started by adding an excess of α-ketoglutarate in 0.1 ml. After a short lag, the absorbance decreased at a rate of 0.04 A unit/min. The cuvette had a light path of 1 cm. Calculate the concentration of GOT in the patient's serum (i.e., the specific activity of the serum in terms of enzyme units/ml).

Solution

$$v = \Delta A / \text{min} = 0.04/\text{min}$$

$$\Delta A = (a_m)(\Delta c)(l) \qquad \text{or} \qquad \Delta c = \frac{\Delta A}{(a_m)(l)}$$

$$\Delta c = \frac{(0.04)}{(6.22 \times 10^3)(1)} = 6.43 \times 10^{-6} \, M = 6.43 \, \mu\text{moles/liter}$$

$$v = 6.43 \, \mu\text{moles} \times \text{liter}^{-1} \times \text{min}^{-1} = 6.43 \times 10^{-3} \, \mu\text{moles} \times \text{ml}^{-1} \times \text{min}^{-1}$$

Thus, the 1.0 ml cuvette contained 6.43×10^{-3} units of enzyme activity. This activity came from 0.1 ml of serum. Therefore, the serum contains:

$$\frac{6.43 \times 10^{-3} \text{ units}}{0.1 \text{ ml}} = \boxed{\mathbf{6.43 \times 10^{-2} \text{ units/ml}}}$$

· **Problem 5-16**

A cell-free extract of *Penicillium chrysogenum* was assayed for β-galactosidase activity. The extract (0.25 ml) was incubated with $3 \times 10^{-3} \, M$ p-nitrophenyl-β-galactoside and buffer in a total volume of 1.0 ml. Periodically, 0.1 ml aliquots were removed and added to 2.9 ml of 0.1 N NaOH. The concentration of free p-nitrophenol was determined by measuring the absorbance at 400 nm against a p-nitrophenyl-β-galactoside + buffer + NaOH blank. (a_m of p-nitrophenol in 0.1 N NaOH is 18,300.) The cell-free extract contained 5 mg protein/ml. The absorbance of the NaOH solution in a 1 cm cuvette is shown below.

Incubation time (minutes)	$A_{400 \, nm}$
2	0.09
4	0.18
6	0.27

Calculate the specific β-galactosidase activity of the cell-free extract.

Solution

$$v = \Delta A / \min = 0.045/\min$$

$$\Delta A = (a_m)(\Delta c)(l) \quad \text{or} \quad \Delta c = \frac{\Delta A}{(a_m)(l)}$$

$$\Delta c = \frac{0.045}{(18.3 \times 10^3)(1)} = 2.46 \times 10^{-6} \, M \times \min^{-1}$$

Thus, the concentration of p-nitrophenol in the NaOH solution increased at a rate of $2.46 \times 10^{-6} \, M/\min = 2.46 \times 10^{-3} \, \mu\text{mole} \times \text{ml}^{-1} \times \min^{-1}$. The volume of the NaOH solution is 3.0 ml. $\therefore$ $(3)(2.46 \times 10^{-3}) = 7.38 \times 10^{-3} \, \mu\text{mole}$ of p-nitrophenol produced/min. The $7.38 \times 10^{-3} \, \mu\text{mole}$ were produced each minute in 0.1 ml of assay mixture. In the assay mixture:

$$v = 7.38 \times 10^{-2} \, \mu\text{mole} \times \text{ml}^{-1} \times \min^{-1}$$

The 1.0 ml of assay mixture contained 0.25 ml of cell-free extract. The protein concentration in the assay was:

$$(0.25 \, \text{ml/ml})(5 \, \text{mg protein/ml}) = 1.25 \, \text{mg/ml}$$

$$\therefore \quad \text{S.A.} = \frac{7.38 \times 10^{-2}}{1.25} = \boxed{\textbf{0.059 units/mg protein}}$$

Note that by 6 min, $(7.38 \times 10^{-2})(6) = 0.4428 \, \mu\text{moles}$ of substrate had been utilized. This represents 14.8% of the initial substrate concentration. The fact that the appearance of product is still linear with time indicates that [S] remains $\gg K_m$ and $v = V_{\max}$.

B. FLUOROMETRY

Many compounds absorb light and then immediately reemit some of the energy as light of a longer wavelength. This *fluorescence* phenomenon can be used with an instrument called a fluorometer to measure very low concentrations of certain compounds. A fluorometer differs from a spectrophotometer in that (a) the emitted fluorescence light is observed at 90° to the incident light and (b) two wavelength selectors are required—one to transmit the desired excitation λ and one to select the desired emission λ. Usually filters (singly or in combination) are used to select the desired wavelengths.

Fluorometry can be extremely selective since only certain wavelengths of light will excite a given compound. Similarly, the fluorescence will occur only at certain wavelengths. In other words, fluorescent compounds have a characteristic excitation spectrum and a characteristic fluorescence spectrum. Two compounds with sufficiently different excitation and/or fluorescence spectra may be determined in the presence of each other in much the same manner as described earlier for spectrophotometry. A compound that does not fluoresce can often be chemically or enzymatically converted to another that does fluoresce.

At low concentrations of a fluorescent compound, the intensity of the fluorescence is directly proportional to concentration. Thus, fluorescence intensity is analogous to absorbance in spectrophotometry, (the linear parameter) not transmission (the logarithmic parameter). The fluorescence emitted by one substance may be absorbed or *quenched* by other substances in the sample. For this reason it is necessary to include an internal ("recovery") standard in each assay. For example, 25 μg of pure compound X might yield a fluorescence intensity of 53 arbitrary units all by itself. However, if 25 μg of pure compound X is added to the sample (which might be diluted urine, or serum, or an enzyme assay mixture containing organic buffers) we may observe an increase of only 29 units above that of the sample itself. Clearly, only 29/53 = 54.7% of the standard's fluorescence is observed under the assay conditions. Thus, under the assay conditions, 25 μg of X is equivalent to 29 units of fluorescence (not 53 units). It is also necessary to subtract from the readings the fluorescence of a blank. The blank should contain the same substances as the sample. For example, if the sample is treated chemically to induce fluorescence, the blank should contain the same chemicals, although added in an order that does not convert the compound being measured to a fluorescent product. The readings can be diagrammed as shown below.

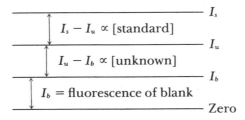

where I_b = fluorescence of blank
             I_u = fluorescence of unknown plus blank
             I_s = fluorescence of internal standard plus unknown plus blank
∴ $I_u - I_b$ = fluorescence of unknown under assay conditions
          $I_s - I_u$ = fluorescence of standard under assay conditions

and $\dfrac{\text{fluorescence of unknown}}{\text{fluorescence of standard}} = \dfrac{\text{amount of unknown}}{\text{amount of standard}}$

or $\boxed{\dfrac{I_u - I_b}{I_s - I_u} = \dfrac{\textbf{amount of unknown}}{\textbf{amount of standard}}}$ (6)

· **Problem 5-17**

Catecholamines (epinephrine and norepinephrine) can be converted to fluorescent compounds (called lutines) by oxidation and treatment with alkali. The concentration of total catecholamine can then be determined by excitation at 405 nm and fluorescence intensity measurements at 495 nm. (These two wavelengths do not discriminate between epinephrine and norepinephrine.) The following data were recorded:

I_u = fluorescence intensity of 1.5 ml of urine, treated
 to induce fluorescence of catecholamines
 = 52 units

I_b = fluorescence intensity of 1.5 ml of urine treated
 with ferricyanide, NaOH, and the other reag-
 ents in the wrong order (so as not to convert
 catecholamines to fluorescent products)
 = 6 units

I_s = fluorescence intensity of 1.5 ml of treated urine
 to which 0.25 μg of epinephrine standard had
 been added
 = 85

Calculate (a) the concentration of catecholamines in the urine sample and (b) the total daily catecholamine excretion if 900 ml of urine were collected in 24 hours.

Solution

(a)   Amount of catecholamines in sample $= \dfrac{I_u - I_b}{I_s - I_u} \times 0.25 \ \mu$g

$$= \frac{52 - 6}{85 - 52} \times 0.25 = \frac{(46)(0.25)}{(33)}$$

> **Amount of catecholamines = 0.348 μg**

$$\text{conc} = \frac{0.348 \ \mu\text{g}}{1.5 \ \text{ml}} = \boxed{\textbf{0.232 } \mu\textbf{g/ml}}$$

(b)   Total excretion $= (0.232 \ \mu\text{g/ml})(900 \ \text{ml}) = \boxed{\textbf{209 } \mu\textbf{g/day}}$

C. OPTICAL ROTATION—POLARIMETRY

Optical rotation refers to the ability of certain compounds (solid or in solution) to rotate the plane of polarized light. Such "optically active" compounds contain at least one asymmetric center and no plane of symmetry. The optical rotation of a solution at a given temperature and wavelength is given by:

$$A° = [\alpha]_{\lambda}^{T} \times C \times l \qquad (7)$$

where $A°$ = the observed rotation in degrees

$[\alpha]_{\lambda}^{T}$ = the specific rotation of the compound in solution at a fixed temperature (generally 20 or 25°C) and wavelength (generally the D line of sodium, 5893 Å)

C = the concentration of the solution in g/ml

l = the light path length (through the solution) in decimeters (dm)

Note that Equation 7 is analogous to Equation 3 for absorbance. The relationship is also frequently given as:

$$[\alpha]_{\lambda}^{T} = \frac{A° \times 100}{l \times C} \qquad (8)$$

where now C = concentration in % w/v (i.e., g/100 ml). The "molar rotation," $[\alpha]_{M}$, is $[\alpha] \times$ MW.

Optical rotation is measured with an instrument called a polarimeter. The essential parts of a polarimeter are shown in Figure 5-3.

The polarizing and analyzing lenses both transmit light that is plane polarized. In the absence of an optically active sample, the analyzing lens can be adjusted so that the light intensity as seen by the viewer is minimal. (This is accomplished by rotating the analyzing lens until its transmission plane is perpendicular to the transmission plane of the polarizing lens.) An optically active sample rotates the plane of the polarized light. The analyzing lens then must be rotated to a new position in order to minimize the light intensity. The angle through which the analyzing lens is

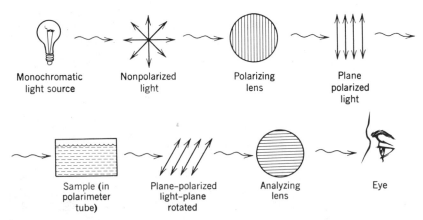

Monochromatic   Nonpolarized   Polarizing   Plane
light source   light   lens   polarized
                                                           light

Sample (in   Plane-polarized   Analyzing   Eye
polarimeter   light-plane   lens
tube)   rotated

Figure 5-3 Essential components of a polarimeter. Polarized light vibrates in a single plane perpendicular to the path of propagation. Nonpolarized light vibrates in an infinite number of planes, all of which are perpendicular to the path of propagation.

rotated is the observed rotation (in degrees) from which $[\alpha]$ may be calculated. If the analyzing lens is rotated clockwise, the substance is said to be dextrorotary (d or $+$). If the analyzing lens is rotated counterclockwise, the substance is said to be levorotary (l or $-$). The symbols d or l are not to be confused with the symbols D or L.

Optical rotation can be used to identify unknown compounds, to determine concentrations of known compounds, and to follow the course of a reaction where the substrate and product have different specific rotations. Optical rotary dispersion measurements (i.e., the optical rotation at several different wavelengths) can provide information concerning the structure and asymmetry of a compound (e.g., percent α-helix content of a protein).

· Problem 5-18

A solution of L-leucine (3.0 g/50 ml of 6 N HCl) had an observed rotation of $+1.81°$ in a 20 cm polarimeter tube. Calculate (a) the specific rotation, $[\alpha]$, and (b) the molar rotation, $[\alpha]_M$, of L-leucine in 6 N HCl.

Solution

(a) $[\alpha] = \dfrac{A°}{l_{dm} \times C_{g/ml}}$ $\qquad l = 20 \text{ cm} = 2 \text{ dm} \qquad C = 3 \text{ g/50 ml} = 0.06 \text{ g/ml}$

$\qquad [\alpha] = \dfrac{+1.81}{2 \times 0.06}$ $\quad$ or $\quad \boxed{[\alpha] = +15.1°}$

(b) $\qquad\qquad [\alpha]_M = [\alpha] \times MW = (+15.1°)(131.2)$

$$\boxed{[\alpha]_M = +1980°}$$

· Problem 5-19

A solution of L-arabinose (containing an equilibrium mixture of α and β forms) has an observed rotation of $+23.7°$ in a 10 cm polarimeter tube at 25°C. Calculate the concentration of L-arabinose in the solution. The $[\alpha]_D^{25}$ for an equilibrium mixture of α- and β-L-arabinose is $+105°$.

Solution

$$A° = [\alpha] \times l \times C \qquad C = \frac{A°}{[\alpha] \times l} = \frac{23.7}{105}$$

$$\boxed{C = 0.225 \text{ g/ml}}$$

· Problem 5-20

An equilibrium mixture of α- and β-D-glucose has an $[\alpha]_D^{25}$ of $+52.7°$. Pure α-D-glucose has an $[\alpha]_D^{25}$ of $+112°$. Pure β-D-glucose has an $[\alpha]_D^{25}$ of $+18.7°$. Calculate the proportions of α- and β-D-glucose in the equilibrium mixture.

Solution

The contribution of each anomer toward the total $A°$ or $[\alpha]_D^{25}$ is directly proportional to the concentrations of each anomer present. Let:

$$X = \% \beta \qquad \therefore \quad (100 - X) = \% \alpha$$

$$\therefore \quad (+18.7)(X) + (+112)(100 - X) = (100)(+52.7)$$

$$18.7X + 11{,}200 - 112X = 5270$$

$$5930 = 93.3X$$

$$X = \frac{5930}{93.3} = \% \beta \qquad \boxed{\beta = 63.5\%}$$

$$\therefore \quad \alpha = (100 - 63.5)\% \qquad \boxed{\alpha = 36.5\%}$$

· **Problem 5-21**

Thirty grams of a polysaccharide containing only D-mannose and D-glucose were acid hydrolyzed. The hydrolysate was diluted to 100 ml. The observed rotation of the solution was $+9.07°$ in a 10 cm polarimeter tube. Calculate the ratio of D-mannose/D-glucose in the polysaccharide. The specific rotations of α/β-D-glucose and α/β-D-mannose are $+52.7$ and $+14.5°$, respectively.

Solution

After acid hydrolysis, the two sugars are present as the equilibrium mixture of their α and β forms. Furthermore, the total weight of monosaccharides after hydrolysis is 33.3 g as a result of adding 1 mole of water (18 g) per mole of monosaccharide residue (162 g).

Let

$$X = \text{g/ml D-glucose present}$$

$$\therefore \quad (0.333 - X) = \text{g/ml D-mannose present}$$

$$A^°_{\text{glucose}} = [\alpha](l)(C) = (52.7)(1)(X) = 52.7X$$

$$A^°_{\text{mannose}} = [\alpha](l)(C) = (14.5)(1)(0.333 - X) = 4.83 - 14.5X$$

$$A^°_{\text{glucose}} + A^°_{\text{mannose}} = A^°_{\text{total}}$$

$$(52.7X) + (4.83 - 14.5X) = 9.07$$

$$38.2X = 4.24$$

$$X = \text{g/ml D-glucose} = \frac{4.24}{38.2}$$

$$\boxed{\text{D-glucose} = 0.111 \text{ g/ml}}$$

$$\therefore \quad \text{D-mannose} = 0.333 - 0.111 \text{ g/ml}$$

$$\boxed{\text{D-mannose} = 0.222 \text{ g/ml}}$$

Because mannose and glucose have the same molecular weights, the molar ratio of the two sugars is the same as the weight ratio.

$$\text{D-mannose/D-glucose} = 2$$

PRACTICE PROBLEMS

Answers to Practice Problems are given on page 429.

1. Calculate the absorbance and the transmission at 260 nm and 340 nm of the following solutions in a 1 cm cuvette: (a) $2.2 \times 10^{-5} M$ NADH, (b) $7 \times 10^{-6} M$ NADH plus $4.2 \times 10^{-5} M$ ATP. The a_m of NADH is 15,000 $M^{-1} \times cm^{-1}$ at 260 nm and 6220 at 340 nm. The a_m of ATP is 15,400 $M^{-1} \times cm^{-1}$ at 260 nm and zero at 340 nm.

2. Calculate the concentrations of ATP and NADPH in solutions with absorbances (in a 1 cm cuvette) of (a) 0.15 at 340 nm and 0.90 at 260 nm, (b) zero at 340 nm and 0.75 at 260 nm, and (c) 0.22 at 340 nm and 0.531 at 260 nm. The a_m values of NADPH at the two wavelengths are the same as those of NADH (problem 1).

3. Calculate the concentrations of two absorbing compounds, A and B, if the absorbance of this solution in a 3 cm cuvette is 0.62 at 450 nm and 0.54 at 485 nm. Compound A has an a_m of 12,000 $M^{-1} \times cm^{-1}$ at 450 nm and 4000 $M^{-1} \times cm^{-1}$ at 485 nm. Compound B has an a_m of 5000 $M^{-1} \times cm^{-1}$ at 450 nm and 11,600 $M^{-1} \times cm^{-1}$ at 485 nm.

4. A standard solution of bovine serum albumin containing 1.0 mg/ml had an absorbance of 0.58 at 280 nm. (a) What is the protein concentration in a partially purified enzyme preparation if the absorbance is 0.12 at 280 nm? (b) Why might the calculated value be in error even if the preparation is free of nucleic acids?

5. The preparation described in practice problem 4 was diluted fivefold and the absorbance at 215 nm and at 225 nm was determined. The ΔA value was 0.31. (a) What is the protein concentration of the undiluted preparation? (b) Is this value more reliable than that obtained from the absorbance at 280 nm?

6. Suppose that 1.5 ml of a $2 \times 10^{-4} M$ solution of NADPH were added to 1.5 ml of a solution containing an unknown concentration of oxidized glutathione and a catalytic amount of the enzyme glutathione reductase. The final absorbance of the solution at 340 nm was 0.25 in a 1 cm cuvette. Calculate the concentration of oxidized glutathione in the original 1.5 ml. The reaction catalyzed by glutathione reductase is GSSG + NADPH + $H^+ \rightarrow 2GSH + NADP^+$ and goes essentially to completion.

7. Devise a spectrophotometric assay based on light absorption at 340 nm by which the concentrations of fructose-1,6-diphosphate, glyceraldehyde-3-phosphate, and dihydroxyacetone phosphate in a mixture may be determined.

8. Calculate the concentrations of citric acid and isocitric acid in a mixture, given the following information: (a) After the addition of 1.5 ml of solution containing excess NAD^+ and the enzyme isocitric dehydrogenase to 2.0 ml of the original solution, the absorbance at 340 nm in a 1 cm cuvette increased to 0.48. (b) After an additional 3.5 ml containing the enzyme aconitase were added, the absorbance remained constant at 0.48.

9. A commercial sample of adenosine-5'-phosphosulfate (APS) is known to be contaminated with 5'-AMP. A solution of the APS that had an $A_{260\,nm}$ of 0.90 was prepared. Exactly 0.9 ml of this solution was mixed with 0.1 ml of a solution containing excess inorganic pyrophosphate, glucose, $NADP^+$, Mg^{2+}, and the enzymes ATP sulfurylase, hexokinase, and glucose-6-phosphate dehydrogenase. The $A_{340\,nm}$ increased to 0.262. Calculate the purity of the APS sample. (Assume that APS and AMP have a_m values of

15,400 $M^{-1} \times cm^{-1}$ at 260 nm.) ATP sulfurylase catalyzes the reaction $APS + PP_i \rightleftharpoons SO_4^{2-} + ATP$ with a K_{eq} of 10^8 as written.

10. If the extract described in problem 5-16 is reassayed using o-nitrophenyl-β-galactoside, what would the $A_{420 nm}$ of the NaOH solution be for the 2-min sample? The a_m of o-nitrophenol at 420 nm (in base) is 21,300 $M^{-1} \times cm^{-1}$. (Assume that the velocity with o-nitrophenyl-β-galactoside is the same as that with p-nitrophenyl-β-galactoside.)

11. Lactate dehydrogenase was assayed by following the appearance of NADH spectrophotometrically in a 1 cm cuvette. The assay mixture (3.0 ml total volume) contained excess lactate, buffer, 0.1 ml of enzyme preparation, and semicarbazide (to trap the pyruvate and pull the reaction to completion). The enzyme preparation contained 120 μg protein/ml. The $A_{340 nm}$ increased at a rate of 0.048/min. (a) What is the lactate dehydrogenase content of the preparation (units/ml)? (b) What is the specific activity of lactate dehydrogenase in the preparation (units/mg protein)?

12. The cardiologist mentioned in problem 4-24 decides to check the activity of the commercial glycerol kinase used for the glycerol assay. An assay mixture is prepared containing excess MgATP, PEP, pyruvic kinase, lactate dehydrogenase, glycerol, and $3 \times 10^{-4} M$ NADH in a total volume of 0.9 ml. The reaction is started by adding 0.1 ml of glycerol kinase solution. If the specific activity of the enzyme is as it should be, the 0.1 ml should have added 0.03 units of activity. After a short lag, the absorbance at 365 nm decreased at a rate of 0.08/min. (The assay was performed at 365 nm rather than 340 nm in order to keep the absorbance readings below 1.0. The a_m of NADH at 365 nm is $3.11 \times 10^3 M^{-1} \times cm^{-1}$.) Does the stock glycerol kinase solution really contain 0.3 units/ml?

13. Lead poisoning causes an increased daily excretion of coproporphyrin, which can be measured fluorometrically in urine (excitation at 405 nm, fluorescence measurements at 595 nm). A 0.5 ml sample of urine from a suspected lead poisoning case was processed appropriately and diluted to 25 ml for fluorescence measurements. The following data were recorded: $I_u = 40$, $I_b = 5$, $I_s = 80$. The internal standard added 1.25 μg of coproporphyrin to the 25 ml assay volume. The total volume of urine excreted per 24 hours was 800 ml. What is the daily coproporphyrin excretion of the patient? (Normal values are 75 to 300 μg/day.) Keep in mind that I_u refers to fluorescence of the unknown plus that of the blank; I_s refers to the fluorescence of the internal standard plus that of the unknown and the blank.

14. A solution of D-histidine (4 g/100 ml of 1 M HCl) had an observed rotation of $-0.41°$ in a 20 cm polarimeter tube. Calculate (a) the specific rotation $[\alpha]$ and (b) the molar rotation $[\alpha]_M$ of D-histidine in 1 M HCl.

15. A solution of L-ribulose (containing an equilibrium mixture of α and β forms) has an observed rotation of $-3.75°$ in a 10 cm polarimeter tube. Calculate the concentration of L-ribulose in the solution. The $[\alpha]_D^{25}$ for an equilibrium mixture of α- and β-L-ribulose is $-16.6°$.

16. An equilibrium mixture of α- and β-D-mannose has an $[\alpha]_D^{25}$ of $+14.5°$. Pure α-D-mannose has an $[\alpha]_D^{25}$ of $+29.3°$. Pure β-D-mannose has an $[\alpha]_D^{25}$ of $-16.3°$. Calculate the proportion of α- and β-D-mannose in the equilibrium mixture.

17. The $[\alpha]_D^{25}$ of α-D-mannose is $+29.30°$. The $[\alpha]_D^{25}$ of β-D-mannose is $-16.30°$. A freshly prepared solution of α-D-mannose had an observed rotation of $+14.65°$ in a 10 cm polarimeter tube. After 10 min, the observed rotation decreased to $+11.0°$. Calculate the overall net rate of mutarotation.

6

ISOTOPES IN BIOCHEMISTRY

A. ISOTOPES AND RADIOACTIVE DECAY

ISOTOPES

"Isotopes" are atoms that contain the same number of protons (have the same atomic number) but different numbers of neutrons (have different atomic weights). Most naturally occurring elements exist as mixtures of isotopes. For example, magnesium exists as Mg^{24}, Mg^{25}, and Mg^{26}, which account for about 78.6, 10.11, and 11.29%, respectively, of the total magnesium in nature. Because of the mass distribution, the weighted average atomic weight of magnesium is 24.31. The chemical properties of an element are determined by its atomic number, not its atomic weight. Consequently, all three isotopes of magnesium react identically. "Isotope effects" caused by the slight mass differences are generally negligible in biochemical studies. Both radioactive and stable isotopes are used in biological research.

MODES OF RADIOACTIVE DECAY

Many of the naturally occurring and man-made isotopes are unstable. The nuclei of these isotopes decay to more stable forms by one or more of the processes shown in Table 6-1. Such isotopes are called "radioactive

Table 6-1 Modes of Radioactive Decay

Decay Process	Nuclear Transformation	Net Equation
Beta particle emission	$_0n^1 \rightarrow {_{+1}}p^1 + {_{-1}}\beta^0$	$_{AN}I^{AW} \rightarrow {_{AN+1}}I^{AW} + {_{-1}}\beta^0$
Positron particle emission	$_{+1}p^1 \rightarrow {_0}n^1 + {_{+1}}\beta^0$	$_{AN}I^{AW} \rightarrow {_{AN-1}}I^{AW} + {_{+1}}\beta^0$
Alpha particle emission	Loss of $_{+2}He^4(\alpha)$	$_{AN}I^{AW} \rightarrow {_{AN-2}}I^{AW-4} + {_{+2}}He^4$
Electron capture (EC)	$_{+1}p^1 + {_{-1}}e^0 \rightarrow {_0}n^1$	$_{AN}I^{AW} + {_{-1}}e^0 \rightarrow {_{AN-1}}I^{AW} + \gamma$
Particle emission followed by isomeric transition of still unstable nucleus		$_{AN}^{*}I^{AW} \rightarrow {_{AN}}I^{AW} + \gamma$

isotopes." Because of the decay process many isotopes have long since disappeared from nature. For example, Mg^{23} and Mg^{27} no longer constitute a significant proportion of the magnesium in nature, although they can be produced by appropriate nuclear reactions for use in research.

Most radioisotopes used in biochemical studies are beta and/or gamma emitters (Appendix XII).

· Problem 6-1

C^{14}, P^{32}, S^{35}, and H^3 are radioisotopes commonly used in biological research. All are β emitters. Write the nuclear reactions by which they decay.

Solution

In all four cases, the ejection of a β particle will result in a stable isotope with an atomic number one higher than the original radioisotope and an atomic weight identical to the parent.

$$_6C^{14} \rightarrow {_{-1}}\beta^0 + {_7}N^{14}$$
$$_{15}P^{32} \rightarrow {_{-1}}\beta^0 + {_{16}}S^{32}$$
$$_{16}S^{35} \rightarrow {_{-1}}\beta^0 + {_{17}}Cl^{35}$$
$$_1H^3 \rightarrow {_{-1}}\beta^0 + {_2}He^3$$

EQUATIONS OF RADIOACTIVE DECAY

The decay of radioactive isotopes is a simple exponential (first-order) process.

$$-\frac{dN}{dt} = \lambda N \tag{1}$$

where $-\dfrac{dN}{dt}$ = the number of atoms decaying per small increment of time (i.e., the count rate)

N = the total number of radioactive atoms present at any given time

λ = a decay constant, different for each isotope

The negative sign indicates that the number of radioactive atoms *decreases* with time.

Although λ is a proportionality constant, we can see its physical significance by rearranging the above equation.

$$-\frac{dN}{dt} = \lambda N \qquad \therefore \quad \lambda = -\frac{dN}{N\,dt}$$

$$\lambda = \frac{-dN/N}{dt} \tag{2}$$

In other words, λ is the fraction of the radioactive atoms that decays per small increment of time.

The differential decay equation may be integrated to obtain a far more useful relationship.

$$-\frac{dN}{dt} = \lambda N \qquad \frac{dN}{N} = -\lambda\,dt \qquad \int\frac{dN}{N} = -\lambda\int dt$$

Integrating between the limits of N_0 (the original number of radioactive atoms) and N (the number of radioactive atoms at any other time) and between the limits of zero time and any other time:

$$\int_{N_0}^{N}\frac{dN}{N} = -\lambda\int_{t=0}^{t} dt \qquad \ln\frac{N}{N_0} = -\lambda t$$

$$\boxed{\ln\frac{N_0}{N} = \lambda t}$$

or $$\boxed{2.3\log\frac{N_0}{N} = \lambda t}$$ or $$\boxed{N = N_0 e^{-\lambda t}} \qquad (3)$$

or, in linear form, $$\boxed{\log N = -\frac{\lambda}{2.3}\,t + \log N_0} \qquad (4)$$

N_0 and N can be expressed in any consistent manner. For example, $N_0 = 100\%$, $N = \%$ remaining after time interval t; $N_0 = 1.00$, $N =$ fraction remaining (as a decimal) after time interval t; $N_0 =$ original CPM in sample, $N =$ CPM remaining after time interval t; $N_0 =$ S.A. of sample at a certain time, $N =$ S.A. of the sample after an elapsed time, t.

HALF-LIFE

The "half-life" ($t_{1/2}$) of a radioactive isotope is the time required for half of the original number of atoms to decay. The relationship between $t_{1/2}$ and λ is shown below.

$$\ln\frac{N_0}{N} = \lambda t \qquad\qquad\text{or}\qquad\qquad 2.303\log\frac{N_0}{N} = \lambda t$$

$$\ln\frac{1}{0.5} = \lambda t_{1/2} \qquad\qquad\qquad 2.303\log\frac{1}{0.5} = \lambda t_{1/2}$$

$$\ln 2 = \lambda t_{1/2} \qquad\qquad\qquad 2.303\log 2 = \lambda t_{1/2}$$

$$0.693 = \lambda t_{1/2} \qquad\qquad\qquad (2.303)(0.301) = \lambda t_{1/2}$$

$$0.693 = \lambda t_{1/2}$$

$$\boxed{\lambda = \frac{0.693}{t_{1/2}} \qquad\text{or}\qquad t_{1/2} = \frac{0.693}{\lambda}} \qquad (5)$$

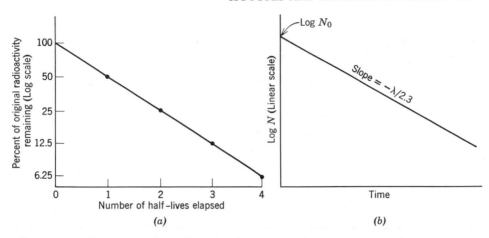

Figure 6-1   The decay of a radioactive isotope is a first-order process.

The amount of radioactivity remaining in a sample can easily be determined by constructing a semilog plot as shown in Figure 6-1.   A single point (50% at one half-life) is sufficient to construct the curve.

THE CURIE

The curie, abbreviated Ci, is a standard unit of radioactive decay.   It was originally defined as the rate at which 1 g of radium^{226} decays.   Because of the relatively long half-life of Ra^{226}, the isotope served as a convenient standard.   The curie is now defined as the quantity of any radioactive substance in which the decay rate is 3.700×10^{10} disintegrations per second (2.22×10^{12} DPM).   Because the efficiency of most radiation detection devices is less than 100%, a given number of curies almost always yields a lower than theoretical count rate.   Hence, there is the distinction between DPM and CPM.   For example, a sample containing 1 μCi of radioactive material has a decay rate of 2.22×10^6 DPM.   If only 30% of the disintegrations are detected, the observed count rate is 6.66×10^5 CPM.

· Problem 6-2

Ca^{45} has a half-life of 163 days.   Calculate (a) the decay constant (λ) in terms of day^{-1} and sec^{-1}, and (b) the percent of the initial radioactivity remaining in a sample after 90 days.

Solution

$$\lambda = \frac{0.693}{t_{1/2}} = \frac{0.693}{163 \text{ days}} = \frac{6.93 \times 10^{-1}}{1.63 \times 10^2} \text{ day}^{-1}$$

$$\boxed{\lambda = 4.25 \times 10^{-3} \text{ day}^{-1}}$$

$$\lambda = \frac{0.693}{163 \text{ days} \times 24 \text{ hr/day} \times 60 \text{ min/hr} \times 60 \text{ sec/min}}$$

$$= \frac{0.693}{163 \text{ days} \times 86{,}400 \text{ sec/day}} = \frac{0.693}{(1.63 \times 10^2)(8.64 \times 10^4)}$$

$$= \frac{69.3 \times 10^{-2}}{14.1 \times 10^6} \text{ sec}^{-1} \qquad \boxed{\lambda = 4.92 \times 10^{-8} \text{ sec}^{-1}}$$

(b) $$2.3 \log \frac{N_0}{N} = \lambda t$$

Let $N_0 = 100\%$.

$$2.3 \frac{100}{N} = (4.26 \times 10^{-3})(90) = 0.3834$$

$$\log \frac{100}{N} = \frac{0.3834}{2.3} = 0.167$$

$$\log 100 - \log N = 0.0167$$

$$\log N = 2.000 - 0.167 = 1.833$$

$$\boxed{N = 68.1\%}$$

· **Problem 6-3**

C^{14} has a half-life of 5700 years. Calculate the fraction of the C^{14} atoms that decays (a) per year (b) per minute.

Solution

(a) Calculate λ:

$$\lambda = \frac{0.693}{5700 \text{ yr}} = \frac{6.93 \times 10^{-1}}{5.7 \times 10^3} \text{ yr}^{-1} \qquad \boxed{\lambda = 1.216 \times 10^{-4} \text{ yr}^{-1}}$$

That is, 1.216×10^{-4} atoms per atom decays per year or 1 atom out of $1/1.216 \times 10^{-4}$ atoms decays per year.

$$\frac{1}{1.216 \times 10^{-4}} = 0.8225 \times 10^4 = 8.225 \times 10^3$$

$$\boxed{\therefore \quad \textbf{1 out of every 8225 radioactive atoms decays per year}}$$

(b)
$$\lambda = \frac{1.216 \times 10^{-4} \, yr^{-1}}{(365)(24)(60)} = 2.31 \times 10^{-10} \, min^{-1}$$

$$\frac{1}{2.31 \times 10^{-10}} = 4.32 \times 10^{9}$$

---

1 out of 4.32×10^{9} radioactive atoms decays per minute

· Problem 6-4

K^{40} ($t_{1/2} = 1.3 \times 10^{9}$ yr) constitutes 0.012% of the potassium in nature. The human body contains about 0.35% potassium by weight. Calculate the total radioactivity resulting from K^{40} decay in a 75 kg human.

Solution

$$\text{total } K^{40} = 0.012\% \times 0.35\% \times 75 \times 10^{3} \, g$$
$$= (1.2 \times 10^{-4})(3.5 \times 10^{-3})(7.5 \times 10^{4})$$
$$= 3.15 \times 10^{-2} \, g$$

$$\text{number of } K^{40} \text{ atoms} = \frac{3.15 \times 10^{-2} \, g}{40 \, g/g\text{-atom}} \times 6.023 \times 10^{23} \text{ atoms/g-atom}$$
$$= 4.74 \times 10^{20} \text{ atoms}$$

$$\lambda = \frac{0.693}{1.3 \times 10^{9} \times 365 \times 24 \times 60} = \frac{6.93 \times 10^{-1}}{6.83 \times 10^{14}} \, min^{-1}$$

$$\lambda = 1.014 \times 10^{-15} \, min^{-1}$$

$$\text{DPM} = -\frac{dN}{dt} = \lambda N$$

$$= (1.014 \times 10^{-15})(4.74 \times 10^{20}) = \boxed{\textbf{4.81} \times \textbf{10}^{5} \textbf{ DPM}}$$

or
$$\frac{4.81 \times 10^{5} \, DPM}{2.22 \times 10^{6} \, DPM/\mu Ci} = \boxed{\textbf{0.217} \, \pmb{\mu}\textbf{Ci}}$$

SPECIFIC ACTIVITY

It is not necessary for every molecule of a compound to be radioactive for us to use the radioactivity as a measure of concentration (and, thereby, determine reaction rates, and so on). All that is necessary is that the sample contain enough radioactive molecules to count accurately and that we know the *specific activity* of the compound. Specific activity (S.A.) refers to the amount of radioactivity per unit amount of substance. It is, in fact, a way of designating the fraction of the total molecules present that is radioactive. Specific activity may be given in terms of curies per gram (Ci/g), millicuries per milligram (mCi/mg), millicuries per millimole (mCi/mmole), disintegrations per minute per millimole (DPM/mmole), counts per minute per micromole (CPM/μmole), or in any other convenient way. Once the specific activity of a compound is known, any given count rate can be equated to the amount of the compound in a sample.

In most studies with radioactive compounds, it is assumed that there are no *isotope effects*, that is, it is assumed that the radioactive molecules are randomly distributed among the total molecules of the compound and behave identically to the nonradioactive molecules. This is a reasonable assumption for most of the isotopes used in biology. The one exception is H^3, which has a mass three times that of normal H. However, for most biologically important H^3-labeled compounds, the total molecular weight is not much different from that of the unlabeled compound. For example, H_2^3O has a molecular weight only 19% greater than that of H_2^1O.

· Problem 6-5

C^{14} is produced continuously in the upper atmosphere by the bombardment of N^{14} with neutrons of cosmic radiation. The reaction is $_7N^{14} + _0n^1 \rightarrow _6C^{14} + _1H^1$. As a result, all carbon-containing compounds currently being biosynthesized on the earth contain sufficient C^{14} to yield 13 DPM/g carbon. After death of an organism the C^{14} decays with a half-life of 5700 years. Calculate (a) the abundance of C^{14} in the carbon that is participating in the carbon cycle on the surface of the earth today and (b) the age of a sample of biological material that contains 3 DPM/g carbon.

Solution

(a)
$$\text{DPM/g} = -\frac{dN}{dt} = \lambda N$$

where N = the number of C^{14} atoms per gram of carbon and λ = the decay constant, 2.31×10^{-10} min^{-1} for C^{14}.

$$13 = 2.31 \times 10^{-10} N$$

$$N = \frac{13}{2.31 \times 10^{-10}} = \boxed{\textbf{5.63} \times \textbf{10}^{10} \textbf{ atoms C}^{14}\textbf{/g carbon}}$$

1 g of carbon contains

$$\frac{1 \text{ g}}{12 \text{ g/g-atom}} \times 6.023 \times 10^{23} \text{ atoms/g-atom} =$$

$$\boxed{\textbf{5.02} \times \textbf{10}^{22} \textbf{ total atoms of carbon}}$$

$$\text{abundance} = \frac{5.63 \times 10^{10} \text{ atoms C}^{14}}{5.02 \times 10^{22} \text{ total atoms carbon}} \times 100\% = \boxed{\textbf{1.12} \times \textbf{10}^{-10}\%}$$

(b)
$$2.3 \log \frac{13}{3} = \lambda t$$

where
$$\lambda = 1.216 \times 10^{-4} \text{ yr}^{-1} \text{ for } C^{14}$$

$$2.3 \log 4.33 = 1.216 \times 10^{-4} t$$

$$(2.3)(0.636) = 1.216 \times 10^{-4} t$$

$$t = \frac{(2.3)(0.636)}{1.216 \times 10^{-4}} \qquad \boxed{\textbf{age} = t = \textbf{12,029 yr}}$$

Check: The C^{14} activity has decayed to about $\frac{1}{5}$ of its original level. After 5700 years it would decay to $\frac{1}{2}$. After 11,400 years (2 half-lives), it would decay to $\frac{1}{4}$. After 17,100 years (3 half-lives), it would decay to $\frac{1}{8}$. Thus, the sample is between 2 and 3 half-lives old.

SPECIFIC ACTIVITY OF CARRIER-FREE ISOTOPES

· Problem 6-6

(a) What is the specific activity of pure C^{14} in terms of DPM/g, Ci/g, and Ci/g-atom? (b) What is the theoretical maximum specific activity (Ci/mole) at which L-phenylalanine-C^{14} (uniformly labeled) could be prepared? (c) What proportion of the molecules is actually labeled in a preparation of L-phenylalanine-C^{14} that has a specific activity of 200 mCi/mmole? $\lambda = 2.31 \times 10^{-10} \text{ min}^{-1}$.

Solution

(a)
$$1 \text{ g } C^{14} = \frac{1 \text{ g}}{14 \text{ g/g-atom}} = 0.0714 \text{ g-atom}$$

$$N = 0.0714 \text{ g-atom} \times 6.023 \times 10^{23} \text{ atoms/g-atom}$$

$$N = 4.3 \times 10^{22} \text{ atoms}$$

$$\frac{\text{DPM}}{\text{g}} = -\frac{dN}{dt} = \lambda N$$

where N = the number of atoms in 1 g of C^{14}.

$$\text{DPM} = 2.31 \times 10^{-10} \text{ min}^{-1} \times 4.3 \times 10^{22} \text{ atoms}$$

$$\boxed{\textbf{specific activity} = \textbf{9.94} \times \textbf{10}^{12} \text{ \textbf{DPM/g}}}$$

or
$$\frac{9.94 \times 10^{12} \text{ DPM/g}}{2.22 \times 10^{12} \text{ DPM/Ci}} = \boxed{\textbf{4.48 Ci/g}}$$

or
$$4.48 \text{ Ci/g} \times 14 \text{ g/g-atom} = \boxed{\textbf{62.7 Ci/g-atom}}$$

(b)   L-phenylalanine contains 9 g-atoms of carbon per mole. As shown above, pure C^{14} has a specific activity of 62.7 Ci/g-atom.

$$\therefore \quad \text{maximum specific activity} = 9 \text{ g-atoms/mole} \times 62.7 \text{ Ci/g-atom}$$

$$= \boxed{\textbf{564 Ci/mole}}$$

(c) $\dfrac{200 \text{ Ci/mole}}{564 \text{ Ci/mole}} \times 100 = \% \ C^{14} \text{ labeled molecules} = \boxed{\textbf{35.5\%}}$

· Problem 6-7

Calculate (a) the number of radioactive atoms and (b) the weight in grams of phosphorous in 1 Ci of pure P^{32}.   (c) Calculate the specific activity of pure P^{32}.   The half-life of P^{32} is 14.3 days.

Solution

(a)   1 Ci $= 2.22 \times 10^{12}$ DPM.   First calculate λ in terms of min^{-1}.

$$\lambda = \frac{0.693}{14.3 \times 24 \times 60} = \frac{0.693}{2.06 \times 10^4} = \frac{6.93 \times 10^{-1}}{2.06 \times 10^4} \ \text{min}^{-1}$$

$$\lambda = 3.36 \times 10^{-5} \ \text{min}^{-1}$$

$$\text{DPM} = -\frac{dN}{dt} = \lambda N$$

$$2.22 \times 10^{12} = 3.36 \times 10^{-5} \ N$$

$$N = \frac{2.22 \times 10^{12}}{3.36 \times 10^{-5}} = 0.66 \times 10^{17} \qquad \boxed{\textbf{N} = \textbf{6.6} \times \textbf{10}^{\textbf{16}} \textbf{ atoms/Ci}}$$

(b)   1 g-atom of P^{32} (i.e., 32 g) contains 6.023×10^{23} atoms. $\therefore$ 6.6×10^{16} atoms weigh:

$$\frac{6.6 \times 10^{16}}{6.023 \times 10^{23}} \times 32 \text{ g} = \boxed{\textbf{3.51} \times \textbf{10}^{\textbf{-6}} \textbf{ g}}$$

(c)   Pure P^{32} contains 6.6×10^{16} atoms/Ci, or:

$$\frac{1 \text{ Ci}}{6.6 \times 10^{16} \text{ atoms}} = 1.515 \times 10^{-17} \text{ Ci/atom}$$

$$1.515 \times 10^{-17} \text{ Ci/atom} \times 6.023 \times 10^{23} \text{ atoms/g-atom} = \boxed{\textbf{9.125} \times \textbf{10}^{\textbf{6}} \textbf{ Ci/g-atom}}$$

or $\dfrac{9.125 \times 10^6 \text{ Ci/g-atom}}{32{,}000 \text{ mg/g-atom}} = \boxed{\textbf{285.2 Ci/mg}}$

A general equation relating the specific activity of a pure isotope ("carrier-free" or isotope at "100% enrichment") to its half-life can be derived easily:

$$\text{DPM} = -\frac{dN}{dt} = \lambda N = \frac{0.693}{t_{\frac{1}{2}(min)}} N$$

$$\frac{\text{DPM}}{\text{g-atom}} = \frac{0.693}{t_{\frac{1}{2}(min)}} \times 6.023 \times 10^{23}$$

If $t_{\frac{1}{2}}$ is given in days and the specific activity in Ci/g-atom:

$$\text{S.A.} = \frac{(0.693)(6.023 \times 10^{23})}{(t_{\frac{1}{2}})(24)(60)(2.22 \times 10^{12})}$$

$$\boxed{\text{S.A.}_{\text{Ci/g-atom}} = \frac{1.305 \times 10^8}{t_{\frac{1}{2}(days)}}} \quad \text{or} \quad \boxed{\text{S.A.}_{\text{Ci/g}} = \frac{1.305 \times 10^8}{\text{AW} \times t_{\frac{1}{2}(days)}}} \quad (6)$$

where AW = the atomic weight of the isotope.

B. SOLUTIONS OF RADIOACTIVE COMPOUNDS

· Problem 6-8

A bottle contains 1 mCi of L-phenylalanine-C^{14} (uniformly labeled) in 2.0 ml of solution. The specific activity of the labeled amino acid is given as 150 mCi/mmole. Calculate (a) the concentration of L-phenylalanine in the solution and (b) the activity of the solution in terms of CPM/ml at a counting efficiency of 80%.

Solution

(a) Since 1 mmole is equivalent to 150 mCi, we can calculate the number of mmoles that corresponds to 1 mCi:

$$\frac{1 \text{ mmole}}{150 \text{ mCi}} = 0.00667 \text{ mmole/mCi}$$

The 1 mCi is dissolved in 2.0 ml.

$$\therefore \quad \text{concentration} = \frac{6.67 \times 10^{-3} \text{ mmole}}{2.0 \text{ ml}} = 3.335 \times 10^{-3} \text{ mmole/ml}$$

$$\boxed{\textbf{concentration} = \textbf{3.335} \times \textbf{10}^{-3} \textbf{ M}}$$

(b) $$1 \text{ Ci} = 2.22 \times 10^{12} \text{ DPM}$$

$$\therefore \quad 1 \text{ mCi} = 2.22 \times 10^9 \text{ DPM}$$

$$\text{total activity} = (0.80)(2.22 \times 10^9) \text{ CPM}$$
$$= 1.775 \times 10^9 \text{ CPM in 2.0 ml}$$

$$\frac{1.775 \times 10^9 \text{ CPM}}{2.0 \text{ ml}} = 0.888 \times 10^9 \text{ CPM/ml}$$

$$\boxed{\textbf{activity} = \textbf{8.88} \times \textbf{10}^8 \textbf{ CPM/ml}}$$

· Problem 6-9

A solution of L-glutamic acid-C^{14} (uniformly labeled) contains 1.0 mCi and 0.25 mg of glutamic acid per milliliter.   Calculate the specific activity of the labeled amino acid in terms of (a) mCi/mg, (b) mCi/mmole, (c) DPM/μmole, and (d) CPM/μmole of carbon at a counting efficiency of 70%.

Solution

(a) $$\text{S.A.} = \frac{1.0 \text{ mCi}}{0.25 \text{ mg}} = \boxed{\textbf{4.0 mCi/mg}}$$

(b) $$\text{S.A.} = 4.0 \text{ mCi/mg} \times 147.1 \text{ mg/mmole}$$

$$\boxed{\textbf{S.A.} = \textbf{588 mCi/mmole}}$$

(c) $$\text{S.A.} = \frac{588 \text{ mCi}}{1 \text{ mmole}} = \frac{588 \text{ mCi}}{1000 \text{ } \mu\text{mole}} = 0.588 \text{ mCi}/\mu\text{mole}$$

$$\text{S.A.} = 0.588 \text{ mCi}/\mu\text{mole} \times 2.22 \times 10^9 \text{ DPM/mCi}$$

$$\boxed{\textbf{S.A.} = \textbf{1.305} \times \textbf{10}^9 \textbf{ DPM/}\mu\textbf{mole}}$$

(d)   One micromole of L-glutamic acid contains 5 μmoles of carbon.

$$\therefore \quad \text{S.A.} = \frac{1.305 \times 10^9}{5} \text{ DPM}/\mu\text{mole carbon}$$

$$= 0.261 \times 10^9 \text{ DPM}/\mu\text{mole carbon}$$

At 70% efficiency:

$$\text{S.A.} = (0.70)(2.61 \times 10^8)$$

$$\boxed{\textbf{S.A.} = \textbf{1.83} \times \textbf{10}^8 \textbf{ CPM/}\mu\textbf{mole carbon}}$$

· Problem 6-10

Describe preparation of 100 ml of a 10^{-2} M solution of L-methionine-S^{35} in which the amino acid has a specific activity of 1.5×10^5 DPM/μmole.   Assume that you have available a 0.1 M solution of unlabeled L-methionine and a stock solution of L-methionine-S^{35} (30 mCi/mmole and 1 mCi/ml).

Solution

First calculate the amount of radioactivity needed.

$$10^{-2} M = 10 \text{ } \mu\text{moles/ml}$$
$$10 \text{ } \mu\text{moles/ml} \times 100 \text{ ml} = 1000 \text{ } \mu\text{moles}$$

$$1000 \text{ } \mu\text{moles} \times 1.5 \times 10^5 \text{ DPM}/\mu\text{mole} = \boxed{\textbf{1.5} \times \textbf{10}^8 \textbf{ DPM}}$$

Next calculate the amount of the radioactive stock solution that is needed to provide 1.5×10^8 DPM.

$$1 \text{ mCi/ml} \times 2.22 \times 10^9 \text{ DPM/mCi} = 2.22 \times 10^9 \text{ DPM/ml}$$

$$\frac{1.5 \times 10^8 \text{ DPM}}{2.22 \times 10^9 \text{ DPM/ml}} = 0.676 \times 10^{-1} \text{ ml} = \boxed{\textbf{67.6 } \mu\textbf{l}}$$

Thus, $67.6 \mu l$ of the radioactive stock provide the radioactivity required.   Next, calculate whether $67.6 \mu l$ also provides any significant amount of L-methionine.   Stock solution:

$$\frac{1 \text{ mmole}}{30 \text{ mCi}} = 0.0333 \text{ mmole/mCi}$$

Because the stock contains 1 mCi/ml, its concentration is 0.0333 mmole/ml or 33.3μmoles/ml.   In the $67.6 \mu l$, we have:

$$0.0676 \text{ ml} \times 33.3 \mu\text{moles/ml} = 2.25 \mu\text{moles}$$

For most applications the amount of L-methionine added from the radioactive stock (2.25μmoles) is so small compared to the total (1000μmoles) that it can be ignored.   The radioactive stock is treated as if it were "carrier-free"— as if it contained only radioactivity and no mass.

∴   **Take 67.6 μl of radioactive L-methionine-S^{35} solution, add 10.0 ml of 0.1 M (1000 μmoles) nonradioactive L-methionine solution, and then add sufficient water to make 100 ml final volume.**

· Problem 6-11

Ten milliliters of a $10^{-3} M$ unlabeled L-methionine solution and $100 \mu l$ of the L-methionine-S^{35} stock solution described in the previous problem were mixed and diluted to a final volume of 100 ml.   What are (a) the concentration and (b) the specific activity of the L-methionine-S^{35} in the final solution?

Solution

(a)   The unlabeled L-methionine solution contains:

$$0.010 \text{ liter} \times 0.001 \, M = 10^{-5} \text{ mole} = 10 \, \mu\text{moles}$$

The radioactive solution contains:

$$33.3 \, \mu\text{moles/ml} \times 0.1 \text{ ml} = 3.33 \, \mu\text{moles}$$

The final solution will contain:

$$10 + 3.33 = 13.33 \, \mu\text{moles/100 ml} = 133.3 \, \mu\text{moles/liter}$$
$$= 133.3 \times 10^{-6} \, M$$

$$= \boxed{\textbf{1.333} \times \textbf{10}^{-4} \, \textbf{\textit{M}}}$$

(b)   The radioactive stock solution provides:

$$1 \text{ mCi/ml} \times 0.1 \text{ ml} = 0.1 \text{ mCi}$$

$$\text{S.A.} = \frac{0.1 \text{ mCi}}{13.33 \text{ } \mu\text{moles}} = \frac{100 \text{ } \mu\text{Ci}}{13.33 \text{ } \mu\text{moles}}$$

> **S.A. = 7.5 μCi/μmole (or 7.5 mCi/mmole or 7.5 Ci/mole)**

In terms of DPM:

$$\text{S.A.} = 7.5 \text{ } \mu\text{Ci/}\mu\text{mole} \times 2.22 \times 10^6 \text{ DPM/}\mu\text{Ci}$$

> **S.A. = 16.65 $\times$ 10^6 DPM/μmole**

Note that in problem 6-11 the amount of L-methionine provided by the radioactive stock solution was significant compared to the amount provided by the nonradioactive solution.

C. ASSAYS USING RADIOACTIVE SUBSTRATES

The following problems illustrate some of the applications of radioactive assays in the biochemical laboratory.

DETERMINATION OF UNKNOWN VOLUMES

· Problem 6-12

Ten microcuries of C^{14}-labeled inulin were added to 15.0 ml of a yeast suspension.   The suspension was then centrifuged and the supernatant fluid carefully drawn off.   The pellet of packed yeast occupied 0.2 ml and contained 10,000 CPM.   The counting efficiency was 25%.   Calculate the proportion of the packed yeast pellet that is interstitial space, assuming that the yeast cells were completely impermeable to the inulin and that the inulin did not adsorb to the cell surface.

Solution

The "specific activity" of the original suspension is 10 μCi/15 ml or 0.667 μCi/ml.   Because the yeast cells occupy such a small proportion of the total volume, we could assume that each milliliter of extracellular fluid contains 0.667 μCi.   However, for a more exact calculation, we can assume that the volume of the suspension occupied by the yeast cells is at least the same volume as the packed yeast pellet.   The "specific activity" of the extracellular fluid then becomes:

$$\frac{10 \text{ } \mu\text{Ci}}{15.0 - 0.2 \text{ ml}} = \frac{10}{14.8} = \boxed{0.676 \text{ } \mu\text{Ci/ml}}$$

Under the given counting conditions (e.g., 0.2 ml of packed yeast cells spread out and dried on a planchet or suspended in a given volume of scintillation fluid), the efficiency of counting is 25%.

∴ One milliliter of extracellular fluid is equivalent to:

$$0.676 \ \mu\text{Ci/ml} \times 2.22 \times 10^6 \ \text{DPM}/\mu\text{Ci} \times 0.25 = \boxed{\textbf{3.75} \times \textbf{10}^5 \ \textbf{CPM/ml}}$$

$$\therefore \quad \text{interstitial volume} = \frac{10 \times 10^3 \ \text{CPM}}{3.75 \times 10^5 \ \text{CPM/ml}} = \boxed{\textbf{2.67} \times \textbf{10}^{-2} \ \textbf{ml}}$$

The packed yeast pellet contains:

$$\frac{0.0267 \ \text{ml interstitial space}}{0.2 \ \text{ml total volume}} = \boxed{\textbf{0.134 ml interstitial space/ml packed cells}}$$

or $\qquad \boxed{\textbf{13.4\% interstitial volume}}$

DETERMINATION OF INTRACELLULAR CONCENTRATIONS

· Problem 6-13

A microorganism was grown in a synthetic medium containing $S^{35}O_4^{2-}$ as the sole sulfur source. The initial concentration of $S^{35}O_4^{2-}$ in the medium was $7 \times 10^{-3} \ M$. One milliliter of the medium contained 2×10^6 CPM of radioactivity. After several days of growth the cells were harvested, washed, and extracted with boiling water. The extract was fractionated by ion-exchange chromatography. One gram wet weight of cells contained 53,000 CPM of S^{35} in the L-methionine fraction. Calculate the intracellular concentration of L-methionine in the organism, assuming that the 1 g wet weight contained 0.2 g of dry cell constituents and 0.8 ml of intracellular water.

Solution

To convert CPM to moles, we must first know the specific activity of the S^{35}. We know that the original medium contained $7 \times 10^{-3} \ M \ S^{35}O_4^{2-}$ (i.e., $7 \ \mu\text{moles/ml}$) and 2×10^6 CPM/ml.

$$\text{S.A.} = \frac{2 \times 10^6 \ \text{CPM/ml}}{7 \ \mu\text{moles/ml}} = 0.286 \times 10^6 \ \text{CPM}/\mu\text{mole}$$

$$\boxed{\textbf{S.A.} = \textbf{2.86} \times \textbf{10}^5 \ \textbf{CPM}/\mu\textbf{mole}}$$

All sulfur compounds in the organism are derived from the $S^{35}O_4^{2-}$. Consequently, the specific activity of all sulfur compounds containing 1 atom of sulfur per molecule also is 2.86×10^5 CPM/μmole.

$$\therefore \quad \text{amount of L-methionine} = \frac{53,000 \text{ CPM}}{2.86 \times 10^5 \text{ CPM/}\mu\text{mole}}$$

$$= \frac{5.3 \times 10^4}{2.86 \times 10^5} \, \mu\text{mole}$$

$$= 0.185 \, \mu\text{mole}$$

The L-methionine came from 0.8 ml of intracellular water.

$$\therefore \quad \text{intracellular concentration of L-methionine} = \frac{1.85 \times 10^{-1} \, \mu\text{moles}}{0.8 \text{ ml}}$$

$$= 2.31 \times 10^{-1} \, \mu\text{moles/ml}$$

$$= 2.31 \times 10^{-4} \text{ moles/liter}$$

$$\boxed{\textbf{intracellular concentration of L-methionine} = \textbf{2.31} \times \textbf{10}^{-4} \, \boldsymbol{M}}$$

ENZYME AND TRANSPORT ASSAYS

· Problem 6-14

Glucose-1-phosphate-C^{14} (uniformly labeled, specific activity 16,000 CPM/ μmole) was incubated with glycogen in the presence of a cell-free extract containing the enzyme glycogen phosphorylase. Radioactivity was incorporated into the glycogen primer at an initial velocity of 2550 CPM/min. (a) Calculate the rate of the enzymic reaction in terms of μmoles glucose incorporated/min. (b) Calculate the rate in terms of μmoles $\times$ liter$^{-1} \times$ min^{-1} if the reaction volume was 0.2 ml. (c) Calculate the rate in terms of μmoles $\times$ mg protein$^{-1} \times$ min^{-1} if the incubation mixture contained 0.35 mg of protein. (d) Calculate the specific (enzyme) activity of the preparation.

Solution

(a)
$$v = \frac{2550 \text{ CPM/min}}{16,000 \text{ CPM/}\mu\text{mole}} = \frac{2.55 \times 10^3}{1.6 \times 10^4} \, \mu\text{mole/min}$$

$$v = 1.59 \times 10^{-1} \, \mu\text{mole/min}$$

$$\boxed{v = \textbf{0.159} \, \boldsymbol{\mu}\textbf{mole/min}}$$

(b)
$$v = \frac{0.159 \, \mu\text{mole/min}}{0.2 \text{ ml}} = \frac{15.9 \times 10^{-2} \, \mu\text{mole/min}}{2 \times 10^{-4} \text{ liter}}$$

$$\boxed{v = \textbf{795} \, \boldsymbol{\mu}\textbf{moles} \times \textbf{liter}^{-1} \times \textbf{min}^{-1}}$$

(c)
$$v = \frac{0.159 \ \mu\text{mole/min}}{0.35 \text{ mg protein}}$$

$$\boxed{v = 0.454 \ \mu\text{mole} \times \text{mg protein}^{-1} \times \text{min}^{-1}}$$

(d)
$$\boxed{\text{S.A.} = 0.454 \text{ units/mg protein}}$$

· Problem 6-15

One gram wet weight of *Penicillium chrysogenum* mycelium (an ATP-sulfurylase negative mutant) was suspended in 100 ml of buffer.   One milliliter of a $10^{-3} \ M$ solution of $K_2S^{35}O_4$ was added at zero-time.   (Ten microliters of the stock $K_2S^{35}O_4$ solution contained 1.2×10^4 CPM under standard counting conditions.)   Four 5 ml aliquots of the mycelial suspension were filtered at 30-sec intervals.   The mycelial pads were washed and then counted in scintillation fluid.   The samples counted as follows: 10,600 CPM at 30 sec; 21,000 CPM at 60 sec; 31,200 at 90 sec; 41,900 CPM at 120 sec.   A preliminary experiment established that the mycelial pads contained 85% water and 15% dry matter.   (a) Calculate the sulfate transport rate in terms of μmoles $\times$ g dry wt$^{-1} \times$ min^{-1}.   (b) The final samples were taken after 3 hr and 4 hr of incubation.   At these times, the mycelium was too "hot" to count accurately, but the medium contained 220 CPM/ml at both times.   Estimate the K_{eq} of the sulfate transport system.

Solution

(a)   The stock $K_2S^{35}O_4$ solution ($10^{-3} \ M$) contained 1 μmole/ml.

$$\therefore \quad 10 \ \mu l = 0.01 \text{ ml} \backsimeq 0.01 \ \mu\text{mole}$$

$$\text{S.A.} = \frac{1.2 \times 10^4 \text{ CPM}}{0.01 \ \mu\text{mole}} = 1.2 \times 10^6 \text{ CPM}/\mu\text{mole}$$

The mycelium transports $S^{35}O_4^{2-}$ at an initial rate of about 21,000 CPM $\times$ min$^{-1} \times$ sample^{-1}.   Each 5 ml sample contained:

$$\frac{(1.0 \text{ g wet wt})(0.15 \text{ g dry wt/g wet wt})(5 \text{ ml})}{(100 \text{ ml})} = 0.0075 \text{ g dry wt mycelium}$$

The initial transport rate on a dry weight basis is:

$$\frac{(21,000 \text{ CPM} \times \text{min}^{-1} \times \text{sample}^{-1})}{(1.2 \times 10^6 \text{ CPM} \times \mu\text{mole}^{-1})(7.5 \times 10^{-3} \text{ g} \times \text{sample}^{-1})} =$$

$$\boxed{2.33 \ \mu\text{moles} \times \text{g}^{-1} \times \text{min}^{-1}}$$

(b)   One gram wet weight of mycelium contains approximately 0.85 ml of intracellular water.   Thus, the external (i.e., medium)/internal (i.e., cellular) volume ratio in the suspension is about 118:1.   A decrease of 1 μmole/ml in the external solution corresponds to an increase of 118 μmoles/ml in the internal volume.   By 3 or 4 hr, the external medium contained:

$$\frac{220 \text{ CPM/ml}}{1.2 \times 10^6 \text{ CPM}/\mu\text{mole}} = 1.83 \times 10^{-4} \ \mu\text{mole/ml} = 1.83 \times 10^{-7} \ M \ S^{35}O_4^{2-}$$

At zero-time, the medium contained 0.01 μmole/ml $= 10^{-5} \ M \ S^{35}O_4^{2-}$.   The medium has been depleted of:

$$(1.000 \times 10^{-5}) - (0.0183 \times 10^{-5}) = 0.9817 \times 10^{-5} \ M \ S^{35}SO_4^{2-}$$

The mycelium has accumulated:

$$(0.9817 \times 10^{-5})(118) = 1.158 \times 10^{-3} \ M \ ^{35}SO_4^{2-}$$

$$K_{eq} = \frac{1.158 \times 10^{-3}}{1.83 \times 10^{-7}} = \boxed{6330}$$

ISOTOPE COMPETITION IN ENZYME ASSAYS

· Problem 6-16

An enzyme catalyzing the reaction $S \rightarrow P$ is competitively inhibited by I.   The assay is based on the incorporation of label from radioactive S into P. $K_m = 2.3 \times 10^{-5} \ M$ and $V_{max} = 290$ nmoles/min under standard assay conditions. $K_i = 2.3 \times 10^{-5} \ M$.   The specific activity of the labeled S is 4.5×10^5 CPM/μmole.   (a) What is the observed velocity in terms of CPM incorporated into P per minute when $[S] = 2.3 \times 10^{-5} \ M$?   (b) What would the observed rate be (CPM/min) if a fivefold excess of unlabeled S were added?   (c) What would the observed rate be (CPM/min) if a fivefold excess of I were added?

Solution

(a)   When $[S] = 2.3 \times 10^{-5} \ M = K_m$, $v = 0.5 \ V_{max}$.

$$v = (0.5)(290 \text{ nmoles/min})(4.5 \times 10^2 \text{ CPM/nmole})$$

$$\boxed{v = 62{,}250 \text{ CPM/min}}$$

(b)   The addition of $(5)(2.3 \times 10^{-5} \ M)$ unlabeled S reduces the specific activity of the substrate to $\frac{1}{6}$ of the former value:

$$\text{S.A.} = \frac{4.5 \times 10^2 \text{ CPM/nmole}}{6} = 75 \text{ CPM/nmole}$$

The increase in $[S]$ to $6 \ K_m$ increases the actual velocity to $\frac{6}{7}$ of $V_{max} = 0.857 \ V_{max}$:

$$v = (0.857)(290) = 248.6 \text{ nmoles/min}$$

The observed velocity in terms of CPM/min is:

$$v = (248.6)(75) = \boxed{18{,}643 \text{ CPM/min}}$$

Compared to the original v of 62,250, we observe:

$$\frac{v_i}{v_0} = \frac{18,645}{62,250} = 0.3 \quad \text{or} \quad \boxed{i_\% = 70}$$

Thus, we observe an apparent 70% inhibition. (The same apparent inhibition is observed if v is calculated in terms of nmoles/min using the observed CPM/min and the original, undiluted specific activity of S.)

(c) $v_i = \dfrac{[S]\,V_{max}}{K_m\left(1 + \dfrac{[I]}{K_i}\right) + [S]}$

$$v_i = \frac{(2.3 \times 10^{-5})(290)}{(2.3 \times 10^{-5})\left(1 + \dfrac{11.5 \times 10^{-5}}{2.3 \times 10^{-5}}\right) + 2.3 \times 10^{-5}} = \frac{6.67 \times 10^{-3}}{(2.3 \times 10^{-5})(6) + 2.3 \times 10^{-5}}$$

$$= \frac{6.67 \times 10^{-3}}{1.61 \times 10^{-4}} = 41.43 \text{ nmoles/min}$$

$$v_i = (41.43)(4.5 \times 10^2) = \boxed{\textbf{18,643 CPM/min}}$$

We see that the addition of unlabeled substrate yields the same apparent degree of inhibition as an equal concentration of a competitive inhibitor where $K_i = K_m$. If the degree of inhibition by I was less than that caused by the same concentration of unlabeled S, then we can conclude that $K_i > K_m$. If the degree of inhibition by I was greater than that caused by the same concentration of unlabeled S, the $K_i < K_m$.

Problem 6-17

As shown above, unlabeled substrate behaves as a competitive inhibitor with respect to labeled substrate. This displacement phenomenon can be used to determine an unknown concentration of substrate in a solution *known to be free of true inhibitors*. For example, suppose 0.1 ml of a column fraction containing unlabeled S was included in a 1.0 ml standard reaction volume containing $[S^*] = 2.3 \times 10^{-5} M$ and an $[E]_t$ sufficient to yield a V_{max} of 290 nmoles/min. The observed initial velocity was 48,000 CPM/min. What is the concentration of unlabeled S in the column fraction?

Solution

The unlabeled substrate acts as a competitive inhibitor whose K_i equals K_m. Thus, from Equation 4-42 we can write the following expression for the relative activity:

$$\frac{v_i}{v_0} = a = \frac{K_m + [S^*]}{K_m\left(1 + \dfrac{[S]}{K_m}\right) + [S^*]} = \frac{K_m + [S^*]}{K_m + [S] + [S^*]}$$

where $[S^*]$ = the concentration of labeled substrate of a given specific activity.
and $[S]$ = the concentration of unlabeled substrate.

Solving for [S]:

$$[S] = \frac{(K_m + [S^*])(1 - a)}{a} \tag{7}$$

$$a = \frac{48,000}{62,250} = 0.771 \qquad (1 - a) = 0.229$$

$$[S] = \frac{(4.6 \times 10^{-5})(0.229)}{(0.771)} = 1.37 \times 10^{-5} \, M$$

The incubation contained $1.37 \times 10^{-5} \, M$ unlabeled S along with $2.3 \times 10^{-5} \, M$ labeled S of specific activity $4.5 \times 10^{5} \, \text{CPM}/\mu\text{mole}$. The unlabeled S was originally present in 0.1 ml of a column fraction. The concentration of S in the fraction then is

$$\boxed{1.37 \times 10^{-4} \, M}$$

EQUILIBRIUM BINDING ASSAYS

· Problem 6-18

An amino acid binding protein (MW = 35,000) was purified from *E. coli*. A solution of the protein (0.5 mg/ml) was placed in one compartment (0.10 ml) of an equilibrium dialysis chamber (the "plus" compartment). A solution of L-leucine-C^{14} (S.A. = $2 \times 10^{7} \, \text{CPM}/\mu\text{mole}$) was placed in the other ("minus") compartment (0.10 ml). At equilibrium, $10 \, \mu\text{l}$ of solution from the "plus" compartment contained 4600 CPM. Ten microliters of solution from the "minus" compartment contained 3400 CPM. Calculate the concentration of bound L-leucine-C^{14}, and the K_S for the protein-leucine complex. Assume one binding site per protein molecule.

Solution

The concentration of bound + free L-leucine-C^{14} is that in the "plus" compartment:

$$\frac{4600 \, \text{CPM}/10 \, \mu\text{l}}{2 \times 10^{7} \, \text{CPM}/\mu\text{mole}} = 2.3 \times 10^{-4} \, \mu\text{mole}/10 \, \mu\text{l} = 2.3 \times 10^{-2} \, \mu\text{mole/ml}$$

$$= 2.3 \times 10^{-5} \, \text{mmole/ml} = 2.3 \times 10^{-5} \, M$$

The concentration of free L-leucine-C^{14} is that in the "minus" compartment:

$$\frac{3400 \, \text{CPM}/10 \, \mu\text{l}}{2 \times 10^{7} \, \text{CPM}/\mu\text{mole}} = 1.7 \times 10^{-4} \, \mu\text{mole}/10 \, \mu\text{l} = 1.7 \times 10^{-2} \, \mu\text{mole/ml}$$

$$= 1.7 \times 10^{-5} \, \text{mmole/ml} = 1.7 \times 10^{-5} \, M$$

The concentration of bound L-leucine-C^{14} is the difference between the bound + free L-leucine-C^{14} in the "plus" compartment and the free L-leucine-

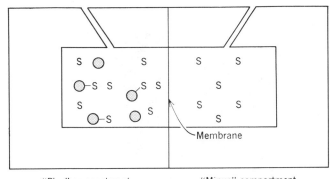

"Plus" compartment

$V_{(+)} = 0.10$ ml

$[S] + [PS] = 46{,}000$ CPM

$= 2.3 \times 10^{-5} \, M$

$[P] + [PS] = 1.43 \times 10^{-5} \, M$

$[PS] = 0.6 \times 10^{-5} \, M$

$[P] = 0.83 \times 10^{-5} \, M$

"Minus" compartment

$V_{(-)} = 0.10$ ml

$[S] = 34{,}000$ CPM

$= 1.7 \times 10^{-5} \, M$

Figure 6-2 Equilibrium dialysis chamber. The protein is introduced into one compartment. The labeled substrate is introduced into either or both compartments. The membrane restricts the protein to one compartment, but the substrate can freely diffuse across the membrane. At equilibrium, the free [S] will be the same in both compartments, but the total concentration of S will be greater in the compartment containing the protein because the bound substrate, PS, cannot equilibrate across the membrane.

C^{14} in the "minus" compartment:

$$[\text{L-leucine-}C^{14}]_b = (2.3 \times 10^{-5} \, M) - (1.7 \times 10^{-5} \, M)$$

$$\boxed{[\text{L-leucine-}C^{14}]_b = 0.6 \times 10^{-5} \, M}$$

The rest of the calculations are shown in Problem 4-11. Figure 6-2 shows the distribution of label and protein.

D. DOUBLE-LABEL ANALYSIS

The β particles from a given radioactive isotope are emitted with a continuous energy distribution extending up to some maximum value (e.g., 0.0176 mev for H^3, 0.155 mev for C^{14}, 1.701 mev for P^{32}). When two isotopes have different emission energy spectra, the amount of each present in a mixture can be determined by selectively measuring the radioactivity at different energy levels. This can be accomplished easily with a two-channel scintillation counter. Sometimes, it is impossible to attain 100% discrimination and, consequently, the activity measured in one channel (or both) will result from both isotopes (Fig. 6-3). Nevertheless, the amount of each

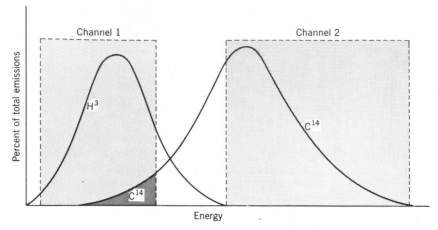

Figure 6-3 Energy distribution spectra of H^3 and C^{14}. By setting the discriminators of a dual-channel scintillation counter as shown, Channel 1 records mostly H^3 (with a little C^{14}), while Channel 2 records only C^{14}.

isotope present can still be calculated.   If a dual-channel scintillation counter is unavailable, a Geiger-Muller counter can be used.   In this case a piece of filter paper or aluminum foil of appropriate thickness (placed between the sample and the detector) can be used to differentiate energy levels.

· **Problem 6-19**

A sample of labeled RNA containing C^{14}-adenine and H^3-uracil yielded 25,000 CPM in Channel 1 and 45,000 CPM in Channel 2 of a dual-channel scintillation counter.   A C^{14} standard, containing 40,000 DPM (or 40,000 CPM under optimum counting conditions) yielded 10,000 CPM in Channel 1 and 20,000 CPM in Channel 2.   An H^3 standard containing 200,000 DPM (or 200,000 CPM under optimum counting conditions) yielded 40,000 CPM in Channel 1 and 100 CPM in Channel 2.   All counts are corrected for background.   Calculate the H^3 DPM and C^{14} DPM in the RNA sample.

Solution

We see from the standards that essentially all the counts in Channel 2 originate from C^{14}, but only half the C^{14} DPM is recorded.   (The very small contribution of H^3 to Channel 2 is ignored.)   Thus, the 45,000 Channel 2 CPM of the sample is equivalent to:

$$\frac{45,000}{0.5} = \boxed{\textbf{90,000 DPM C}^{14}}$$

The 25,000 CPM recorded in Channel 1 represents 25% of the C^{14} DPM plus 20% of the H^3 DPM.

$$(0.25)(90,000) + (0.20)(H^3 \text{ DPM}) = 25,000$$

$$H^3 = \frac{(25,000) - (0.25)(90,000)}{0.2} = \frac{2500}{0.2}$$

$$\boxed{H^3 = 1250 \text{ DPM}}$$

· Problem 6-20

An alga was grown in synthetic medium containing $S^{35}O_4^{2-}$ (2.9×10^6 DPM/μmole) and P_i^{32} (5.2×10^6 DPM/μmole) as sole sulfur and phosphorous sources, respectively.   A cell-free, deproteinated extract of the organism was analyzed by paper chromatography.   A radioactive spot containing S^{35} and P^{32} was eluted from the chromatogram and counted in a dual-channel scintillation counter. S^{35} and P^{32} standards were also counted.   The data are shown below.

	Observed (corrected for background)	
Added	Channel 1	Channel 2
P^{32} standard (58,000 DPM)	39,000 CPM	14,000 CPM
S^{35} standard (110,000 DPM)	18,000 CPM	56,000 CPM
Eluted sample	74,600 CPM	38,500 CPM

Calculate the P/S ratio in the unknown compound.

Solution

$$\text{Channel 1 counts:}\quad \frac{39,000}{58,000} = 67.2\% \text{ of the } P^{32} \text{ DPM}$$

$$\text{and}\quad \frac{18,000}{110,000} = 16.4\% \text{ of the } S^{35} \text{ DPM}$$

$$\text{Channel 2 counts:}\quad \frac{14,000}{58,000} = 24.1\% \text{ of the } P^{32} \text{ DPM}$$

$$\text{and}\quad \frac{56,000}{110,000} = 50.9\% \text{ of the } S^{35} \text{ DPM}$$

Neither channel represents the CPM of only one of the isotopes.   Therefore, we must set up two simultaneous equations to solve for P^{32} DPM and S^{35} DPM.

$$\text{Channel 1:}\quad (0.672)(P^{32}) + (0.164)(S^{35}) = 74,600$$

$$\text{Channel 2:}\quad (0.241)(P^{32}) + (0.509)(S^{35}) = 38,500$$

Solving for P^{32} in terms of S^{35} using the Channel 1 equation:

$$(P^{32}) = \frac{(74,600) - (0.164)(S^{35})}{(0.672)}$$

Substituting for (P^{32}) in the Channel 2 equation:

$$(0.241)\left(\frac{(74,600)-(0.164)(S^{35})}{0.672}\right)+0.509(S^{35})=38,500$$

$$\frac{(0.241)(74,600)}{0.672}-\frac{(0.241)(0.164)(S^{35})}{0.672}+0.509(S^{35})=38,500$$

$$26,753-0.059(S^{35})+0.509(S^{35})=38,500$$

$$0.45(S^{35})=11,747$$

$$\boxed{(S^{35})=26,104 \text{ DPM}}$$

Next, calculate (P^{32}) from the Channel 1 or Channel 2 equation:

$$(0.672)(P^{32})+(0.164)(26,104)=74,600$$

$$(P^{32})=\frac{(74,600)-(0.164)(26,104)}{0.672}=\frac{70,318}{0.672}$$

$$\boxed{(P^{32})=104,641 \text{ DPM}}$$

The amounts of each isotope in the sample are:

$$P^{32}=\frac{104,641 \text{ DPM}}{5.2\times10^6 \text{ DPM}/\mu\text{mole}}=0.02 \ \mu\text{mole}$$

$$S^{35}=\frac{26,104 \text{ DPM}}{2.9\times10^6 \text{ DPM}/\mu\text{mole}}=0.009 \ \mu\text{mole}$$

or

$$\boxed{P/S\approx2}$$

E. BIOLOGICAL HALF-LIFE—TURNOVER

EFFECTIVE HALF-LIFE

When a short-lived radioactive isotope is introduced into a biological system, the observed decay in radioactivity results from a combination of normal radioactive decay and biological turnover (e.g., removal of the isotope from the bloodstream by excretion or transport into tissues). If the biological turnover is a first-order process, then λ_{app}, the apparent first-order rate constant, is the sum of λ_r (radioactive) $+ \lambda_b$ (biological). This is quite understandable since λ represents the fraction of the activity present that disappears per small increment of time. Fractions can be added. The observed radioactivity at any time is given by:

$$2.3 \log \frac{N_0}{N}=\lambda_{app}t=(\lambda_r+\lambda_b)t$$

The effective half-life (when $N_0/N = 1:0.5 = 2$) is given by:

$$\boxed{\frac{0.693}{(\lambda_r + \lambda_b)} = t_{\frac{1}{2}(\text{eff})}} \tag{8}$$

or

$$\frac{1}{t_{\frac{1}{2}(\text{eff})}} = \frac{\lambda_r + \lambda_b}{0.693} = \frac{\lambda_r}{0.693} + \frac{\lambda_b}{0.693}$$

$$\boxed{\frac{1}{t_{\frac{1}{2}(\text{eff})}} = \frac{1}{t_{\frac{1}{2}r}} + \frac{1}{t_{\frac{1}{2}b}}} \tag{9}$$

or

$$\boxed{t_{\frac{1}{2}(\text{eff})} = \frac{(t_{\frac{1}{2}r})(t_{\frac{1}{2}b})}{t_{\frac{1}{2}r} + t_{\frac{1}{2}b}}} \tag{10}$$

· Problem 6-21

A guinea pig was given a single injection of $Na^{24}Cl$. Periodically, blood samples were withdrawn and analyzed immediately for radioactivity. The data are shown below. Calculate (a) the biological half-life of Na^{24} in the bloodstream, (b) the specific activity (CPM/ml) of the 1 hr sample if all samples were counted at 24 hr. The radioactive half-life of Na^{24} is 15 hr.

Time after Injection (hr)	Specific Activity (CPM/ml)
1	3604
3	2928
5	2376
10	1412
16	756
24	329

Solution

(a) First calculate the effective half-life from the specific activities at any two times, for example, 3 and 10 hr:

$$2.3 \log \frac{2928}{1412} = \frac{0.693}{t_{\frac{1}{2}(\text{eff})}} \tag{7}$$

$$t_{\frac{1}{2}(\text{eff})} = \frac{(0.693)(7)}{2.3 \log 2.074} = \frac{4.851}{(2.3)(0.3167)}$$

$$\boxed{t_{\frac{1}{2}(\text{eff})} = 6.66 \text{ hr}}$$

Now, $t_{\frac{1}{2}b}$ can be calculated:

$$\frac{1}{t_{\frac{1}{2}(\text{eff})}} = \frac{1}{t_{\frac{1}{2}r}} + \frac{1}{t_{\frac{1}{2}b}} \qquad \frac{1}{t_{\frac{1}{2}b}} = \frac{1}{t_{\frac{1}{2}(\text{eff})}} - \frac{1}{t_{\frac{1}{2}r}}$$

$$\frac{1}{t_{\frac{1}{2}b}} = \frac{1}{6.66} - \frac{1}{15} = 0.150 - 0.0667 = 0.0833$$

$$t_{\frac{1}{2}b} = \frac{1}{0.0833} \qquad \boxed{t_{\frac{1}{2}b} = \textbf{12 hr}}$$

(b)   At 24 hr, the radioactive decay factor is the same for all samples.   The decrease in specific activity of each sample reflects solely the rate at which Na^{24} was removed from the bloodstream with a half-life of 12 hr.   The 24 hr sample would count 329 CPM/ml.   The specific activity of the 1 hr sample can be calculated as shown below:

$$2.3 \log \frac{N_0}{329} = \frac{0.693}{12} \quad (23)$$

$$\log N_0 - \log 329 = \frac{(0.693)(23)}{(12)(2.3)} = 0.5775$$

$$\log N_0 = 0.5775 + \log 329 = 3.0946$$

$$\boxed{N_0 = \textbf{1243 CPM/ml}}$$

TURNOVER

Many cellular components are in a *dynamic state* whereby they are synthesized and degraded at a constant *turnover rate* while their total concentration remains constant.   The turnover rate can be measured if the specific component can be labeled.

· Problem 6-22

A culture of *E. coli* was grown in synthetic medium containing P_i^{32} as sole phosphorous source.   After several generations, the cells were harvested, washed, and resuspended in fresh medium containing unlabeled phosphate and lacking a nitrogen source (so no further growth could occur).   Periodically, a portion of the stationary cells were harvested, washed, and the lipids extracted with a chloroform-methanol mixture.   The individual phospholipids were separated by thin-layer chromatography on silicic acid plates.   The spot corresponding to phosphatidyl glycerol was eluted, analyzed chemically for total phosphorous, and counted.   The specific activity of the phosphatidyl glycerol is shown below.   All samples were counted within a few minutes of each other, so no correction for decay of P^{32} is needed.

Time after Resuspension (hr)	Specific Activity of Phosphatidyl Glycerol (CPM/μmole P)
0	40,000
0.5	30,306
1.0	22,962
2.0	13,181
2.5	9,987
3.0	7,567
5.0	2,494

Calculate the rate at which phosphatidyl glycerol turns over in *E. coli* (k and $t_{1/2}$).

Solution

We can estimate from the recorded data that $t_{\frac{1}{2}}$ lies somewhere between 1 and 1.5 hr. (Between 0 and 1 hr, less than half the P^{32} had been removed; between 0.5 and 2.0 hr more than half of the P^{32} present at 0.5 hr had been removed.) A plot of log specific activity versus time can be constructed and k, the first-order rate constant, determined from the slope (slope = $-k/2.3$). Alternatively, we can calculate k from any two points (e.g., 1 and 2.5 hr):

$$2.3 \log \frac{22,962}{9987} = k \text{ (90 min)}$$

$$k = \frac{2.3}{90} \log 2.299 = (0.0256)$$

$$\boxed{k = 9.24 \times 10^{-3} \text{ min}^{-1}}$$

$$t_{\frac{1}{2}} = \frac{0.693}{k} = \frac{0.693}{9.24 \times 10^{-3}} \qquad \boxed{t_{\frac{1}{2}} = 75 \text{ min}}$$

PRECURSOR-PRODUCT RELATIONSHIPS

Isotopic tracers have been extremely useful in elucidating metabolic pathways. Precursor-product relationships (i.e., the *sequence* of reactions) can be established by introducing a very small amount of highly labeled suspected precursor into the system and then following the appearance of label in the other intermediates and the final product. The mass in the labeled precursor should be very small so that the steady-state pool sizes of the various intermediates do not change. For example, consider the sequence shown below where we suspect that C is an intermediate in the pathway from A to D.

$$\rightarrow A \rightarrow B \rightarrow C? \rightarrow D \rightarrow E \rightarrow$$

At zero-time, B is made radioactive by injecting or feeding the organism a very small amount of high specific activity B*. What will we observe if C is indeed an intermediate between B and D? We observe an immediate rise in the specific activity of B followed by a first-order decay as labeled B is converted to C and unlabeled A is converted to B. The total pool of B remains constant because for every mole of B converted to C, a mole of A is converted to B. Similarly, under this steady-state condition, the total pools of C, D, E, and so on, remain constant. Only the radioactivity in B, C, and so on, changes with time. The specific activity of an intermediate increases when labeled molecules enter the pool. The specific activity of the intermediate is unaffected by its further metabolism (labeled and unlabeled molecules leave the pool at a constant ratio). The specific activity of an intermediate decreases (by dilution) when the specific activity of the precursor entering the pool is lower than the specific activity of the intermediate in the pool.

Figure 6-4 shows an analogy of the precursor-product relationship. For simplicity, it is assumed that all pool sizes are the same, but this need not be true.

Figure 6-5 shows the speciic activities of B, C, and D as a function of time. Note that the crossover point of the B curve occurs at the maximum of the C curve. Similarly, the C curve crosses the D curve at its maximum. Several conclusions can be drawn:

If C is made from B, the specific activity of C can never rise above S.A. maximum of B. Similarly, the specific activity of D can never be greater than S.A. maximum of C Also, the maximum specific activity of C occurs when the specific activity of C equals the specific activity of B. Similarly, the maximum specific activity of D occurs when the specific activity of D equals the specific activity of C. These observations establish the direct precursor-product relationships between B and C and between C and D.

For the irreversible sequence $A \xrightarrow{k_1} B \xrightarrow{k_2} C \xrightarrow{k_3} D$ it can be shown that the amount of radioactivity in A, B, and C (CPM) is given by:

$$A^* = A_0^* \, e^{-kt} \qquad (11)$$

$$B^* = \frac{k_1 A_0^*}{(k_2 - k_1)} (e^{-k_1 t} - e^{-k_2 t}) \qquad (12)$$

$$C^* = k_1 k_2 A_0^* \left(\frac{e^{-k_1 t}}{(k_2 - k_1)(k_3 - k_1)} + \frac{e^{-k_2 t}}{(k_1 - k_2)(k_3 - k_2)} + \frac{e^{-k_3 t}}{(k_1 - k_3)(k_2 - k_3)} \right)$$

$$(13)$$

where A_0^* = the original CPM in A

k_1, k_2, k_3 = the first-order rate constants for the biological turnover of A, B, and C (also indicated as λ_1, λ_2, and λ_3)

Figure 6-4 Precursor-product relationships in the pathway $A \rightarrow B \rightarrow C \rightarrow D \rightarrow$ and so on. The dots represent radioactively labeled molecules. Since the pool sizes shown are all identical (for simplicity), the number of dots also represents the specific activity of the intermediate. If the pool of C, for example, were twice as large as the pool of B, there would be twice as many dots in C when $S.A._C = S.A._B$. The relationship could be set up as a lab demonstration using a dye in place of radioactivity. The absorbancies of the solutions would be analogous to specific activities.

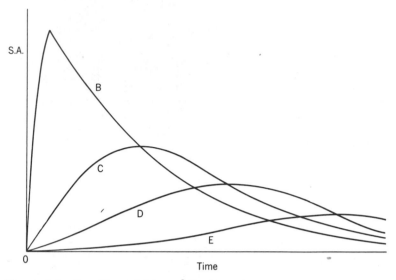

Figure 6-5   Specific activities of B, C, and D as a function of time, where B is an immediate precursor of C, and C is an immediate precursor of D, and so on.

The specific activities of A, B, and C at any time equal the CPM present divided by the total number of μmoles in the pool.   It is assumed that the isotope has a relatively long half-life compared to the biological half-lives of A, B, C, and so on.   By suitable mathematical procedures, a general equation for n intermediates can be derived [e.g., see C. Cappellos and B. H. J. Bielski, *Kinetic Systems*, Ch. 9, Wiley-Interscience (1972)].

· **Problem 6-23**

The biosynthesis of δ-amino levulinic acid (ALA, a chlorophyll precursor) was studied in greening barley leaves.   Preliminary long-term experiments showed that radioactivity from glycine-2-C^{14} (α-carbon labeled) and from glutamate-2,3,-C^{14} (methylene carbons labeled) appeared in ALA.   The specific activity data of a more detailed study is shown in Figure 6-6.   What conclusions can be drawn?

Solution

(a)   Glycine is not a direct precursor of ALA.   The C^{14} originally present in glycine-C^{14} does eventually appear in ALA, but this is not unexpected.   After all, glycine can be converted to serine, and serine to pyruvate.   A great many compounds can arise from pyruvate, one or more of which may be precursors of ALA.

(b)   The crossover point occurs very close to the maximum specific activity of ALA.   Thus, glutamate is a direct precursor of ALA, or glutamate is converted to a direct precursor with a very short half-life and very small pool size.

Note: these results suggest that ALA synthesis in plants may occur by a completely different pathway from that in animals (where glycine condenses with succinyl-S-CoA to form ALA directly).

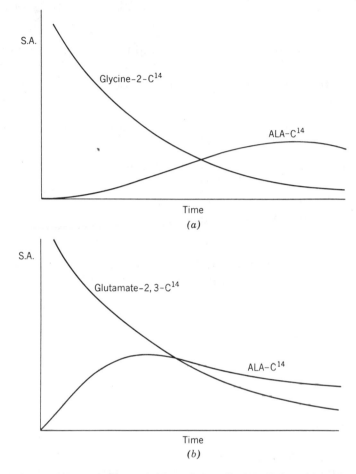

Figure 6-6 Specific activities of δ-aminolevulinic acid and two potential precursors: (*a*) glycine-2-C^{14} and (*b*) glutamate-2,3-C^{14}.

F. RADIOACTIVE TRACER AND DILUTION ANALYSES

DETERMINATION OF UNKNOWN AMOUNTS OF UNLABELED COMPOUNDS—ISOTOPE DILUTION

Radioactive tracers may be used to determine the amount of a single substance in a mixture. The tracer technique is especially useful where quantitative recovery of the substance in question is difficult or impossible. Basically, the technique involves adding a known amount of the radioactive compound to the mixture and then (after thorough equilibration) reisolating a small amount of the compound. The amount of compound recovered is unimportant, provided it is sufficient to weigh and to count. From the specific activity of the reisolated compound as well as a knowledge of the total number of counts originally added, the amount of nonradioactive compound in the mixture may be calculated as shown below. Let:

A_o = CPM of tracer added in a weight very small compared to the weight of the nonradioactive compound in the mixture

M_u = unknown amount of the nonradioactive compound in the mixture

$\therefore \quad \dfrac{A_o}{M_u}$ = S.A.$_u$ = specific activity of the compound in the mixture after addition of the tracer

M_r = amount of pure compound reisolated from the mixture after equilibration with tracer

A_r = CPM in the reisolated sample

$\dfrac{A_r}{M_r}$ = S.A.$_r$ = specific activity of the reisolated compound

It is obvious that the ratio of radioactivity to mass—the specific activity of the compound—is the same in the mixture as it is after isolation.

$$\text{S.A.}_u = \text{S.A.}_r \qquad \frac{A_o}{M_u} = \frac{A_r}{M_r}$$

$$\therefore \quad \boxed{M_u = M_r \frac{A_o}{A_r} \qquad \text{or} \qquad M_u = \frac{A_o}{\text{S.A.}_r}} \qquad (14)$$

The above calculation is based on the assumption that the amount of the added tracer is negligible compared to the amount of the compound in the mixture. The condition is easily met in practice because many compounds of biological interest are available in radioactive form with very high specific activities.

If the amount of the radioactive tracer is significant compared to the amount of the nonradioactive compound in the mixture, this equation must be corrected. Let:

A_o = CPM of tracer added

M_o = amount of tracer added

$\therefore \quad \dfrac{A_o}{M_o}$ = specific activity of tracer added

$\qquad$ = S.A.$_o$

M_u = unknown amount of nonradioactive compound in the mixture

$\therefore \quad \dfrac{A_o}{M_u + M_o}$ = specific activity of compound in the mixture after adding tracer

$\qquad$ = S.A.$_u$

M_r = amount of pure compound reisolated from the mixture after equilibration of radioactive and nonradioactive compounds

A_r = CPM in the reisolated sample

$\therefore \dfrac{A_r}{M_r} =$ specific activity of reisolated sample

$= S.A._r$

As before, the specific activity of the compound in the reisolated sample is the same as it was in the mixture.

$$S.A._u = S.A._r$$

$$\frac{A_o}{M_u + M_o} = \frac{A_r}{M_r}$$

$$A_r M_u + A_r M_o = A_o M_r$$

$$A_r M_u = A_o M_r - A_r M_o$$

$$M_u = \frac{A_o M_r - A_r M_o}{A_r} = \frac{A_o M_r}{A_r} - \frac{A_r M_o}{A_r}$$

$$\boxed{M_u = M_r \frac{A_o}{A_r} - M_o \quad \text{or} \quad M_u = \frac{A_o}{S.A._r} - M_o} \tag{15}$$

We can see that as M_o becomes small compared to M_u, the equation reduces to Equation 14.

If we concern ourselves only with specific activities, then several points are obvious. First, the specific activity of the reisolated compound is less than the specific activity of the original tracer. Second, the degree to which the specific activity decreases (i.e., the dilution factor) is directly related to the amount of unlabeled compound in the mixture. In other words, the dilution of specific activity is identical to the dilution of radioactive tracer.

$$\text{dilution} = \frac{S.A._r}{S.A._o} = \frac{M_o}{M_o + M_u}$$

$$S.A._r M_o + S.A._r M_u = S.A._o M_o$$

$$S.A._r M_u = S.A._o M_o - S.A._r M_o$$

$$M_u = \frac{S.A._o M_o - S.A._r M_o}{S.A._r}$$

$$M_u = \left(\frac{S.A._o - S.A._r}{S.A._r}\right) M_o = \left(\frac{S.A._o}{S.A._r} - \frac{S.A._r}{S.A._r}\right) M_o$$

$$\boxed{M_u = \left(\frac{S.A._o}{S.A._r} - 1\right) M_o} \tag{16}$$

· Problem 6-24

Carrier-free $S^{35}O_4^{2-}$ (5×10^8 CPM) was added to a sample containing an unknown amount of unlabeled sulfate. After equilibration, a small sample of the sulfate was reisolated as $BaS^{35}O_4$. A 2 mg sample of the $BaS^{35}O_4$ contained 2.9×10^5 CPM. Calculate M_u, the amount of unlabeled SO_4^{2-} in the samples as Na_2SO_4.

Solution

The specific activity of the reisolated $BaS^{35}O_4$, S.A.$_r$, is:

$$S.A._r = S.A._{\cdot BaS^{35}O_4} = \frac{2.9 \times 10^5 \text{ CPM}}{2 \text{ mg}} = \boxed{\textbf{1.45} \times \textbf{10}^5 \textbf{ CPM/mg}}$$

or $\quad 1.45 \times 10^5 \text{ CPM/mg} \times 233.5 \text{ mg/mmole} = \boxed{\textbf{3.39} \times \textbf{10}^7 \textbf{ CPM/mmole}}$

One mmole of $BaSO_4$ contains 1 mmole of SO_4^{2-}.   Therefore, the specific activity of the SO_4^{2-} is also 3.39×10^7 CPM/mmole.   We can now calculate the amount of unlabeled SO_4^{2-} in the sample.

$$M_u = \frac{A_o}{S.A._r}$$

where $\quad A_o = $ total activity added $= 5 \times 10^8$ CPM
$\qquad\quad$ S.A.$_r = 3.39 \times 10^7$ CPM/mmole

$$M_u = \frac{5 \times 10^8 \text{ CPM}}{3.39 \times 10^7 \text{ CPM/mmole}} \qquad \boxed{M_u = \textbf{14.74 mmoles}}$$

Expressed as Na_2SO_4:

$$M_u = 14.74 \text{ mmoles} \times 142 \text{ mg/mmole} = 2093 \text{ mg}$$

$$\boxed{M_u = \textbf{2.09 g}}$$

· Problem 6-25

Twenty milligrams of C^{14}-labeled glycogen (6.7×10^4 CPM/mg) were added to a solution containing an unknown amount of unlabeled glycogen.   A small amount of glycogen was then reisolated from the solution and reprecipitated with ethanol to constant specific activity (2.8×10^4 CPM/mg).   Calculate the amount of unlabeled glycogen in the solution.

Solution

Because S.A.$_r$ is of the same order of magnitude as S.A.$_o$, the amount of unlabeled glycogen in the solution obviously was significant compared to the amount of labeled material added.   We must use Equation 16:

$$M_u = \left(\frac{S.A._o}{S.A._r} - 1\right)M_o \qquad M_u = \left(\frac{6.7 \times 10^4}{2.8 \times 10^4} - 1\right)20 \text{ mg}$$

$$= (2.39 - 1)20 - (1.39)(20)$$

$$\boxed{M_u = \textbf{27.9 mg}}$$

DETERMINATION OF UNKNOWN AMOUNTS OF RADIOACTIVE COMPOUNDS—REVERSE ISOTOPE DILUTION

Dilution analysis can also be used to determine how much of a radioactive compound is present in a mixture, if the specific activity of the compound is known. For example, if an organism is grown on labeled $C^{14}O_2$ or uniformly labeled glucose-C^{14} as the sole carbon source, then all carbon compounds in the organism have the same specific activity (CPM/μmole of C) as the original radioactive carbon source. Let:

M_u = unknown amount of a radioactive compound in a mixture

A_u = CPM in the above unknown weight

$\therefore \quad \dfrac{A_u}{M_u}$ = specific activity of the radioactive compound

$\qquad = \text{S.A.}_u$

M_o = amount of nonradioactive compound added to the mixture

$\therefore \quad \dfrac{A_u}{M_u + M_o}$ = specific activity of the compound in the mixture after adding the nonradioactive tracer

$\qquad = \text{S.A.}_o$

M_r = amount of compound reisolated from the mixture after equilibration of radioactive and nonradioactive compound

A_r = CPM in recovered sample

$\therefore \quad \dfrac{A_r}{M_r}$ = specific activity of recovered compound

$\qquad = \text{S.A.}_r$

$\text{S.A.}_r = \text{S.A.}_o$

$\text{S.A.}_r = \dfrac{A_u}{M_u + M_o}$

If the amount of the added nonradioactive compound ("carrier") is very large compared to the amount of the radioactive compound in the mixture, the equation may be simplified:

$$\text{S.A.}_r = \dfrac{A_u}{M_o} \qquad A_u = \text{S.A.}_r M_o$$

By definition:

$$\text{S.A.}_u = \dfrac{A_u}{M_u} \qquad \therefore \quad M_u = \dfrac{A_u}{\text{S.A.}_u}$$

Substituting:

$$\boxed{M_u = \dfrac{\text{S.A.}_r}{\text{S.A.}_u} M_o} \qquad\qquad (17)$$

If the amount of radioactive compound is significant compared to the amount of carrier added, then suitable corrections must be made.

$$S.A._r = \frac{A_u}{M_u + M_o}$$

$$S.A._r M_u + S.A._r M_o = A_u$$

$$A_u = S.A._u M_u$$

$$S.A._r M_u + S.A._r M_o = S.A._u M_u$$

$$S.A._r M_u - S.A._u M_u = -S.A._r M_o$$

$$S.A._u M_u - S.A._r M_u = S.A._r M_o$$

$$M_u[S.A._u - S.A._r] = S.A._r M_o$$

$$M_u = \frac{S.A._r}{S.A._u - S.A._r} M_o \qquad (18)$$

· Problem 6-26

A plant was grown in an atmosphere containing $C^{14}O_2$ (3×10^8 CPM/μmole). After several weeks a leaf extract was prepared for glucose-1-phosphate determination via reverse isotope dilution analysis. To 20 ml of the extract, 1.5 mmoles of unlabeled dipotassium glucose-1-phosphate were added. A small amount of dipotassium glucose-1-phosphate was reisolated from the extract and recrystallized to constant specific activity from aqueous ethanol. The recrystallized salt had a specific activity of 2.6×10^5 CPM/μmole. Calculate the concentration of labeled glucose-1-phosphate in the extract.

Solution

All carbon compounds in the plant have the same specific activity (on a per mole of carbon basis) as the $C^{14}O_2$ provided. Glucose-1-phosphate contains 6 g-atoms of carbon per mole.

$$\therefore \quad S.A._u = \text{specific activity of the unknown amount of G-1-P}$$
$$= 6 \times 3 \times 10^8 = 18 \times 10^8 \text{ CPM/}\mu\text{mole}$$

$$M_u = \frac{S.A._r}{S.A._u - S.A._r} M_o$$

where M_o = amount of nonradioactive carrier added to the extract
        = 1.5 mmoles

$$M_u = \frac{2.6 \times 10^5}{1.8 \times 10^9 - 2.6 \times 10^5} \, 1.5 \text{ mmoles}$$

The 2.6×10^5 is insignificant compared to the 1.8×10^9 and may be discarded in the denominator. The fact that $S.A._r$ is so small compared to $S.A._u$ immediately shows that the amount of carrier added was very large compared to the amount of radioactive glucose-1-phosphate in the extract.

$$M_u = \frac{2.6 \times 10^5}{1.8 \times 10^9} 1.5 = 2.167 \times 10^{-4} \, \text{mmole}$$

$$\boxed{M_u = 0.2167 \, \mu\text{mole}}$$

$$\text{concentration} = \frac{0.2167 \, \mu\text{mole}}{20 \, \text{ml}} = \frac{21.65 \times 10^{-2}}{20}$$

$$= 1.084 \times 10^{-2} \, \mu\text{mole/ml}$$

$$\boxed{\textbf{concentration} = 1.084 \times 10^{-5} \, \textbf{\textit{M}}}$$

RADIOACTIVE DERIVATIVE ANALYSIS

An unknown amount of an unlabeled compound in a mixture may be determined, even though the labeled compound is unavailable, if a suitable radioactive derivative can be prepared and isolated.   By reacting the mixture with a suitable radioactive reagent of known specific activity, the compound in question is converted to a derivative of the same specific activity.   The amount of the derivative in the mixture can be quantitated by the calculations outlined above.

· **Problem 6-27**

A mixture of amino acids was reacted with p-iodobenzene sulfonyl chloride ("pipsyl" chloride) labeled with I^{131} to produce the radioactive pipsyl derivatives of the individual amino acids.   The specific activity of the pipsyl chloride was $4.23 \times 10^5 \, \text{CPM}/\mu\text{mole}$.   After the reaction, 250 mg of unlabeled pipsyl derivative of leucine were added to the mixture.   A small amount of the leucine derivative was reisolated and purified to a constant specific activity of $1700 \, \text{CPM}/\mu\text{mole}$.   Calculate the amount of unlabeled leucine in the original mixture.

Solution

After the reaction with pipsyl chloride the pipsyl derivatives of all the amino acids present have the same specific activity ($4.23 \times 10^5 \, \text{CPM}/\mu\text{mole}$).

$$M_u = \frac{\text{S.A.}_r}{\text{S.A.}_u - \text{S.A.}_r} \qquad M_o = \frac{17 \times 10^2}{4.23 \times 10^5 - 0.017 \times 10^5} 250 \, \text{mg}$$

$$= \frac{17 \times 10^2}{4.21 \times 10^5} 250 = (4.04 \times 10^{-3})(250) \, \text{mg}$$

$$\boxed{\textbf{\textit{M}}_u = \textbf{1.01 mg}}$$

MOLECULAR WEIGHT OF ENZYMES BY AFFINITY LABELING

· Problem 6-28

Exactly 3.4 mg of a purified proteolytic enzyme (with esterase activity) was treated with excess diisopropylfluorophosphate-P^{32} (1 μCi/mmole). After 24 hr, KOH was added to the solution to destroy the unreacted (extremely toxic) DFP^{32}. The P^{32}-labeled enzyme was precipitated with TCA, washed free of soluble P^{32}, and then redissolved in 0.5 ml of dilute KOH. The 0.5 ml was transferred quantitatively to a scintillation vial and counted. The sample counted 7380 counts per hour above background at 80% efficiency. What is the minimum molecular weight of the enzyme?

Solution

The DFP^{32} reacts with serine residues at the active site. The minimum molecular weight of the enzyme is that which contains one mole of active serine residue. The number of moles of P^{32} bound to the enzyme equals the number of moles of active serine (assuming the reaction went to completion).

First, note that after 24 hr, the specific activity of the P^{32} had decayed slightly:

$$2.3 \log \frac{\text{S.A.}_o}{\text{S.A.}} = \lambda t \qquad \log 1 - \log \text{S.A.} = \frac{(0.693)}{(14.3)(2.3)}$$

$$\log \text{S.A.} = \log 1 - 0.0210 \qquad \log \text{S.A.} = -0.0210$$

$$\text{S.A.} = 0.953 \ \mu\text{Ci/mmole}$$

At the given counting efficiency:

$$\text{S.A.} = (0.953 \ \mu\text{Ci/mmole})(2.22 \times 10^6 \ \text{DPM}/\mu\text{Ci})(0.80 \ \text{CPM/DPM})$$

$$= 16.93 \times 10^5 \ \text{CPM/mmole} = 16.93 \times 10^8 \ \text{CPM/mole}$$

The labeled enzyme contains:

$$\frac{7380}{60} = 123 \ \text{CPM of } P^{32}$$

$$\frac{123 \ \text{CPM}}{16.93 \times 10^8 \ \text{CPM/mole}} = 7.265 \times 10^{-8} \ \text{mole of } P^{32}$$

Using the procedure outlined in Problem 2-11:

$$\frac{7.265 \times 10^{-8} \ \text{mole of } P^{32}}{3.4 \times 10^{-3} \ \text{g}} = \frac{1 \ \text{mole of } P^{32}}{\text{MW}_\text{g}}$$

$$\text{MW} = \frac{3.4 \times 10^{-3}}{7.265 \times 10^{-8}} \qquad \boxed{\textbf{MW = 46,800}}$$

G. COUNTING ERRORS

SELF-ABSORPTION

When radioactive samples are counted on planchets, some of the radiation is absorbed by the sample itself and never reaches the counting tube.   The best way to avoid errors caused by self-absorption is to count all samples (including standards) at a constant density (mg material/cm^2).   If the samples are in solution, they can be mixed with some inert material (e.g., a dilute gelatin solution) so that, after drying, each planchet receives the same amount of total mass.   Small variations in mass (determined by weighing the planchet empty and after the sample + gelatin has dried) can be corrected for with a self-absorption curve.   The curve is prepared by counting a constant amount of radioactive standard on a planchet containing different amounts of inert diluent.   Figure 6-7 shows the correction curve where 10 mg/planchet is taken as the standard density.   The count rate at any other density is corrected by dividing the observed count rate by the self-absorption factor.   The inert diluent should have the same self-absorption characteristics as the materials in the sample.   Thus, gelatin is a reasonable diluent for solutions of organic compounds, gelatin plus NaCl for salt-containing samples, unlabeled $BaCO_3$ for $BaC^{14}O_3$, and so on.

If sufficient material is available, the sample can be counted at *infinite thickness*.   An infinitely thick sample is one in which the addition of more sample to the planchet does not increase the count rate.   At infinite thickness, the radiation from the bottom of the sample is completely absorbed; only the radiation from the top 1.2 mm, for example, of the sample reaches the counting tube.   The addition of more sample will not increase the count

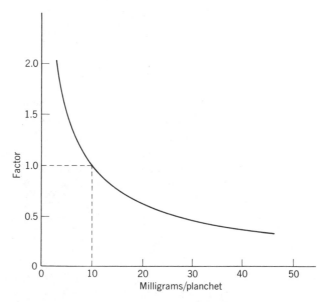

Figure 6-7   Self-absorption correction curve: 10 mg/ planchet is taken as the standard density.

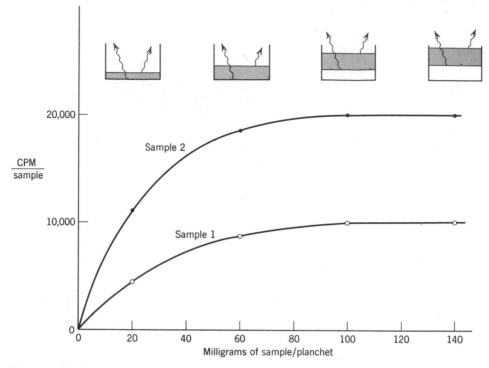

Figure 6-8 Plot of CPM/sample versus amount of sample in the planchet (i.e., sample thickness) illustrating that a constant activity is observed at "infinite thickness."

rate because, even though the amount of labeled sample under the tube is greater, only the radiation emitted from the top 1.2 mm is detected. Thus, the count rate observed at infinite thickness is directly proportional to the specific activity of the sample. This is illustrated in Figure 6-8 where the specific activity of Sample 2 is twice that of Sample 1.

QUENCHING IN SCINTILLATION COUNTING

The presence of inert material or colored material in a sample may reduce the radioactivity observed in scintillation counting. This *quenching* effect can be corrected for quite easily by means of an internal or added standard. The first thing to be sure of is that the samples are counted under the same conditions as the standards. For example, if the samples are 0.5 ml of an enzyme assay mixture (in aqueous buffer) that are added to 5 ml of scintillation fluid, then the specific activity of the standard should be determined in 0.5 ml of the same buffer, counted in 5 ml of scintillation fluid.

· Problem 6-29

A column eluate was analyzed for H^3-cyclic AMP. Exactly 0.2 ml of each fraction was counted in 0.5 ml of scintillation fluid. The column was eluted with a linear KCl gradient, so that the salt concentration is not constant in each fraction. To correct for quenching, 10 μl of H^3-cyclic AMP were added to each scintillation vial after counting the sample. The results are shown below. Calculate the unquenched H^3 activity of each sample. The blank was 0.2 ml of water.

| | CPM Above Background | |
Sample	Sample Alone	Sample + Added H^3-Cyclic AMP
Blank	5	1786
Fraction 50	9	1610
Fraction 75	115	1575
Fraction 80	5250	6630
Fraction 82	1589	2829
Fraction 85	125	1285

Solution

If we take the blank value as our standard, then the 10 μl of H^3-cyclic AMP added 1781 CPM above background. In Fraction 50, the 10 μl of H^3-cyclic AMP yields $1610 - 9 = 1601$ CPM. Thus, only $1601/1781 = 89.9\%$ of the added counts were recovered. The corrected count rate of fraction 50 is:

$$\frac{9}{0.899} = 10 \text{ CPM} \qquad \text{or} \qquad \frac{1781}{1601} \times 9 = 10 \text{ CPM}$$

Similarly, the corrected count rates of the other fractions are given by:

$$\frac{(CPM)_{added}}{(CPM)_{recovered}} \times (CPM)_{sample\ alone}$$

Fraction 75: $\dfrac{1781}{1575 - 115} \times 115 =$ **140 CPM**

Fraction 80: $\dfrac{1781}{6630 - 5250} \times 5250 =$ **6775 CPM**

Fraction 82: $\dfrac{1781}{2829 - 1589} \times 1589 =$ **2282 CPM**

Fraction 85: $\dfrac{1781}{1285 - 125} \times 125 =$ **191 CPM**

H. STABLE ISOTOPES

Isotopes that are not radioactive are called "stable isotopes." If a particular isotope contains a greater number of neutrons than the most common form of the element, the isotope is frequently referred to as a "heavy isotope." Stable isotopes commonly used in biochemical research include: H^2 (deuterium), C^{13}, N^{15}, O^{18}, O^{17}, S^{34}, Ca^{44}, Fe^{54}, Fe^{57}, Zn^{66}, and Zn^{68}.

ATOM PERCENT EXCESS

The degree to which a radioactive compound is labeled is given in terms of its specific activity (CPM/μmole, mCi/mmole, and so on). The degree of labeling of a compound with a stable isotope is expressed in terms of "atom percent excess." This term represents the proportion of stable isotope *over and above that normally present in nature*. For example, if P_n is the normal abundance (%) of a given isotope in nature and P_e is the abundance (%) in a labeled (enriched) compound, then $P_e - P_n$ is the atom percent excess (APE).

Stable isotopes may be used for enzyme assays, precursor-product determinations, and carrier dilution analyses in much the same manner as radioactive tracers. The calculations are based on the atom percent excess per mole instead of specific radioactivities.

GENERAL REFERENCES

Hendee, W. R., *Radioactive Isotopes in Biological Research.* Wiley (1973).

PRACTICE PROBLEMS

Answers to Practice Problems are given on page 430.

1. Write the nuclear reaction showing how (a) Ca^{45}, (b) Cl^{36}, (c) K^{42}, and (d) P^{33} decay by β emission.

2. An isotope has a half-life of 4 yr. Calculate (a) the decay constant, λ, in terms of yr^{-1}, day^{-1}, hr^{-1}, min^{-1}, and sec^{-1} and (b) the fraction of the original activity remaining after 13 months.

3. I^{131} has a half-life of 8.1 days. Calculate (a) the fraction of the I^{131} atoms that decays per day and per minute and (b) the specific activity of pure I^{131} in terms of Ci/g, Ci/g-atom, and DPM/g.

4. A sample of organic matter from a stream near a petroleum-refining plant contains a level of C^{14} sufficient to provide a count of 10 DPM/g carbon. What fraction of the total carbon in the stream is contamination from the plant? (Note: the organic matter in petroleum is millions of years old; essentially all the C^{14} has decayed.)

5. (a) What is the theoretical maximum specific activity (mCi/mmole) at which fructose-1,6-diphosphate-P^{32} could be pre-

pared? (b) What proportion of the molecules is actually labeled in a sample of FDP^{32} that has a specific activity of 2×10^6 DPM/μmole?

6. Calculate the weight in grams of calcium-45 in 1 mCi of carrier-free Ca^{45}. The half-life of Ca^{45} is 163 days.

7. A bottle of serine-C^{14} (uniformly labeled) contains 2.0 mCi in 3.5 ml of solution. The specific activity is given as 160 mCi/mmole. Calculate (a) the concentration of serine in the solution and (b) the activity of the solution in terms of CPM/ml at a counting efficiency of 68%.

8. A solution of L-lysine-C^{14} (uniformly labeled) contains 1.2 mCi and 0.77 mg of L-lysine per ml. Calculate the specific activity of the lysine in terms of (a) mCi/mg, (b) mCi/mmole, (c) DPM/μmole, and (d) CPM/μmole of carbon at a counting efficiency of 80%.

9. Describe the preparation of 75 ml of a 10^{-2} M solution of L-cysteine-S^{35} hydrochloride in which the amino acid has a specific

activity of 3.92×10^4 DPM/μmole. Assume that you have available solid unlabeled L-cysteine hydrochloride and a stock solution of L-cysteine-S^{35} (14 mCi/mmole and 1.2 mCi/ml).

10. Describe the preparation of 50 ml of a 10^{-3} M solution of glucose-C^{14} in which the sugar has a specific activity of 3000 DPM/μmole. Assume that you have available a 10^{-2} M stock solution containing 0.02 μCi/ml and solid glucose.

11. An aliquot of a cell-free extract of *Neurospora crassa*, containing 0.72 mg of protein, was incubated with O-acetylhomoserine and labeled methylmercaptan (C^{14}H$_3$-SH, specific activity 2.4×10^6 CPM/μmole) in a total

dialysis cell had a volume of 0.1 ml. At equilibrium, 10 μl of the solution in the chamber containing the protein gave 2570 CPM; 10 μl of the solution from the chamber without the protein gave 1870 CPM. Calculate [S], [PS], [P]$_t$, [P], and K_S.

14. A sample of C^{14}- and P^{32}-labeled AMP was prepared with a specific activity of 9500 CPM/μmole (75% of the activity results from decay of P^{32}, 25% from decay of C^{14}). Calculate the specific activity of the sample after (a) 5 days, (b) 10 days, and (c) 25 days.

15. The following data were obtained with a windowless, gas-flow planchet counter.

(Data for Practice Problem 15)

Addition	CPM Above Background	
	Without Shield	With Aluminum Shield
C^{14} Standard 250,000 DPM	75,000	12,520
P^{32} Standard 125,000 DPM	80,000	50,000
Sample	217,000	113,900

volume of 1.5 ml. The C^{14} was enzymatically incorporated into L-methionine at a rate of 2240 CPM/min. Calculate the rate of the reaction in terms of (a) μmoles/min, (b) μmoles $\times$ liter^{-1} $\times$ min^{-1}, and (c) μmoles $\times$ mg protein^{-1} $\times$ min^{-1}.

12. A 10 ml suspension of Cr^{51}-labeled red blood cells, containing 3×10^8 CPM total radioactivity, was injected into a subject. After 10 min a small blood sample was taken and found to contain 5×10^4 CPM/ml. Calculate the total blood volume of the individual.

13. A cyclic GMP binding protein was purified to homogeneity from lymphosarcoma cells. The protein binds cyclic GMP reversibly. The molecular weight of the binding protein is 60,000. A solution of the binding protein containing 6 μg/ml was placed in one side of a dialysis membrane. A solution containing H^3-cyclic GMP (S.A. = 10^9 CPM/μmole) was placed on the other side of the membrane. Each chamber of the

Calculate the true C^{14} and P^{32} DPM in the sample.

16. A small amount of Cu^{64} ($t_{\frac{1}{2}}$ = 12.8 hr) was injected into the bloodstream of an animal. Blood samples were taken periodically and counted immediately. The specific activities of the samples are shown below. (a) From the data, calculate the effective half-life and the biological half-life of the isotope in the bloodstream. (b) What was the zero-time S.A. (CPM/ml)?

Sample Time (hours)	Specific Activity (CPM/ml)
2	7120
4	5070
6	3610
10	1830
18	470

17. A solution containing 0.5 mg of D-mannose-C^{14} (uniformly labeled, specific activity 3.3×10^6 CPM/μmole) was added to 50 ml of a solution containing an unknown amount of unlabeled mannose. After mixing, the D-mannose was reisolated as the osazone. The osazone had a specific activity of 14,280 CPM/μmole. Calculate the concentration of unlabeled D-mannose in the original solution.

18. Fifty-six micrograms of Co^{60}-labeled vitamin B$_{12}$, containing 7.39×10^5 CPM, were added to a sample containing an unknown amount of unlabeled vitamin B$_{12}$. The sample was then extracted and the vitamin B$_{12}$ purified by chromatography. The final product contained 49 μg of vitamin B$_{12}$ and 1.58×10^5 CPM of radioactivity. Calculate the amount of unlabeled vitamin B$_{12}$ in the sample.

19. A yeast culture was grown in a synthetic medium containing S^{35}O$_4^{2-}$ (specific activity 4.78×10^7 CPM/μmole) as the sulfur source. After several days of growth, the cells were harvested and extracted. To 50 ml of extract, 500 mg of unlabeled reduced glutathione were added. Glutathione was then reisolated from the mixture. The reisolated compound had a specific activity of 6.97×10^6 CPM/μmole. Calculate the concentration of glutathione in the extract.

20. A mixture of fatty acids was treated with C^{14}-labeled diazomethane (specific activity 1.93×10^5 CPM/μmole) to produce the methyl-C^{14}-labeled esters of each acid pres-ent. Unlabeled methyl stearate (2 mmoles) was then added to the mixture. A small amount of methyl stearate was reisolated from the mixture and found to have a specific activity of 4.87×10^3 CPM/μmole. Calculate the amount of stearic acid in the mixture.

21. Exactly 1.7 mg of a purified enzyme (MW = 55,000) was incubated with an excess of iodoacetamide-C^{14} (S.A. = 2 μCi/mmole). The carboxymethylated protein was then precipitated, washed free of unreacted iodoacetamide-C^{14}, dissolved in a small amount of buffer, and the entire solution counted in a scintillation counter operating at 80% efficiency. In one hour, the sample gave 13,190 counts above background. Calculate the number of reactive SH groups per molecule of protein.

22. The enzyme ADPG phosphorylase (ADPG synthetase) was assayed by following the incorporation of P^{32} from PP$_i^{32}$ into ATP: ADPG + PP$_i^{32}$ $\rightleftharpoons$ G-1-P + ATP^{32}. The ATP^{32} formed was adsorbed into charcoal, washed free of occluded PP$_i^{32}$, resuspended in 1.0 ml of aqueous ethanol-NH$_3$, and 0.5 ml counted in a scintillation counter. A sample gave 21,550 CPM above background. To check for quenching, 10,000 CPM of PP$_i^{32}$ (counted earlier in the absence of charcoal) was added to the sample. The sample plus added PP$_i^{32}$ now counted 28,270 CPM above background. (a) What is the true (unquenched) activity of the sample? (b) How can corrections resulting from quenching be made unnecessary?

APPENDICES

PROPERTIES OF COMMERCIAL CONCENTRATED SOLUTIONS OF ACIDS AND BASES

Compound	MW	Specific Gravity	Percent w/w	g/100 ml	Approximate N	Milliliters Required for 1 liter of 1 N Solution
HCl	36.5	1.19	37	44	12.1	82.5
HNO$_3$	63.0	1.42	70	91	15.8	63.5
H$_2$SO$_4$	98.1	1.84	96	173	35.2	29
H$_3$PO$_4$	98.0	1.71	85	146	44.5	22.5
HClO$_4$	100.5	1.66	70	116.2	11.6	86.5
HCOOH	46.0	1.20	88	105.6	24	41.6
CH$_3$COOH	60.0	1.06	100	106	17.4	57.5
NH$_3$	17.0	0.91	28	22.8	14.8	67.5

FRACTIONATION WITH SOLID AMMONIUM SULFATE

Final concentration of ammonium sulfate—% saturation at 0°C

solid ammonium sulfate to add to 100 ml of solution

Initial conc.	20	25	30	35	40	45	50	55	60	65	70	75	80	85	90	95	100
0	10.6	13.4	16.4	19.4	22.6	25.8	29.1	32.6	36.1	39.8	43.6	47.6	51.6	55.9	60.3	65.0	69.7
5	7.9	10.8	13.7	16.6	19.7	22.9	26.2	29.6	33.1	36.8	40.5	44.4	48.4	52.6	57.0	61.5	66.2
10	5.3	8.1	10.9	13.9	16.9	20.0	23.3	26.6	30.1	33.7	37.4	41.2	45.2	49.3	53.6	58.1	62.7
15	2.6	5.4	8.2	11.1	14.1	17.2	20.4	23.7	27.1	30.6	34.3	38.1	42.0	46.0	50.3	54.7	59.2
20	0	2.7	5.5	8.3	11.3	14.3	17.5	20.7	24.1	27.6	31.2	34.9	38.7	42.7	46.9	51.2	55.7
25		0	2.7	5.6	8.4	11.5	14.6	17.9	21.1	24.5	28.0	31.7	35.5	39.5	43.6	47.8	52.2
30			0	2.8	5.6	8.6	11.7	14.8	18.1	21.4	24.9	28.5	32.3	36.2	40.2	44.5	48.8
35				0	2.8	5.7	8.7	11.8	15.1	18.4	21.8	25.4	29.1	32.9	36.9	41.0	45.3
40					0	2.9	5.8	8.9	12.0	15.3	18.7	22.2	25.8	29.6	33.5	37.6	41.8
45						0	2.9	5.9	9.0	12.3	15.6	19.0	22.6	26.3	30.2	34.2	38.3
50							0	3.0	6.0	9.2	12.5	15.9	19.4	23.0	26.8	30.8	34.8
55								0	3.0	6.1	9.3	12.7	16.1	19.7	23.5	27.3	31.3
60									0	3.1	6.2	9.5	12.9	16.4	20.1	23.9	27.9
65										0	3.1	6.3	9.7	13.2	16.8	20.5	24.4
70											0	3.2	6.5	9.9	13.4	17.1	20.9
75												0	3.2	6.6	10.1	13.7	17.4
80													0	3.3	6.7	10.3	13.9
85														0	3.4	6.8	10.5
90															0	3.4	7.0
95																0	3.5
100																	0

Initial concentration of ammonium sulfate, % saturation at 0°C

Reprinted by permission of the Oxford University Press (Oxford) from *Data for Biochemical Research*, 2nd ed. Edited by R. M. C. Dawson, D. C. Elliott, W. H. Elliott, and K. M. Jones. © Oxford University Press (1969). See *Methods in Enzymology*, Vol. 1, p. 76 (1955) for a similar table prepared for 25°C.

Note: The pH of the solution may decrease significantly on addition of ammonium sulfate.

FRACTIONATION WITH SATURATED AMMONIUM SULFATE SOLUTION

The table gives the milliliters of saturated ammonium sulfate solution to be added to one liter of solution to produce the desired change in percent saturation. The volume changes on mixing are negligible. The pH of a saturated ammonium sulfate solution is about 5.5. The pH can be adjusted to 7 by adding a few drops of concentrated NH_4OH.

Initial concentration of ammonium sulfate in the preparation (percent saturation)

Final	0	5	10	15	20	25	30	35	40	45	50	55	60	65	70	75	80	85
5	52.6																	
10	111	55.8																
15	177	118	58.8															
20	250	188	125	62.5														
25	333	267	200	133	66.7													
30	429	357	286	214	143	71.4												
35	559	462	385	308	231	154	76.9											
40	667	583	500	417	333	250	167	83.3										
45	818	727	637	546	455	364	273	182	91.0									
50	1000	900	800	700	600	500	400	300	200	100								
55	1222	1111	1000	889	778	667	556	444	333	222	111							
60	1500	1375	1250	1125	1000	875	750	625	500	375	250	125						
65	1857	1714	1571	1429	1286	1143	1000	857	714	571	429	286	143					
70	2333	2167	2001	1833	1667	1500	1333	1167	1000	833	667	500	333	167				
75	3000	2800	2600	2400	2200	2000	1800	1600	1400	1200	1000	800	600	400	200			
80	4000	3750	3500	3250	3000	2750	2500	2250	2000	1750	1500	1250	1000	750	500	250		
85	5667	5333	5000	4667	4333	4000	3667	3333	3000	2667	2333	2000	1667	1333	1000	667	333	
90	9000	8500	8000	7500	7000	6500	6000	5500	5000	4500	4000	3500	3000	2500	2000	1500	1000	500

Final concentration of ammonium sulfate desired (percent saturation)

Saturated ammonium sulfate solutions at various temperatures

Temperature (°C)	0	10	20	25	30
Moles $(NH_4)_2SO_4$ per 1000 g H_2O	5.35	5.53	5.73	5.82	5.91
Percentage by weight	41.42	42.22	43.09	43.47	43.85
g $(NH_4)_2SO_4$ required to saturate 1000 ml H_2O	706.8	730.5	755.8	766.8	777.5
g $(NH_4)_2SO_4$ per liter of saturated solution	514.8	525.2	536.5	541.2	545.9
Molarity of saturated solution	3.90	3.97	4.06	4.10	4.13

Reprinted by permission of the Oxford University Press (Oxford) from *Data for Biochemical Research*, 2nd ed. Edited by R. M. C. Dawson, D. C. Elliott, W. H. Elliott, and K. M. Jones, © Oxford University Press (1969).

APPENDIX IV

pK_a VALUES OF ACIDS AND BASES USEFUL IN PREPARING BUFFERS

Listed below are some acids and bases that are useful in preparing buffers for enzyme assays. The choice of a particular compound depends on many factors. For example, multicarboxylic acids would be poor choices for reactions involving metal ions as cofactors; amino acids may be poor choices for reactions involving amino acids as substrates. The number of buffer components can be kept to a minimum by using an acid and a base to cover the desired region. For example, maleic acid and Tris can be mixed to produce Tris-maleate buffers of pH 5.7 to 8.6 (rather than using maleic acid-NaOH and Tris-HCl).

The activity coefficients of multivalent ions change markedly with concentration. Consequently, the pH of a stock buffer should be checked after dilution and readjusted if necessary. As a general rule, the pH of a reaction mixture should be checked at the end of an assay period to insure that it remained constant.

Free Acid or Base	MW	pK_a at 25°C
Pyrophosphoric	177.98	0.85 (pK_{a_1})
Oxalic	95.07	1.19 (pK_{a_1})
Glycerophosphoric	172.08	1.47 (pK_{a_1})
Ethylenediamine tetraacetic acid (EDTA)	292.24	1.70 (pK_{a_1})
Histidine	155.16	1.82 (pK_{a_1})
Pyrophosphoric	177.98	1.96 (pK_{a_2})
Maleic	116.07	2.00 (pK_{a_1})
Benzenehexacarboxylic (mellitic)	342.17	2.08 (pK_{a_1})
Phosphoric	98.0	2.12 (pK_{a_1})
Brucine tetrahydrate	466.53	2.30 (pK_{a_1})
Benzenepentacarboxylic	296.18	2.34 (pK_{a_1})
Glycine	75.07	2.34 (pK_{a_1})
Benzene-1,2,4,5-tetracarboxylic (pyromellitic)	254.15	2.43 (pK_{a_1})
Benzenehexacarboxylic (mellitic)	342.17	2.46 (pK_{a_2})
EDTA	292.24	2.6 (pK_{a_2})
Malonic	146.02	2.85 (pK_{a_1})
Phthalic	116.13	2.90
Benzenepentacarboxylic	298.16	2.95 (pK_{a_2})
Salicylic	138.12	2.98
Benzene-1,2,3-tricarboxylic (hemimellitic)	246.18	2.98 (pK_{a_1})
1,4-Piperazinebis-(ethanesulfonic acid) "PIPES"	302.37	3.0 (pK_{a_3})
Tartaric	150.09	3.02 (pK_{a_1})
Fumaric	116.07	3.03 (pK_{a_1})

Free Acid or Base	MW	pK_a at 25°C
Glycylglycine	132.12	3.06
Citric acid	192.12	3.06 (pK_{a_1})
Cyclopentanetetra-1,2,3,4-carboxylic	246.17	3.07 (pK_{a_1})
o-Phthalic	166.13	3.10 (pK_{a_1})
Benzene-1,2,4,5-tetracarboxylic (pyromellitic)	254.15	3.13 (pK_{a_2})
Benzene-1,3,5-tricarboxylic (trimesic)	210.14	3.16 (pK_{a_1})
Benzenehexacarboxylic (mellitic)	342.17	3.24 (pK_{a_3})
Dimethylmalonic	132.12	3.29 (pK_{a_1})
Mandelic	152.15	3.36
Butane-1,2,3,4-tetracarboxylic	234.12	3.36 (pK_{a_1})
Malic	134.09	3.40 (pK_{a_1})
1,1-Cyclohexanediacetic	200.18	3.52 (pK_{a_1})
2-Methylpropane-1,2,3-triscarboxylic (β-methyltricarballylic)	190.15	3.53 (pK_{a_1})
Hippuric	179.18	3.64
Propane-1,2,3-tricarboxylic (tricarballylic)	176.12	3.67 (pK_{a_1})
Formic	46.02	3.75
3,3-Dimethylglutaric	160.17	3.79 (pK_{a_1})
1,1-Cyclopentanediacetic (3,3 tetra-methyleneglutaric acid)	186.21	3.82 (pK_{a_1})
Itaconic	130.1	3.84 (pK_{a_1})
Lactic	90.08	3.86
Benzenepentacarboxylic	298.16	3.94 (pK_{a_3})
Benzene-1,3,5-tricarboxylic (trimesic)	210.14	3.98 (pK_{a_2})
Barbituric	128.09	3.98
Ascorbic	176.12	4.1 (pK_{a_1})
2,2-Dimethylsuccinic	146.14	4.11 (pK_{a_1})
Succinic	118.09	4.19 (pK_{a_1})
Benzoic	122.12	4.20
Oxalic	95.07	4.21 (pK_{a_2})
Benzene-1,2,3-tricarboxylic (hemimellitic)	246.18	4.25 (pK_{a_2})
3,6-Endomethylene-1,2,3,6-tetrahydrophthalic acid "EMTA" (endo-5-norbornene-2,3-dicarboxylic acid "ENDCA")	183.62	4.3 (pK_{a_1})
2,2-Dimethylglutaric	160.17	4.31 (pK_{a_1})
Butane-1,2,3,4-tetracarboxylic	234.12	4.38 (pK_{a_2})
Benzenehexacarboxylic (mellitic)	342.17	4.44 (pK_{a_4})
Benzene-1,2,4,5-tetracarboxylic (pyromellitic)	254.15	4.44 (pK_{a_3})
Fumaric	116.07	4.47 (pK_{a_2})
Cyclopentanetetra-1,2,3,4-carboxylic	246.17	4.48 (pK_{a_2})
Tartaric	150.09	4.54 (pK_{a_2})
Citric	210.14	4.74 (pK_{a_2})
Acetic	60.05	4.76
n-Butyric	88.1	4.82
Propane-1,2,3-tricarboxylic (tricarballylic)	176.12	4.84 (pK_{a_2})
Benzene-1,3,5-tricarboxylic (trimesic)	210.14	4.85 (pK_{a_3})
Propionic	74.08	4.87
2-Methylpropane-1,2,3-triscarboxylic (β-methyltricarballylic)	190.15	5.02 (pK_{a_2})
Malic	134.09	5.05 (pK_{a_2})
Benzenepentacarboxylic	298.16	5.07 (pK_{a_4})
Pyridine	79.1	5.23

Free Acid or Base	MW	pK$_a$ at 25°C
o-Phthalic	116.13	5.27 (pK$_{a_2}$)
Citric	192.12	5.40 (pK$_{a_3}$)
Butane-1,2,3,4-tetracarboxylic	234.12	5.45 (pK$_{a_3}$)
Benzenehexacarboxylic (mellitic)	342.17	5.50 (pK$_{a_5}$)
2,2-Dimethylglutaric	160.17	5.51 (pK$_{a_2}$)
Itaconic	130.1	5.55 (pK$_{a_2}$)
Cyclopentanetetra-1,2,3,4-carboxylic	246.17	5.57 (pK$_{a_3}$)
Succinic	118.09	5.57 (pK$_{a_2}$)
Benzene-1,2,4,5-tetracarboxylic (pyromellitic)	254.15	5.61 (pK$_{a_4}$)
Benzene-1,2,3-tricarboxylic (hemimellitic)	246.18	5.87 (pK$_{a_3}$)
Dimethylmalonic	132.12	5.98 (pK$_{a_2}$)
Histidine	156.16	6.00 (pK$_{a_2}$)
Hydroxylamine	34.0	6.03
Carbonic (H$_2$CO$_3$ + CO$_2$)	62(CO$_2$)	6.10 (pK$'_{a_1}$)
Malonic	104.06	6.10 (pK$_{a_2}$)
2-(N-Morpholino)-ethane sulfonic acid "MES"	195.2	6.15 (pK$_{a_2}$)
Glycerophosphoric	172.08	6.19 (pK$_{a_2}$)
Propane-1,2,3-tricarboxylic (tricarballylic)	176.12	6.20 (pK$_{a_3}$)
Benzenepentacarboxylic	298.16	6.25 (pK$_{a_5}$)
Maleic	116.07	6.26 (pK$_{a_2}$)
2,2-Dimethylsuccinic	146.14	6.29 (pK$_{a_2}$)
EDTA	292.24	6.30 (pK$_{a_3}$)
3,3-Dimethylglutaric	160.17	6.31 (pK$_{a_2}$)
Bis(2-hydroxyethyl)imino-tris(hydroxymethyl)-methane "BIS-TRIS"	209.24	6.46
Benzenehexacarboxylic (mellitic)	342.17	6.59 (pK$_{a_6}$)
N-(2-Acetamido)imino-diacetic acid "ADA"	190.17	6.6 (pK$_{a_3}$)
Butane-1,2,3,4-tetracarboxylic	234.12	6.63 (pK$_{a_4}$)
Pyrophosphoric	177.98	6.68 (pK$_{a_3}$)
1,1-Cyclopentanediacetic (3,3 tetramethylene-glutaric acid)	186.21	6.70 (pK$_{a_2}$)
1,4-Piperazinebis-(ethanesulfonic acid) "PIPES"	302.37	6.8 (pK$_{a_4}$)
N-(2-Acetamido)-2-aminoethanesulfonic acid "ACES"	182.20	6.9 (pK$_{a_2}$)
1,1-Cyclohexanediacetic	200.18	6.94 (pK$_{a_2}$)
3,6-Endomethylene-1,2,3,6-tetrahydrophthalic acid "EMTA" ("ENDCA")	183.62	7.0 (pK$_{a_2}$)
Imidazole	68.08	7.0
2-(Aminoethyl)trimethylammonium chloride "CHOLAMINE"	156.69	7.1
N,N-Bis(2-hydroxyethyl)-2-aminoethanesulfonic acid "BES"	213.25	7.15 (pK$_{a_2}$)
2-Methylpropane-1,2,3-triscarboxylic (β-methyltricarballylic)	190.15	7.20 (pK$_{a_3}$)
2-(N-Morpholino)propane-sulfonic acid "MOPS"	209.27	7.2 (pK$_{a_2}$)
Phosphoric	98.0	7.21 (pK$_{a_2}$)
N-Tris(hydroxymethyl)methyl-2-aminoethane sulfonic acid "TES"	229.28	7.5 (pK$_{a_2}$)
N-2-Hydroxyethylpiperazine-N'-2-ethanesulfonic acid "HEPES"	238.31	7.55 (pK$_{a_2}$)
2-Hydroxyethylimino-tris(hydroxymethyl)methane "MONO-TRIS"	165.18	7.83

Free Acid or Base	MW	pK_a at 25°C
Brucine tetrahydrate	466.53	7.95 (pK_{a_2})
4-(2-Hydroxyethyl)-1-piperazinepropane sulfonic acid "EPPS"	252.23	8.0
Tris(hydroxymethyl)aminomethane "TRIS"	121.14	8.1
N-Tris(hydroxymethyl)methylglycine "TRICINE"	180.18	8.15
Glycinamide	74.04	8.2
N,N-Bis(2-hydroxyethyl)glycine "BICINE"	163.18	8.35
N-Tris(hydroxymethyl)methyl-2-aminopropane sulfonic acid "TAPS"	243.3	8.4 (pK_{a_2})
N-Glycyl-glycine	132.12	8.4
Histidine	155.16	9.17 (pK_{a_3})
Boric	43.82	9.24
Pyrophosphoric	177.98	9.39 (pK_{a_4})
Ethanolamine	61.08	9.44
Glycine	75.07	9.6 (pK_{a_2})
Trimethylamine	59.11	9.74
Cyclopentanetetra-1,2,3,4-carboxylic	246.17	10.06 (pK_{a_4})
Carbonic ($H_2CO_3 + CO_2$)	62(CO_2)	10.25 (pK_{a_2})
3-Cyclohexylamino-1-propanesulfonic acid "CAPS"	221.32	10.40 (pK_{a_2})
EDTA	292.24	10.6 (pK_{a_4})
Methylamine	31.06	10.64
Dimethylamine	45.09	10.72
Ethylamine	45.09	10.75
Triethylamine	101.19	10.76
Diethylamine	73.14	10.98
Ascorbic	176.12	11.79 (pK_{a_2})
Phosphoric	98.0	12.32 (pK_{a_3})

ACTIVITY COEFFICIENTS OF SOME IONS IN AQUEOUS SOLUTION

Ion	Ionic Concentration (M)		
	0.001 M	0.01 M	0.1 M
H^+	0.975	0.933	0.86
OH^-	0.975	0.925	0.805
Acetate$^-$	0.975	0.928	0.82
$H_2PO_4^-$	0.975	0.928	0.744
HPO_4^{2-}	0.903	0.740	0.445
PO_4^{3-}	0.796	0.505	0.16
H_2citrate$^-$	0.975	0.926	0.81
Hcitrate^{2-}	0.903	0.741	0.45
Citrate^{3-}	0.796	0.51	0.18
HCO_3^-	0.975	0.928	0.82
CO_3^{2-}	0.903	0.742	0.445

APPENDIX VI

pK_a CORRECTION FACTORS FOR IONIC STRENGTH

Correction: ΔpK_a

$\Gamma/2$	$Z=+1$	$Z=0$	$Z=-1$	$Z=-2$
0.01	+0.04	−0.04	−0.13	−0.22
0.05	+0.08	−0.08	−0.25	−0.42
0.10	+0.11	−0.11	−0.32	−0.53

Selected values from Bruening, G., Criddle, R., Preiss, J., and Rudert, F. *Biochemical Experiments,* p. 60 Wiley-Interscience (1970).

Values are given for 20°C. They are essentially the same between 0 and 37°C.
$\Gamma/2$ = the ionic strength (total of all ions present)
Z = the charge on the conjugate acid of the buffer
At any total ionic strength: $pK_a' = pK_a + \Delta pK_a$
The corrections are based on the Debye-Hückel equation:

$$\log \gamma = -0.509\, Z^2 \sqrt{\Gamma/2}$$

where γ = the activity coefficient of an ion.

The relationship between pK_a and pK_a' is shown below:

$$\begin{aligned} \mathrm{pH} &= pK_a + \log \frac{\gamma_{\mathrm{C.B.}}\ [\mathrm{C.B.}]}{\gamma_{\mathrm{C.A.}}\ [\mathrm{C.A.}]} \\ &= pK_a + \log \frac{\gamma_{\mathrm{C.B.}}}{\gamma_{\mathrm{C.A.}}} + \log \frac{[\mathrm{C.B.}]}{[\mathrm{C.A.}]} \\ &= pK_a + \Delta pK_a + \log \frac{[\mathrm{C.B.}]}{[\mathrm{C.A.}]} \\ &= pK_a' + \log \frac{[\mathrm{C.B.}]}{[\mathrm{C.A.}]} \end{aligned}$$

where: $pK_a' = pK_a + \Delta pK_a$

and $\Delta pK_a = \log \dfrac{\gamma_{\mathrm{C.B.}}}{\gamma_{\mathrm{C.A.}}}$

APPENDIX **VII**

IONIZATION CONSTANTS, pK_a, pK_b, AND pI VALUES OF SOME COMMON AMINO ACIDS[a]

Compound	MW	Conjugate Acid	K_a	pK_a	Conjugate Base	K_b	pK_b	pI
α-Alanine	89.1	α-COOH	4.47×10^{-3}	2.35	α-COO$^-$	2.24×10^{-12}	11.65	6.02
		α-NH$_3^+$	2.04×10^{-10}	9.69	α-NH$_2$	4.90×10^{-5}	4.31	
β-Alanine	89.1	α-COOH	2.51×10^{-4}	3.60	α-COO$^-$	3.98×10^{-11}	10.40	6.90
		β-NH$_3^+$	6.46×10^{-11}	10.19	β-NH$_2$	1.55×10^{-4}	3.81	
Arginine	174.2	α-COOH	6.76×10^{-3}	2.17	α-COO$^-$	1.48×10^{-12}	11.83	10.76
		α-NH$_3^+$	9.12×10^{-10}	9.04	α-NH$_2$	1.10×10^{-5}	4.96	
		Guanidinium-NH$_2^+$	3.31×10^{-13}	12.48	Guanidinium-NH	3.02×10^{-2}	1.52	
Asparagine	132.1	α-COOH	9.55×10^{-3}	2.02	α-COO$^-$	1.05×10^{-12}	11.98	5.41
		α-NH$_3^+$	1.58×10^{-9}	8.8	α-NH$_2$	6.31×10^{-6}	5.2	
Aspartic acid	133.1	α-COOH	8.13×10^{-3}	2.09	α-COO$^-$	1.23×10^{-12}	11.91	2.98
		β-COOH	1.38×10^{-4}	3.86	β-COO$^-$	7.25×10^{-11}	10.14	
		α-NH$_3^+$	1.51×10^{-10}	9.82	α-NH$_2$	6.61×10^{-5}	4.18	
Citrulline	175.2	α-COOH	3.72×10^{-3}	2.43	α-COO$^-$	2.69×10^{-12}	11.57	5.92
		α-NH$_3^+$	3.89×10^{-10}	9.41	α-NH$_2$	2.57×10^{-5}	4.59	
Cysteine	121.2	α-COOH	1.95×10^{-2}	1.71	α-COO$^-$	5.13×10^{-13}	12.29	5.02
		β-SH	4.68×10^{-9}	8.33	β-S$^-$	2.14×10^{-6}	5.67	
		α-NH$_3^+$	1.66×10^{-11}	10.78	α-NH$_2$	6.03×10^{-4}	3.22	

Compound	MW	Conjugate Acid	K_a	pK_a	Conjugate Base	K_b	pK_b	pI
Cystine	240.3	α-COOH	2.24×10^{-2}	1.65	α-COO⁻	4.47×10^{-13}	12.35	5.06
		α-COOH	5.50×10^{-3}	2.26	α-COO⁻	1.82×10^{-12}	11.74	
		α-NH₃⁺	1.41×10^{-8}	7.85	α-NH₂	7.08×10^{-7}	6.15	
		α-NH₃⁺	1.41×10^{-10}	9.85	α-NH₂	7.08×10^{-5}	4.15	
Glutamic acid	147.1	α-COOH	6.46×10^{-3}	2.19	α-COO⁻	1.55×10^{-12}	11.81	3.22
		γ-COOH	5.62×10^{-5}	4.25	γ-COO⁻	1.78×10^{-10}	9.75	
		α-NH₃⁺	2.14×10^{-10}	9.67	α-NH₂	4.68×10^{-5}	4.33	
Glutamine	146.1	α-COOH	6.76×10^{-3}	2.17	α-COO⁻	1.48×10^{-12}	11.83	5.65
		α-NH₃⁺	7.41×10^{-10}	9.13	α-NH₂	1.35×10^{-5}	4.87	
Glycine	75.1	α-COOH	4.57×10^{-3}	2.34	α-COO⁻	2.19×10^{-12}	11.66	5.97
		α-NH₃⁺	2.51×10^{-10}	9.6	α-NH₂	3.98×10^{-5}	4.4	
Histidine	155.2	α-COOH	1.51×10^{-2}	1.82	α-COO⁻	6.61×10^{-13}	12.18	7.58
		Imidazole-NH⁺	1.0×10^{-7}	6.0	Imidazole-N⁻	1.0×10^{-9}	8.0	
		α-NH₃⁺	6.76×10^{-10}	9.17	α-NH₂	1.48×10^{-5}	4.83	
Homocysteine	135.2	α-COOH	6.03×10^{-3}	2.22	α-COO⁻	1.66×10^{-12}	11.78	5.54
		α-NH₃⁺	1.35×10^{-9}	8.87	α-NH₂	7.41×10^{-6}	5.13	
		γ-SH	1.38×10^{-11}	10.86	γ-S⁻	7.25×10^{-4}	3.14	
Homocystine	268.3	α-COOH	2.57×10^{-2}	1.59	α-COO⁻	3.89×10^{-13}	12.41	5.53
		α-COOH	2.88×10^{-3}	2.54	α-COO⁻	3.47×10^{-12}	11.46	
		α-NH₃⁺	3.02×10^{-9}	8.52	α-NH₂	3.31×10^{-6}	5.48	
		α-NH₃⁺	3.63×10^{-10}	9.44	α-NH₂	2.76×10^{-5}	4.56	
Hydroxylysine	162.2	α-COOH	7.41×10^{-3}	2.13	α-COO⁻	1.35×10^{-12}	11.87	9.15
		α-NH₃⁺	2.40×10^{-9}	8.62	α-NH₂	4.17×10^{-6}	5.38	
		ε-NH₃⁺	2.14×10^{-10}	9.67	ε-NH₂	4.68×10^{-5}	4.33	
Hydroxyproline	131.1	α-COOH	1.20×10^{-2}	1.92	α-COO⁻	8.32×10^{-13}	12.08	5.83
		a-NH₃⁺	1.86×10^{-10}	9.73	α-NH₂	5.37×10^{-5}	4.27	
Isoleucine	131.2	α-COOH	4.37×10^{-3}	2.36	α-COO⁻	2.29×10^{-12}	11.64	6.02
		α-NH₃⁺	2.09×10^{-10}	9.68	α-NH₂	4.78×10^{-5}	4.32	
Leucine	131.2	α-COOH	4.37×10^{-3}	2.36	α-COO⁻	2.29×10^{-12}	11.64	5.98
		α-NH₃⁺	2.51×10^{-10}	9.60	α-NH₂	3.98×10^{-5}	4.40	

Amino acid	MW		K_a	pK_a		K_b	pK_b	pI
Lysine	146.2	α-COOH	6.61×10^{-3}	2.18	α-COO$^-$	1.51×10^{-12}	11.82	9.74
		α-NH$_3^+$	1.12×10^{-9}	8.95	α-NH$_2$	8.91×10^{-6}	5.05	
		ϵ-NH$_3^+$	2.95×10^{-11}	10.53	ϵ-NH$_2$	3.39×10^{-4}	3.47	
Methionine	149.2	α-COOH	5.25×10^{-3}	2.28	α-COO$^-$	1.91×10^{-12}	11.72	5.75
		α-NH$_3^+$	6.17×10^{-10}	9.21	α-NH$_2$	1.62×10^{-5}	4.79	
Ornithine	132.2	α-COOH	1.15×10^{-2}	1.94	α-COO$^-$	8.71×10^{-13}	12.06	9.70
		α-NH$_3^+$	2.24×10^{-9}	8.65	α-NH$_2$	4.47×10^{-6}	5.35	
		δ-NH$_3^+$	1.74×10^{-11}	10.76	δ-NH$_2$	5.76×10^{-4}	3.24	
Phenylalanine	165.2	α-COOH	1.48×10^{-2}	1.83	α-COO$^-$	6.76×10^{-13}	12.17	5.48
		α-NH$_3^+$	7.41×10^{-10}	9.13	α-NH$_2$	1.35×10^{-5}	4.87	
Proline	115.1	α-COOH	1.02×10^{-2}	1.99	α-COO$^-$	9.77×10^{-13}	12.01	6.30
		α-NH$_3^+$	2.51×10^{-11}	10.60	α-NH$_2$	3.98×10^{-4}	3.40	
Serine	105.1	α-COOH	6.17×10^{-3}	2.21	α-COO$^-$	1.62×10^{-12}	11.79	5.68
		α-NH$_3^+$	7.08×10^{-10}	9.15	α-NH$_2$	1.41×10^{-5}	4.85	
Taurine	125.1	-SO$_3$H	3.16×10^{-2}	1.5	-SO$_3^-$	3.16×10^{-13}	12.5	5.12
		α-NH$_3^+$	1.82×10^{-9}	8.74	α-NH$_2$	5.50×10^{-6}	5.26	
Threonine	119.1	α-COOH	2.35×10^{-3}	2.63	α-COO$^-$	4.27×10^{-12}	11.37	6.53
		α-NH$_3^+$	3.72×10^{-11}	10.43	α-NH$_2$	2.69×10^{-4}	3.57	
Tryptophan	204.2	α-COOH	4.17×10^{-3}	2.38	α-COO$^-$	2.40×10^{-12}	11.62	5.88
		α-NH$_3^+$	4.07×10^{-10}	9.39	α-NH$_2$	2.46×10^{-5}	4.61	
Tyrosine	181.2	α-COOH	6.31×10^{-3}	2.20	α-COO$^-$	1.59×10^{-11}	11.80	5.65
		α-NH$_3^+$	7.76×10^{-10}	9.11	α-NH$_2$	1.29×10^{-5}	4.89	
		-OH	8.51×10^{-11}	10.07	-O$^-$	1.18×10^{-4}	3.93	
Valine	117.1	α-COOH	4.79×10^{-3}	2.32	α-COO$^-$	2.09×10^{-12}	11.68	5.97
		α-NH$_3^+$	2.40×10^{-10}	9.62	α-NH$_2$	4.17×10^{-5}	4.38	
			K_{a_1}	pK_{a_1}		K_{b_3}	pK_{b_3}	
			K_{a_2}	pK_{a_2}		K_{b_2}	pK_{b_2}	
			K_{a_3}	pK_{a_3}		K_{b_1}	pK_{b_1}	

[a] K and pK values are numbered as shown on the right.

$\Delta G'$ VALUES FOR THE HYDROLYSIS OF SOME COMPOUNDS OF BIOLOGICAL IMPORTANCE

General Type	*Example*	*Hydrolysis Products*	*Approximate* $\Delta G'$ (pH ~ 7) *(kcal/mole)*
Phosphosulfate anhydride	Adenosine phosphosulfate (APS)	$AMP + SO_4^{2-}$	-18
Pyrophosphate	Inorganic PP_i	$2 P_i$	-4.2
	ATP	$ADP + P_i$	-7.7
	ATP	$AMP + PP_i$	-9.9
	ADP	$AMP + P_i$	-6.4
Acyl phosphate	Acetyl phosphate	$Acetate + P_i$	-10
	1,3-DiPGA	$3\text{-}PGA + P_i$	-12
	Carbamyl phosphate	$Carbamate + P_i$	-12 (pH 9.5)
Acyl adenylate	Amino acid adenylate	$Amino\ acid + AMP$	-13
Amino acid ester	Glycine ethylester	$Glycine + ethanol$	-8
	Glycyl tRNA	$Glycine + t$RNA	-8
Phenolic ester	p-NO_2-Phenylacetate	p-Nitrophenol + acetate	-13
	p-NO_2-Phenylsulfate	p-Nitrophenol + SO_4^{2-}	-13
Nucleotide diphosphosugar	UDPG	$UDP + glucose$	-7.3
Cyclic phosphodiester	$3',5'$-Cyclic AMP	$5'$-AMP	-10
Enolic phosphate	Phosphoenol pyruvate (PEP)	$Ketopyruvate + P_i$	-13.8
Guanidinium phosphate	Creatine phosphate	$Creatine + P_i$	-10.5
	Arginine phosphate	$Arginine + P_i$	-10.5

General Type	Example	Hydrolysis Products	Approximate $\Delta G'$ (pH ~ 7) (kcal/mole)
Thioester	Acetyl-S-CoA	CoASH + acetate	-8.2
	Acetoacetyl-S-CoA	CoASH + acetoacetate	-10.5
Acyl ester	Acetyl carnitine	Carnitine + acetate	-7.2
	Acetyl choline	Choline + acetate	-6
Hemiacetal-1-phosphate	Glucose-1-phosphate	Glucose + P_i	-5
Simple phosphate ester	Glucose-6-phosphate	Glucose + P_i	-3
	α-Glyceryl-phosphate	Glycerol + P_i	-2.5
	AMP	Adenosine + P_i	-2
Amide	Glutamine	Glutamate + P_i	-3.4
	Asparagine	Aspartate + P_i	-3.4
Peptide bond	Glycylglycine	Amino acids	-0.5

STANDARD REDUCTION POTENTIALS OF SOME OXIDATION-REDUCTION HALF-REACTIONS[a]

Reaction	Half-Reaction (Written as a Reduction)	E_0' at pH 7.0 (volts)
1	$\frac{1}{2}O_2 + 2H^+ + 2e^- \rightarrow H_2O$	0.816
2	$Fe^{+3} + 1e^- \rightarrow Fe^{+2}$	0.771
3	$SO_4^{2-} + 2H^+ + 2e^- \rightarrow SO_3^{2-} + H_2O$	0.48
4	$NO_3^- + 2H^+ + 2e^- \rightarrow NO_2^- + H_2O$	0.42
5	$2I^- + 2e^- \rightarrow I_2$	0.536
6	Cytochrome a_3-$Fe^{+3} + 1e^- \rightarrow$ cytochrome-a_3-Fe^{+2}	0.55
7	$\frac{1}{2}O_2 + H_2O + 2e^- \rightarrow H_2O_2$	0.30
8	Cytochrome-a-$Fe^{+3} + 1e^- \rightarrow$ cytochrome-a-Fe^{+2}	0.29
9	Cytochrome-c-$Fe^{+3} + 1e^- \rightarrow$ cytochrome-c-Fe^{+2}	0.25
10	2,6-Dichlorophenolindophenol$_{(ox)}$ + $2H^+ + 2e^- \rightarrow$ 2,6-DCPP$_{(red)}$	0.22
11	Crotonyl-S-CoA + $2H^+ + 2e^- \rightarrow$ butyryl-S-CoA	0.19
12	$Cu^{+2} + 1e^- \rightarrow Cu^+$	0.15
13	Methemoglobin-$Fe^{+3} + 1e^- \rightarrow$ hemoglobin-Fe^{+2}	0.139
14	Ubiquinone + $2H^+ + 2e^- \rightarrow$ ubiquinone-H_2	0.10
15	Dehydroascorbate + $2H^+ + 2e^- \rightarrow$ ascorbate	0.06
16	Metmyoglobin-$Fe^{+3} + 1e^- \rightarrow$ myoglobin-Fe^{+2}	0.046
17	Fumarate + $2H^+ + 2e^- \rightarrow$ succinate	0.030
18	Methylene blue$_{(ox)}$ + $2H^+ + 2e^- \rightarrow$ methylene blue$_{(red)}$	0.011
19	Pyruvate + $NH_3 + 2H^+ + 2e^- \rightarrow$ alanine	-0.13
20	α-Ketoglutarate + $NH_3 + 2H^+ + 2e^- \rightarrow$ glutamate + H_2O	-0.14
21	Acetaldehyde + $2H^+ + 2e^- \rightarrow$ ethanol	-0.163
22	Oxalacetate + $2H^+ + 2e^- \rightarrow$ malate	-0.175
23	$FAD + 2H^+ + 2e^- \rightarrow FADH_2$	-0.18^b
24	Pyruvate + $2H^+ + 2e^- \rightarrow$ lactate	-0.190
25	Riboflavin + $2H^+ + 2e^- \rightarrow$ riboflavin-H_2	-0.200
26	Cystine + $2H^+ + 2e^- \rightarrow$ 2 cysteine	-0.22
27	$GSSG + 2H^+ + 2e^- \rightarrow$ 2 GSH	-0.23
28	$S^0 + 2H^+ + 2e^- \rightarrow H_2S$	-0.23
29	1,3-Diphosphoglyceric acid + $2H^+ + 2e^- \rightarrow$ GAP + P_i	-0.29
30	Acetoacetate + $2H^+ + 2e^- \rightarrow \beta$-hydroxybutyrate	-0.290

Reaction	Half-Reaction (Written as a Reduction)	E^0 at pH 7.0 (volts)
31	Lipoate$_{(ox)}$ + 2H$^+$ + 2e^- → lipoate$_{(red)}$	-0.29
32a	NAD$^+$ + 2H$^+$ + 2e^- → NADH + H$^+$	-0.320
b	NADP$^+$ + 2H$^+$ + 2e^- → NADPH + H$^+$	-0.320
33	Pyruvate + CO$_2$ + 2H$^+$ + 2e^- → malate	-0.33
34	Uric acid + 2H$^+$ + 2e^- → xanthine	-0.36
35	Acetyl-S-CoA + 2H$^+$ + 2e^- → acetaldehyde + CoA	-0.41
36	CO$_2$ + 2H$^+$ + 2e^- → formate	-0.420
37	2H$^+$ + 2e^- → H$_2$	-0.414
38	Ferredoxin-Fe^{+3} + 1e^- → ferredoxin-Fe^{+2}	-0.432
39	Gluconate + 2H$^+$ + 2e^- → glucose + H$_2$O	-0.45
40	3-Phosphoglycerate + 2H$^+$ + 2e^- → glyceraldehyde-3-phosphate + H$_2$O	-0.55
41	Methylviologen$_{(ox)}$ + 2H$^+$ + 2e^- → methylviologen$_{(red)}$	-0.55
42	Acetate + 2H$^+$ + 2e^- → acetaldehyde	-0.60
43	Succinate + CO$_2$ + 2H$^+$ + 2e^- → α-ketoglutarate + H$_2$O	-0.67
44	Acetate + CO$_2$ + 2H$^+$ + 2e^- → pyruvate	-0.70

[a] Standard conditions: Unit activity of all components except H$^+$, which is maintained at 10^{-7} M. Gases are at 1 atm pressure.

[b] The value given is for free FAD/FADH$_2$. The E_0' of the protein-bound coenzyme varies.

ABSORPTION MAXIMA AND ABSORPTION COEFFICIENTS OF SOME COMPOUNDS OF BIOCHEMICAL IMPORTANCE

Compound	λ_{max} (nm)	Molar Absorption Coefficient[a] $(a_m \times 10^{-3})$
Adenine	260.5	13.3
Adenosine, AMP, ADP, ATP	259	15.4
Cytidine	271	8.9
Cytosine	267	6.1
CMP, CDP, CTP	271	9.1
NAD$^+$, NADP$^+$	259	18
NADH, NADPH	339	6.22
	259	15
Flavin adenine dinucleotide		
(FAD)	450	11.3
	375	9.3
	260	37
Guanine	275.5	8.1
	246	10.7
Guanosine, GMP, GDP, GTP	252	13.7
Nicotinamide	260	4.6
Phenylalanine (in 0.1 N HCl)	257.5	0.19
Phenylalanine (in 0.1 N NaOH)	258	0.206
Pyridoxal phosphate	388	4.9
	330	2.5
Riboflavin	450	12.2
	375	10.6
	260	27.7
Riboflavin phosphate (FMN)	450	12.2
	375	10.4
	260	27.1
Thiamine hydrochloride	267	9.0
	235	11.5
Thymidine	267	9.7
	207.5	9.6
Thymine	264	7.9

Compound	λ_{max} (nm)	Molar Absorption Coefficient[a] $(a_m \times 10^{-3})$
Tryptophan (in 0.1 N HCl)	278	5.6
	218	33.5
Tryptophan (in 0.1 N NaOH)	280.5	5.43
	221.5	34.6
Tyrosine (in 0.1 N HCl)	274.5	1.34
	223	8.2
Tyrosine (in 0.1 N NaOH)	293.5	2.33
	240	11.1
Uracil	259.5	8.2
Uridine, UMP, UDP, UTP, UDPG, UDPGal	262	10.0

[a] Absorption coefficients are given for a 1 cm light path.

SPECIFIC ROTATION OF SOME CARBOHYDRATES AND DERIVATIVES

Compound	Specific Rotation $[\alpha]_D^{T=20-25°C}$
β-D-Arabinose	$-175^a \rightarrow -103^a$
α-L-Arabinose	$+55.4 \rightarrow +105$
β-L-Arabinose	$+190.6 \rightarrow +104.5$
β-D-Fructose	$-133.5 \rightarrow -92$
D-Galactonic acid	$-11.2 \rightarrow +57.6$
α-D-Galactosamine	$+121 \rightarrow +80$
α-D-Galactose	$+150.7 \rightarrow +80.2$
β-D-Galactose	$+52.8 \rightarrow +80.2$
β-D-Galacturonic acid	$+27 \rightarrow +55.6$
D-Gluconic acid	$-6.7 \rightarrow +11.9$
α-D-Glucosamine	$+100 \rightarrow +47.5$
α-D-Glucose	$+112 \rightarrow +52.7$
β-D-Glucose	$+18.7 \rightarrow +52.7$
α-L-Glucose	$-95.5 \rightarrow -51.4$
β-D-Glucuronic acid	$+11.7 \rightarrow +36.3$
D-Glyceraldehyde	$+13.5$
α-D-Mannose	$+29.3 \rightarrow +14.5$
β-D-Mannose	$-16.3 \rightarrow +14.5$
β-D-Mannuronic acid	$-47.9 \rightarrow -23.9$
α-L-Rhamnose	$-8.6 \rightarrow +8.2$
α-D-Xylose	$+9.36 \rightarrow +18.8$

[a] The first figure given indicates the $[\alpha]_D^T$ of the original form; the second figure given indicates the $[\alpha]_D^T$ of the equilibrium mixture of α and β forms after mutarotation. The aldonic acids equilibrate with the lactone.

RADIOISOTOPES USED IN BIOLOGICAL RESEARCH

Isotope	Half-life	Decay Energy (MeV)	
		Beta (β^- or β^+)	Gamma
Calcium-45	163 days	0.254	
Carbon-14	5700 yr	0.154	
Cesium-137	33 yr	0.52	0.032
		1.18	0.662
Chlorine-36	4.4×10^5 yr	0.714	
Chromium-51	27.8 days		0.267
			0.32
Cobalt-60	5.3 yr	0.31	1.17
			1.33
Copper-64	12.8 hr	0.573	1.35
		0.654	
Gold-198	2.69 days	0.290	0.411
		0.97	0.676
		1.38	1.087
Hydrogen-3	12.3 yr	0.0179	
Iodine-131	8.1 days	0.250	0.080
		0.31	0.284
		0.608	0.364
			0.638
Iron-55	2.9 yr		K-capture: 0.232
Iron-59	45.1 days	0.27	0.19
		0.47	1.10
		1.57	1.29
Lead-210	25 yr	0.018	0.047
		0.029	
Manganese-54	314 days		0.84
Mercury-203	46.6 days	0.212	0.279
Molybdenum-99	66 hr	<0.2	0.04
		0.445	0.367
		1.23	0.740
			0.780
Nickel-63	85 yr	0.063	
Phosphorus-32	14.3 days	1.718	
Phosphorous-33	25.2 days	0.248	
Potassium-42	12.4 hr	1.98	1.51
		3.58	

Isotope	Half-life	Decay Energy (MeV)	
		Beta (β^- or β^+)	Gamma
Rubidium-86	18.7 days	1.82 0.72	1.08
Selenium-75	128 days		0.025, 0.066, 0.081, 0.097, 0.121, 0.136, 0.199, 0.265, 0.280, 0.305, 0.402
Sodium-22	2.6 yr	0.55 0.58 1.8	0.51 1.27
Sodium-24	15.06 hr	1.390	1.38 2.758
Strontium-90	28 yr	0.54	
Sulfur-35	87.1 days	0.167	
Technetium-99	2.1×10^5 yr	0.293	
Zinc-65	244 days	0.325	0.201 1.11
Zirconium-95	65 days	0.84 0.371	0.72

ANSWERS TO PRACTICE PROBLEMS

CHAPTER 1. AQUEOUS SOLUTIONS AND ACID-BASE CHEMISTRY

Concentrations of Solutions

1. (a) 78.95 g/liter, (b) 7.895% (w/v), (c) 7895 mg %, (d) 0.711 M, (e) 2.134 Osmolar, (f) 2.134

2. (a) 39.52 ml, (b) 18.6% (w/w), (c) 20.24% (w/v), (d) 1.73 m, (e) 1.53 M, (f) 4.59 Osmolar, (g) 0.03, (h) $\Gamma/2 = 4.59$

3. (a) 18.3 g (at 0°C—calculated from Equation 1-17); 18.0 g (at 0°C—calculated from Appendix II), (b) 75 ml (from Equation 1-18 or Appendix III)

4. 17.15 M

Strong Acids and Bases—pH

7. $[H^+] = 0.4\ M$, $[OH^-] = 2.5 \times 10^{-14}\ M$, pH = 0.398, pOH = 13.602

8. pOH = 3.42, pH = 10.58

9. $a_{H^+} = 0.071$, $\gamma_{H^+} = 0.71$

10. pH = 12.86, pOH = 1.14

11. (a) 20.8 ml, (b) 33.3 ml, (c) 68 ml, (d) 180 ml, (e) 4.4×10^{-3} ml

12. (a) 12.2 M, (b) 8.2 ml con HCl/500 ml of solution, (c) 14.3 ml con HCl/350 ml of solution, (d) 666 g (or 560 ml) con HCl + 334 g (or ml) of water, (e) 1.64×10^{-3} ml con HCl/liter of solution

5.

	(a)	(b)	(c)	(d)
pH	2	4	2.3	9.57
pOH	12	10	11.7	4.43
H^+ ions/liter	6.023×10^{21}	6.023×10^{19}	3.0×10^{21}	1.63×10^{14}
OH^- ions/liter	6.023×10^{11}	6.023×10^{13}	1.2×10^{12}	2.23×10^{19}

	(e)	(f)	(g)	(h)	(i)
pH	6.88	11.46	0	-1	4.52
pOH	7.12	2.54	14	15	9.48
H^+ ions/liter	7.9×10^{16}	2.1×10^{12}	6.023×10^{23}	6.023×10^{24}	1.81×10^{19}
OH^- ions/liter	4.88×10^{16}	1.75×10^{21}	6.023×10^{9}	6.023×10^{8}	1.99×10^{14}

6.

	(a)	(b)	(c)	(d)
$[H^+]$	$1.86 \times 10^{-3}\ M$	$5.14 \times 10^{-6}\ M$	$1.66 \times 10^{-7}\ M$	$2.24 \times 10^{-9}\ M$
$[OH^-]$	$5.4 \times 10^{-12}\ M$	$1.95 \times 10^{-9}\ M$	$6.02 \times 10^{-8}\ M$	$4.46 \times 10^{-6}\ M$
H^+ ions/liter	1.12×10^{21}	3.1×10^{18}	1.0×10^{17}	1.35×10^{15}
OH^- ions/liter	3.26×10^{12}	1.18×10^{15}	3.6×10^{16}	2.7×10^{18}

	(e)	(f)	(g)
$[H^+]$	$3.02 \times 10^{-10}\ M$	$3.9 \times 10^{-12}\ M$	1 M
$[OH^-]$	$3.31 \times 10^{-5}\ M$	$2.57 \times 10^{-3}\ M$	$1 \times 10^{-14}\ M$
H^+ ions/liter	1.82×10^{14}	2.3×10^{12}	6.023×10^{23}
OH^- ions/liter	1.99×10^{19}	1.55×10^{21}	6.023×10^{9}

13. (a) 400 g NaOH/5 liters of solution, (b) 0.25 g NaOH/2 liters of solution, (c) 356 g NaOH/500 ml of solution

14. 202 ml

15. 1.06 g

16. 24.5×10^4 ml (24.5 liters)

Weak Acids and Bases—Buffers

17. (a) $K_a = 1.27 \times 10^{-4}$, (b) pH = 2.28, (c) 1210 ml, (d) 1.75×10^{21} ions

18. (a) $[H^+] = 5 \times 10^{-5}$ M, (b) $1.85 \times 10^{-2}\%$, (c) $K_a = 9.25 \times 10^{-9}$

19. (a) $[OH^-] = 1.49 \times 10^{-12}$ M, (b) 4.46%

20. 12 K_a

21. (a) pH = 11.07, (b) 2.36%

22. (a) $pK_a = 3.21$ and $pK_b = 10.79$, (b) $pK_a = 4.54$ and $pK_b = 9.46$, (c) $pK_a = 4.47$ and $pK_b = 9.53$, (d) $pK_a = 5.14$ and $pK_b = 8.86$

23. (a) $pK_b = 4.68$ and $pK_a = 9.32$, (b) $pK_b = 5.51$ and $pK_a = 8.49$, (c) $pK_b = 4.11$ and $pK_a = 9.89$, (d) $pK_b = 3.04$ and $pK_a = 10.96$

24. pH = 11.4

25. (a) 1.46, (b) 4.67, (c) 9.76, (d) 12.74, (e) 9.03, (f) 4.98, (g) 11.59, (h) 9.35, (i) 5.07, (j) 5.07

26. 1080 ml

27. 487.5 ml

28. $[H^+] = 5 \times 10^{-13}$ M, pH = 12.3
29. $[H^+] = 1.66 \times 10^{-9}$ M, pH = 8.78

30. $[H^+] = 7.25 \times 10^{-5}$ M, pH = 4.14

31. $[H^+] = 1.6 \times 10^{-13}$ M, pH = 12.8

32. $[NH_3] = 0.103$ M, $[NH_4Cl] = 0.047$ M

33. (a) pH = 12.32, (b) pH = 12.15, (c) $pK_{a_3}' = 11.79$

34. pH = 8.28

35. pH = 10.15

36. pH = 9.56

37. 28.9 g sodium formate + 75.5 ml 1 M formic acid/2 liters of solution

38. (a) 400 ml 2 M H_3PO_4 + 1070 ml 1 N KOH/40 liters of solution, (b) 1000 ml 0.8 M H_3PO_4 + 42.8 g NaOH/40 liters of solution, (c) 54 ml 14.8 M H_3PO_4 + 1070 ml 1 M KOH/40 liters of solution, (d) 533 ml KH_2PO_4 + 267 ml Na_2HPO_4/40 liters of solution, (e) 72.5 g KH_2PO_4 + 46.5 g K_2HPO_4/40 liters of solution, (f) 139.1 g K_2HPO_4 + 355 ml 1.5 M HCl/40 liters of solution, (g) 666.7 ml K_2HPO_4 + 133.0 ml 2 M H_2SO_4/40 liters of solution, (h) 108.8 g KH_2PO_4 + 133.5 ml 2 M KOH/40 liters of solution, (i) 533.3 ml 1.5 M KH_2PO_4 + 267 ml 1 M NaOH/40 liters of solution, (j) 131.2 g Na_3PO_4 + 1330 ml 1 M HCl/40 liters of solution

39. 21 ml glacial acetic acid and 62.2 g potassium acetate/5 liters of solution

40. (a) pH = 6.96, (b) pH = 2.4, (c) $HPO_4^{2-} + H^+ \rightleftharpoons H_2PO_4^-$

41. $\frac{1}{2} = Tris^0/Tris^+$, (b) $\frac{1}{1} = Tris^0/Tris^+$, (c) pH = 8.1, (d) pH = 12.52, (e) $R-NH^+ \rightleftharpoons R-N^0 + H^+$ replacing a large portion of the H^+ utilized and converting some of the $Tris^+$ to $Tris^0$

42. $[H^+]$ formed = 1.49×10^{-3} M = 1.49 μmoles/ml

43. (a) $\beta = 4.69 \times 10^{-3}$ M in both directions, (b) $BC_a = 2.47 \times 10^{-3}$ M, $BC_b = 5.15 \times 10^{-3}$ M

44. (a) pH = 4.23, (b) pH = 4.23, (c) $H_2A : HA^- : A^{2-} = 1 : 19.95 : 8.90$; $[H_2A] = 1.675 \times 10^{-3}$ M, $[HA^-] = 33.41 \times 10^{-3}$ M, $[A^{2-}] = 14.91 \times 10^{-3}$ M

Amino Acids and Peptides

45. (a) pH = 3.07, (b) pH = 6.02, (c) pH = 10.3

46. (a) 2250 ml, (b) 200 ml, (c) 600 ml, (d) 1200 ml

47. (a) 250 ml, (b) 468.8 ml, (c) 1225 ml, (d) 468.8 ml

48. pH = 9.45

49. (a) AA^0, (b) AA^-, (c) AA^+

50. Dissolve 41.92 g (0.2 moles) of histidine hydrochloride monohydrate (AA^+) in some water. Add 152 ml of 1 M KOH. Dilute to 1.0 liter. The final solution contains 0.152 M AA^0 and 0.048 M AA^+.

Blood Buffers

51. $1.2 \times 10^{-3} M$

52. $[CO_2]:[HCO_3^-]:[CO_3^{2-}] = 1:20:0.0283$

53. pH = 7.34

54. $K'_{O_2} = 0.0275$

55. $0.716 \, H^+/O_2$

CHAPTER 2   CHEMISTRY OF BIOLOGICAL MOLECULES

Amino Acids, Peptides, and Proteins

1. Disregarding hydrophobic interactions, we predict the order to be glu, ser, trp, ala, arg. However, glu is slightly hydrophobic (because of the two methylene groups) and ser is quite hydrophilic, so the order is reversed. Similarly, trp is significantly more hydrophobic than ala, so even though the pI of trp is slightly lower than the pI of ala, ala elutes before trp. The actual order is ser, glu, ala, trp, arg. (Arginine will not elute at all unless the pH is increased.)

2. Serine will not move (pH = pI). Arginine will move the fastest toward the negative pole, followed by alanine and then tryptophan. Glutamate will move toward the positive pole.

3. lys-met-tyr-ser-phe-ala-gly

4. glu-phe-lys-pro-lys

5. met-asp-phe-thr-ser

6. (a) 6.4×10^7, (b) 2.79×10^7, (c) 38,760, (d) 60,459, (e) 177,100

7. (a) 183 Å, (b) 439.2 Å, (c) 14,640

8. (a) First calculate the weight fraction of each amino acid *residue*: ser = 0.279, pro = 0.311, ala = 0.228, and gly = 0.183. Then calculate $\bar{v}$ of the lipoprotein from Σ (wt. fraction of amino acid) ($\bar{v}$ of amino acid). $\bar{v} = 0.700 \, cm^3/g$, $\rho = 1/\bar{v} = 1.429 \, g/cm^3$, (b) volume = $9.077 \times 10^{-21} \, cm^3 = 9077 \, Å^3$, diameter = 25.88 Å

9. 25,700

10. Use the law of cosines: $a^2 = b^2 + c^2 - 2bc \cos A$ (where a, b, and c are the sides of a triangle and A is the angle opposite a) to obtain $x = 2.397$ Å and $y = 2.151$ Å.

11. 13 [GS, GS-AMP, GS(AMP)$_2$ .... GS(AMP)$_{12}$]

12. MW = 33,459

13. The enzyme is probably composed of four subunits of MW 22,000, each subunit containing one atom of selenium.

14. (a) 39,901, (b) 41,649 at 2.5 mg/ml → 43,397 at infinite dilution

15. MW = 13,921

Carbohydrates

16. (a) 4, (b) 8

17. 11 [$1 \rightarrow 2, 3, 4, 6$; α or β, plus three 1-1 linked disaccharides: α-α, α-β, (same as β-α), and β-β]

18. 90% (75 mg of pure cellulose should yield 83.33 mg of glucose on hydrolysis)

19. glucose $(1 \rightarrow 2)$ glucose

20. (a) Glucose (α or $\beta 1 \rightarrow 3$) glucose, or glucose ($\beta 1 \rightarrow 4$) glucose. The nonreducing residue utilizes two moles of periodate and releases one mole of formic acid. The sugar alcohol formed from the reducing residue utilizes three moles of periodate and liberates two moles of formaldehyde (from carbons 1 and 6) and one mole of formic acid (from carbon 5 for the 1-3 possibility; from carbon 2 for the 1-4 possibility). (b) The polysaccharide (called nigeran or mycodextran) consists of alternating 1-4 and 1-3 linked glucose residues.

21. There are a total of 200 μmoles of glucose in 32.4 mg of amylopectin. (a) The products are 10 μmoles of 2,3,4,6-tetramethylglucose. ∴ 10 μmoles of 2,3-dimethylglucose (from branch points), and 180 μmoles of 2,3,6-trimethylglucose. (b) 5%, (c) 370

22. (a) 5, (b) 2, (c) 2,3,4,6-tetramethylgalactose; 2,3,4-trimethylglucose; and 1,3,4,6-tetramethylfructose

23. (a) 438,727, (b) 1.477

Lipids

24. (a) 64, (48, discounting like-ended molecules), (b) 40, (c) 20

25. 556.3

26. 730.4

27. 2

28. Elemental analysis yields $C_{10.49}H_{14.90}O$ (or possibly $C_{10}H_{15}O$), or $C_{21}H_{30}O_2$. The freezing point depression calculation yields a MW of 313.8, which agrees with $C_{21}H_{30}O_2$ (actual MW = 314.45).

29. Use $\bar{v}_{lipoprotein} = \Sigma$ (wt. fraction of component) $\times$ ($\bar{v}$ of component) to obtain $\bar{v} = 0.833$ cm^3/g. $\rho = 1/\bar{v} = 1.20$ g/cm^3.

Nucleotides and Nucleic Acids

30. If T = 32.8%, A = 32.8%, G = 17.2%, and C = 17.2%

31. (a) 3.56×10^6, (b) 1.21 mm, (c) volume of DNA = 3.8×10^{-15} cm^3 or 0.24% of the total cell volume

32. (a) $(3.56 \times 10^6) \times 2 = 7.12 \times 10^6$ nucleotides/40 min = 1.78×10^5 nucleotide bonds/min, (b) 0.03025 mm/min = 30.25 μm/min, (c) 356,000 turns/40 min = 8,900 turns/min

33. 40,000

34. 75% of 3.56×10^6 coding nucleotides = 2.67×10^6 nucleotides = 890,000 codons. A protein of MW 60,000 contains 500 amino acids. $\therefore$ 1,780 different proteins can be made.

35. 49,440 (Assuming double-stranded DNA)

36. (a) 57.5% G+C, (b) % G+C = 2.4 $(T_m - 49.3)$

37. 2.5×10^7

CHAPTER 3   BIOCHEMICAL ENERGETICS

1. $\Delta G' = +500$ cal/mole

2. (a) $\Delta G = -4092$ cal/mole ($\Delta G' = 0$). (b) The reaction proceeds in the direction of ADPG + PP$_i$ synthesis.

3. The hydrolysis of glucose-6-sulfate will proceed further to the right (higher K'_{eq} and more negative $\Delta G'$) because the product, HSO_4^-, is a strong acid (p$K_{a_2} \simeq 1.9$) that spontaneously ionizes at pH 7 to $SO_4^{2-} + H^+$. The product of glucose-6-phosphate hydrolysis, $HPO_4^{2-} + H_2PO_4^-$, is a weak acid (p$K_{a_3} \simeq 12.5$) that does not ionize further.

4. $K'_{eq} = 64.26$, $\Delta G' = -2466$ cal/mole

5. $\Delta G' = -12,481$ cal/mole

6. $\Delta G_{ion} = -1972$ cal/mole (when $K_{a_2} = 2.69 \times 10^{-6}$)

7. (a) The overall synthesis can be considered as the sum of two reactions:

ATP + H$_2$O $\rightleftharpoons$ AMP + PP$_i$	$\Delta G' = -8000$ cal/mole
NH$_4^+$ + aspartate $\rightleftharpoons$ asparagine + H$_2$O	$\Delta G' = +3400$ cal/mole

Sum: aspartate + ATP + NH$_4^+$ $\rightleftharpoons$ asparagine + AMP + PP$_i$ $\Delta G' = -4600$ cal/mole

A more rapid mental calculation says: ATP (worth -8000 cal/mole) is used to make asparagine (worth -3400 cal/mole). Therefore, the overall $\Delta G'$ (the *difference*) is -4600 cal/mole.

(b)   (1) aspartate + ATP $\rightleftharpoons$ β-aspartyladenylate + PP$_i$ $\Delta G'_1 = +2000$ cal/mole

(2) β-aspartyladenylate + NH$_4^+$ $\rightleftharpoons$ asparagine + AMP $\Delta G'_2 = -6600$ cal/mole

Overall: (3) aspartate + ATP + NH$_4^+$ $\rightleftharpoons$ asparagine + PP$_i$ + AMP $\Delta G'_3 = -4600$ cal/mole

ATP (worth -8000 cal/mole) is used to make β-aspartyladenylate (worth $-10,000$ cal/mole). Therefore, $\Delta G'_1 = +2000$ cal/mole. The second reaction must have a $\Delta G'$ such that the overall $\Delta G'$ is -4600 cal/mole: $\Delta G'_1 + \Delta G'_2 = \Delta G'_3$. Therefore, $\Delta G'_2 = \Delta G'_3 - \Delta G'_1 = (-4600) - (+2000) = -6600$ cal/mole.

8. $[\text{G-6-P}] > 0.221\ M$

9. $\Delta G' = -18{,}912$ cal/mole. (The $\Delta G'$ of the ATP sulfurylase reaction is $+10{,}912$ cal/mole. ATP is worth -8000. Therefore, APS must be worth $-18{,}912$.)

10. $\text{ATP} \rightleftharpoons \text{cyclic AMP} + \text{PP}_i$ $K'_{eq} = 0.065$

$\therefore$ $\text{cyclic AMP} + \text{PP}_i \rightleftharpoons \text{ATP}$ $K'_{eq} = 15.38$ $\Delta G' = -1619$ cal/mole

$\text{ATP} + \text{H}_2\text{O} \rightleftharpoons \text{AMP} + \text{PP}_i$ $\Delta G' = -8000$ cal/mole

---

$\text{cyclic AMP} + \text{H}_2\text{O} \rightleftharpoons \text{AMP}$ $\Delta G' = -9619$ cal/mole

11.   (a) ~ 0, (b) -3415 cal/mole, (c) ~ 0, (d) -1700 cal/mole

12.   (a) [glucose-6-P] = 0.95 M, [glucose-1-P] = 0.05 M, [glucose-6-P]/[glucose-1-P] = 19
(b) [glucose-6-P] = 0.095 M, [glucose-1-P] = 0.005 M, [glucose-6-P]/[glucose-1-P] = 19
(c) [glucose-6-P] = $9.5 \times 10^{-3}\ M$, [glucose-1-P] = $5 \times 10^{-4}\ M$, [glucose-6-P]/[glucose-1-P] = 19
(d) [glucose-6-P] = $9.5 \times 10^{-4}\ M$, [glucose-1-P] = $5 \times 10^{-5}\ M$, [glucose-6-P]/[glucose-1-P] = 19
(e) [glucose-6-P] = $9.5 \times 10^{-5}\ M$, [glucose-1-P] = $5 \times 10^{-6}\ M$, [glucose-6-P]/[glucose-1-P] = 19

13.   (a) isocitrate = 0.845 M, glyoxylate = 0.155 M, succinate = 0.155 M

$\dfrac{\text{isocitrate}}{\text{glyoxylate}} = 5.45$ $\dfrac{\text{glyoxylate}}{\text{succinate}} = 1.0$

(b) isocitrate = 0.0591 M, glyoxylate = 0.0409 M, succinate = 0.0409 M

$\dfrac{\text{isocitrate}}{\text{glyoxylate}} = 1.44$ $\dfrac{\text{glyoxylate}}{\text{succinate}} = 1.0$

(c) isocitrate = 0.0022 M, glyoxylate = 0.0078 M, succinate = 0.0078 M

$\dfrac{\text{isocitrate}}{\text{glyoxylate}} = 0.282$ $\dfrac{\text{glyoxylate}}{\text{succinate}} = 1.0$

(d) isocitrate = $3.3 \times 10^{-5}\ M$, glyoxylate = $9.67 \times 10^{-4}\ M$, succinate = $9.67 \times 10^{-4}\ M$

$\dfrac{\text{isocitrate}}{\text{glyoxylate}} = 0.034$ $\dfrac{\text{glyoxylate}}{\text{succinate}} = 1.0$

(e) isocitrate = $\sim 0\ M$, glyoxylate = $\sim 1 \times 10^{-4}\ M$, succinate = $\sim 1 \times 10^{-4}\ M$

$\dfrac{\text{isocitrate}}{\text{glyoxylate}} = \sim 0$ $\dfrac{\text{glyoxylate}}{\text{succinate}} = 1.0$

14.   (a) pyruvate + β-hydroxybutyrate $\rightarrow$ lactate + acetoacetate; (b) pyruvate is reduced to lactate, β-hydroxybutyrate is oxidized to acetoacetate, pyruvate is the oxidizing agent, β-hydroxybutyrate is the reducing agent; (c) $\Delta E'_0 = +0.100$ v, $\Delta G' = -4612$ cal/mole, $K'_{eq} = 2408$

15.   (a) ubiquinone + succinate $\rightarrow$ ubiquinone-H$_2$ + fumarate;   (b) $\Delta E = +0.070$ v, $\Delta G' = -3228.8$ cal/mole, $K'_{eq} = 233$

16.   (a) -0.149 v, (b) -0.081 v, (c) -0.06 v, (d) -0.046 v, (e) -0.019 v, (f) $+0.017$ v

17.   (a) $[\text{NADH}]/[\text{NAD}^+] = 1.22 \times 10^{-5}$ (equation 33), (b) $[\text{NADH}]/[\text{NAD}^+] = 1.22 \times 10^{-3}$

18.   (a) $E'_{0_{\text{pH }9}} = E'_{0_{\text{pH }7}} - (0.059)(2)$ $\therefore$ $E'_{0_{\text{pH }9}} = -0.293$ v, (b) $E'_{0_{\text{pH }9}} = E'_{0_{\text{pH }7}}$ because H$^+$ is not involved in the reaction $\therefore$ $E'_{0_{\text{pH }9}} = +0.771$ v

19.   (a) ~ 6 ATP/mole S oxidized, (b) ~ 3 ATP/mole NH$_4^+$ oxidized

20.   The fatty acid because it is more highly reduced than the sugar (more hydrogens per mole to burn). The $\Delta G'$ of hexanoic acid oxidation will be more negative than the $\Delta G'$ of fructose oxidation.

21.   18 moles ATP/mole ethanol

22. $\Delta G = (2)(1364)\Delta\text{pH}$.   Under the given conditions, $\Delta G = 6336$ cal/mole. $\therefore$ $\Delta\text{pH} = 2.32$

23.   (a) $\mathcal{E} = 109{,}808$ cal/einstein, (b) $\mathcal{E} = 38{,}067$ cal/einstein

24. ~ 2 ATP at 40 to 50% efficiency

25.   (a) $\Delta G = -543$ cal/mole, (b) $\Delta G = +543$ cal/mole (i.e., 543 cal are required to move a mole of Cl$^-$ outward).

26.   At 37°C, $\Delta G = 1419\ \Delta\text{pH}$ $\therefore$ $\Delta G = 7518$ cal/mole

27.   (a) $\Delta G = Z\mathcal{F}\Delta\Psi$ $\therefore$ $\Delta G = -13{,}838$ cal/mole, (b) $\sim 10^{10}$

28. $\Delta H = -70,000$ cal/mole, $\Delta G' = -65,314$ cal/mole, $T\Delta S = -4686$ cal/mole, $\Delta S = -15.7$ e.u. at 25°C (298°K)

29. (a) $\Delta H = -11,092$ cal/mole, $\Delta G' = -2568$ cal/mole, $T\Delta S = -8524$ cal/mole, $\Delta S = -27.5$ e.u., (b) K'_{eq} at 28°C = 111.4

30. v at 37°C is 3.77 times greater than v at 15°C

CHAPTER 4   ENZYMES

1. (a) $K_m = 10^{-5}$ M, $V_{max} = 120 \times 10^{-9}$ M/min. (b) Verify by plotting v versus [S], showing a hyperbolic curve, and by plotting $1/v$ versus $1/[S]$ showing a straight line. You can also show that the original v versus [S] data yield the same K_m value regardless of which values are substituted into the Michaelis-Menten equation. (c) 0.012 min^{-1}.

2. (a) 0.107 nmole × liter^{-1} × min^{-1}, (b) 26.6 nmoles × liter^{-1} × min^{-1}, (c) 37.7 nmoles × liter^{-1} × min^{-1}, (d) 114.2 nmoles × liter^{-1} × min^{-1}, (e) 128 nmoles × liter^{-1} × min^{-1} = V_{max}

3. Each of the indicated values of v would be five times greater.

4. (a) $k_f = V_{max_f}/K_{m_S} = 0.168$ min^{-1}, (b) $k_r = k_f/K_{eq} = 8.4 \times 10^{-5}$ min^{-1}, (c) $V_{max_r}/K_{m_p} = k_r = 8.4 \times 10^{-5}$ min^{-1}

5. $v_{net} = -5.4$ μmoles × liter^{-1} × min^{-1} (P → S)

6. (a) 5.4 μmoles/liter, (b) 8.55 μmoles/liter, (c) 16.2 μmoles/liter, (d) 28.62 μmoles/liter, (e) 2.7×10^{-2}%, 4.28×10^{-2}%, 8.1×10^{-2}%, 14.31×10^{-2}%

7. (a) 2.1×10^{-6} M at 5 min, (b) 3.9×10^{-6} M at 10 min

8. (a) 22.4%, (b) 22.4%, (c) 1.52×10^{-5} moles × liter^{-1} × min^{-1}, (d) 13.7 min, (e) 27.4 min ($k = 0.051$ min^{-1})

9. (a) 361, (b) 16, (c) 9, (d) 3

10. $1/v = $ "2" $= 0.02$, $V_{max} = 1/0.02 = 50$ nmoles × liter^{-1} × min^{-1}, $1/[S] = $ "4" $= 4 \times 10^4$, $K_m = 1/(4 \times 10^4) = 2.5 \times 10^{-5}$ M

11. (a) [PS] $= 0.65 \times 10^{-6}$ M, (b) [P] $= 0.50 \times 10^{-6}$ M, [P]$_t = 1.15 \times 10^{-6}$ M, (c) $K_S = 6 \times 10^{-7}$ M. A Scatchard plot using data at several different concentrations of [S]$_f$ would yield a more reliable estimate and also establish whether only one type of binding site is present.

12. A reciprocal plot will establish that the kinetic constants for the adult liver extract are $K_m = 3 \times 10^{-4}$ M, $V_{max} = 20$ μmoles/mg protein. The embryonic liver extract has kinetic constants of $K_m = 5 \times 10^{-5}$ M and $V_{max} = 20$ μmoles/mg protein. The different K_m values suggest that the adult and embryonic enzymes are not kinetically identical. They may be the products of two different genes, or, alternately, the adult enzyme might be a modified form of the embryonic enzyme. The identical V_{max} values may be coincidental since $V_{max} = k_p[E]_t$. Thus, one extract might contain a higher concentration of enzyme with a lower catalytic rate constant. Another possibility is that the two enzymes are indeed identical and present at the same concentrations, but that the extract of the adult tissue contains a competitive inhibitor that increases the apparent K_m.

13. A reciprocal plot shows that the K_m of the enzyme present in the serum is 3×10^{-4} M. Thus, liver damage is a more likely diagnosis than strenuous exercise (assuming the absence of inhibitors or activators that might alter the K_m value).

14. (a) 17.6 nmoles × liter^{-1} × min^{-1}, 87%; (b) $v_i = 3.7$ nmoles × liter^{-1} × min^{-1}, 91.8%; (c) $v_i = 210.2$ nmoles × liter^{-1} × min^{-1}, 1.3%

15. (a) 3.7×10^{-4} M, (b) 2.93×10^{-2} M

16. 6.66×10^{-5} M

17. (a) 20.7 nmoles × liter^{-1} × min^{-1}, (b) 89.3%

18. $[S]_i = \left(1 + \dfrac{[I]}{K_i}\right)[S]_0$

19. The control data (or a reciprocal plot) yields $K_m = 1 \times 10^{-3}$ M and $V_{max} = 100$ nmoles/min. I is a competitive inhibitor. (The reciprocal plot intersects the control plot at V_{max}.) $K_i = 3 \times 10^{-6}$ M (calculated from either the slope of the reciprocal plot or from $K_{m_{app}}$). X is a noncompetitive

inhibitor. (The reciprocal plot intersects the control plot on the $1/[S]$ axis at $-1/K_m$.) $K_i = 1.5 \times 10^{-5}\ M$ (calculated from the slope of the reciprocal plot or from the $1/v$-axis intercept). Y is an uncompetitive inhibitor. (The reciprocal plot is parallel to the control plot.) $K_i = 1 \times 10^{-3}\ M$ (calculated from the $1/v$-axis intercept of the reciprocal plot or from $K_{m_{app}}$). Z is a mixed-type inhibitor. (The reciprocal plot intersects the control plot to the left of the $1/v$ axis, above the $1/[S]$ axis.) $K_i = 2 \times 10^{-4}\ M$ (calculated from the slope of the reciprocal plot). $\alpha K_i = 8 \times 10^{-4}\ M$ (calculated from the $1/v$-axis intercept or the $1/[S]$ intersection coordinate. Thus, $\alpha = 4$.

20. The v versus $[S]$ plot is shown in Figure PP-4-1. This is an example of an "energy-charge" type of response. S and P might be ATP and ADP, respectively. ATP is converted to ADP in the reaction described in this problem, but the total nucleotide pool remains constant. Thus, as the ATP concentration increases, there is a *simultaneous* reduction of product inhibition resulting in the unusual velocity curve. If the $[S]/[P]$ ratio is poised at about $9 \times 10^{-4}\ M$, v would be extremely responsive to small changes in the ratio. The energy charge response is seen only when $K_P < K_S$; that is, when the enzyme has a higher affinity for the product than for the substrate.

21. $\mathrm{pH_{opt}} = \frac{1}{2}(\mathrm{p}K_a + \mathrm{p}K_e) = 5.5$

22. The situation is identical to that described in Practice Problem 21 except now both ionizable groups are on the enzyme. (a)

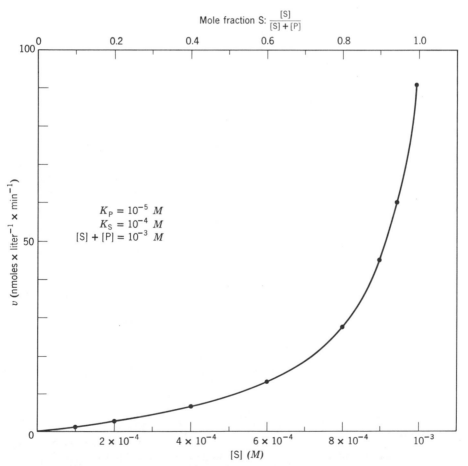

Mole fraction S: $\dfrac{[S]}{[S]+[P]}$

$K_P = 10^{-5}\ M$
$K_S = 10^{-4}\ M$
$[S] + [P] = 10^{-3}\ M$

v (nmoles $\times$ liter^{-1} $\times$ min^{-1})

$[S]$ (M)

Figure PP4-1 (Practice Problem 20) Energy charge response where $[S]+[P]$ is constant and P is a product inhibitor.

$pH_{opt} = 5.5$. (b) The reactions are:

$$
\begin{array}{c}
E^{n+1} \\
\Big\updownarrow K_{e_1} \\
H^+ \\
+ \\
E^n + S \underset{K_S}{\rightleftharpoons} ES \xrightarrow{k_p} E + P \\
\Big\updownarrow K_{e_2} \\
H^+ \\
+ \\
E^{n-1}
\end{array}
$$

where $E^n = {}^+HN\text{-}E\text{-}COO^-$, $E^{n+1} = {}^+HN\text{-}E\text{-}COOH$, $E^{n-1} = N\text{-}E\text{-}COO^-$. $[H^+]$ acts as a competitive inhibitor as its concentration approaches K_{e_1} (i.e., as the pH decreases toward pK_{e_1}). An identical effect on the slope of the reciprocal plot is observed as $[H^+]$ *decreases* (i.e., as the pH increases toward pK_{e_2} and the dead-end E^{n-1} form accumulates). The velocity equation can be written according to the usual rules.

$$
\frac{v}{V_{max}} = \frac{\dfrac{[S]}{K_S}}{1 + \dfrac{[S]}{K_S} + \dfrac{[H^+]}{K_{e_1}} + \dfrac{K_{e_2}}{[H^+]}}
$$

or

$$
\frac{v}{V_{max}} = \frac{[S]}{K_S\left(1 + \dfrac{[H^+]}{K_{e_1}} + \dfrac{K_{e_2}}{[H^+]}\right) + [S]}
$$

where the $K_{e_2}/[H^+]$ term represents the concentration of E^{n-1} relative to E^n.

23. (a) $E_a = 11,828$ cal/mole, (b) $Q_{10} = 1.92$

24. (a) 1.6×10^{-3} μmole/min, (b) 16 μmoles $\times$ liter^{-1} $\times$ min^{-1}, (c) 3.33×10^{-3} μmole $\times$ mg protein^{-1} $\times$ min^{-1}, (d) 0.08 unit/ml, (e) 0.0033 unit/mg protein

25. (a) 88.5%, (b) 2.95-fold

26. (a) 43,200 (moles S$\rightarrow$P) $\times$ mole enzyme^{-1} $\times$ min^{-1}, (b) 2.3×10^{-5} min

27. (a) $V_{max} = 3.6 \times 10^{-4}$ M/min $= 3.6 \times 10^{-1}$ μmole $\times$ ml^{-1} $\times$ min$^{-1} = 0.36$ unit/ml, (b) $[E] \times t = $ constant. Therefore, $t = 9.0$ min

28. (a) 4.72 nmoles $\times$ liter^{-1} $\times$ min^{-1}, (b) 29.79 nmoles $\times$ liter^{-1} $\times$ min^{-1}

29. (a)

$$
\frac{v}{V_{max}} = \frac{[A]}{K_{m_A}\left(1 + \dfrac{K_{ia}K_{m_B}}{K_{m_A}[B]}\right) + [A]\left(1 + \dfrac{K_{m_B}}{[B]}\right)}
$$

$$
= \frac{[A]}{K_{m_A}(\text{slope factor}) + [A](\text{intercept factor})}
$$

(b)

$$
\frac{v}{V_{max}} = \frac{[B]}{K_{m_B}\left(1 + \dfrac{K_{ia}}{[A]}\right) + [B]\left(1 + \dfrac{K_{m_A}}{[A]}\right)}
$$

$$
= \frac{[B]}{K_{m_B}(\text{slope factor}) + [B](\text{intercept factor})}
$$

30. The equations for initial velocity studies of a rapid equilibrium ordered bireactant system are:

$$
\frac{v}{V_{max}} = \frac{[A]}{K_A\left(\dfrac{K_B}{[B]}\right) + [A]\left(1 + \dfrac{K_B}{[B]}\right)}
$$

$$
= \frac{[A]}{K_A(\text{slope factor}) + [A](\text{intercept factor})}
$$

and

$$
\frac{v}{V_{max}} = \frac{[B]}{K_B\left(1 + \dfrac{K_A}{[A]}\right) + [B]}
$$

$$
= \frac{[B]}{K_B(\text{slope factor}) + [B]}
$$

Unlike the steady-state system, the slope of the $1/v$ versus $1/[A]$ plot for the rapid equilibrium system goes to zero as $[B]$ approaches infinity. (As $[B]$ increases, the $K_B/[B]$ term of the slope factor becomes very small.) Also, unlike the steady-state system, the plots of $1/v$ versus $1/[B]$ intersect on the vertical axis at $1/V_{max}$. (There is no intercept factor—the denominator $[B]$ term is not multiplied by an $[A]$-containing term.)

31. $a < 1$, $b > 1$, $c < 1$. $b > 1$ (negative cooperativity) indicates that the second molecule of S that binds makes it more difficult for the third molecule to bind.

32. $V_{max} = 100$ μmoles $\times$ liter^{-1} $\times$ min^{-1}. $[S]_{0.5} = 5 \times 10^{-3}$ M. The v versus $[S]$ plot is sigmoidal. The slope of the Hill plot = 2 (i.e., $n_{app} = 2$). $K' = 2.5 \times 10^{-5}$ M^2.

33. (a) $n_{app} = 2$, (b) $[S]_{0.9}/[S]_{0.1} = 3$

34. $L = 49.7$

35. The rules of equilibrium require that $L \times K_{S_R} = K_{S_1} \times L_1$. $\therefore$ $L_1 = (L)(K_{S_R})/K_{S_1}$ or $L_1 = Lc$. Similarly, $L_2 = Lc^2$, $L_3 = Lc^3$, and $L_4 = Lc^4$.

36. (a) 45 units $\times$ g tissue$^{-1} \times$ min^{-1}, (b) 0.053 min^{-1}, (c) 25.44 units $\times$ g tissue$^{-1} \times$ min^{-1}

CHAPTER 5   SPECTROPHOTOMETRY AND OTHER OPTICAL METHODS

1. (a) At 260 nm: $A = 0.33$, $I = 0.468$. At 340 nm: $A = 0.137$, $I = 0.730$. (b) At 260 nm: $A = 0.752$, $I = 0.177$. At 340 nm: $A = 0.044$, $I = 0.905$

2. (a) $[NADPH] = 2.41 \times 10^{-5} M$, $[ATP] = 3.50 \times 10^{-5} M$. (b) $[NADPH] = 0$, $[ATP] = 4.87 \times 10^{-5} M$. (c) $[NADPH] = 3.54 \times 10^{-5} M$, $[ATP] = 0$

3. $[A] = 1.26 \times 10^{-5} M$, $[B] = 1.11 \times 10^{-5} M$

4. (a) 0.207 mg/ml. (b) The proteins in the preparation may not have the same average aromatic amino acid composition as bovine serum albumin.

5. (a) 223.2 μg/ml = 0.223. (b) Yes—the absorbance at 215 nm and 225 nm results from the peptide bond.

6. $[GSSG] = 1.196 \times 10^{-4} M$ in the original 1.5 ml

7. To the unknown solution, add a solution containing excess NAD^+, P_i, and glyceraldehyde-3-phosphate dehydrogenase. The GAP present will be converted to 1,3-DiPGA. One mole of NADH will be produced for every mole of GAP originally present. Then add triosephosphate isomerase to convert the DHAP to GAP that will then be converted to 1,3-DiPGA producing another mole of NADH for every mole of DHAP originally present. Finally, add aldolase to convert the FDP to GAP and DHAP which, upon conversion to 1,3-DiPGA, will yield 2 moles of NADH per mole of FDP originally present. The use of arsenate in place of phosphate will insure that the glyceraldehyde-3-phosphate dehydrogenase reaction goes to completion. (The triose-1-arsenate anhydride spontaneously hydrolyzes.)

8. [citrate] $= 1.35 \times 10^{-4} M$, [isocitrate] $= 1.35 \times 10^{-4} M$

9. The $A_{340\,nm}$ of 0.262 corresponds to [NADPH] of $4.21 \times 10^{-5} M$. Therefore, the APS concentration in the assay was $4.21 \times 10^{-5} M$. The original solution contains $(4.21 \times 10^{-5})(1.0)/0.9 = 4.68 \times 10^{-5} M$ APS. The total nucleotide concentration of the original solution is $5.84 \times 10^{-5} M$. Therefore, the preparation is 80.1% APS.

10. $A_{420\,nm} = 0.105$

11. (a) 0.232 units/ml, (b) 1.93 units/mg protein

12. $\Delta A_{365\,nm}$ of 0.08/min $\backsimeq \Delta C$ of $2.57 \times 10^{-5} M \times$ min$^{-1} = 25.7 \times 10^{-6} M \times$ min$^{-1} = 25.7 \mu$moles $\times$ liter$^{-1} \times$ min$^{-1} = 0.0257 \mu$mole $\times$ ml$^{-1} \times$ min$^{-1} = 0.0257$ unit/ml. The glycerol kinase solution contains 0.257 unit/ml (85.7% of the stated activity).

13. The assay mixture contained 1.094 μg/25 ml. The urine contained 2.19 μg/ml. The amount excreted is 1750 μg/24 hours, which is considerably above normal.

14. (a) $-5.12°$, (b) $-794.6°$

15. 0.226 g/ml

16. $\alpha = 67.5\%$, $\beta = 32.5\%$

17. $A° = [\alpha] \times c \times l$; $A°_{TOT} = A°_\alpha + A°_\beta$. $c = 0.5$ g/ml. At 10 min, the β form represents 16.0% of the total (0.080 g/ml). $\therefore$ $v = 8.0$ mg $\times$ ml$^{-1} \times$ min^{-1}.

CHAPTER 6   ISOTOPES IN BIOCHEMISTRY

1.   (a) $_{20}Ca^{45} \rightarrow _{-1}\beta^0 + _{21}Sc^{45}$, (b) $_{17}Cl^{36} \rightarrow _{-1}\beta^0 + _{18}A^{36}$, (c) $_{19}K^{42} \rightarrow _{-1}\beta^0 + _{20}Ca^{42}$, (d) $_{15}P^{33} \rightarrow _{-1}\beta^0 + _{16}S^{33}$

2.   (a) 0.173 yr^{-1}, $4.75 \times 10^{-4} \text{ day}^{-1}$, $1.98 \times 10^{-5} \text{ hr}^{-1}$, $3.30 \times 10^{-7} \text{ min}^{-1}$, $5.49 \times 10^{-9} \text{ sec}^{-1}$, (b) 82.9%

11.   (a) $9.33 \times 10^{-4} \mu\text{mole/min}$, (b) $0.622 \mu\text{mole} \times \text{liter}^{-1} \times \text{min}^{-1}$, (c) $1.30 \times 10^{-3} \mu\text{moles} \times \text{mg protein}^{-1} \times \text{min}^{-1}$

12.   6 liters

13.   The data and calculations are shown below.

(Data for Practice Problem 13)

"Minus" chamber without protein = [S]	$1870 \text{ CPM}/10 \mu\text{l} = 1.87 \times 10^{-6} \mu\text{mole}/10 \mu\text{l} = 1.87 \times 10^{-7} \mu\text{moles}/\mu\text{l} = 1.87 \times 10^{-7} M$
"Plus" chamber with protein = [S] + [PS]	$2570 \text{ CPM}/10 \mu\text{l} = 2.57 \times 10^{-6} \mu\text{mole}/10 \mu\text{l} = 2.57 \times 10^{-7} \mu\text{mole}/\mu\text{l} = 2.57 \times 10^{-7} M$
[PS] = difference	$700 \text{ CPM}/10 \mu\text{l} = 0.7 \times 10^{-6} \mu\text{mole}/10 \mu\text{l} = 0.7 \times 10^{-7} \mu\text{mole}/\mu\text{l} = 0.7 \times 10^{-7} M$
$[P]_t$	$6 \times 10^{-6} \text{ g/ml} = 6 \times 10^{-3} \text{ g/liter}$ $\dfrac{6 \times 10^{-3} \text{ g/liter}}{60 \times 10^3 \text{ g/mole}} = 0.1 \times 10^{-6} \text{ mole/liter} = 1 \times 10^{-7} M$
$[P] = [P]_t - [PS]$	$[P] = (1 \times 10^{-7}) - (0.7 \times 10^{-7}) = 0.3 \times 10^{-7} M$
K_S	$K_S = \dfrac{[P][S]}{[PS]} = \dfrac{(3 \times 10^{-8})(1.87 \times 10^{-7})}{(7 \times 10^{-8})} = 8.01 \times 10^{-8} M$

3.   (a) One atom out of every 11.7 radioactive atoms present decays per day; one out of every 16,831 decays per minute; (b) 12.3×10^4 Ci/g, 16.11×10^6 Ci/g-atom, 2.73×10^{17} DPM/g

4.   23%

5.   (a) 18.25×10^6 Ci/mole $(18.25 \times 10^3$ mCi/μmole), (b) 4.93×10^{-6}%

6. 5.62×10^{-8} g/mCi

7.   (a) $3.57 \times 10^{-3} M$, (b) 8.63×10^8 CPM/ml

8.   (a) 1.56 mCi/mg, (b) 227.7 mCi/mmole, (c) 5.05×10^8 DPM/μmole, (d) 6.74×10^7 CPM/μmole carbon

9.   Take 11 μl of radioactive L-cysteine-S^{35} stock solution plus 0.1189 g solid, unlabeled, anhydrous L-cysteine hydrochloride and dissolve in sufficient water or buffer to make 75 ml of solution.

10.   Take 3.38 ml of radioactive glucose-C^{14} stock solution plus 2.92 mg solid, unlabeled glucose and dissolve in sufficient water to make 50 ml of solution.

14.   (a) 7965 CPM/μmole, (b) 6761 CPM/μmole, (c) 4494 CPM/μmole

15. $C^{14} = 158,000$ DPM, $P^{32} = 265,000$ DPM

16.   (a) $t_{\frac{1}{2}(\text{eff})} \approx 4.085$ hr, $t_{\frac{1}{2}b} = 6$ hr, (b) S.A. at zero time = 10,000 CPM/ml

17.   0.0128 M

18.   173.2 μg

19.   85.35 mg/50 ml = 1.71 mg/ml

20.   51.8 μmoles

21.   There is 6.19×10^{-8} mole of reactive SH per 3.09×10^{-8} mole of enzyme, or 2 SH/molecule

22.   (a) 6720 CPM of the known 10,000 CPM (67.2%) was recovered. The unquenched activity of the sample is 32,068 CPM. (b) Determine the specific activity of the PP_i^{32} under the same quenching conditions that the samples are counted (i.e., in the presence of charcoal).

LOGARITHMS

Natural numbers	0	1	2	3	4	5	6	7	8	9	Proportional Parts								
											1	2	3	4	5	6	7	8	9
10	0000	0043	0086	0128	0170	0212	0253	0294	0334	0374	4	8	12	17	21	25	29	33	37
11	0414	0453	0492	0531	0569	0607	0645	0682	0719	0755	4	8	11	15	19	23	26	30	34
12	0792	0828	0864	0899	0934	0969	1004	1038	1072	1106	3	7	10	14	17	21	24	28	31
13	1139	1173	1206	1239	1271	1303	1335	1367	1399	1430	3	6	10	13	16	19	23	26	29
14	1461	1492	1523	1553	1584	1614	1644	1673	1703	1732	3	6	9	12	15	18	21	24	27
15	1761	1790	1818	1847	1875	1903	1931	1959	1987	2014	3	6	8	11	14	17	20	22	25
16	2041	2068	2095	2122	2148	2175	2201	2227	2253	2279	3	5	8	11	13	16	18	21	24
17	2304	2330	2355	2380	2405	2430	2455	2480	2504	2529	2	5	7	10	12	15	17	20	22
18	2553	2577	2601	2625	2648	2672	2695	2718	2742	2765	2	5	7	9	12	14	16	19	21
19	2788	2810	2833	2856	2878	2900	2923	2945	2967	2989	2	4	7	9	11	13	16	18	20
20	3010	3032	3054	3075	3096	3118	3139	3160	3181	3201	2	4	6	8	11	13	15	17	19
21	3222	3243	3263	3284	3304	3324	3345	3365	3385	3404	2	4	6	8	10	12	14	16	18
22	3424	3444	3464	3483	3502	3522	3541	3560	3579	3598	2	4	6	8	10	12	14	15	17
23	3617	3636	3655	3674	3692	3711	3729	3747	3766	3784	2	4	6	7	9	11	13	15	17
24	3802	3820	3838	3856	3874	3892	3909	3927	3945	3962	2	4	5	7	9	11	12	14	16
25	3979	3997	4014	4031	4048	4065	4082	4099	4116	4133	2	3	5	7	9	10	12	14	15
26	4150	4166	4183	4200	4216	4232	4249	4265	4281	4298	2	3	5	7	8	10	11	13	15
27	4314	4330	4346	4362	4378	4393	4409	4425	4440	4456	2	3	5	6	8	9	11	13	14
28	4472	4487	4502	4518	4533	4548	4564	4579	4594	4609	2	3	5	6	8	9	11	12	14
29	4624	4639	4654	4669	4683	4698	4713	4728	4742	4757	1	3	4	6	7	9	10	12	13
30	4771	4786	4800	4814	4829	4843	4857	4871	4886	4900	1	3	4	6	7	9	10	11	13
31	4914	4928	4942	4955	4969	4983	4997	5011	5024	5038	1	3	4	6	7	8	10	11	12
32	5051	5065	5079	5092	5105	5119	5132	5145	5159	5172	1	3	4	5	7	8	9	11	12
33	5185	5198	5211	5224	5237	5250	5263	5276	5289	5302	1	3	4	5	6	8	9	10	12
34	5315	5328	5340	5353	5366	5378	5391	5403	5416	5428	1	3	4	5	6	8	9	10	11
35	5441	5453	5465	5478	5490	5502	5514	5527	5539	5551	1	2	4	5	6	7	9	10	11
36	5563	5575	5587	5599	5611	5623	5635	5647	5658	5670	1	2	4	5	6	7	8	10	11
37	5682	5694	5705	5717	5729	5740	5752	5763	5775	5786	1	2	3	5	6	7	8	9	10
38	5798	5809	5821	5832	5843	5855	5866	5877	5888	5899	1	2	3	5	6	7	8	9	10
39	5911	5922	5933	5944	5955	5966	5977	5988	5999	6010	1	2	3	4	5	7	8	9	10

Natural numbers	0	1	2	3	4	5	6	7	8	9	Proportional Parts								
											1	2	3	4	5	6	7	8	9
40	6021	6031	6042	6053	6064	6075	6085	6096	6107	6117	1	2	3	4	5	6	8	9	10
41	6128	6138	6149	6160	6170	6180	6191	6201	6212	6222	1	2	3	4	5	6	7	8	9
42	6232	6243	6253	6263	6274	6284	6294	6304	6314	6325	1	2	3	4	5	6	7	8	9
43	6335	6345	6355	6365	6375	6385	6395	6405	6415	6425	1	2	3	4	5	6	7	8	9
44	6435	6444	6454	6464	6474	6484	6493	6503	6513	6522	1	2	3	4	5	6	7	8	9
45	6532	6542	6551	6561	6571	6580	6590	6599	6609	6618	1	2	3	4	5	6	7	8	9
46	6628	6637	6646	6656	6665	6675	6684	6693	6702	6712	1	2	3	4	5	6	7	7	8
47	6721	6730	6739	6749	6758	6767	6776	6785	6794	6803	1	2	3	4	5	5	6	7	8
48	6812	6821	6830	6839	6848	6857	6866	6875	6884	6893	1	2	3	4	4	5	6	7	8
49	6902	6911	6920	6928	6937	6946	6955	6964	6972	6981	1	2	3	4	4	5	6	7	8
50	6990	6998	7007	7016	7024	7033	7042	7050	7059	7067	1	2	3	3	4	5	6	7	8
51	7076	7084	7093	7101	7110	7118	7126	7135	7143	7152	1	2	3	3	4	5	6	7	8
52	7160	7168	7177	7185	7193	7202	7210	7218	7226	7235	1	2	2	3	4	5	6	7	7
53	7243	7251	7259	7267	7275	7284	7292	7300	7308	7316	1	2	2	3	4	5	6	6	7
54	7324	7332	7340	7348	7356	7364	7372	7380	7388	7396	1	2	2	3	4	5	6	6	7
55	7404	7412	7419	7427	7435	7443	7451	7459	7466	7474	1	2	2	3	4	5	5	6	7
56	7482	7490	7497	7505	7513	7520	7528	7536	7543	7551	1	2	2	3	4	5	5	6	7
57	7559	7566	7574	7582	7589	7597	7604	7612	7619	7627	1	2	2	3	4	5	5	6	7
58	7634	7642	7649	7657	7664	7672	7679	7686	7694	7701	1	1	2	3	4	4	5	6	7
59	7709	7716	7723	7731	7738	7745	7752	7760	7767	7774	1	1	2	3	4	4	5	6	7
60	7782	7789	7796	7803	7810	7818	7825	7832	7839	7846	1	1	2	3	4	4	5	6	6
61	7853	7860	7868	7875	7882	7889	7896	7903	7910	7917	1	1	2	3	4	4	5	6	6
62	7924	7931	7938	7945	7952	7959	7966	7973	7980	7987	1	1	2	3	3	4	5	6	6
63	7993	8000	8007	8014	8021	8028	8035	8041	8048	8055	1	1	2	3	3	4	5	5	6
64	8062	8069	8075	8082	8089	8096	8102	8109	8116	8122	1	1	2	3	3	4	5	5	6
65	8129	8136	8142	8149	8156	8162	8169	8176	8182	8189	1	1	2	3	3	4	5	5	6
66	8195	8202	8209	8215	8222	8228	8235	8241	8248	8254	1	1	2	3	3	4	5	5	6
67	8261	8267	8274	8280	8287	8293	8299	8306	8312	8319	1	1	2	3	3	4	5	5	6
68	8325	8331	8338	8344	8351	8357	8363	8370	8376	8382	1	1	2	3	3	4	4	5	6
69	8388	8395	8401	8407	8414	8420	8426	8432	8439	8445	1	1	2	2	3	4	4	5	6
70	8451	8457	8463	8470	8476	8482	8488	8494	8500	8506	1	1	2	2	3	4	4	5	6
71	8513	8519	8525	8531	8537	8543	8549	8555	8561	8567	1	1	2	2	3	4	4	5	5
72	8573	8579	8585	8591	8597	8603	8609	8615	8621	8627	1	1	2	2	3	4	4	5	5
73	8633	8639	8645	8651	8657	8663	8669	8675	8681	8686	1	1	2	2	3	4	4	5	5
74	8692	8698	8704	8710	8716	8722	8727	8733	8739	8745	1	1	2	2	3	4	4	5	5
75	8751	8756	8762	8768	8774	8779	8785	8791	8797	8802	1	1	2	2	3	3	4	5	5
76	8808	8814	8820	8825	8831	8837	8842	8848	8854	8859	1	1	2	2	3	3	4	5	5
77	8865	8871	8876	8882	8887	8893	8899	8904	8910	8915	1	1	2	2	3	3	4	4	5
78	8921	8927	8932	8938	8943	8949	8954	8960	8965	8971	1	1	2	2	3	3	4	4	5
79	8976	8982	8987	8993	8998	9004	9009	9015	9020	9026	1	1	2	2	3	3	4	4	5
80	9031	9036	9042	9047	9053	9058	9063	9069	9074	9079	1	1	2	2	3	3	4	4	5
81	9085	9090	9096	9101	9106	9112	9117	9122	9128	9133	1	1	2	2	3	3	4	4	5
82	9138	9143	9149	9154	9159	9165	9170	9175	9180	9186	1	1	2	2	3	3	4	4	5
83	9191	9196	9201	9206	9212	9217	9222	9227	9232	9238	1	1	2	2	3	3	4	4	5
84	9243	9248	9253	9258	9263	9269	9274	9279	9284	9289	1	1	2	2	3	3	4	4	5

Natural numbers	0	1	2	3	4	5	6	7	8	9	Proportional Parts								
											1	2	3	4	5	6	7	8	9
85	9294	9299	9304	9309	9315	9320	9325	9330	9335	9340	1	1	2	2	3	3	4	4	5
86	9345	9350	9355	9360	9365	9370	9375	9380	9385	9390	1	1	2	2	3	3	4	4	5
87	9395	9400	9405	9410	9415	9420	9425	9430	9435	9440	0	1	1	2	2	3	3	4	4
88	9445	9450	9455	9460	9465	9469	9474	9479	9484	9489	0	1	1	2	2	3	3	4	4
89	9494	9499	9504	9509	9513	9518	9523	9528	9533	9538	0	1	1	2	2	3	3	4	4
90	9542	9547	9552	9557	9562	9566	9571	9576	9581	9586	0	1	1	2	2	3	3	4	4
91	9590	9595	9600	9605	9609	9614	9619	9624	9628	9633	0	1	1	2	2	3	3	4	4
92	9638	9643	9647	9652	9657	9661	9666	9671	9675	9680	0	1	1	2	2	3	3	4	4
93	9685	9689	9694	9699	9703	9708	9713	9717	9722	9727	0	1	1	2	2	3	3	4	4
94	9731	9736	9741	9745	9750	9754	9759	9763	9768	9773	0	1	1	2	2	3	3	4	4
95	9777	9782	9786	9791	9795	9800	9805	9809	9814	9818	0	1	1	2	2	3	3	4	4
96	9823	9827	9832	9836	9841	9845	9850	9854	9859	9863	0	1	1	2	2	3	3	4	4
97	9868	9872	9877	9881	9886	9890	9894	9899	9903	9908	0	1	1	2	2	3	3	4	4
98	9912	9917	9921	9926	9930	9934	9939	9943	9948	9952	0	1	1	2	2	3	3	4	4
99	9956	9961	9965	9969	9974	9978	9983	9987	9991	9996	0	1	1	2	2	3	3	4	4

ATOMIC NUMBERS AND ATOMIC WEIGHTS OF THE ELEMENTS

Element	Symbol	Atomic Number	Atomic Weight	Element	Symbol	Atomic Number	Atomic Weight
Aluminum	Al	13	26.97	Neodymium	Nd	60	144.27
Antimony	Sb	51	121.76	Neon	Ne	10	20.183
Argon	A	18	39.944	Nickel	Ni	28	58.69
Arsenic	As	33	74.91	Niobium	Nb	41	92.91
Barium	Ba	56	137.36	Nitrogen	N	7	14.008
Beryllium	Be	4	9.02	Osmium	Os	76	190.2
Bismuth	Bi	83	209.00	Oxygen	O	8	16.000
Boron	B	5	10.82	Palladium	Pd	46	106.7
Bromine	Br	35	79.916	Phosphorus	P	15	30.98
Cadmium	Cd	48	112.41	Platinum	Pt	78	195.23
Calcium	Ca	20	40.08	Potassium	K	19	39.096
Carbon	C	6	12.01	Praseodymium	Pr	59	140.92
Cerium	Ce	58	140.13	Protactinium	Pa	91	231
Cesium	Cs	55	132.91	Radium	Ra	88	226.05
Chlorine	Cl	17	35.457	Radon	Rn	86	222
Chromium	Cr	24	52.01	Rhenium	Re	75	186.31
Cobalt	Co	27	58.94	Rhodium	Rh	45	102.91
Copper	Cu	29	63.57	Rubidium	Rb	37	85.48
Dysprosium	Dy	66	162.46	Ruthenium	Ru	44	101.7
Erbium	Er	68	167.2	Samarium	Sm	62	150.43
Europium	Eu	63	152.0	Scandium	Sc	21	45.10
Fluorine	F	9	19.00	Selenium	Se	34	78.96
Gadolinium	Gd	64	156.9	Silicon	Si	14	28.06
Gallium	Ga	31	69.72	Silver	Ag	47	107.880
Germanium	Ge	32	72.60	Sodium	Na	11	22.997
Gold	Au	79	197.2	Strontium	Sr	38	87.63
Hafnium	Hf	72	178.6	Sulfur	S	16	32.06
Helium	He	2	4.003	Tantalum	Ta	73	180.88
Holmium	Ho	67	164.94	Tellurium	Te	52	127.61
Hydrogen	H	1	1.0081	Terbium	Tb	65	159.2
Indium	In	49	114.76	Thallium	Tl	81	204.39
Iodine	I	53	126.92	Thorium	Th	90	232.12
Iridium	Ir	77	193.1	Thulium	Tm	69	169.4
Iron	Fe	26	55.84	Tin	Sn	50	118.70
Krypton	Kr	36	83.7	Titanium	Ti	22	47.90
Lanthanum	La	57	138.92	Tungsten	W	74	183.92
Lead	Pb	82	207.21	Uranium	U	92	238.07
Lithium	Li	3	6.940	Vanadium	V	23	50.95
Lutecium	Lu	71	175.00	Xenon	Xe	54	131.3
Magnesium	Mg	12	24.32	Ytterbium	Yb	70	173.04
Manganese	Mn	25	54.93	Yttrium	Y	39	88.92
Mercury	Hg	80	200.61	Zinc	Zn	30	65.38
Molybdenum	Mo	42	95.95	Zirconium	Zr	40	91.22

INDEX